MICROBIAL GENETICS APPLIED TO BIOTECHNOLOGY

PRINCIPLES AND TECHNIQUES OF GENE TRANSFER AND MANIPULATION

MICROBIAL GENETICS APPLIED TO BIOTECHNOLOGY

PRINCIPLES AND TECHNIQUES OF GENE TRANSFER AND MANIPULATION

VENETIA A. SAUNDERS AND JON R. SAUNDERS

CROOM HELM
London & Sydney

Croom Helm Ltd, Provident House, Burrell Row,
Beckenham, Kent BR3 1AT

Croom Helm Australia, 44-50 Waterloo Road,
North Ryde, 2113, New South Wales

British Library Cataloguing in Publication Data

Saunders, Venetia A.
Microbial genetics applied to biotechnology.
1. Microbial genetic engineering
2. Biotechnology
I. Title II. Saunders, Jon R.
660'.62 TP248.6

ISBN 0-7099-2365-1
ISBN 0-7099-4435-7 Pbk

Typeset in Times Roman by Leaper & Gard Ltd, Bristol, England
Printed and bound in Great Britain

For our son
MARK

Contents

Preface

This book describes techniques of microbial genetics and how they may be applied to biotechnology. The text is concerned largely with the application of these techniques to microbial technology. We have therefore utilised illustrative material that is given in our own courses in applied microbiology. The book assumes in the reader a basic knowledge of microbial genetics and industrial microbiology. We hope it will prove useful to undergraduates, postgraduates and others taking courses in applied microbiology.

We would like to thank various colleagues, including John Carter, Julian Davies, Gordon Dougan, David Hopwood, Gwyn Humphreys, Alan McCarthy, David O'Connor, Tony Hart, Steve Oliver, Roger Pickup, Hilary Richards, Bob Rowlands, David Sherratt, Peter Strike, Richard Sykes and Liz Wellington, all of whom provided information at various stages during the writing of this book. Many thanks are also due to Linda Marsh for patiently typing the many drafts of the manuscript.

1

Introduction

Natural genetic variation has always been exploited by man to improve the properties of microbial strains. Spontaneous mutations that arise in microbial populations and that have properties advantageous to man have been gradually selected over centuries of use. However, it is only since the development of modern genetic techniques that more rational approaches have been possible. Such newer technologies have permitted the tailoring of microorganisms, plant or animal cells to manufacture specific products of commercial or social benefit and to manage the environment.

It is the intention of this book to demonstrate how fundamental principles of microbial genetics have been applied to biotechnology. Microorganisms have a wide range of physiological capabilities not encountered in higher organisms. Bacteria and fungi can also normally be grown in culture under controlled conditions. Furthermore, microorganisms possess relatively simple genetic organisations and are more amenable to manipulation than higher organisms. Genetics is a discipline *par excellence* for breaking down and analysing complex biological problems and for using the information so obtained for the rational manipulation of biological processes. Fundamental studies in genetics have provided both direct and indirect benefits to biotechnology by permitting a better understanding of the biology of microorganisms. This has generated an infrastructure of knowledge, particularly for bacteria, such as *Escherichia coli*, and fungi, such as *Saccharomyces cerevisiae*, that has, in turn, promoted the development of modern manipulative techniques. During recent years much emphasis has been placed on the role of *in vitro* recombinant DNA technology (see Chapters 3, 4 and 5) in stimulating a resurgence of biotechnology. Although it is true that *in vitro* techniques do provide a radical new ingredient in the exploitation of biological systems, it should not be forgotten that many commercially important applications of microbial genetics involve the use of traditional *in vivo* gene manipulations (see Chapters 6, 7, 8 and 9). Moreover, the development of much *in vitro* genetic manipulation technology has been strictly dependent on 'conventional' *in vivo* techniques (Chapters 2 and 4).

The ability to exploit microbial genetics has depended to a great extent on gene transfer and other evolutionary mechanisms found in natural populations of microorganisms. Crucial to the development of genetic technologies has been an understanding of the biology of plasmid and virus genomes. Such extrachromosomal elements, which are widespread in both bacteria and fungi, provide important vehicles for the natural exchange of genes both within and between species (see Chapter 2). These elements are relatively small, autonomously replicating, nucleic acid molecules, and are particularly amenable to manipulation by both *in vivo* and *in vitro* genetic techniques. Furthermore, many extrachromosomes are present in host cells at copy numbers greater than chromosomes. This provides a means of amplifying genes that the element happens to carry. Plasmids may be classified into incompatibility (Inc) groups based on the ability of pairs of plasmids to coexist in the same cell. Those plasmids that fail to coexist belong to the same Inc group. Incompatible plasmids share homologous systems for controlling replication and hence copy number (see Chapter 5). The diversity of Inc groups in different bacteria and fungi has permitted the construction of a variety of useful genetic vectors, of both broad and narrow host range, for gene manipulation (see Chapters 2, 3 and 5).

This book describes *in vivo* and *in vitro* genetic manipulation methodologies and considers examples of their application to biotechnology. Genetics can be used for the deliberate enhancement of existing attributes of microorganisms (see Chapter 6) and for their manipulation to provide novel properties or products (see, for example, Chapters 6 and 7). Examples of the application of the principles of microbial genetics to veterinary and human medicine (Chapter 7), to agriculture (Chapter 8) and to the management of the environment (Chapter 9) are also considered.

2

In Vivo Genetic Manipulation

Genes can be shuffled within and between microbial species by a variety of mechanisms. Such mechanisms lead to the formation of new genotypes by bringing together and reassorting genes from different organisms. In bacteria new combinations of genes may be generated by using one of the three natural processes of gene transfer, namely **transformation** (section 2.2), **conjugation** (section 2.3) or **transduction** (section 2.4). In fungi, genetic exchange can be effected through the agency of the **sexual** (section 2.5) or **parasexual** (section 2.6) cycle. Transformation systems are also available for gene transfer in yeasts and certain filamentous fungi. **Protoplast fusion** (section 2.7) provides a further route for combining groups of genes from different strains and modifying the genetic constitution of microorganisms. These various processes for manipulating genomes enable the formation of innumerable genetic combinations, in turn producing variability within microbial populations.

Transposable genetic elements (section 2.1) provide another source of variability among microorganisms. These elements can insert into and excise from a variety of replicons and promote a number of genome rearrangements. Transposable elements can cause mutation when they interrupt the coding sequence of a gene. Various properties of these elements make them useful for manipulating genomes.

This chapter describes mechanisms for gene manipulation *in vivo.*

2.1 TRANSPOSABLE GENETIC ELEMENTS

2.1.1 Properties of transposable elements

Transposable genetic elements are segments of DNA that are capable of inserting as discrete nonpermuted DNA sequences at various sites within a genome. These elements have been found in the genomes of a variety of microorganisms, including bacteria, fungi and bacteriophages, as well as in the genomes of higher organisms. This section is primarily concerned with the properties of prokaryotic transposable elements, some of which have been studied in detail and have been used widely in genetic manipulation. However, transposable elements of fungi, for example *Ty* elements, have some similar properties.

The smallest prokaryotic transposable elements are **insertion sequences (IS)** or IS-like elements. These encode determinants involved in promoting/regulating transposition. Larger genetic entities (originally termed **transposons (Tn)**)[1] encode accessory determinants (for example, antibiotic resistance, lactose fermentation), in addition to transposition functions. Transposable elements can be divided into a number of classes according to genetic organisation and transposition mechanism (see Table 2.1). Class I includes IS-like modules and composite (compound) elements which are formed from them. Composite transposons comprise a central DNA segment flanked on either side by a copy of an IS (Figure 2.1). Available evidence indicates that the information required for transposition of composite transposons is encoded by the IS constituents. Furthermore, the ISs are capable of transposing independently. Class II (complex transposons) comprises Tn*3* and its relatives and Class III transposing bacteriophages, for example Mu. Certain elements cannot be categorised into these classes. Such 'unclassified' elements include Tn*7* and Tn*916*. The properties of transposable elements have been extensively documented; see for example, Calos and Miller (1980), Cohen and Shapiro (1980), Bennett (1985), Campbell (1981), Kleckner (1981), Shapiro (1983), Cullum (1985) and Grindley and Reed (1985).

Transposition (and associated events discussed in section 2.1.3) in *E. coli* is independent of *recA*-mediated homologous recombination, implying reliance on transposon-encoded functions and/or other host recombination processes. Extensive DNA sequence homology between the element and the site of insertion is not required for transposition. Integrity of the ends of the transposable element is crucial for transposition. Normally the ends are inverted repeats (IRs) of one another. Phage Mu (and D108), Tn *7* and Tn*554* are exceptions in that they do not have true terminal repeats. In the case of IS*1* the minimal terminal sequence required for transposition is 21 to 25 bp. A specific sequence at base pairs 13 to 23, found at each extremity of IS*1*, appears to be an essential site. It has been suggested (Gamas *et al.*, 1985) that a host DNA binding protein, such as integration host factor (IHF) or a similar protein, could be involved in the transposition reaction by binding to such sites. For Tn*10* the outer 13 to 27 bp of the terminal IS*10* sequence, at each end, are essential for transposition, and sequences at 27 to 70 bp are important in the process (Way and Kleckner,

[1] The terms transposable element and transposon are now often used synonymously.

Table 2.1: Selected prokaryotic transposable elements and their properties

	Element	Size (kb)	Source	Accessory determinants*
Class IA (Individual IS modules)	IS*1*	0.768	*E. coli* and other entero-bacterial genomes, plasmid R100	none
	IS*2*	1.327	*E. coli* K12 genome, plasmid F	none
	IS*21*	2.1	*Pseudomonas aeruginosa* plasmid R68.45	none
Class IB (composite Tn elements)	Tn*5*	5.7 (IR.IS*50*)	*Klebsiella* plasmid JR67	Km
	Tn*9*	2.5 (DR.IS*1*)	plasmid R100	Cm
	Tn*10*	9.3 (IR.IS*10*)	plasmid R100	Tc
	Tn*903* (Tn*601*)	3.1 (IR.IS*903*)	plasmid R6	Km
Class II (Tn*3* family)	Tn*3*	4.957	plasmid R1	Ap
	Tn *1000* (γδ)	5.8	plasmid F	unknown
	Tn*501*	8.2	*P. aeruginosa* plasmid VS1	Hg
	Tn*551*	5.3	*Staphylococcus aureus* plasmid pl258	Em
	Tn*917*	5.3	*Streptococcus faecalis* plasmid pAD2	Em
Class III (transposing bacteriophages)	Mu	37.0	*E. coli* K12 genome	phage life cycle
Unclassified	Tn*7*	13.5	plasmid R483	Tp, Sm

* Ap, Cm, Em, Hg, Km, Sm, Tc, Tp: resistance to ampicillin, chloramphenicol, erythromycin, mercuric ions, kanamycin (and other aminoglycosides including G418), streptomycin, tetracycline and trimethoprim respectively. IR, inverted repeat; DR, direct repeat

1984). Transposons of the Tn*3* family (such as Tn*1/3*, Tn*21* and Tn*1721*) can transpose when only a single end (either right or left IR) is present (Arthur *et al.*, 1984; Avila *et al.*, 1984; Motsch and Schmitt, 1984). Such one-ended transposition occurs at low frequency and results in the formation of cointegrates when the appropriate transposase enzyme (see section 2.1.2.a) is present in the cell.

Transposable elements can transpose either within the same DNA molecule (**intramolecular transposition**), or from one DNA molecule to another (**intermolecular transposition**). Insertion of the element into the target DNA is generally accompanied by duplication of host DNA sequences at the target site (Tn*554* being a notable exception). The size of

Figure 2.1: Structures of composite (compound) transposons. (i) Insertion sequence (IS) in inverted repeat; (ii) IS in direct repeat

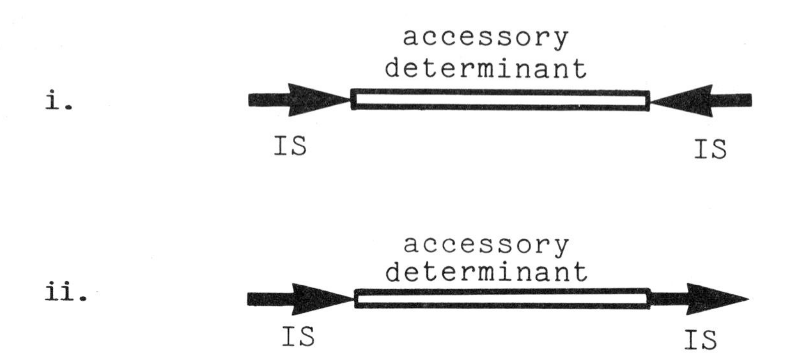

the duplication (typically 3 to 13 base pairs) is characteristic of a particular element. Furthermore, the degree of insertion specificity depends upon the element. Some transposons (for example Tn*5*, Tn*10*) show preference for particular target sites; whereas others (for example Tn*1*, Tn*3*) exhibit regional specificity, inserting efficiently at a number of different sites within preferred regions of a target molecule. Generally, AT-rich regions (where denaturability is greater) are preferred sites for integration of Tn*3*. Furthermore insertions of Tn *3* have been found near sequences resembling its ends.

Members of the Tn*3* family exhibit **transposition immunity** whereby the presence of a Tn on a plasmid can inhibit transposition of a second copy of that Tn on to that plasmid, but not on to another Tn-free plasmid residing in the same cell (Robinson *et al.*, 1977). There is, however, no barrier to introducing a second Tn by homologous recombination. Transposition immunity thus limits the number of copies of a Tn that may transpose on to a replicon. Transposition of the Tn*3* family is also regulated by resolvase acting at the level of transcription (see section 2.1.2.a). Different modes of regulation are employed by different transposons. Transposition of composite Tns, such as Tn*10*, can be regulated by copies of their component IS elsewhere in the cell. Regulation of Tn*10* transposition may involve RNA species encoded by IS*10*, which forms the terminal repeats of Tn*10*. One of the two copies of IS*10*, IS*10*-right (IS*10*-R), encodes a long open reading frame (ORF) specifying a function (presumptive transposase) essential for transposition. An inwardly directed promoter, pIN, which is located just upstream of this ORF, is responsible for its expression. A second outwardly directed promoter, pOUT, is located just inside the start of the ORF. The region of overlap includes the ATG translation start codon for the ORF. It is proposed that the transcript obtained from pOUT pairs with the transcript for the putative transposase from pIN, in turn inhibiting translation of the transposase gene (Simons and Kleckner, 1983; Way and Kleckner, 1984; Kleckner, 1986).

The frequency of transposition is generally within the range 10^{-4} to 10^{-7}. However, transposition frequencies may be influenced by environmental stimuli (for example, temperature or conditions of stress for the host). The transposition of most transposable elements probably depends upon host replication functions. In *E. coli*, mutations in genes such as *polA* (for DNA polymerase I) and *gyrB* (for DNA gyrase, B subunit) have resulted in decreased transposition frequencies for a number of elements (Isberg and Syvanen, 1982; Syvanen *et al.*, 1982). By contrast, strains defective in the *dam* gene product (for DNA methylation) show increased frequencies of transposition (see section 4.12).

2.1.2 Transposition models

There are two general classes of transposition models, **replicative** and **conservative** (non-replicative). The underlying features of replicative models are (i) the precise joining of transposon sequences to target DNA at the insertion site (break/join process), and (ii) replication of the transposing element. The outcome of intermolecular replicative transposition is that one copy of the element remains at the original locus of the donor DNA molecule while a second copy integrates at the target site:

$$\text{Replicon 1 : : Tn + Replicon 2} \rightarrow \text{Replicon 1 : : Tn + Replicon 2 : : Tn}$$

However, transposable elements appear to differ in the manner in which they resolve transposition intermediates formed between donor and target molecules. In some cases, for example Tn*3* and Tn*1000* ($\gamma\delta$), resolution of intermediates (to give final separation of donor and target DNA each carrying a copy of the transposon) can occur after completion of element replication, through a site-specific recombination event involving a site internal to the element. In other cases, where no internal resolution system appears to operate, separation probably occurs as an integral part of the break/join and replication processes themselves, or does not occur at all.

There are various molecular models for replicative transposition (see, for example, Arthur and Sherratt, 1979; Shapiro, 1979; Bukhari, 1981; Galas and Chandler, 1981; Harshey *et al.*, 1982; Craigie and Mizuuchi, 1985), all of which account for break/join and replication processes, but differ in the order and details of these events. Such models can be divided into two classes, **symmetric** and **asymmetric**, depending upon whether transposition is initiated at both or only one end of the transposon. According to the Shapiro model (Figure 2.2A), an archetype for numerous other models, intermolecular transposition is initiated by separation of

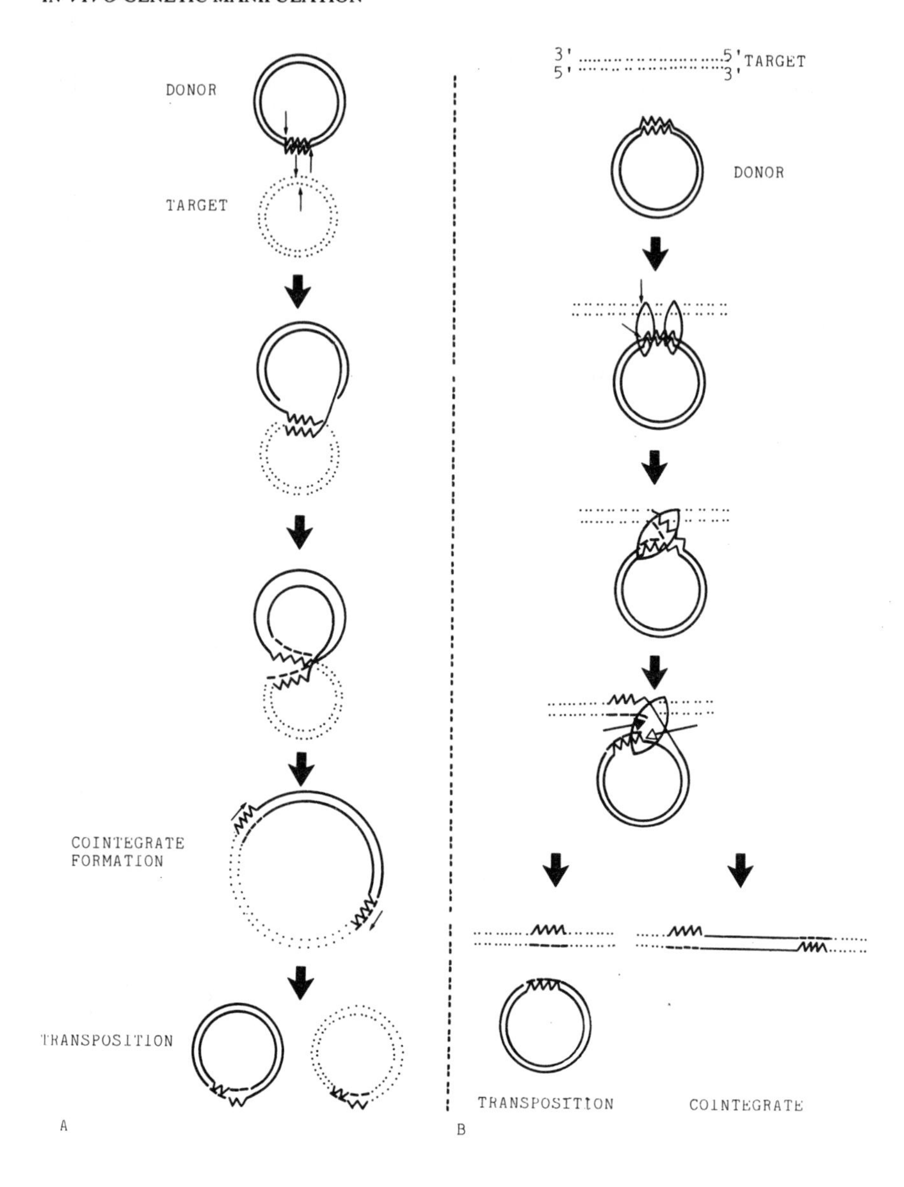

transposon sequences from adjacent donor DNA sequences by a single-stranded nick at the boundaries of the transposon. Both ends of the element become ligated to the target DNA. Replication of the element results in the formation of a cointegrate structure comprising two directly repeated copies of the transposon, one at each junction of donor and target molecules. Resolution of the cointegrate occurs by recombination between the two copies of the element. This may be a site-specific recombination (as

Figure 2.2: Models for transposition. (A) The Shapiro model. Transposition is initiated by single-stranded nicks (↓) at the ends of the transposon on the donor molecule. Nicks also occur at the target site on the recipient molecule and the ends of the Tn become ligated to the recipient DNA. Donor and recipient molecules are fused, but four nicks are present in the structure. Replication of the element results in a cointegrate as an obligate intermediate. Resolution of the cointegrate is effected by recombination between the directly repeated copies of the Tn. (B) Harshey and Bukhari model. A protein-mediated association is brought about between the donor and target DNA molecules. The target site undergoes a double-stranded cleavage. Cleavage also occurs at one end of the Tn in one strand that becomes attached to the target DNA. (Cleavage sites are indicated by arrows, ↓.) Replication of the element (roll-in replication) occurs. Where the 3′ end of the newly synthesised strand (indicated by the arrow ⟶ at the left) is recognised for ligation, a simple (direct) transposition results. Where a nick is made in the parental strand (indicated by the arrow ⟶ at the right), and is ligated to the free end of the target strand, insertion of the entire donor molecule (carrying the Tn) into the target site results, generating a cointegrate structure. ⁓⁓, Transposon; ---, new DNA synthesis; O, protein

can occur in Tn*3* and Tn*1000*), or may be a *recA*-promoted homologous recombination event. This symmetric model invokes cointegrates as obligatory intermediates in the transposition process.

Asymmetric replicative models invoking branching pathways to explain transposition and cointegrate formation have also been proposed (Galas and Chandler, 1981; Harshey and Bukhari, 1981; Harshey *et al.*, 1982). A salient feature of these models is that replication starts from one end of the element that is ligated to the target DNA (roll-in-type replication) while the other end remains unligated until the final stage in the transposition process (see Figure 2.2B). The end product may be either a cointegrate or a simple (direct) transposition of the element, depending upon the precise nature of the break/join processes at the target site when replication is terminated. In 1% to 5% of Tn*3* transposition events a cointegrate is not involved as intermediate. Accordingly, direct transposition of Tn*3* might proceed by mechanisms such as these (Bennett *et al.*, 1983). One-ended transposition might also occur by such mechanisms (Arthur *et al.*, 1984; Avila *et al.*, 1984). For composite transposons, transposition may involve read-through replication, in which replication begins at the outside end of a flanking IS and proceeds through the central region of the transposon to the outside end of the second copy of the IS, such that the composite transposon can function as a unit (Galas and Chandler, 1981).

Although some transposition events appear to be replicative, others apparently do not involve replication of the entire transposon. In such cases the transposon may be excised from the donor site without replication. A model for such conservative mechanisms (Berg and Berg, 1983) proposes that the transposable element is excised from the donor DNA by double-stranded cleavages at both ends of the element and is inserted into a target molecule. There is no DNA synthesis, except for that required to fill the gaps in the target DNA. The remainder of the donor replicon is lost from the host cell. A second copy of the donor molecule that did not participate in transposition is over-replicated to compensate for the loss of the first copy.

The significance of replicative and conservative modes of transposition would appear to vary from element to element. Some elements, such as IS*1*, might transpose in either mode: utilising the replicative mode where cointegrates are involved, or the conservative mode for direct transposition.

*(a) Structure of Tn*1000 *(γδ)*

The genetic map of Tn*1000* is given in Figure 2.3. The *tnpA* gene product, transposase, is required for the generation of cointegrate structures. The

Figure 2.3: Genetic map and sequenced promoter regions of Tn*1000* (γδ). (a) The element is drawn with the γ end towards the right. Transcription of *tnpA* and *tnpR* (⟶) is divergent from the *tnpA*-*tnpR* intercistronic region. The intercistronic region is enlarged to show DNA sites protected by resolvase binding. Site I contains cleavage sites (↓) for resolvase. These cleavage sites are within the −10 regions for *tnpA* and *tnpR*. *tnpA* gene product is transposase; *tnpR* gene product is resolvase; *res*, resolution site in the intercistronic space. (b) The −10 regions of the *tnpA* and *tnpR* promoters overlap. ↱, nucleotides at which *tnpA* and *tnpR* transcription is initiated (adapted from Reed *et al.*, 1982)

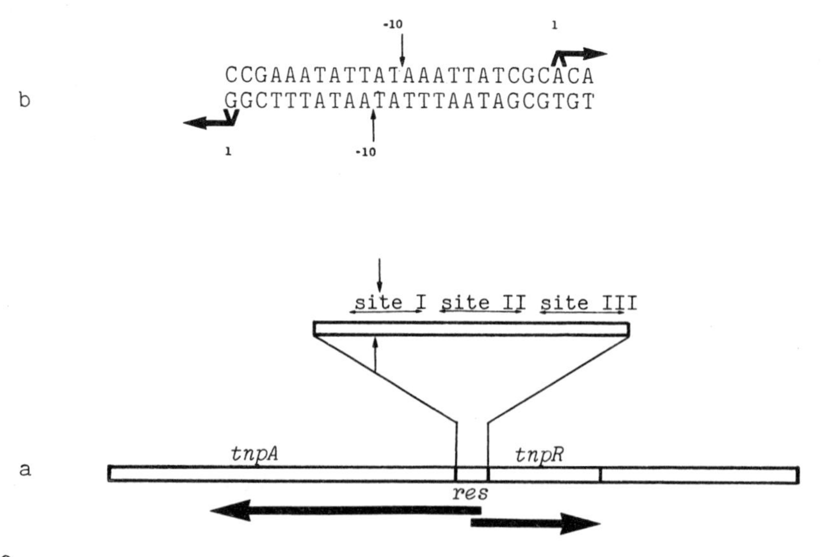

tnpR gene product, resolvase, mediates site-specific recombination at *res*, a site located in the intercistronic region between *tnpA* and *tnpR* genes (Reed, 1981). In addition, resolvase appears to regulate the frequency of transposition by repressing transcription of both the *tnpA* gene and its own gene. There are three sites (sites I, II and III) in the *tnpA-tnpR* intercistronic region that bind resolvase. Site I contains the recombinational cross-over point and promoters for the divergent transcription of *tnpA* and *tnpR* genes. It is likely that resolvase regulates expression of transposition functions and resolves transposition intermediates by acting at these sites (Grindley *et al.*, 1982; Reed *et al.*, 1982; Grindley and Reed, 1985).

2.1.3 Transposon-mediated genome rearrangements

Transposable genetic elements can promote a number of genome rearrangements, including the inversion, deletion and duplication of DNA. Such rearrangements have a natural role in modulating gene expression and in promoting evolution. Transposable elements have been introduced into various Gram-negative and some Gram-positive organisms, including species with poorly developed genetic systems, and are used for a variety of genetic manipulations and analyses. Examples of transposon-mediated genome rearrangements in prokaryotes are described in Figure 2.4. These genetic rearrangements can arise independently of *recA* function, presumably as a consequence of intra- or intermolecular transpositions. Alternatively, combinations of transposition and homologous recombination events may be involved in which the transposon serves as a region of portable homology. By providing portable regions of homology, transposons can be used to mobilise the chromosome and nonconjugative plasmids (section 2.3), to fuse unrelated replicons and to mediate gene duplications. Furthermore, transposons can provide a source of mobile restriction sites to facilitate gene cloning.

Transposons can cause mutations when they insert into genes (section 4.7). Many of these mutations are strongly polar with respect to expression of those genes in the operon that lie distal to the insertion site. The mechanics of such polar effects are not completely clear. Transcription termination mechanisms that have been implicated in other types of polarity in prokaryotic operons (see Adhya and Gottesman, 1978) may be involved. Strong polarity may result at least in part from the presence of multiple translational stop codons within a transposable element and the absence of nearby translational reinitiation signals.

There are a number of cases where insertion of an IS or transposon (for example IS*2* or Tn*10*) can activate expression of adjacent genes. This is presumably due to the presence of promoter sequences at the ends of the element. In some instances the juxtaposition of determinants within/at the

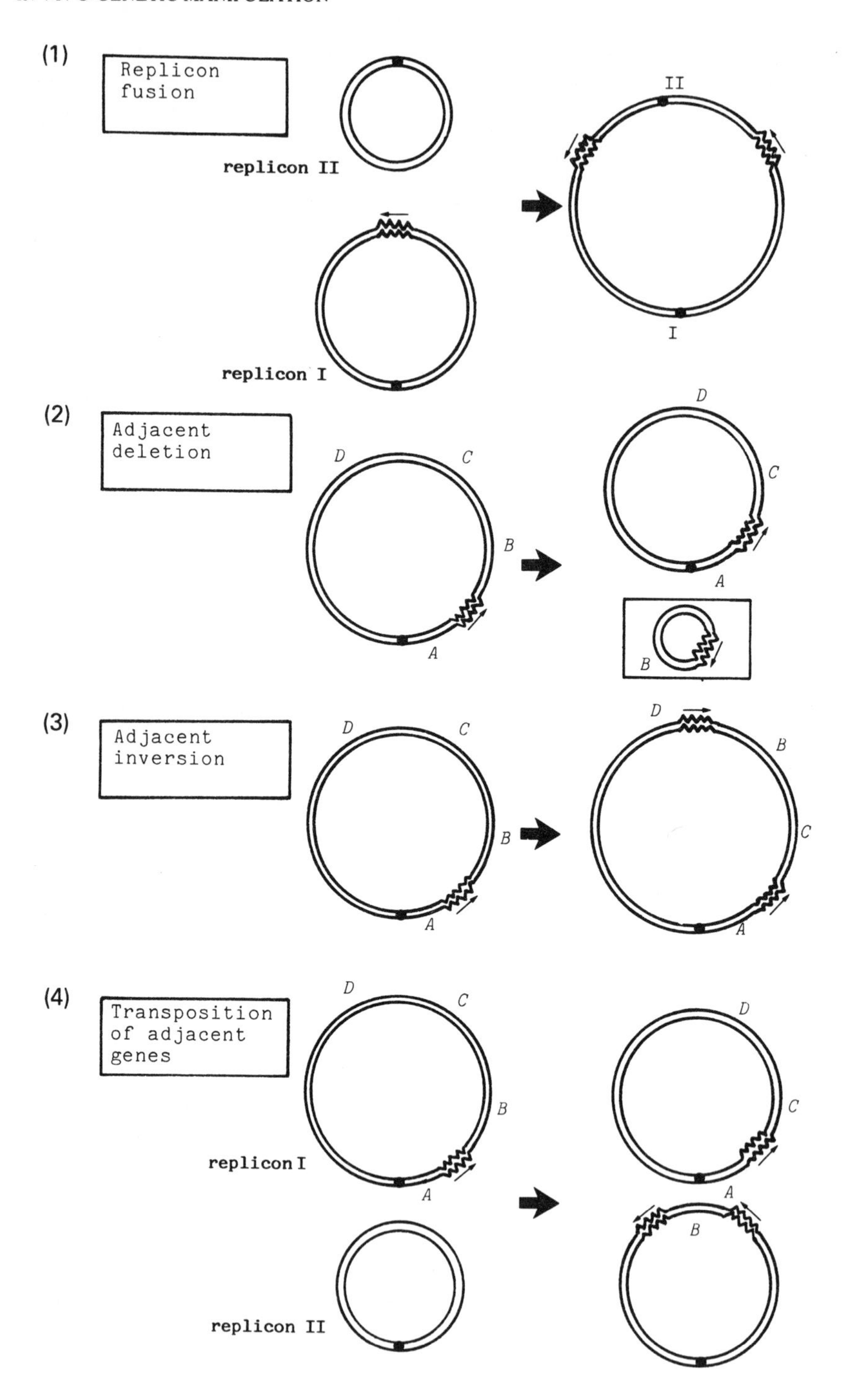
(1)
Replicon fusion
replicon II
replicon I
II
I
(2)
Adjacent deletion
D
C
B
A
D
C
A
B
(3)
Adjacent inversion
D
C
B
A
D
B
C
A
(4)
Transposition of adjacent genes
D
C
B
A
replicon I
replicon II
D
C
A
B

Figure 2.4: Some transposon-mediated genome rearrangements. (1) Two replicons (I and II) can be fused through intermolecular transposition. The cointegrate formed contains directly repeated copies of the transposon at the fusion junctions. (2) Deletion of genetic material (B) adjacent to the Tn effected through intramolecular transposition. (Deleted DNA (B) would be lost, since it does not contain an origin of vegetative replication, unless it is rescued by transposition to a replicon.) (3) Intramolecular transposition effects inversion of intervening genetic material (BC) between oppositely oriented copies of the Tn. (4) Transposition of adjacent genetic material (B) to replicon II between directly repeated copies of the Tn. Formally this is adjacent deletion followed by cointegrate formation. ~~, Transposon; •; origin of vegetative replication. Note: Although these products arise through *recA*-independent transpositions, suitable combinations of transposition and *recA*-dependent homologous recombination could generate such products.

ends of the element with target DNA sequences may generate appropriate signals to activate expression of distal genes. Indeed, activation of genes by transposable elements is often correlated with insertion of the element at or near a pre-existing regulatory site in the target DNA where suitable sequences for transcription/translation may be available. One of the two copies of IS*10* (IS*10*-R) that form the terminal repeats of Tn*10* contains inwardly and outwardly directed promoters (pIN and pOUT respectively). The stronger pOUT can activate genes distal to the site of IS*10* or Tn*10* insertion (Simons *et al.*, 1983).

Transposon-mediated events analogous to those described above for prokaryotes, including the inversion and deletion of genetic material, have also been reported in eukaryotic microorganisms, such as yeast (Roeder and Fink, 1983).

2.2 TRANSFORMATION

Transformation is the process by which microorganisms take up naked DNA and subsequently acquire an altered genotype. Transformation has been reported in various bacteria, in yeast and in some filamentous fungi, for example *Neurospora crassa* (Table 2.2). The process involves the binding of DNA to competent cells, uptake of the DNA and its establishment within recipients (either as a replicon itself or by recombination with a resident replicon). The transforming DNA may be chromosomal, plasmid or viral (uptake and infection of cells with naked viral DNA is generally termed **transfection**). In naturally transformable species, **competence** (the ability of a cell to take up DNA into a DNase-resistant form) normally develops under specific growth conditions. However, some species, including a number of industrial importance, do not become transformable

Table 2 2: Selected transformation systems

Competence induction	Organism	Comments*
Natural	*Bacillus subtilis*	Require oligomeric plasmid DNA or recombinational rescue
	Streptococcus sanguis	
	Anacystis nidulans	
	Haemophilus influenzae	Uptake sequence required
	Methylobacterium organophilum	
	Rhodopseudomonas sphaeroides	Requires helper phage
	Thiobacillus thioparus	
	Saccharomyces cerevisiae	
Artificial: (a) Intact cells		
1. Divalent cation systems	*Azotobacter vinelandii*	Ca^{2+}
	Staphylococcus aureus	Ca^{2+} and helper phage
	Streptomyces griseus	Ca^{2+}
	Alcaligenes eutrophus	Ca^{2+} with Mg^{2+}
	Erwinia carotovora	Ca^{2+}
	Escherichia coli	Ca^{2+}; Ca^{2+} with Rb^{+}; Ca^{2+} with Mg^{2+}
	Flavobacterium sp.	Protease pretreatment, then Ca^{2+}
	Pseudomonas aeruginosa	Mg^{2+}; Mg^{2+} with $Ca^{2+} \pm Rb^{+}$
	Pseudomonas putida	Ca^{2+}
	Rhizobium meliloti	Ca^{2+}
	Rhodopseudomonas sphaeroides	Tris and Ca^{2+}
	Serratia marcescens	Ca^{2+}
2. Monovalent cation systems	*Saccharomyces cerevisiae*	PEG with Cs^{+} or Li^{+} or Rb^{+}
3. Freeze-thaw systems	*Agrobacterium tumefaciens*	
	Escherichia coli	
	Rhizobium leguminosarum	
	Rhizobium meliloti	

4. Triton treatment	*Saccharomyces cerevisiae*	
Artificial: (b) PEG/protoplasts or spheroplasts		
	Bacillus cereus	
	Bacillus subtilis	
	Bacillus stearothermophilus	
	Bacillus thuringiensis	
	Brevibacterium lactofermentum	
	Clostridium acetobutylicum	
	Staphylococcus carnosus	
	Streptococcus faecalis	
	Streptococcus lactis	
	Streptomyces antibioticus	
	Streptomyces coelicolor	
	Streptomyces griseus	
	Streptomyces lividans	
	Streptomyces parvulus	
	Alcaligenes eutrophus	
	Aspergillus nidulans	
	Cephalosporium acremonium (Acremonium chrysogenum)	Dimethyl sulphoxide
	Neurospora crassa	
	Saccharomyces cerevisiae	
	Schizosaccharomyces pombe	
Artificial: (c) Liposome systems		
1. Intact cells	*Escherichia coli*	
	Rhodopseudomonas sphaeroides	
2. Protoplasts	*Streptomyces lividans*	
	Neurospora crassa	
	Saccharomyces cerevisiae	

* Ca^{2+}, Cs^{+}, Li^{+}, Mg^{2+}, Rb^{+}: calcium, caesium, lithium, magnesium, rubidium ions; PEG, polyethylene glycol; Tris, Tris (hydroxymethyl) aminomethane.

naturally. Competence must be induced by artificial treatments that render the cells permeable to transforming DNA.

Transformation and transfection are valuable genetic tools because they permit the introduction of defined DNA sequences into microorganisms. DNA can therefore be subjected to manipulations *in vitro* prior to reintroduction and subsequent maintenance *in vivo.*

2.2.1 Bacterial transformation systems

In naturally transformable Gram-positive bacteria, for example *Streptococcus pneumoniae* and *Bacillus subtilis,* a number of components, including competence factors, DNA-binding proteins, nucleases and autolysins, may be involved in the binding and processing of DNA for its subsequent intracellular establishment (Smith *et al.*, 1981). Linear chromosomal DNA is transported in a single-stranded form, with concomitant degradation of the complementary strand.

Typically bacteria transformable by linear chromosomal DNA can also be transformed by circular plasmid DNA (or transfected by phage DNA), although the efficiencies of the processes differ. This is due, in part, to the different topologies of the molecules involved. In *B. subtilis* monomeric plasmid molecules are generally ineffective in transformation, whereas oligomeric forms transform with high efficiencies (Canosi *et al.*, 1978). This is explained by the mode of DNA processing for transformation, whereby circular plasmid DNA is converted into a single-stranded linear form during uptake by *B. subtilis* (Figure 2.5). Incoming monomeric molecules are consequently unable to recircularise. However, complementary single strands derived from oligomers can reanneal within the cell to generate partially double-stranded circular molecules. These can be converted into completely double-stranded replicons by DNA (repair) synthesis. The requirement for oligomeric plasmids can, however, be circumvented if the recipient harbours a replicon containing sequences homologous to the donor plasmid, thereby enabling **recombinational rescue** of the incoming plasmid markers.

The ability to introduce plasmids into cells by transformation is an important prerequisite for gene cloning *in vitro* (see Chapter 3). Accordingly, transformation systems are now being developed for a wide range of microorganisms that have not previously been explored genetically (see Saunders *et al.*, 1984). Most of these systems depend upon the artificial induction of competence. Transformation of *E. coli* by plasmid DNA was first achieved by subjecting recipient cells to high concentrations (25 mM and above) of Ca^{2+} at 0°C, followed by a heat pulse at 42°C (Cohen *et al.*, 1972). Ca^{2+}-treated *E. coli* will take up circular or linear, single- or double-stranded DNA molecules. Since the transforming DNA does not appear to

Figure 2.5: A model for the uptake of oligomeric plasmid DNA by *Bacillus subtilis.* (i) Competent cells are exposed to trimeric plasmid DNA. (ii) A double-stranded nick is introduced. (iii)-(iv) DNA is taken up in single-stranded form; the other strand is degraded. The single strands, which can be damaged on entry and intracellularly, are converted into acid-soluble products (‹) and fragments smaller than monomeric length. (v) The fragments form partially double-stranded molecules. (vi) Partially double-stranded molecules are converted into completely double-stranded molecules by new DNA synthesis (---). CCC monomeric plasmid DNA is found in recipient cells (adapted from de Vos *et al.*, 1981)

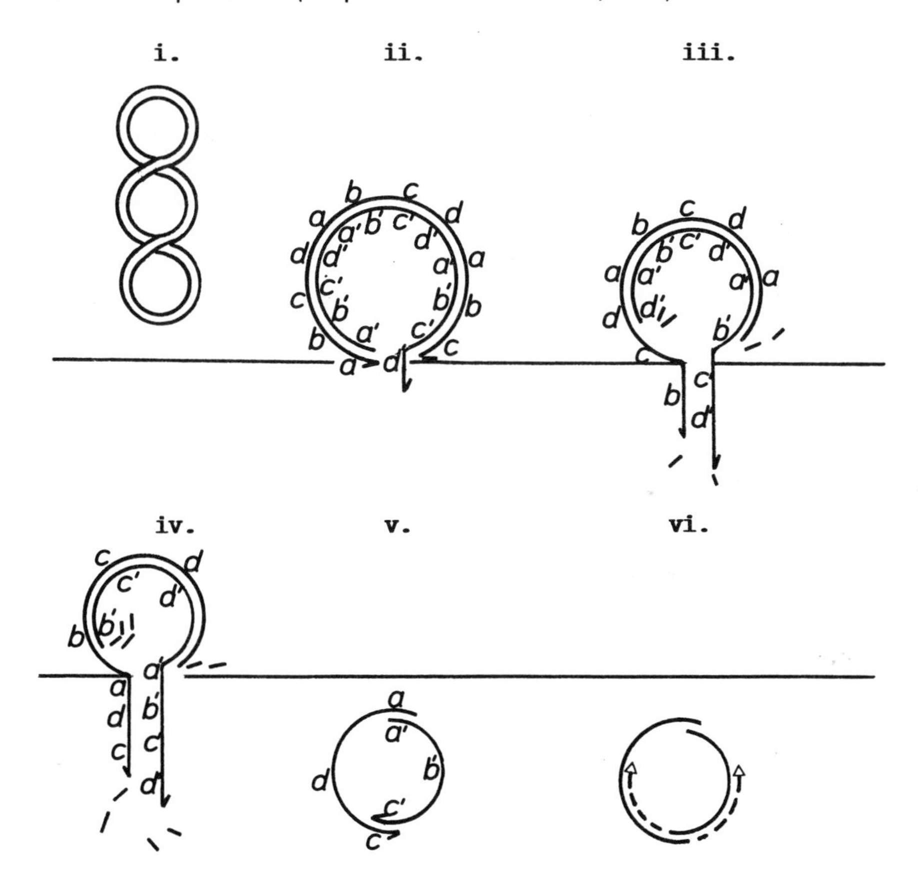

be damaged during transport, circular monomeric plasmid molecules are effective in transformation. Linearised plasmid DNA transforms *E. coli* with low efficiency. Plasmids in the transformants recovered often contain deletions (see section 4.9.1). Modifications to the calcium chloride-temperature shock procedure, for example, by employing other divalent cations, such as Mg^{2+} or Ba^{2+}, have been developed for transforming other bacterial species (see Table 2.2). Freezing and thawing cells in the presence of DNA, a method originally devised for transforming *E. coli* (Dityatkin and Ilyashenko, 1979), has also been successfully adapted for transforming a number of bacteria, including *Agrobacterium tumefaciens* and *Rhizobium* species.

A method of increasing importance and wide applicability is the transformation of protoplasts in the presence of polyethylene glycol (PEG) and divalent cations. Protoplasts are generated by the complete removal of the rigid cell wall by, for example, the action of lysozyme, in the presence of osmotic stabilisers. After transformation the cell wall is regenerated by growth in appropriate media. Protoplast transformation was first developed for *Streptomyces* (Bibb *et al.*, 1978). The technique has since been applied effectively to a number of organisms. In *B. subtilis* it is more efficient than transformation of intact competent cells. Furthermore, transformation of *Bacillus* protoplasts requires neither oligomeric plasmid DNA nor recombinational rescue, since the transforming DNA is not converted into single strands during uptake.

The use of DNA entrapped in liposomes provides an alternative method for transforming certain organisms. Liposome-mediated transformation was first described in Ca^{2+}-treated *E. coli*, albeit at lower frequencies than those attained with free DNA (Fraley *et al.*, 1979). However, transformation and transfection of PEG/protoplasts of *Streptomyces* can be enhanced by encasing the DNA in liposomes (Makins and Holt, 1981; Rodicio and Chater, 1982). The entrapment of DNA in liposomes may overcome certain barriers to transformation (such as the activity of extracellular nucleases) that might exist in some species. Interestingly, liposomes free of DNA can also stimulate transformation of PEG/protoplasts of *Streptomyces* by phage or plasmid DNA.

Most transformable bacterial species are able to take up heterospecific (foreign) DNA in addition to their own DNA. However, bacteria of the genera *Haemophilus* and *Neisseria* discriminate against foreign DNA. In *Haemophilus influenzae* uptake specificity is determined by membrane proteins that recognise the nucleotide sequence 5′-AAGTGCGGTCA-3′ in the DNA (Smith *et al.*, 1981; Pifer and Smith, 1985). The 11 base pair sequence occurs frequently (approximately once in every 4000 base pairs) in *Haemophilus* DNA, but rarely in DNA from unrelated species. The occurrence of analogous uptake specificity in other bacteria may be a potential barrier to the development of new transformation systems. In such cases heterospecific transfer would dictate that the transforming DNA either contains the appropriate uptake sequence intrinsically, or can be engineered to do so. Alternatively, induction of competence by artificial means might circumvent sequence specificity. In *Haemophilus*, uptake specificity may be avoided if cells are rendered competent by Ca^{2+} treatment, rather than relying upon natural competence. It is also noteworthy that Ca^{2+}-treated *E. coli* shows no apparent uptake specificity.

2.2.2 Fungal transformation systems

Natural but low-efficiency transformation has been reported in yeasts. More effective transformation can be achieved by using PEG/protoplasts or spheroplasts (where only part of the cell wall is removed) of yeasts (and of other fungi). Fungal protoplasts may be generated by the action of glucanases, which remove the cell wall. Intact cells of *S. cerevisiae* that have been treated with caesium or lithium in the presence of PEG can also be transformed by plasmid DNA (Ito *et al.*, 1983). Transformation efficiencies comparable to those achieved using protoplast methods can be obtained with some yeast plasmids. Deletion of plasmid sequences occurs during transformation (at a frequency several orders of magnitude higher than spontaneous mutation frequencies). Linearisation of the plasmid molecules by restriction endonuclease cleavage prior to transformation has been found to stimulate deletion formation (Clancy *et al.*, 1984).

In *S. cerevisiae*, efficiencies of transformation are highest if the transforming DNA contains sequences capable of directing autonomous replication in the cell. Furthermore, the topology of the transforming DNA can influence transformation efficiency. Linearisation of yeast integrating (YIp) vectors (see section 3.5.5.a), by restriction endonuclease cleavage, produces 5 to 20 times higher transformation frequencies, presumably as a consequence of stimulating recombination with endogenous DNA (Hinnen and Meyhack, 1982). In contrast, transformation frequencies are reduced two- to three-fold by using linearised yeast extrachromosomal (YEp) vectors (see section 3.5.5.b). Single-stranded circular plasmid DNA molecules have been shown to transform yeast 10 to 30 times more efficiently than double-stranded molecules of identical sequence (Singh *et al.*, 1982).

Various procedures have been employed in order to develop efficient transformation systems for the filamentous fungus *Neurospora crassa*, including treatment with Ca^{2+} (Mishra, 1979), spheroplasting (Case *et al.*, 1979; Schweizer *et al.*, 1981) and the use of DNA encapsulated in liposomes (Radford *et al.*, 1981). Transformation of protoplasts of *Aspergillus nidulans* and of *Cephalosporium acremonium* (*Acremonium chrysogenum*) in the presence of PEG and Ca^{2+} has also been achieved (Ballance *et al.*, 1983; Queener *et al.*, 1985).

2.2.3 Cotransformation

Selection for the uptake of DNA molecules that do not encode readily identifiable phenotypes is generally impracticable, due to the relatively low efficiency of most transformation systems. However, **cotransformation** (or cotransfection) can be performed using a sub-saturating concentration of a DNA species that carries a selectable marker together with a large excess (10^2- to 10^4-fold) of DNA carrying the nonselectable phenotype (see, for example, Kretschmer *et al.*, 1975; Colbère-Garapin *et al.*, 1982). Selection for the marker yields transformants in which a high proportion (up to 80%) have coinherited DNA for the nonselectable phenotype.

2.3 BACTERIAL CONJUGATION

Conjugation is the process of gene transfer that requires cellular contact between donors and recipients. The transfer (*tra*) genes required for conjugation are usually encoded by **conjugative plasmids**. Such plasmids are found in a wide variety of Gram-negative bacteria and in some Gram-positive bacteria. In addition to transferring copies of themselves by conjugation, many conjugative plasmids can mediate transfer of host chromosomal genes between bacteria (a property designated **chromosome mobilisation ability (Cma),** Holloway, 1979). This may involve some form of covalent association of the plasmid with the chromosome as in the formation of **Hfr** (High frequency of recombination) strains or **prime** plasmids. Conjugative plasmids can also mobilise for transfer certain **nonconjugative plasmids** that are resident in the same cell.

2.3.1 Conjugation in Gram-negative bacteria

The conjugative system of F and F-like plasmids has been studied extensively and serves as a model system for understanding the genetic and biochemical basis of conjugation in Gram-negative species.

(a) The F conjugation system

Figure 2.6 describes the main stages in the mating cycle (for reviews, see Manning and Achtman, 1980; Willetts and Skurray, 1980; Willetts and Wilkins, 1984; Willetts, 1985). The F pilus specified by the F plasmid enables donor cells to recognise and contact recipients. Aggregates (comprising differing proportions of donors and recipients) form, are subsequently stabilised, and conjugal DNA metabolism is triggered. Transfer of the plasmid DNA is initiated at a specified site, the **origin of transfer** (*oriT*), and a pre-existing single strand of the plasmid is transferred to the recipient. The transferred strand is replaced in the donor by *de novo* DNA synthesis, and its complement is synthesised in the recipient. Recircularisation of DNA molecules occurs in both cells. Donors and recipients then actively disaggregate.

The organisation and control of the *tra* genes are shown in Figure 2.7. Most of the *tra* genes, including those governing DNA metabolism, pilus

Figure 2.6: A model for the mating cycle of *Escherichia coli*. The F pilus, synthesised by the donor, enables donors to recognise and bind to recipients. Pilus retraction may bring cells into wall-to-wall contact, ultimately to form stable mating pairs. Mating-pair formation triggers conjugal DNA metabolism. A specific strand of plasmid DNA is transferred to the recipient beginning at *oriT*. The transferred strand is replaced in the donor by new DNA synthesis and its complement is synthesised in the recipient. Plasmid recircularisation occurs in both donor and recipient. Donor and recipient actively disaggregate and expression of *tra* genes can occur. The precise timing of certain molecular events in plasmid transfer requires clarification. **O**, Conjugative plasmid DNA; ⌒, pilus. Note: mating pairs or mating aggregates (involving groups of cells) may form during conjugation (adapted from Willetts and Skurray, 1980)

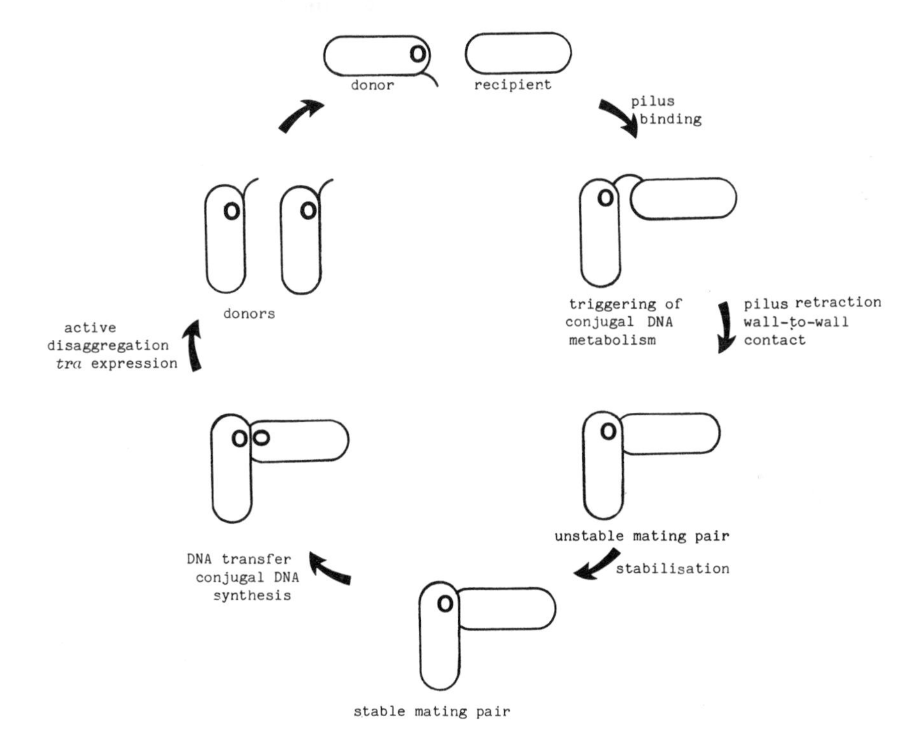

formation and entry (surface) exclusion, are contained within the *tra Y→Z* operon. Transcription of this operon and of *traM* relies upon the *traJ* product, which may be required (either directly or indirectly) to initiate transcription or to act as an antiterminator. Two chromosomal genes, *sfrA* and *sfrB*, appear to be involved in preventing premature termination of transcription of *traJ* and of the *tra Y→Z* operon respectively (Gaffney *et al*., 1983). Whether transcription from promoter pI is also dependent upon *traJ* has yet to be resolved.

Figure 2.7: Organisation and control of the *tra* genes of the F plasmid. Transcripts originate from promoters p*M*, p*J*, p*YZ* and p*I*. Transcription from promoters for *traM* and for the *traY→Z* operon is dependent on the product of *traJ*, which is negatively controlled by the *finOP* products. Pilus formation involves *traA, L, E, K, B, V, W/C, U, F, Q, H* and *G*; stabilisation of mating pairs, *traN* and *G*; conjugative DNA metabolism, *traM, Y, D, I* and *Z*; surface exclusion *traS* and *T*; *oriT*, origin of transfer; o*J*, o*YZ*, operator regions for *traJ* and *traY→Z* respectively (from Willetts and Skurray, 1980; Willetts and Wilkins, 1984)

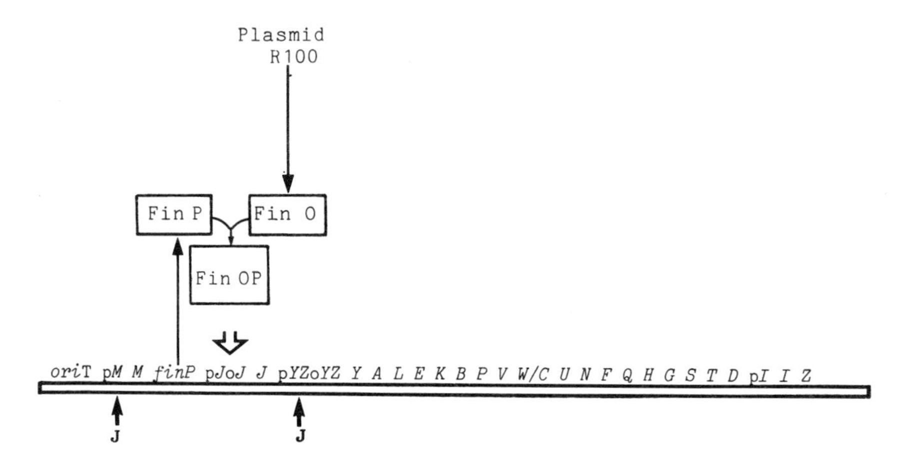

R100 and other F-like plasmids have similar genetic organisation and specify similar pili to the F plasmid. However, unlike F, these plasmids are normally repressed for transfer due to the concerted action of the products of their *finO* and *finP* genes (FinOP fertility inhibition system). The F plasmid is *finO*$^-$ *finP*$^+$ and is therefore derepressed for transfer. F transfer may, however, be repressed when F coresides with another F-like (FinO$^+$) plasmid. The *finO* product of the F-like plasmid together with the F *finP* product inhibit expression of *traJ* and in turn prevent expression of *traM* and of the *tra Y→Z* operon. *finP*, which is plasmid-specific, may encode an antisense RNA that interacts with *traJ* mRNA to prevent its translation. FinO may stabilise the interaction. Such a control mechanism would thus be similar to that for control of ColE1 and R1 replication: FinO being regarded as analogous to Rom and FinP to RNA I (see section 5.4.1) (Sherratt, 1986).

Conjugative systems specified by plasmids other than F and its relatives are less well characterised. However, the broad host range conjugative plasmids (particularly those of the IncP-1 group) have received attention in view of their utility in genetic manipulation. The sex pili produced by IncP-1 plasmids are morphologically distinct from F pili and are able to recognise cell surface components in a wide range of unrelated Gram-negative bacteria. IncP-1 plasmids are also transferred more efficiently when mating takes place on a solid substratum rather than in a liquid medium, which is normally employed in matings with cells harbouring F-like plasmids. The arrangement of genes for conjugation in the prototype IncP-1 plasmid, RP4 (alias RP1), differs from F in that identified genes are located in two separate regions (Figure 2.8) (Willetts, 1985). There is also no fertility inhibition system, and transfer functions are expressed constitutively (Schmidt *et al.*, 1980).

(b) Interaction of conjugative plasmids with the chromosome

Conjugative plasmids can interact with the chromosome in various ways and as a consequence are able to mobilise regions of that chromosome to other bacteria.

Transposable elements (IS*2*, IS*3* and Tn*1000*) present on the F plasmid (see Figure 2.8) have a role in its insertion into and excision from the chromosome. Stable integration of F into the chromosome to form Hfr strains can occur in either $recA^+$ or $recA^-$ backgrounds, although, the majority of Hfrs result from *recA*-dependent events (Cullum and Broda, 1979). The presence of copies of a transposable element in both the F plasmid and the chromosome may afford the necessary homology for *recA*-mediated integration. In $recA^-$ hosts the transposition function of the element may promote F insertion. Hfr formation enables oriented chromosome transfer, which can be exploited in the construction of genetic maps (Bachmann, 1984). The F plasmid can mobilise not only the *E. coli* chromosome, but also the chromosomes of related enterobacteria, including *Erwinia, Klebsiella, Salmonella* and *Shigella.*

Prime plasmids are plasmids that have acquired chromosomal sequences (Low, 1972). F prime (F′) plasmids arise as a result of aberrant excision of an integrated F plasmid from the host chromosome, such that adjacent chromosomal DNA is removed along with F. F′ plasmids can be divided broadly into two classes, type I and type II, depending upon the mode of F excision involved in their formation (Figure 2.9). A type I F′ is formed when excision involves a site on the chromosome and a site within the F plasmid. Residual F DNA sequences remaining in the chromosome as a result of this event constitute a sex factor affinity (*sfa*) locus, which provides homology with sequences on an incoming F plasmid enabling its integration at that locus. A type II F′ is formed when excision involves chromosomal sequences (sometimes these are known insertion sequences, for example IS*5*, (Timmons *et al.*, 1983)) that border both sides of the integrated F plasmid.

Upon transfer of an F′ to an F^- $recA^+$ recipient the plasmid can integrate into the chromosome at that site homologous with the chromosomal sequences carried by the prime plasmid. In the absence of such an homologous site within the recipient chromosome, the F′ may integrate at other

Figure 2.8: Maps of plasmids F and RP4. *tra*, Tra, transfer region; IS*2*, IS*3*, IS*8*, Tn*1*, Tn*1000* (γδ), transposable sequences; *oriV*, origin of vegetative replication; *oriT*, origin of transfer; Ap^r, ampicillin resistance; Km^r, kanamycin resistance; Tc^r, tetracycline resistance. Kilobase coordinates are shown inside the circles

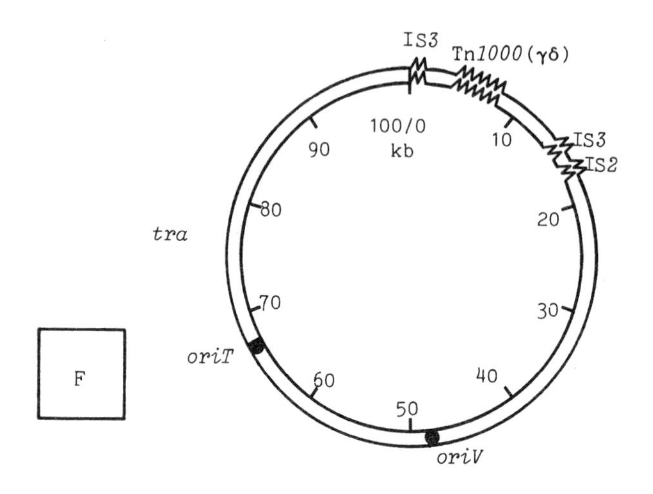

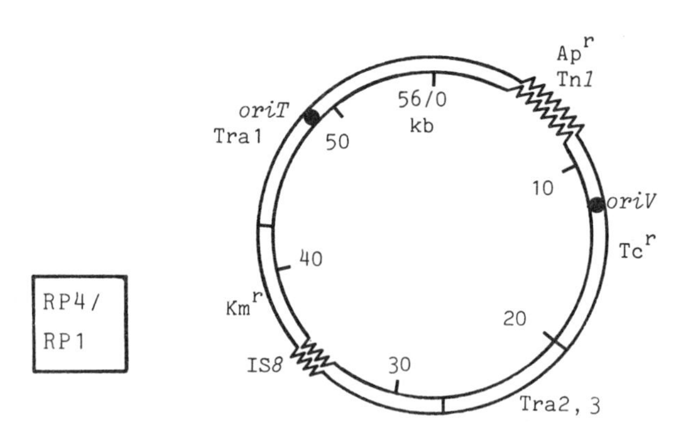

sites. This results in transposition of chromosomal genes, carried by the prime plasmid, to new sites.

F′ plasmids tend to be unstable. They often acquire and lose sequences of DNA either due to recombination events between sites within the plasmid, or by repeated integration into and excision from the chromosome. In these ways novel F′ plasmids can be generated and chromosomal genes transposed to other locations.

Figure 2.9: Modes of aberrant excision of F from the chromosome. Aberrant excision generates substituted plasmids that contain sequences of bacterial DNA. Type I plasmids are formed when excision involves a site on the chromosome and a site within the F plasmid. A type I plasmid thus carries host DNA sequences from only one side of the integrated plasmid and lacks some F plasmid DNA. Type II plasmids are formed when excision involves sites on the chromosome that border both sides of the integrated F plasmid. A type II plasmid thus carries host DNA sequences from both sides of the F insertion and the entire F plasmid

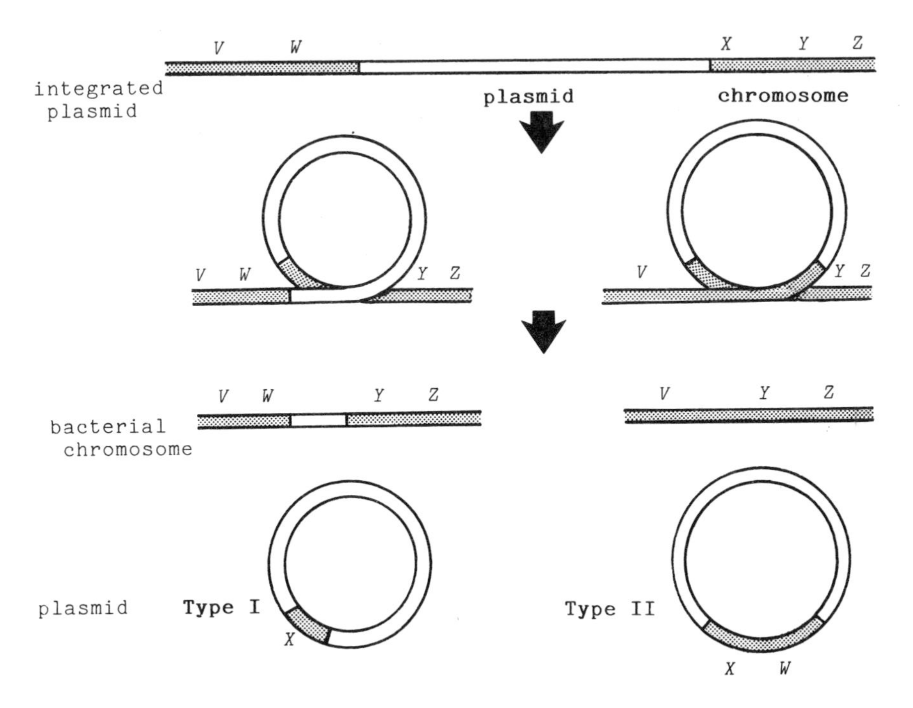

Conjugative plasmids from a number of incompatibility groups, notably IncP, can form primes. This property of P group plasmids, allied with their wide host range, provides scope for the genetic manipulation of a number of bacteria, including *E. coli, Klebsiella pneumoniae, Pseudomonas* spp., *Rhizobium* spp. and *Rhodopseudomonas capsulata.* Various derivatives of P group plasmids have been isolated that possess an enhanced ability to transfer chromosomal genes. The best known of these plasmids is R68.45 (a derivative of R68) isolated from *Pseudomonas aeruginosa* (Haas and Holloway, 1978). This plasmid is able to mobilise the chromosome of a number of bacteria including *Agrobacterium* spp., *Azospirillum* spp., *Pseudomonas putida, Rhizobium* spp. and *Rhodopseudomonas sphaeroides,* in addition to that of *P. aeruginosa.* The presence of IS*21* on R68.45 probably confers enhanced Cma on this plasmid (Willetts *et al.,*

1981). Formation of a cointegrate between the chromosome and plasmid during transposition of IS*21* from R68.45 to the chromosome presumably promotes chromosome mobilisation.

The creation of a region of homology between a specific conjugative plasmid and the host chromosome can endow that plasmid with enhanced Cma. In practice this may be achieved either by inserting a copy of a transposable element (for example Tn*501* or phage Mu) into both chromosome and plasmid, or by incorporating specific duplicated sequences of the host chromosome into the plasmid. The plasmid may then integrate into the chromosome (of a *recA*$^+$ host) through the homologous regions so generated and promote chromosome transfer.

Another procedure for promoting Cma is to utilise the ability of certain conjugative plasmids to suppress temperature sensitive (Ts) defects (normally *dnaA*[2] mutations) in the initiation of chromosome replication. At the restrictive temperature, plasmids that have integrated into the chromosome provide DNA initiation functions *in cis*, permitting chromosome replication. Plasmids exhibiting such **integrative suppression** can promote high-frequency chromosome transfer and can form primes.

(c) Transfer of nonconjugative plasmids

Nonconjugative plasmids can often be transferred between bacteria through the agency of coresident conjugative plasmids. Such transfers can be achieved by a number of mechanisms, of which mobilisation is the most efficient.

1. Mobilisation. Certain nonconjugative plasmids, for example ColE1, can be mobilised for transfer by a coresident conjugative plasmid in a process that does not involve covalent attachment between the participating replicons. The conjugative plasmid provides the effective cellular contact required for mating, but is not necessarily cotransferred. Nonconjugative plasmids participate actively in the mobilisation process. ColE1 possesses a mobility (*mob*) region that specifies *trans*-acting proteins required for mobilisation and has an origin of transfer site (*oriT*) from which transfer can be initiated (see Figure 2.10) (Warren *et al.*, 1978). The *oriT* site is probably the same as the *nic* site (in the *b*asis *o*f *m*obility (*bom*) region) and is required *in cis* for mobilisation. ColE1 can exist as a **relaxation complex** (a supercoiled DNA-protein complex), in which the protein component (relaxation protein(s)) can be induced *in vitro*, by, for example, treatment with ionic detergents, to nick the DNA in a specific strand at the *nic* (*oriT*) site.

When donors harbouring the conjugative and nonconjugative plasmids

---

[2] *dnaA* gene product is required for the initiation of a round of chromosome replication in *E. coli.*

Figure 2.10: Map of ColE1. *cea*, production of colicin E1; *imm*, immunity to colicin E; *mob*, mobilisation; *oriV*, origin of vegetative replication; *oriT*, origin of transfer. Kilobase coordinates are shown inside the circle

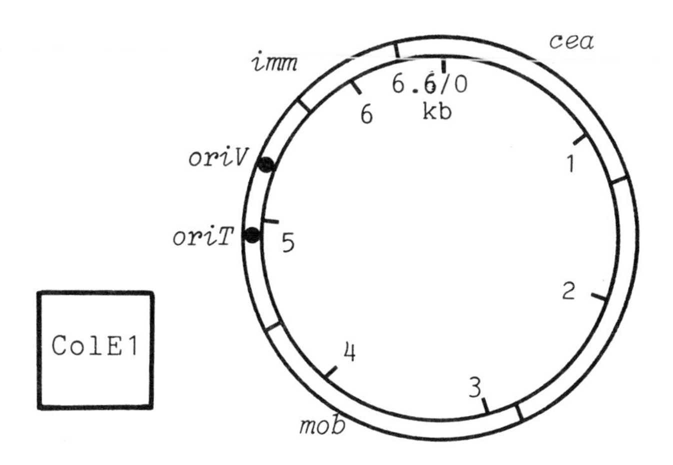

mate with recipients, a mating signal promotes activation of the relaxation proteins. This results in the nicking of a single strand of the DNA-protein complex, such that a 5′ terminus covalently attached to relaxation protein and a 3′ terminus are generated (see Figure 2.11). The relaxation protein then probably acts as a pilot protein leading the 5′ end of the DNA strand into the recipient. The mobilisation genes, which include those encoding relaxation proteins, may be responsible for both nicking and triggering events required for transfer of ColE1. The precise mode of DNA processing during transfer is not completely clear. Several models have been proposed to explain mobilisation: see, for example, Warren *et al.* (1978), Broome-Smith (1980) and Nordheim *et al.* (1980).

The efficiency of mobilisation depends upon the plasmid combinations involved. For instance, ColE1 can be efficiently mobilised by IncI and IncP plasmids, but not by IncW plasmids; whereas RSF1010 is efficiently mobilised by IncP plasmids, but poorly mobilised by plasmids of IncI and IncW groups and not at all by IncF plasmids (Willetts, 1981; Willetts and Wilkins, 1984).

*2. Transposon-mediated transfer (*recA-*independent).* This may occur when one of the plasmids contains a transposable element. *recA*-independent fusion between the conjugative and nonconjugative plasmid can occur during transposition of the transposable element from one plasmid to the other (Figure 2.12). The cointegrate formed may be transferred to recipient cells by utilising the transfer functions specified by the conjugative plasmid. Resolution of the cointegrate in the recipient results in separation

Figure 2.11: Model for mobilisation of nonconjugative plasmids. (i) The donor cell harbours conjugative and nonconjugative plasmids (only the nonconjugative plasmid with relaxation proteins is shown). (ii) Mating promotes activation of relaxation proteins and results in a single-stranded nick at the relaxation site generating a 5′ terminus attached to relaxation protein and a 3′ terminus. (iii) The relaxed single strand with relaxation protein attached to the 5′ leading end is transferred to the recipient. (iv) The transferred strand is recircularised in the recipient. The timing of complementary strand synthesis for the generation of double-stranded molecules is not completely clear. - - - , New DNA synthesis

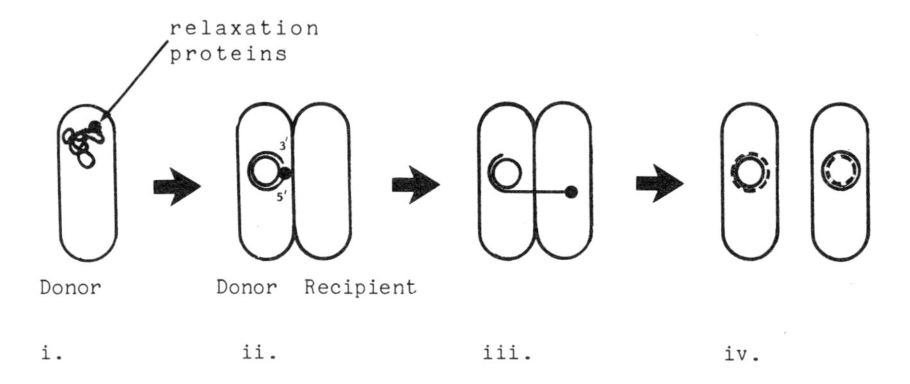

of the two plasmids, each now carrying a copy of the transposable element. Examples of this kind of transfer include transfer of pBR322 by F, accompanied by transposition of Tn*1000* to pBR322 (Guyer, 1978) and transfer of pBR325 by R68.45, accompanied by transposition of IS*21* to pBR325 (Willetts *et al.*, 1981). Such transposon-mediated transfer can be exploited to introduce mutations into cloned genes by transposition mutagenesis (see section 4.7).

3. recA$^+$*-dependent transfer.* A conjugative and nonconjugative plasmid containing homologous DNA regions may become transiently fused by *recA*-dependent homologous recombination. The resultant cointegrate can then be transferred by using the functions of the conjugative plasmid. Resolution of the cointegrate in the recipient by homologous recombination leads to separation of the two plasmids.

2.3.2 Conjugation in Gram-positive bacteria

Gene transfer requiring cell-to-cell contact between donors and recipients and having genetic consequences characteristic of Gram-negative conjugation systems has been described for a number of Gram-positive genera,

Figure 2.12: Transposon-mediated transfer of nonconjugative plasmids. (i) A conjugative plasmid that carries a transposon coresides with a nonconjugative plasmid. (ii) A cointegrate is formed between the conjugative and nonconjugative plasmids during intermolecular transposition of the Tn. (iii) The cointegrate may be transferred from donor to recipient cells (by using functions specified by the conjugative plasmid), where resolution of the cointegrate occurs. Both plasmids now carry a Tn

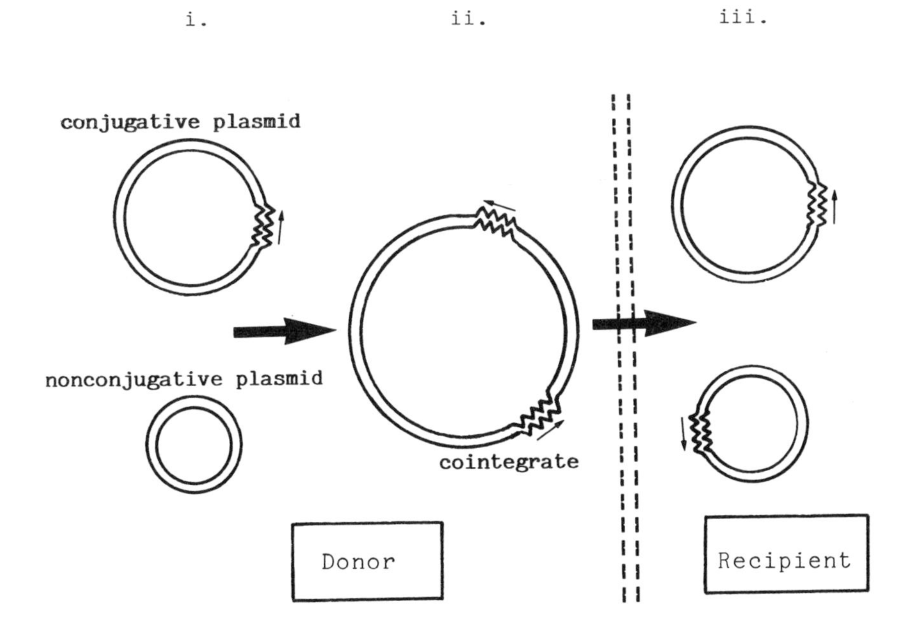

including *Clostridium, Staphylococcus, Streptococcus* and *Streptomyces.* The mechanics of the conjugation processes are poorly defined and may well differ from those in Gram-negative organisms.

(a) Streptomyces

Many *Streptomyces* conjugative plasmids have been found (for example, SCP1 and SCP2 in *S. coelicolor*; SLP2 in *S. lividans 66*; pIJ101 in *S. lividans* ISP 5434; SRP1 in *S. rimosus*) that are capable of promoting generalised chromosome recombination (Hopwood and Chater, 1984). Plasmid SCP1 is similar, in a number of ways, to the F plasmid, existing autonomously (SCP1$^+$), integrated into the chromosome (SCP1) or as a prime plasmid (SCP1′). The chromosome of *S. coelicolor* contains a number of cryptic plasmids. These are only detected after mating with species, such as *S. lividans,* where the plasmids are maintained extrachromosomally.

A phenomenon often accompanying the transfer of conjugative *Strepto-*

myces plasmids to plasmid-free recipients is the formation of **pocks**. These are zones of recipient growth inhibition in areas where plasmid transfer has occurred (see Figure 2.13). The inhibition reaction has been called 'lethal zygosis' (Ltz), although the precise mechanism involved is not yet known. Production of pocks provides a useful visual aid for detecting clones carrying conjugative plasmids (see Chater and Hopwood, 1984).

(b) Streptococcus

A number of conjugative plasmids have been found in strains of *Streptococcus faecalis* and other streptococci. In addition to self-transfer, these plasmids are capable of mobilising certain nonconjugative plasmids and chromosomal genes (Franke *et al.*, 1978; Clewell, 1981). Two types of conjugative streptococcal plasmid have been distinguished on the basis of transfer ability. One class transfers poorly in broth (but relatively efficiently on solid substrata) and has a host range that includes various streptococci

Figure 2.13: Pocks produced when plasmid-containing *Streptomyces lavendulae* grows in a background of plasmid-free individuals (magnification × 4). Courtesy of E.M. Wellington

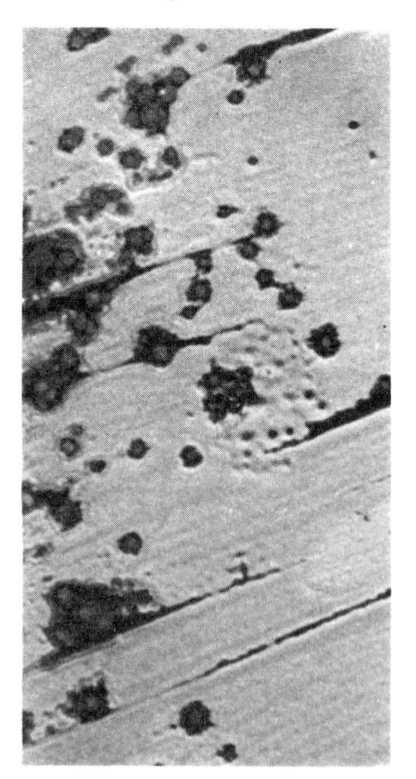

and other Gram-positive species. The other class transfers efficiently in broth, but has a host range apparently limited to *S. faecalis* (see Clewell, 1981). High frequency transfer of these plasmids involves specific peptide sex pheromones, designated clumping-inducing agents (CIAs), which facilitate mating aggregate formation in *S. faecalis* (Clewell, 1981; Clewell *et al.*, 1982, 1984; Tortorello and Dunny, 1985). CIAs are produced by recipient cells and induce certain donor cells to synthesise a proteinaceous adhesin and aggregate (with themselves and with recipients). These pheromones additionally induce a function(s) related to plasmid DNA transfer *per se* (Clewell and Brown, 1980). Various recipient strains produce multiple CIAs, each of which is specific for donors harbouring a particular class of conjugative plasmid. Recipients that have acquired a conjugative plasmid no longer produce the particular CIA to which that plasmid determines a response. The CIA molecule appears to be inactivated by a plasmid-determined modification enzyme (Ike *et al.*, 1983). The production of different CIAs by a specific recipient enables that strain to acquire conjugative plasmids from a range of donors (such a cell is thus 'promiscuously receptive').

Transfer of chromosomal determinants in the absence of detectable conjugative plasmids has been reported in a number of streptococci (see, for example, Buu-hoi and Horodniceanu, 1980; Franke and Clewell, 1981). The precise nature of the process is not clear. In some systems at least, the conjugative elements may represent **conjugative transposons** (Gawron-Burke and Clewell, 1982; Smith and Guild, 1982; Clewell *et al.*, 1985; Fitzgerald and Clewell, 1985). Transfer may thus involve a transposition event in which the donor and recipient replicons are in separate cells.

2.4 TRANSDUCTION

Transduction is the process of gene transfer that is mediated by a bacteriophage (Zinder and Lederberg, 1952). In **generalised transduction** random or quasi-random tracts of donor DNA are packaged in phage capsids and transferred to recipient cells. In **specialised transduction** specific regions of bacterial DNA in covalent union with phage DNA are packaged and transferred. Both types of transduction are normally mediated by temperate phages, although certain virulent phages can mediate generalised transduction. Transducing phages are found in a number of bacteria including *E. coli* (for example phages λ and P1); *Salmonella typhimurium* (for example phage P22); *Bacillus subtilis* (for example phage SPP1), *Pseudomonas aeruginosa* (for example phage F116L) and *Streptomyces venezuelae* (for example actinophage ϕSV1).

2.4.1 Generalised transduction

Coliphage P1 has been widely used in generalised transduction and is described here as a model system for phage-mediated gene transfer. Production of generalised transducing P1 phages occurs by accidental packaging of host DNA into virions during phage morphogenesis. DNA is packaged by the headful mechanism, and the maximum size of transducing DNA is generally equivalent to the size of the mature phage genome. An ability to degrade host DNA upon infection allied with weak packaging specificity is important in the formation of transducing phages. In those cases where transducing DNA does not contain a replication origin, for example fragments of chromosomal DNA, inheritance of the transferred DNA will depend upon recombination with a resident replicon in the recipient. High-frequency transducing mutants, which give an increased proportion of transducing phages, have been isolated. Lysates of such P1 mutants transduce markers with increased efficiency (Masters, 1985).

Chromosomal and extrachromosomal DNA can be transferred by generalised transducing phages. Since the amount of DNA that can be transduced is limited by the capacity of the phage capsid, **transductional shortening** of extrachromosomal elements that are larger than the phage genome can occur (Low, 1972). In other cases, where plasmids are considerably smaller than the phage genome, oligomers or plasmid cointegrates (comprising two or more different plasmids) of sufficient size to exceed the minimum packageable length may be transduced (Novick *et al.*, 1981).

2.4.2 Specialised transduction

The temperate phage lambda (λ) is the classical specialised transducing phage of *E. coli* (see Hendrix *et al.*, 1983). When phage λ infects a sensitive host, the λ genome circularises by the annealing of its cohesive ends (the site of the fused ends is termed the cohesive end (*cos*) site) and may either initiate the lytic cycle or establish lysogeny. Lysogeny normally involves the insertion of a λ prophage into the *E. coli* chromosome at a specific site, designated *attB*, which lies between the *galE* gene and the *bioA* gene. This site shares a 15 base pair region of homology (a core region, designated 0) with the *attP* site on the λ genome. A consequence of λ integration is the formation of a linear prophage with terminal hybrid attachment sites, *attL* (BOP′) and *attR* (POB′), that separate it from adjacent bacterial DNA (see Figure 2.14). λ-encoded genes for lytic functions are repressed by λ*c*I repressor. The host cell harbouring the λ prophage is termed a lysogen. The prophage can be induced to enter the lytic cycle by subjecting the lysogen to treatments, such as UV irradiation, that damage DNA and result in the inactivation of the λ*c*I repressor. During induction the λ genome is excised

Figure 2.14: Model for integration of bacteriophage λ. The viral genome circularises by the annealing of its cohesive ends (*cos* site is formed). Recombination involves the *att*λ*P* site on the λ chromosome and the *att*λ*B* site on the bacterial chromosome. These sites have a region of sequence homology (core region, 0). Integration generates two hybrid attachment sites BOP′ (*attL*) and POB′ (*attR*) flanking the λ prophage. *J*, *A*, *R*, *int*, Genes on λ chromosome; *gal*, bacterial gene for galactose metabolism; *bio*, bacterial gene for biotin synthesis

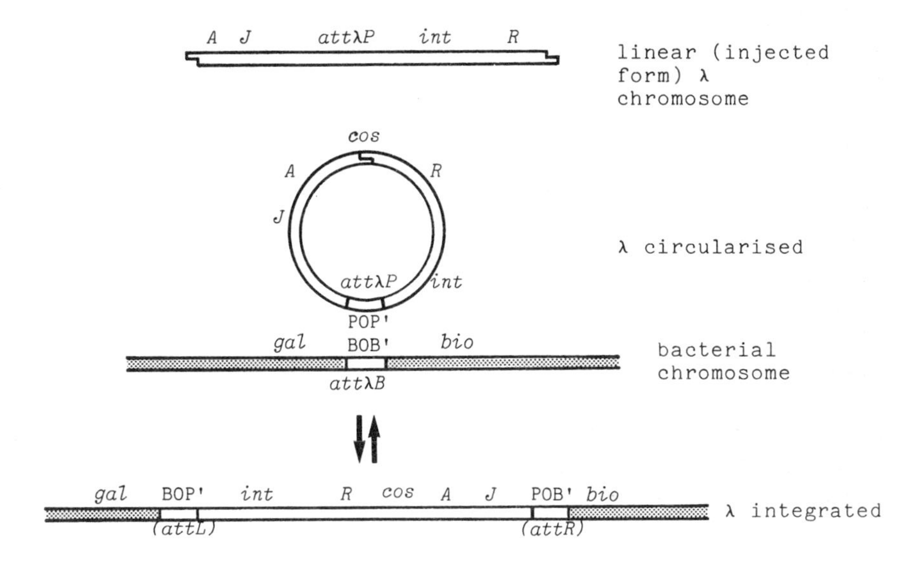

from the host chromosome (for a detailed description of the genetic and biochemical events involved in λ induction see Hendrix *et al.*, 1983).

Integration and excision of the λ genome are promoted by the phage-encoded proteins Int and Xis. Int is required for integration and excision, whereas Xis is required for excision only. Usually, but not invariably, excision of the genome is precise, involving specific recombination between the terminal attachment sites, in turn generating the complete vegetative λ genome with *attP* intact. Occasionally aberrant excision occurs, when recombination involves sequences within the prophage and bacterial sequences that flank the attachment sites. Such excision events result in the removal of bacterial DNA that is contiguous with one or other of the attachment sites, together with phage DNA, leading to the production of specialised transducing phages (for example, λ*gal* and λ*bio*) (Figure 2.15). Phage DNA sequences (normally sequences in the prophage most distal to those bacterial genes included in the specialised transducing phage) are often deleted during these events. For example, λ*gal* phages lack certain genes that are essential to λ morphogenesis. Such phages are therefore

Figure 2.15: Model for aberrant excision of bacteriophage λ. When excision of λ involves sites within the prophage and sites in the bacterial chromosome that flank the attachment sites, specialised transducing phages carrying bacterial DNA are generated. Recombination involving bacterial sites (i) to the left of the attachment site generates λ*gal*, and (ii) to the right generates λ*bio* (see Figure 2.14 for abbreviations)

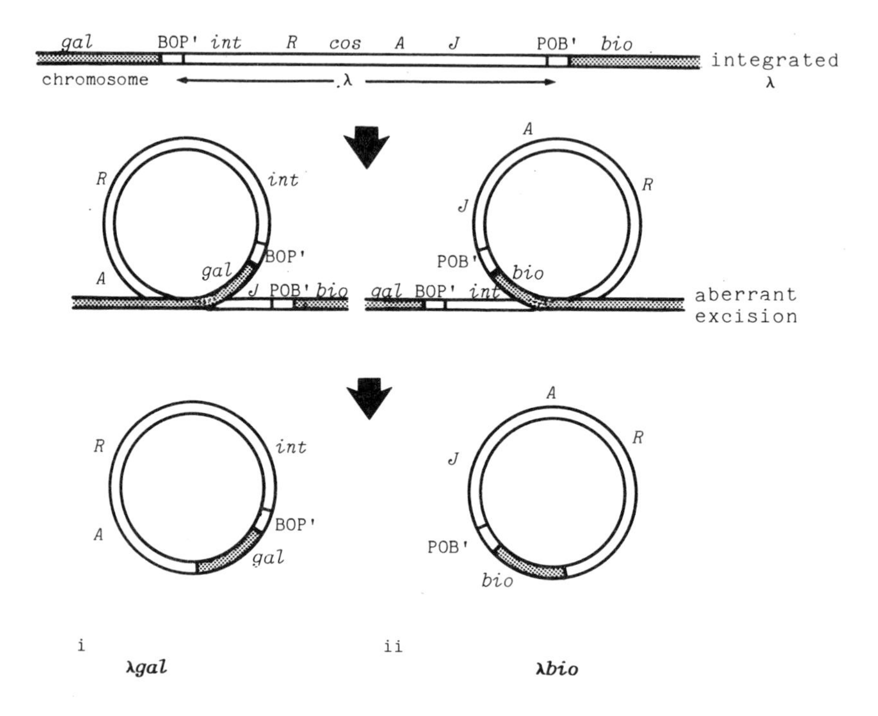

defective. Erroneous excision of λ is a rare event, thus phage lysates obtained from λ lysogens, following induction, contain only a small proportion of specialised transducing phages (for either the *gal* or *bio* region of the *E. coli* chromosome). Such lysates are termed **low-frequency transducing (LFT)** lysates. Infection of a Gal^- recipient with an LFT lysate and selection for Gal^+ transductants commonly yields transductants that are merodiploid and double lysogens, of genotype $gal^-/gal^+/\lambda$. Induction of such lysogens generates a phage lysate containing approximately equal numbers of λgal^+ and λ phages. Although the λgal^+ genome is defective, the presence of the wild-type λ genome in the double lysogen enables both genomes to participate in the lytic cycle and contribute phages to the lysate. The lysate is thus a **high-frequency transducing (HFT)** lysate.

λ transduction is normally limited to those markers bordering the primary attachment site (*attB*) on the *E. coli* chromosome. However, it is possible to extend the range of markers transferred by deleting *attB* and exploiting the ability of λ to integrate, albeit inefficiently, at a number of secondary attachment sites that show partial homology with *attB* (see, for example, Shimada *et al.*, 1972; McIntire and Willetts, 1978). Alternatively, insertion could be achieved by *recA*-dependent recombination between the desired integration site and an homologous region incorporated into the λ genome. Phage Mu::λ hybrids are particularly powerful in this respect. Such hybrids can integrate by using the homology between Mu sequences of the Mu::λ hybrid and Mu sequences in the chromosome (at those sites where Mu has previously been integrated). These hybrids enable various manipulations by utilising the specific properties of the transposable element, Mu, in conjunction with the specialised transducing ability of λ (see, for example, Casadaban, 1976; MacNeil *et al.*, 1980; Bremer *et al.*, 1985).

The amount of bacterial DNA that can be transduced by phage λ can be increased by using λ derivatives that carry deletions in non-essential regions of the λ genome. λ deletion mutants can be used to transduce nonconjugative plasmids, provided that the phage or the plasmid DNA carries a transposable element (Piffaretti and Fayet, 1981). The cointegrate formed during transposition of the transposable element from one replicon to the other can be packaged in λ phage heads and transferred, as long as the amount of DNA does not exceed the maximum packageable length. Resolution of the cointegrate in the recipient enables transduction of the plasmid. A similar molecular mechanism involving cointegrates has been proposed to account, at least in part, for Pl-mediated transduction of small plasmids (Iida *et al.*, 1981).

2.4.3 Uses of transduction in genetic manipulation

Although transduction has classically been used for fine structure genetic mapping (Hayes, 1968), the process can be used in a variety of other genetic manipulations including:

(i) *Strain construction.* A particular gene or cluster of genes can be transferred between strains. Cotransduction can be exploited in the transfer of markers that have no readily identifiable phenotype by selecting for transfer of a neighbouring gene(s).
(ii) *Creation of partial diploids.* Partially diploid strains enable complementation analyses to be performed.
(iii) *Gene transpositions.* By directing the insertion of λ into sites other than *attB*, transposition of bacterial genes can be effected.
(iv) *Gene enrichment and purification.* The induction of double λ lysogens yielding HFT lysates (section 2.4.2) results in increased dosage of specific bacterial genes that are carried on the specialised transducing

phages. Gene dosage can be further amplified by employing λ mutants that are defective in late functions of the lytic cycle, for example *Sam* mutants. Amber (*am*) mutations in the *S* gene prevent cell lysis in nonsuppressing hosts. In this way purified bacterial DNA can be obtained in large quantities from infected cells.

(v) *Enhanced expression of bacterial genes.* This can be accomplished with specialised transducing phages by increasing gene dosage (as described in (iv)) in conjunction with efficient transcription initiated from promoters located within the phage genome.

(vi) *Localised mutagenesis.* Mutagenesis of transducing phages and subsequent transfer of the DNA to unmutagenised hosts enables induction of genetic lesions to be limited to those regions of a genome carried by the phage (Hong and Ames, 1971).

(vii) *Isolation of deletion mutants of plasmids.* Such mutant plasmids can be obtained through transductional shortening (section 2.4.1), where the size of the 'wild-type' plasmid exceeds the maximum packageable length for the particular transducing phage involved. Induction of the prophage of a plasmid::prophage cointegrate can also provide a source of deletion mutants. Deletants of the F plasmid have been obtained by induction of F*lac*::λ hybrids (where λ prophage is integrated at various secondary attachment sites on F). Improper excision of λ results in the formation of specialised transducing phages and F deletion mutants as survivors to induction (McIntire and Willetts, 1978).

2.5 THE SEXUAL CYCLE OF FUNGI

Basic stages in the sexual cycle of fungi are outlined in Figure 2.16. Fusion of haploid (n) gamete nuclei (**karyogamy**), which normally occurs in specialised cells, generates the diploid chromosome number ($2n$). In **heterothallic** fungi, gamete nuclei must be contributed by strains of different mating type, whereas in **homothallic** fungi each strain is self-fertile and consequently can undergo karyogamy on its own. Diploidy gives rise to haploidy via the process of meiosis. The immediate products of meiotic division are a tetrad of haploid nuclei. Meiosis may or may not be followed by a mitotic division producing eight n nuclei. These haploid nuclei, which are normally enclosed in spores, will possess combinations of genes contributed by the parents of the cross and assorted at meiosis. Sexual reproduction thus normally involves alternation between haploid and diploid phases, and the relative duration of each phase varies with different fungi. Where meiosis follows immediately after karyogamy, the fungi are predominantly haploid (for example, the Basidiomycotina and the majority of the Ascomycotina). Where karyogamy occurs immediately or soon after

Figure 2.16: Basic stages in the sexual cycle of fungi. Haploid (*n*) and diploid (2*n*) phases are shown.

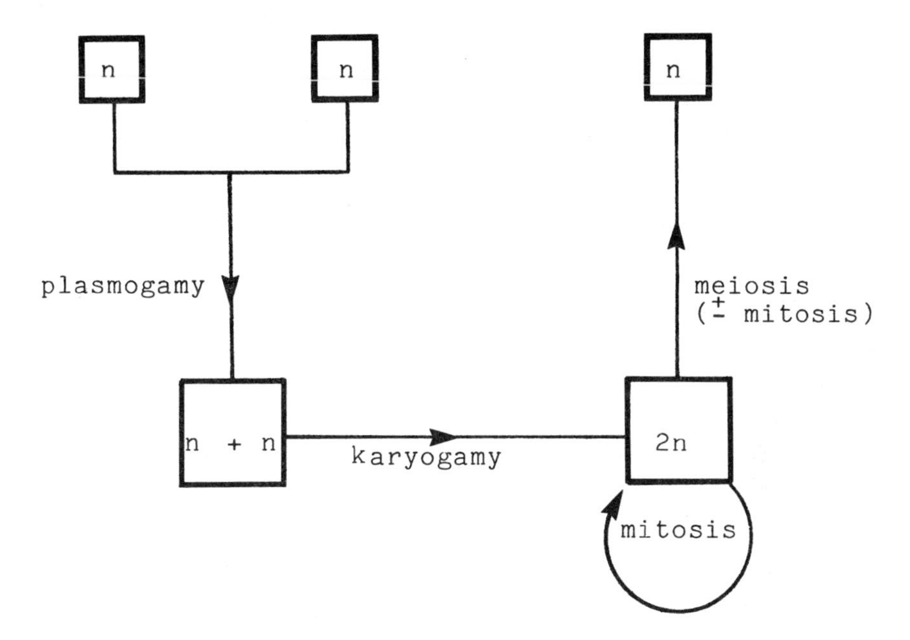

meiosis, the fungi are predominantly diploid under normal conditions (for example, certain yeasts, including *Saccharomyces cerevisiae*).

The potential of the sexual cycle for generating novel genotypes in fungi depends both upon mechanisms that minimise self-fertilisation and upon mechanisms that promote outbreeding. **Sexual dimorphism** (dioecism), in which male and female sex organs are produced on different individuals, and heterothallism both afford barriers to self-fertilisation.

2.5.1 Heterothallism

Heterothallism may involve two mating types encoded by alternate alleles at a specific locus, or numerous mating types determined by multiple alleles of distinct mating-type genes (for explanations of the different nomenclature used for heterothallic systems, refer to Burnett, 1975; Fincham *et al.*, 1979). Where two mating types are involved, there is no discrimination against mating of close relatives: the mean breeding compatibility among sisters of a zygote will be the same as it is among unrelated haploids of the population. However, in multiple-allele heterothallism, compatibility depends upon the two haploid individuals involved differing with respect to each of the mating-type factors. Thus where there are large numbers of different factors in the breeding population the probability of compatibility between sisters of a zygote will be less than between unrelated haploids. Multiple allele systems therefore not only prevent self-fertilisation, as do systems involving two mating types, but additionally encourage outbreeding as opposed to inbreeding (Mather, 1942). In certain groups of fungi, conversion of one mating type to another or to a state permitting self-fertility can occur. This can effectively impose inbreeding on fundamentally outbreeding systems.

2.5.2 Mating-type switching in *Saccharomyces*

The genetic information controlling *a* and α mating types in *Saccharomyces* is carried on transposable elements referred to as *a* and α 'cassettes' (Hicks *et al.*, 1977). These cassettes reside at three distinct loci (*HML*, *HMR* and mating-type locus *MAT*) on yeast chromosome III (see Figure 2.17). Cryptic 'cassettes', repressed by *SIR* (silent information regulator) gene products, are located at *HMR* and *HML* (normally strains carry *HMRa* and *HML*α, to the right and left of *MAT* respectively), whereas the 'cassette' at *MAT* (either *a* or α) is normally expressed. Thus the mating type of the cells is controlled by the information at *MAT* (*MATa* or *MAT*α). In heterothallic strains, which carry the recessive homothallism allele (*ho*), *MATa* and *MAT*α are rarely interconverted, whereas in homothallic strains, which carry the dominant homothallism allele (*HO*), *MATa* and *MAT*α are interconvertible. This enables mating-type switching at virtually every cell division until conjugation between cells of *a* and α mating types leads to the formation of *a*/α diploids, whereupon switching ceases. Interconversion of mating type is due to transposition of a copy of the heterologous mating-type 'cassette' from the appropriate *HM* locus to *MAT*. The transposed 'cassette' is expressed, whereas the 'cassette' previously at *MAT* is discarded. (Transposition of mating-type information is, therefore, a non-reciprocal event.) The sequences at *HML* or *HMR*, copied for transposition, are preserved at the *HM* locus and can be used again in subsequent switching events.

Mating-type 'cassettes' apparently determine cell type by controlling expression of unlinked *a* or α specific genes. *MAT*α performs this through the action of at least two genes: *MAT*α*1* activates α-specific genes, *MAT*α*2* represses *a*-specific genes, which are otherwise expressed constitutively. In *MATa*/*MAT*α diploids, *MATa* in conjunction with *MAT*α*2* prevents expression of mating functions, while permitting sporulation (Nasmyth, 1982; Haber, 1983; Sprague *et al.*, 1983). Other functions, such as resistance to mutagens and budding pattern, also appear to be governed by the *MAT* locus. Thorough characterisation of this locus should, therefore, facilitate the construction of novel yeast strains.

Figure 2.17: Model for mating-type conversion in *Saccharomyces*. Mating-type depends upon expression of the mating-type cassette at the *MAT* locus on chromosome III. Switching occurs as a result of transposition of a copy of the opposite mating-type information from one of the two *HM* loci to *MAT*. In most laboratory strains a silent copy of α information is on the left (*HML*) and a silent copy of *a* on the right (*HMR*) arm of chromosome III (as shown). Transposition of cassette *a* to *MAT* and excision of the α sequences originally at that site converts the α cell, with α cassette at *MAT* (i), to the *a* cell, with *a* cassette at *MAT* (ii). In the α cell *MAT*α*l* positively regulates α-specific genes, *MAT*α*2* negatively regulates *a* specific genes. In the *a* cell, *a*-specific genes are constitutive. ●, Centromere; HO, dominant homothallism allele; *SIR*, silent information regulator; *MAT*, mating-type locus

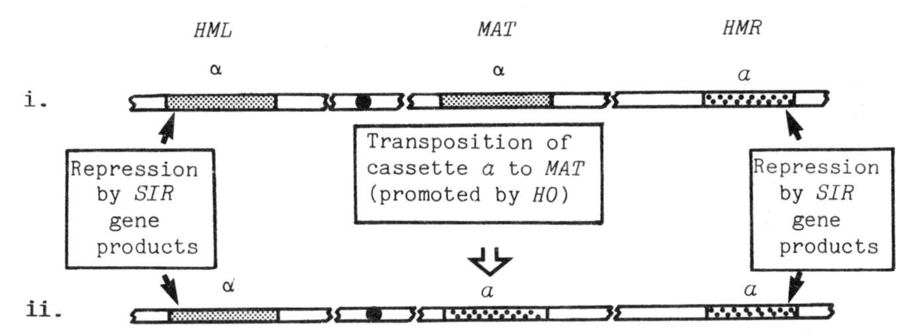

2.5.3 Polyploidy

Polyploidy refers to the condition where an individual possesses more than two whole sets of chromosomes. Triploids ($3n$) with three sets, tetraploids ($4n$) with four sets, and higher polyploids have been reported in fungi. (In **autopolyploids** all the chromosome sets are derived from the same species and fully homologous with each other, and in **allopolyploids** different chromosome sets are derived from different species.) Polyploidy is common in yeasts and normally arises as a result of fusion of haploid and diploid or diploid and diploid of opposite mating type, for triploid and tetraploid, respectively.

In tetraploids (and other even-numbered polyploids) there is a reasonable frequency of viable meiotic products (given a reasonable frequency of regular chromosome disjunction). However, in triploids the viability of meiotic products is greatly reduced. This is primarily due to disjunction at meiosis resulting in aneuploids possessing one copy of some chromosomes and two copies of others. Most of these meiotic products do not survive. Only those aneuploids with chromosome numbers close to the haploid or diploid number are reasonably viable. Other strains of higher odd ploidy also produce many inviable spores and some aneuploid viable spores.

2.6 THE PARASEXUAL CYCLE IN FUNGI

The **parasexual cycle** is found in a variety of fungi, notably the Ascomycotina and Deuteromycotina (for reviews refer to Pontecorvo, 1956; Roper, 1966; Caten, 1981). An essential preliminary to the parasexual crossing of strains is the establishment of **heterokaryosis**. The **heterokaryon** is characterised by the presence of genetically distinct nuclei in a common cytoplasm and is normally formed by the fusion of hyphae of genetically different strains and subsequent mixing of the nuclei (Figure 2.18). (Where formation of the heterokaryon proves difficult due to failure of anastomosis, induced fusion of fungal protoplasts can often be used to initiate the cycle (see section 2.7).) In the parasexual cycle, fusion of unlike haploid nuclei occurs in the heterokaryon to give heterozygous diploid nuclei. The frequency of such fusions varies for different species but normally occurs rarely. Such diploid nuclei can divide repeatedly by mitosis producing clones of diploid descendants. Haploid segregants can arise from the diploids by the process of **haploidisation**. This occurs as a consequence of **nondisjunction** at mitosis. Nondisjunction involves the abnormal segregation of homologous chromosomes during nuclear division, such that aneuploids (hypodiploids and hyperdiploids) are generated. Loss of a single chromosome from the hyperdiploid by nondisjunction results in the

Figure 2.18: Heterokaryon. Formation of the heterokaryon can be effected by the fusion of hyphae carrying genetically distinct nuclei (● ○) and their subsequent mixing

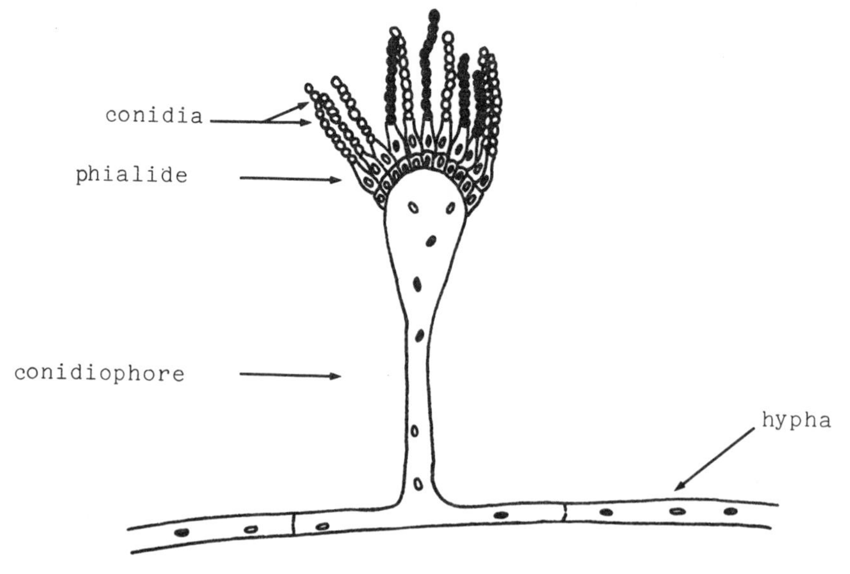

diploid number. This is described as nondisjunctional diploid formation. Sequential loss of chromosomes from the hypodiploid, by repeated nondisjunction, ultimately produces the haploid chromosome number, a process termed haploidisation. The basic stages in the parasexual cycle are summarised in Figure 2.19.

Genetic variation depends upon the underlying processes of mitotic recombination, which occurs during multiplication of the diploid nuclei, nondisjunction and haploidisation. Mitotic recombination permits the rearrangement of alleles between homologous chromosomes, whereas haploidisation enables the reassortment of whole chromosomes. Haploidisation thus not only facilitates the assignment of genes to linkage (haploidisation) groups, but also engineers the exchange of whole chromosomes

Figure 2.19: Basic stages in the parasexual cycle of fungi

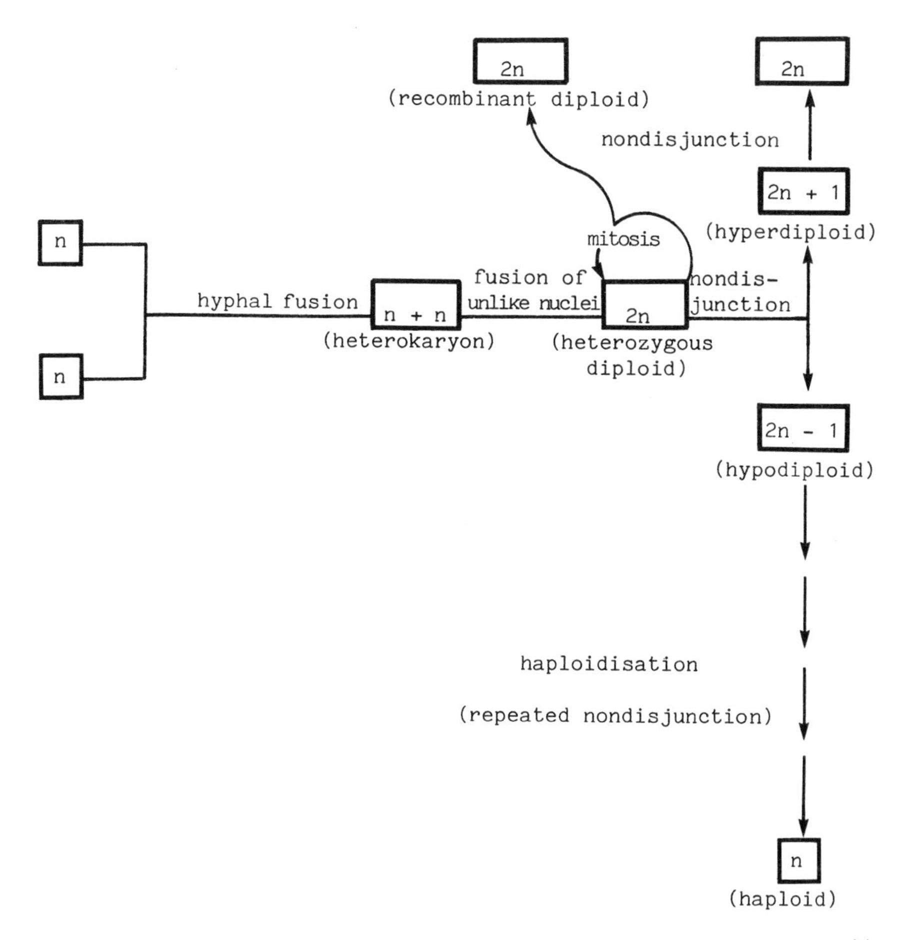

between closely related strains. Generally these processes occur rather infrequently. There are, however, a number of chemical and physical agents that can induce mitotic crossing-over and haploidisation. For example, in *Penicillium chrysogenum*, parafluorophenylalanine induces haploidisation, and 5-fluorouracil and gamma-rays both increase mitotic crossing-over.

A drawback in utilising the parasexual cycle in breeding programmes has been the lack of recombination generally found. This has been attributed largely to the use of strains late in their industrial history, such that the genetic structure of homologous chromosomes of the parents has diverged through repeated mutations and chromosome rearrangements. As a consequence many of the products of recombination events would be inviable and haploids of parental genotype would be preferentially recovered (a phenomenon termed **parental genotype (genome) segregation**). This can be circumvented by the use of closely related strains, differing from each other in only a few mutational steps (see section 6.2.2.a).

2.7 PROTOPLAST FUSION

Induced protoplast fusion has been reported in both prokaryotic and eukaryotic microorganisms (for reviews see Peberdy, 1979; Ferenczy, 1981; Hopwood, 1981; Peberdy and Ferenczy, 1985). It is a particularly useful technique for achieving gene transfer and genetic recombination in organisms with no efficient, natural gene-transfer mechanism. Furthermore, it provides a means of intraspecific, interspecific and intergeneric transfer (see Table 2.3). Interestingly, more than two strains can be combined in one fusion, generating recombinants that have inherited genes from all parents in the fusion (see, for example, Hopwood and Wright, 1978). Polyethylene glycol (PEG) (of varying molecular weight and concentration depending upon the organism) is typically used as the fusogenic agent, normally in the presence of Ca^{2+}. The basic procedure involves PEG-induced fusion of protoplasts followed by their regeneration as normal cells (Figure 2.20). Electrofusion may be used as an alternative to PEG (see, for example, Halfmann *et al.*, 1983). It is noteworthy that the act of protoplasting and regeneration may result in the loss (curing) of resident plasmids.

An important feature of bacterial protoplast fusion is that it enables establishment of a transient diploid or quasi-diploid state during fusion. This permits recombination between complete chromosomes, as opposed to fragments of the donor chromosome (normally transferred during transformation, conjugation or transduction) and the recipient chromosome. The fusion-induced recombinants are, however, normally haploid.

Protoplast fusion in fungi often results in the formation of a hetero-

Table 2.3: Microbial protoplast fusions

Intraspecific

Prokaryotes:
Bacillus spp., for example *B. megaterium, B. subtilis*
Brevibacterium spp., for example *B. flavum, B. lactofermentum*
Corynebacterium glutamicum
Escherichia coli
Micromonospora spp., for example *M. echinospora, M. rosaria*
Providencia alcalifaciens
Streptomyces spp., for example *S. antibioticus, S. coelicolor, S. fradiae, S. griseus, S. lividans, S. parvulus, S. viridosporus*

Eukaryotes:
Aspergillus spp., for example *A. flavus, A. nidulans, A. niger*
Candida tropicalis
Cephalosporium acremonium (*Acremonium chrysogenum*)
Kluyveromyces lactis
Penicillium spp., for example *P. chrysogenum, P. cyaneo-fulvum, P. frequentans, P. roquefortii*
Saccharomyces cerevisiae
Schizosaccharomyces pombe
Yarrowia lipolytica

Interspecific

Prokaryotes:
Bacillus cereus × *B. thuringiensis*
Brevibacterium flavum × *B. lactofermentum*
Streptomyces fradiae × *S. narbonensis*
S. viridosporus × *S. setonii*

Eukaryotes:
Aspergillus nidulans × *A. rugulosus*
Penicillium chrysogenum × *P. cyaneo-fulvum*
P. chrysogenum × *P. notatum*
P. cyaneo-fulvum × *P. citrinum*

Intergeneric

Prokaryotes:
Bacillus subtilis × *Escherichia coli*
Brevibacterium flavum × *Corynebacterium glutamicum*

Eukaryotes:
Candida tropicalis × *Saccharomycopsis fibuligero*
Hansenula wingei × *Saccharomyces cerevisiae*

Figure 2.20: Fungal protoplast fusion. ○, ●, Genetically distinct nuclei of parental strains; PEG, polyethylene glycol

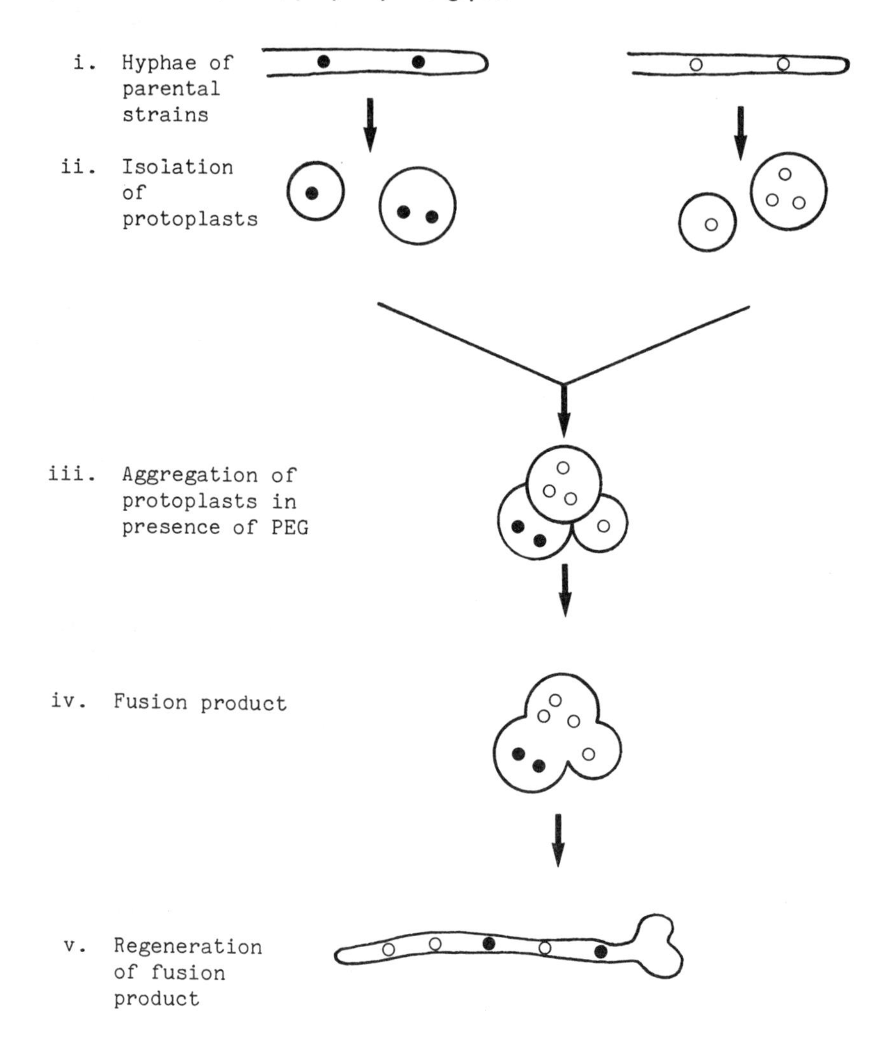

karyon with the occasional occurrence of stable or partial diploids. Such events constitute initial stages in the parasexual cycle. Haploid recombinants may be obtained from the fusion depending upon how far the cycle has progressed (refer to section 2.6). In some cases fusion results in the production of unstable heterokaryons, in which the nucleus of one parent or the other ultimately prevails. Importantly, fungal protoplast fusion provides a means of overcoming certain compatibility barriers between strains. In some yeasts fusion of protoplasts of identical mating type has been achieved (see, for example, Svoboda, 1978; Halfmann *et al.*, 1983). However, the first detectable products of such fusions are normally diploids in which spore formation is either dramatically reduced or absent.

Protoplast fusion affords a means of transferring cytoplasmic organelles, plasmid DNA or viral nucleic acid between microbial cells. Fusion of mini-protoplasts, derived from anucleate fungal cells, with normal (nucleated) protoplasts enables transfer of cytoplasmic components in the absence of nuclei. This mini-protoplast fusion technique has facilitated intergeneric mitochondrial transfer from *Hansenula wingei* to petite (respiration-deficient) strains of *Saccharomyces cerevisiae* (Yamashita *et al.*, 1981).

Plasmid DNA can be transferred from bacteria to eukaryotic cells by protoplast fusion. Schaffner (1980) has transferred a recombinant plasmid of pBR322 containing Simian virus (SV40) DNA from *E. coli* to mammalian cells by PEG-induced fusion. This technique thus provides a means of transferring cloned genes directly from prokaryotes to eukaryotes and vice versa.

2.7.1 Fusions with nonviable protoplasts

Fusions involving nonviable protoplasts find application particularly where counterselection of one (or both) parental type(s) in the fusion is required in order to enrich for recombinants that cannot be selected directly.

One strategy is to inactivate one of the partners in the fusion by, for example, heat treatment or UV irradiation. The use of heat does not bias the genotypes of recombinants against the treated parent (Fodor *et al.*, 1978), whereas the use of UV irradiation does (Hopwood and Wright, 1981). Most of the recombinants acquire individual markers (with complete loss of linkage) from the UV-treated parent. UV treatment might thus be applied specifically to fusions where particular short tracts of the chromosome of one partner need to be introduced into the recipient. On the other hand, heat treatment might be preferred where temporary inactivation of heat-sensitive components (for example a restriction system) of one partner is required.

The inactivation of both partners can also be used to enrich for recombinants. UV-irradiated *Streptomyces* protoplasts can be fused to yield viable recombinants by recombining out UV lesions in the parental genomes (Hopwood and Wright, 1979). Although this technique enhances the recombination frequency, the genetic damage inherited by the recombinant survivors may be at an unacceptable level. One of the advantages of the high frequency of recombination of chromosomal genes that can be attained in *Streptomyces* using such protoplast fusion techniques is that it eliminates the need to introduce selectable markers into parental strains

Table 2.4: Some barriers to genetic exchange

Barrier	Comments
External barriers	
Differential growth requirements	Growth conditions required for donor and recipient may be antagonistic
Inactivation of genetic vector by for example:	
UV irradiation, X-rays extracellular protease activity	Inactivate phage vectors
extracellular nuclease activity	Inactivates transforming DNA. May be overcome by encasing DNA in liposomes
Inactivation of parental strain(s) by, for example: production of antibiotics, bacteriocins, killer particles	May be overcome by protoplast fusion using nonviable parents
Lack of surface receptors, for example: phage attachment sites (for transduction) sex pilus receptor sites (for conjugation) DNA binding sites (for transformation)	May be overcome by using protoplasts
Failure of anastomoses for mycelial fungi	May be overcome by protoplast fusion
Incompatibility of mating-type	May be overcome by protoplast fusion
Lack of appropriate sexual cells/organs	
Entry (surface) exclusion for conjugative plasmids	May be overcome by using stationary phase recipients (phenocopies)
Internal barriers	
Intracellular nuclear activity: non-specific exo- and endonucleases, restriction endonucleases	Inactivates incoming DNA. May be overcome by using nuclease-deficient recipients or by heat inactivation of restriction systems
Incompatibility of plasmids	Overcome by the use of compatible plasmids or by recombining the incompatible plasmid with a compatible replicon
Incompatibility of fungal nuclei genic incompatibility (for alleles at different loci) allelic incompatibility (for pairs of alleles at the same locus)	Limits nuclear intermingling and can result in hyphal degeneration
Lack of appropriate replication apparatus for maintenance of extrachromosomal elements	May be overcome by directing integration of the elements into resident replicons or by incorporating appropriate replicating sequences into the elements
Recombination barriers, for example: lack of appropriate recombination sites/ genetic homology lack of suitable enzymes for recombination	Can result in unilinear inheritance
Expression barriers, for example: lack of appropriate transcriptional and/or translational signals	Genes are silent or poorly expressed. May be overcome by insertion of the genes downstream of appropriate expression signals

(which is, in itself, a laborious task). This in turn avoids the potentially deleterious effects such genetic marking can have on desirable traits.

2.8 BARRIERS TO GENETIC EXCHANGE

Gene transfer among microorganisms is not indiscriminate. Various barriers exist that limit the traffic of genes (and associated recombination events) within and between species and across generic barriers. Table 2.4 outlines some of these barriers.

BIBLIOGRAPHY

Adhya, S. and Gottesman, M. (1978) Control of transcription termination. *Ann. Rev. Biochem.* **47**, 967-96

Arthur, A. and Sherratt, D.J. (1979) Dissection of the transposition process: a transposon-encoded site-specific recombination system. *Mol. Gen. Genet.* **175**, 267-74

Arthur, A., Nimmo, E., Hettle, S. and Sherratt, D. (1984) Transposition and transposition immunity of different transposon Tn*3* derivatives having different ends. *EMBO J.* **3**, 1723-9

Avila, P., de la Cruz, F., Ward, E. and Grinsted, J. (1984) Plasmids containing one inverted repeat of Tn*21* can fuse with other plasmids in the presence of Tn*21* transposase. *Mol. Gen. Genet.* **195**, 288-93

Bachmann, B.J. (1984) Linkage map of *Escherichia coli* K-12. In *Genetic maps 1984. A compilation of linkage and restriction maps of genetically studied organisms,* vol 3 (S.J. O'Brien, ed.), pp. 145-61. Cold Spring Harbor Laboratory, Cold Spring Harbor, NY

Ballance, D.J., Buxton, F.P. and Turner, G. (1983) Transformation of *Aspergillus nidulans* by the orotidine-5′-phosphate decarboxylase gene of *Neurospora crassa. Biochem. Biophys. Res. Commun.* **112**, 284-9

Bennett, P. (1985) Bacterial transposons. In *Genetics of bacteria* (J. Scaife, D. Leach and A. Galizzi, eds). pp. 97-115. Academic Press, New York

Bennett, P.M., de la Cruz, F. and Grinsted, J. (1983) Cointegrates are not obligatory intermediates in transposition of Tn*3* and Tn*21. Nature (London)* **305**, 743-4

Berg, D.E. and Berg, C.M. (1983) The prokaryotic transposable element Tn*5. Biotechnology* **1**, 417-35

Bibb, M.J., Ward, J.M. and Hopwood, D.A. (1978) Transformation of plasmid DNA into *Streptomyces* at high frequency. *Nature (London)* **274**, 398-400

Bremer, E., Silhavy, T.J. and Weinstock, G.M. (1985) Transposable λp*lac*Mu bacteriophages for creating *lacZ* operon fusions and kanamycin resistance insertions in *Escherichia coli. J. Bacteriol.* **162**, 1092-9

Broome-Smith, J. (1980) *RecA* independent, site specific recombination between ColE1 or ColK and a miniplasmid they complement for mobilization and relaxation: implications for the mechanism of DNA transfer during mobilization. *Plasmid* **4**, 51-63

Bukhari, A.I. (1981) Models of DNA transposition. *Trends Biochem. Sci.* **6**, 56-60

Burnett, J.H. (1975) *Mycogenetics.* John Wiley, London

Buu-hoi, A. and Horodniceanu, T. (1980) Conjugative transfer of multiple antibiotic resistance markers in *Streptococcus pneumoniae. J. Bacteriol.* **143**, 313-20
Calos, M.P. and Miller J.H. (1980) Transposable elements. *Cell* **20**, 579-96
Campbell, A. (1981) Evolutionary significance of accessory DNA elements in bacteria. *Ann. Rev. Microbiol.* **35**, 55-83
Canosi, U., Morelli, G. and Trautner, T.A. (1978) The relationship between molecular structure and transformation efficiency of some *S. aureus* plasmids isolated from *B. subtilis. Mol. Gen. Genet.* **166**, 259-67
Casadaban, M.J. (1976) Transposition and fusion of the *lac* genes to selected promoters in *Escherichia coli* using bacteriophage lambda and Mu. *J. Mol. Biol.* **104**, 541-55
Case, M.E., Schweizer, M., Kushner, S.R. and Giles, N.H. (1979) Efficient transformation of *Neurospora crassa* by utilizing hybrid plasmid DNA. *Proc. Natl Acad. Sci. USA* **76**, 5259-63
Caten, C.E. (1981) Parasexual processes in fungi. In *The fungal nucleus* (K. Gull and S.G. Oliver, eds), pp. 191-214. Cambridge University Press, Cambridge
Chater, K.F. and Hopwood, D.A. (1984) *Streptomyces* genetics. In *The biology of the actinomycetes* (M. Goodfellow, M. Mordarski and S.T. Williams, eds), pp. 229-86. Academic Press, London and New York
Clancy, S., Mann, C., Davis, R.W. and Calos, M.P. (1984) Deletion of plasmid sequences during *Saccharomyces cerevisiae* transformation. *J. Bacteriol.* **159**, 1065-7
Clewell, D.B. (1981) Plasmids, drug resistance and gene transfer in the genus *Streptococcus. Microbiol. Rev.* **45**, 409-36
Clewell, D. and Brown, B. (1980) Sex pheromone cAD1 in *Streptococcus faecalis*: induction of a function related to plasmid transfer. *J. Bacteriol.* **143**, 1063-5
Clewell, D.B., Yagi, Y., Ike, Y., Craig, R.A., Brown, B.L. and An, F. (1982) Sex pheromones in *Streptococcus faecalis*: multiple pheromone systems in strain DS5, similarities of pAD1 and pAMγ1, and mutants of pAD1, altered in conjugative properties. In *Microbiology — 1982* (D. Schlessinger, ed.), pp. 97-100. American Society for Microbiology, Washington, DC
Clewell, D.B., White, B.A., Ike, Y. and An, F., (1984) Sex pheromones and plasmid transfer in *Streptococcus faecalis.* In *Microbial development* (R. Losick and L. Shapiro, eds), pp. 133-49. Cold Spring Harbor Laboratory, Cold Spring Harbor, N Y
Clewell, D.B., An, F.Y., White, B.A. and Gawron-Burke, C. (1985) *Streptococcus faecalis* sex pheromone (cAM373) also produced by *Staphylococcus aureus* and identification of a conjugative transposon (Tn *918*). *J. Bacteriol.* **162**, 1212-20
Cohen, S.N. and Shapiro, J.A. (1980) Transposable genetic elements. *Scient. Amer.* **242**, 36-45
Cohen, S.N., Chang, A.C.Y. and Hsu, L. (1972) Non-chromosomal antibiotic resistance in bacteria: genetic transformation of *Escherichia coli* by R-factor DNA. *Proc. Natl Acad. Sci. USA* **69**, 2110-14
Colbère-Garapin, F., Garapin, A. and Kourilsky, P. (1982) Selectable markers for the transfer of genes into mammalian cells. *Curr. Top. Microbiol. Immunol.* **96**, 145-57
Craigie, R. and Mizuuchi, K. (1985) Mechanism of transposition of bacteriophage Mu: structure of a transposition intermediate. *Cell* **41**, 867-76
Cullum, J. (1985) Insertion sequences. In *Genetics of bacteria* (J. Scaife, D. Leach and A. Galizzi, eds), pp. 85-95. Academic Press, New York.
Cullum, J. and Broda, P. (1979) Chromosome transfer and Hfr formation in *rec*$^+$ and *recA* strains of *Escherichia coli* K12. *Plasmid* **2**, 358-65

de Vos, W.M., Venema, G., Canosi, U. and Trautner, T.A. (1981) Plasmid transformation in *Bacillus subtilis*: fate of plasmid DNA. *Mol. Gen. Genet.* **181**, 424-33

Dityatkin, S.Y.A. and Ilyashenko, B.N. (1979) Plasmid transformation of frozen-thawed bacteria. *Genetika* **15**, 220-5

Ferenczy, L. (1981) Microbial protoplast fusion. In *Genetics as a tool in microbiology* (31st Symp. Soc. Gen. Microbiol.), (S.W. Glover and D.A. Hopwood, eds), pp. 1-34. Cambridge University Press, Cambridge

Fincham, J.R.S., Day, P.R. and Radford, A. (1979) *Fungal genetics*. Blackwell Scientific Publications, Oxford

Fitzgerald, G.F. and Clewell, D.B. (1985) A conjugative transposon (Tn*919*) in *Streptococcus sanguis*. *Infect. Immun.* **47**, 415-20

Fodor, K., Demiri, E. and Alfoldi, L. (1978) Polyethylene glycol-induced fusion of heat-inactivated living protoplasts of *Bacillus megaterium*. *J. Bacteriol.* **135**, 68-70

Fraley, R.T., Fornari, C.S. and Kaplan, S. (1979) Entrapment of a bacterial plasmid in phospholipid vesicles. Potential for gene transfer. *Proc. Natl Acad. Sci. USA* **76**, 3348-52

Franke, A. and Clewell, D.B. (1981) Evidence for a chromosome-borne resistance transposon in *Streptococcus faecalis* capable of 'conjugal' transfer in the absence of a conjugative plasmid. *J. Bacteriol*, **145**, 494-502

Franke, A.E., Dunny, G.M., Brown, B.L., An, F., Oliver, D.R., Damle, S.P. and Clewell, D.B. (1978) Gene transfer in *Streptococcus faecalis*: evidence for the mobilization of chromosomal determinants by transmissible plasmids. In *Microbiology — 1978* (D. Schlessinger, ed.), pp. 45-7. American Society for Microbiology, Washington, DC

Gaffney, D., Skurray, R. and Willetts, N. (1983) Regulation of the F conjugation genes studied by hybridization and *tra-lacZ* fusion. *J. Mol. Biol.* **168**, 103-22

Galas, D.J. and Chandler, M. (1981) On the molecular mechanisms of transposition. *Proc. Natl Acad. Sci. USA* **78**, 4858-62

Gamas, P., Galas, D. and Chandler, M. (1985) DNA sequence at the ends of IS*1* required for transposition. *Nature (London)* **317**, 458-60

Gawron-Burke, M.C. and Clewell, D.B. (1982) Tn*916* (Tc^r), a transferable non-plasmid element in *Streptococcus faecalis*. In *Microbiology — 1982* (D. Schlessinger, ed.), pp. 93-6. American Society for Microbiology, Washington, DC

Grindley, N.D.F. and Reed, R.R. (1985) Transpositional recombination in prokaryotes. *Ann. Rev. Biochem.* **54**, 863-96

Grindley, N.D.F., Lauth, M.R., Wells, R.G., Wityk, R.J., Salvo, J.J. and Reed, R.R. (1982) Transposon-mediated site specific recombination: identification of three binding sites for resolvase at the *res* sites of $\gamma\delta$ and Tn*3*. *Cell* **30**, 19-27

Guyer, M.S. (1978) The $\gamma\delta$ sequence of F is an insertion sequence. *J. Mol. Biol.* **126**, 347-65

Haas, D. and Holloway, B.W. (1978) Chromosome mobilization by the R plasmid R68.45: a tool in *Pseudomonas* genetics. *Mol. Gen. Genet.* **158**, 229-37

Haber, J.E. (1983) Mating-type genes of *Saccharomyces cerevisiae*. In *Mobile genetic elements* (J.A. Shapiro, ed.), pp. 559-619. Academic Press, New York

Halfmann, H.J., Emeis, C.C. and Zimmermann, U. (1983) Electro-fusion of haploid *Saccharomyces* yeast cells of identical mating type. *Arch. Microbiol.* **134**, 1-4

Harshey, R.M. and Bukhari, A.I. (1981) A mechanism of DNA transposition. *Proc. Natl Acad. Sci. USA* **78**, 1090-4

Harshey, R.M., McKay, R. and Bukhari, A.I. (1982) DNA intermediates in transposition of phage Mu. *Cell* **29**, 561-71

Hayes, W. (1968) *The genetics of bacteria and their viruses.* Blackwell Scientific Publications, Oxford

Hendrix, R.W., Roberts, J.W., Stahl, F.W. and Weisberg, R.A. (1983) *Lambda II.* Cold Spring Harbor Laboratory, Cold Spring Harbor, NY

Hicks, J.B., Strathern, J.N. and Herskowitz, I. (1977) The cassette model of mating-type interconversion. In *DNA insertion elements, plasmids and episomes* (A.I. Bukhari, J.A. Shapiro and S.L. Adhya, eds), pp. 457-62. Cold Spring Harbor Laboratory, Cold Spring Harbor, NY

Hinnen, A. and Meyhack, B. (1982) Vectors for cloning in yeast. *Curr. Top. Microbiol. Immunol.* **96**, 101-17

Holloway, B.W. (1979) Plasmids that mobilize bacterial chromosome. *Plasmid* **2**, 1-19

Hong, J-S. and Ames, B.N. (1971) Localization mutagenesis of any specific region of the bacterial chromosome. *Proc. Natl Acad. Sci. USA* **68**, 3158-62

Hopwood, D.A. (1981) Genetic studies with bacterial protoplasts. *Ann. Rev. Microbiol.* **35**, 237-72

Hopwood, D.A. and Chater, K.F. (1984) Streptomycetes. In *Genetics and breeding of industrial microorganisms* (C. Ball, ed.), pp. 7-42. CRC Press, Boca Raton, Fla

Hopwood, D.A. and Wright, H.M. (1978) Bacterial protoplast fusion: recombination in fused protoplasts of *Streptomyces coelicolor. Mol. Gen. Genet.* **162**, 307-17

Hopwood, D.A. and Wright, H.M. (1979) Factors affecting recombinant frequency in protoplast fusions of *Streptomyces coelicolor. J. Gen. Microbiol.* **111**, 137-43

Hopwood, D.A. and Wright, H.M. (1981) Protoplast fusions in *Streptomyces*: fusions involving ultraviolet-irradiated protoplasts. *J. Gen. Microbiol.* **126**, 21-7

Iida, S., Meyer, J. and Arber, W. (1981) Cointegrates between bacteriophage P1 DNA and plasmid pBR322 derivatives suggest molecular mechanisms for P1-mediated transduction of small plasmids. *Mol. Gen. Genet.* **184**, 1-10

Ike, Y., Craig, R.A., White, B.A., Yagi, Y. and Clewell, D.B. (1983) Modification of *Streptococcus faecalis* sex pheromones after acquisition of plasmid DNA. *Proc. Natl Acad. Sci. USA* **80**, 5369-73

Isberg, R.R. and Syvanen, M. (1982) DNA gyrase is a host factor required for transposition of Tn*5*. *Cell* **30**, 9-18

Ito, H., Fukuda, Y., Murata, K. and Kimura, A. (1983) Transformation of intact yeast cells treated with alkali cations. *J. Bacteriol.* **153**, 163-8

Kleckner, N. (1981) Transposable elements in prokaryotes. *Ann. Rev. Genet.* **15**, 341-404

Kleckner, N. (1986) Mechanism and regulation of Tn*10* and IS*10* transposition. In *Regulation of gene expression — 25 years on* (39th Symp. Soc. Gen. Microbiol.) (I.R. Booth and C.F. Higgins eds), pp. 221-37, Cambridge University Press, Cambridge

Kretschmer, P.J., Chang, A.C.Y. and Cohen, S.N. (1975) Indirect selection of bacterial plasmids lacking identifiable phenotypic properties. *J. Bacteriol.* **124**, 225-31

Low, K.B. (1972) *Escherichia coli* K-12 F-prime factors, old and new. *Bacteriol. Rev.* **36**, 587-607

MacNeil, D., Howe, M.H. and Brill, W.J. (1980) Isolation and characterisation of lambda specialized transducing bacteriophages carrying *Klebsiella pneumoniae nif* genes. *J. Bacteriol.* **141**, 1264-71

McIntire, S. and Willetts, N. (1978) Plasmid cointegrates of F*lac* and lambda

prophage. *J. Bacteriol.* **134**, 184-92
Makins, J.F. and Holt, G. (1981) Liposome-mediated transformation of streptomycetes by chromosomal DNA. *Nature (London)* **293**, 671-2
Manning, P.A. and Achtman, M. (1980) Cell-to-cell interactions in conjugating *Escherichia coli*: the involvement of the cell envelope. In *Bacterial outer membranes. Biogenesis and functions* (M. Inouye, ed.), pp. 409-47. John Wiley, New York
Masters, M. (1985) Generalized transduction. In *Genetics of bacteria* (J. Scaife, D. Leach and A. Galizzi, eds), pp. 197-215. Academic Press, New York
Mather, K. (1942) Heterothally as an outbreeding mechanism in fungi. *Nature (London)* **142**, 54-6
Mishra, N.C. (1979) DNA-mediated genetic changes in *Neurospora crassa. J. Gen. Microbiol.* **113**, 255-9
Motsch, S. and Schmitt, R. (1984) Replicon fusion mediated by a single-ended derivative of transposon Tn*1721*. *Mol. Gen. Genet.* **195**, 281-7
Nasmyth, K.A. (1982) Molecular genetics of yeast mating type. *Ann. Rev. Genet.* **16**, 439-500
Nordheim, A., Hashimoto-Gotoh, T. and Timmis, K.N. (1980) Location of two relaxation nick sites in R6K and single sites in pSC101 and RSF1010 close to origins of vegetative replication: implication for conjugal transfer of plasmid deoxyribonucleic acid. *J. Bacteriol.* **144**, 923-32
Novick, R.P., Iordanescu, S., Surdeanu, M. and Edelman, I. (1981) Transduction-related cointegrate formation between staphylococcal plasmids: a new type of site-specific recombination. *Plasmid* **6**, 159-72
Peberdy, J.F. (1979) Fungal protoplasts: isolation, reversion, and fusion. *Ann. Rev. Microbiol.* **33**, 21-39
Peberdy, J.F. and Ferenczy, L. (1985) *Fungal protoplasts, applications in biochemistry and genetics* (Mycology Series, vol. 6). Marcel Dekker, New York
Pifer, M.L. and Smith, H.O. (1985) Processing of donor DNA during *Haemophilus influenzae* transformation: analysis using a model plasmid system. *Proc. Natl Acad. Sci. USA* **101**, 72-9
Piffaretti, J-C. and Fayet, O. (1981) Phage-lambda-mediated transduction of non-conjugative plasmids is promoted by transposons. *Gene* **13**, 319-25
Pontecorvo, G. (1956) The parasexual cycle in fungi. *Ann. Rev. Microbiol.* **10**, 393-400
Queener, S.W., Ingolia, T.D., Skatrud, P.L., Chapman, J.L. and Kastar, K.R. (1985) A system for genetic transformation of *Cephalosporium acremonium.* In *Microbiology — 1985* (L. Leive, ed.), pp. 468-72. American Society for Microbiology, Washington, DC
Radford, A., Pope, S., Sazci, A., Fraser, M.J. and Parish, J.H. (1981) Liposome-mediated genetic transformation of *Neurospora crassa. Mol. Gen. Genet.* **184**, 567-9
Reed, R.R. (1981) Resolution of cointegrates between transposons γδ and Tn*3* defines the recombination site. *Proc. Natl Acad. Sci. USA* **78**, 3428-32
Reed, R.R., Shibuya, G.I. and Steitz, J.A. (1982) Nucleotide sequence of γδ resolvase gene and demonstration that its gene product acts as a repressor of transcription. *Nature (London)* **300**, 381-3
Robinson, M.K., Bennett, P.M. and Richmond, M.H. (1977) Inhibition of Tn*A* translocation by Tn*A*. *J. Bacteriol.* **129**, 407-14
Rodicio, M.R. and Chater, K.F. (1982) Small DNA-free liposomes stimulate transfection of *Streptomyces* protoplasts. *J. Bacteriol.* **151**, 1078-85
Roeder, G.S. and Fink, G.R. (1983) Transposable elements in yeast. In *Mobile*

genetic elements (J.A. Shapiro, ed.), pp. 299-328. Academic Press, New York
Roper, J.A. (1966) The parasexual cycle. In *The Fungi*, vol. 2 (G.C. Ainsworth and A.S. Sussman, eds), pp. 589-617. Academic Press, New York
Saunders, J.R., Docherty, A. and Humphreys, G.O. (1984) Transformation of bacteria by plasmid DNA. In *Methods in microbiology*, vol. 17 (P.M. Bennett and J. Grinsted, eds), pp. 61-95. Academic Press, London and New York
Schaffner, W. (1980) Direct transfer of cloned genes from bacteria to mammalian cells. *Proc. Natl Acad. Sci. USA* **77**, 2163-7
Schmidt, L., Watson, J. and Willetts, N. (1980) Genetic analysis of conjugation by RP1. In *Plasmids and transposons. Environmental effects and maintenance mechanisms* (C. Stuttard and K.R. Rozee, eds), pp. 287-92. Academic Press, New York
Schweizer, M., Case, M.E., Dykstra, C.C., Giles, N.H. and Kushner, S.R. (1981) Identification and characterization of recombinant plasmids carrying the complete *qa* gene cluster from *Neurospora crassa* including the *qa*-1^+ regulatory gene. *Proc. Natl Acad. Sci. USA* **78**, 5086-90
Shapiro, J.A. (1979) Molecular model for the transposition and replication of bacteriophage Mu and other transposable elements. *Proc. Natl Acad. Sci. USA* **76**, 1933-7
Shapiro, J.A. (1983) *Mobile genetic elements.* Academic Press, New York
Sherratt, D. (1986) Control of plasmid maintenance. In *Regulation of gene expression — 25 years on* (39th Symp. Soc. Gen. Microbiol.) (I.R. Booth and C.F. Higgins, eds), pp. 239-50. Cambridge University Press, Cambridge
Shimada, K., Weisberg, R.A. and Gottesman, M.E. (1972) Prophage λ at unusual locations. I. Location of the secondary attachment sites and the properties of the lysogens. *J. Mol. Biol.* **63**, 483-503
Simons, R.W. and Kleckner, N. (1983) Translational control of IS*10* transposition. *Cell* **34**, 683-91
Simons, R.W., Hoopes, B.C., McClure, W.R. and Kleckner, N. (1983) Three promoters near the termini of IS*10*: pIN, pOUT and pIII. *Cell* **34**, 673-82
Singh, H., Bieker, J.J. and Dumas, L.B. (1982) Genetic transformation of *Saccharomyces cerevisiae* with single-stranded DNA vectors. *Gene* **20**, 441-9
Smith, H.O., Danner, D.B. and Deich, R.A. (1981) Genetic transformation. *Ann. Rev. Biochem.* **50**, 41-68
Smith, M.D. and Guild, W.R. (1982) Evidence for transposition of the conjugative R determinants of *Streptococcus agalactiae* B109. In *Microbiology — 1982* (D. Schlessinger, ed.), pp. 109-11. American Society for Microbiology, Washington, DC
Sprague, G.F. Jr, Blair, L.C. and Thorner, J. (1983) Cell interactions and regulation of cell type in the yeast *Saccharomyces cerevisiae. Ann. Rev. Microbiol.* **37**, 623-60
Svoboda, A. (1978) Fusion of yeast protoplasts induced by polyethylene glycol. *J. Gen. Microbiol.* **109**, 169-75
Syvanen, M., Hopkins, J.D. and Clements, M. (1982) A new class of mutants in DNA polymerase I that affects gene transposition. *J. Mol. Biol.* **158**, 203-12
Timmons, M.S., Bogardus, A.M. and Deonier, R.C. (1983) Mapping of chromosomal IS*5* elements that mediate type II F-prime plasmid excision in *Escherichia coli* K-12. *J. Bacteriol.* **153**, 395-407
Tortorello, M.L. and Dunny, G.M. (1985) Identification of multiple cell surface antigens associated with the sex pheromone response of *Streptococcus faecalis. J. Bacteriol.* **162**, 131-7
Warren, G.J., Twigg, A.J. and Sherratt, D.J. (1978) ColE1 plasmid mobility and

relaxation complex. *Nature (London)* **274**, 259-61

Way, J.C. and Kleckner, N. (1984) Essential sites at transposon Tn*10* termini. *Proc. Natl Acad. Sci. USA* **81**, 3452-6

Willetts, N. (1981) Sites and systems for conjugal DNA transfer in bacteria. In *Molecular biology, pathogenicity and ecology of bacterial plasmids* (S.B. Levy, R.C. Clowes and E.L. Koenig, eds), pp. 207-15. Plenum Press, New York

Willetts, N. (1985) Plasmids. In *Genetics of bacteria* (J. Scaife, D. Leach and A. Galizzi, eds), pp. 165-95. Academic Press, New York

Willetts, N. and Skurray, R. (1980) The conjugation system of F-like plasmids. *Ann. Rev. Genet.* **14**, 41-76

Willetts, N. and Wilkins, B. (1984) Processing of plasmid DNA during bacterial conjugation. *Microbiol. Rev.* **48**, 24-41

Willetts, N.S., Crowther, C. and Holloway, B. (1981) The insertion sequence IS*21* of R68.45 and the molecular basis for mobilization of the bacterial chromosome. *Plasmid* **6**, 30-52

Yamashita, K., Fukuda, H., Murata, K. and Kimura, A. (1981) Transfer of mitochondria of *Hansenula wingei* into protoplasts of *Saccharomyces cerevisiae* by mini-protoplast fusion. *FEBS Letters* **132**, 305-7

Zinder, N.D. and Lederberg, J. (1952) Genetic exchange in *Salmonella*. *J. Bacteriol.* **64**, 679-99

3

In Vitro Genetic Manipulation

Manipulation of genes by *in vitro* recombinant DNA techniques (molecular or gene cloning) provides a number of distinct advantages over traditional methods of genetic analysis and strain improvement. Taxonomic barriers can generally be overcome, permitting the construction of hybrids that may not occur naturally. Furthermore, *in vitro* techniques are more rapid and provide a much greater degree of control over the final gene product than can be achieved by most *in vivo* methods. Recombinant DNA technology makes it possible to purify particular genes in sufficient quantity to obtain complete nucleotide sequences and to determine regulatory mechanisms. This can be achieved even with organisms that are refractory to conventional genetic techniques or where the genetics and physiology of the organism are not well understood. Fundamental knowledge of the organisation of particular genes, gained by recombinant techniques, may be exploited directly to improve the existing capabilities of microorganisms. Alternatively, this technology permits the manipulation, propagation and expression of particularly useful genes heterospecifically in amenable microorganisms or cultured animal or plant cells.

In vitro gene manipulation generally involves a number of distinct stages:

(i) A source of the desired gene(s), or at least the protein coding sequence must be obtained.
(ii) Purified DNA encoding the gene is obtained as manageable-sized fragments.
(iii) The DNA fragments are inserted *in vitro* into a cloning vector.
(iv) A population of vector molecules carrying inserted passenger DNA is reintroduced into viable host cells wherein the replication functions of the vector generally ensure that the inserted sequence is propagated and maintained. The unique event of introducing into a single cell a single recombinant DNA molecule and its subsequent replication

during growth of the host produces a **clone** of that molecule. The host is normally a microorganism, although cultivated animal or plant cells can be used.

(v) Selection is imposed for clones carrying recombinant (hybrid or chimaeric) DNA molecules comprising vectors with insertions of any or, ideally, specific DNA sequences. The presence of desired sequences is determined by detection either of the sequence *per se* or of expressed gene products. Stages (i) to (v) constitute **primary cloning**.

(vi) Selected hybrid molecules are reisolated from the primary clones and subjected to physical and functional characterisation. This normally involves a series of **subcloning** procedures.

(vii) Where expression as a protein product is desired, the cloned coding sequence of the gene of interest must either have strong endogenous transcriptional and translational signals appropriate to the host, or be placed downstream of such signals by subcloning into 'expression' vectors (see Chapter 5).

3.1 SOURCES OF GENES

3.1.1 Genomic DNA

The ease with which particular genes can be cloned from the genome of an organism depends largely upon the genetic organisation and phenotype involved. Products specified by a single gene or set of linked genes constituting an operon are relatively easy to manipulate. Certain proteins, however, are aggregates of several nonidentical polypeptide subunits, whose coding regions may not be closely linked. Some operons comprise genes that are distantly linked (for example, the *arg* genes of *E. coli*). Furthermore, secondary metabolites may be the end product of multiple operons. Such factors may complicate genetic manipulation.

Fundamental differences between the genetic organisation of prokaryotes and eukaryotes also affect the ease of cloning and expression of particular genes. In prokaryotes the coding DNA, mRNA and polypeptide product are, in most cases, colinear (Figure 3.1). By contrast, eukaryotic genes are generally split: the coding regions (exons) are interspersed with noncoding DNA (introns) (for a review see Nevins, 1983) and the RNA transcripts from such genes have to be cut and spliced to produce functional mRNA (Figure 3.2). Prokaryotes do not apparently contain the necessary biochemical machinery for splicing eukaryotic RNA transcripts. Consequently, most eukaryotic genes would not be correctly translated in prokaryotes. Such basic differences in gene structure mean that the amount

Figure 3.1: General features of prokaryotic genes that encode proteins. RBS, ribosome binding site. N, amino terminus; C, carboxy terminus

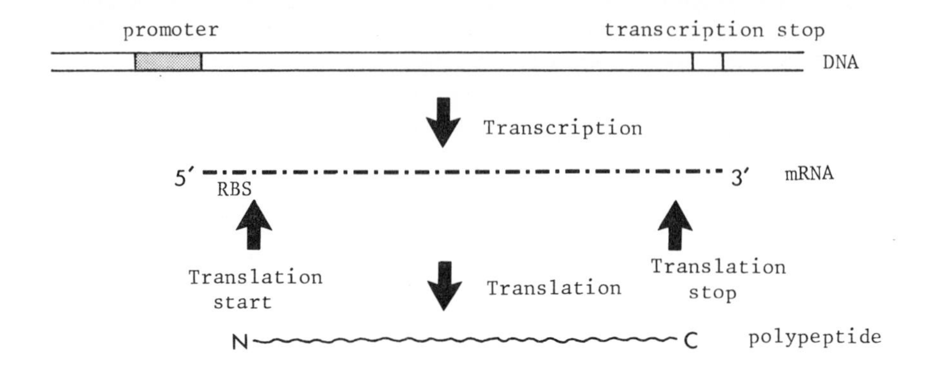

Figure 3.2: General features of eukaryotic genes that encode proteins. Bases are numbered from transcription start (+1)

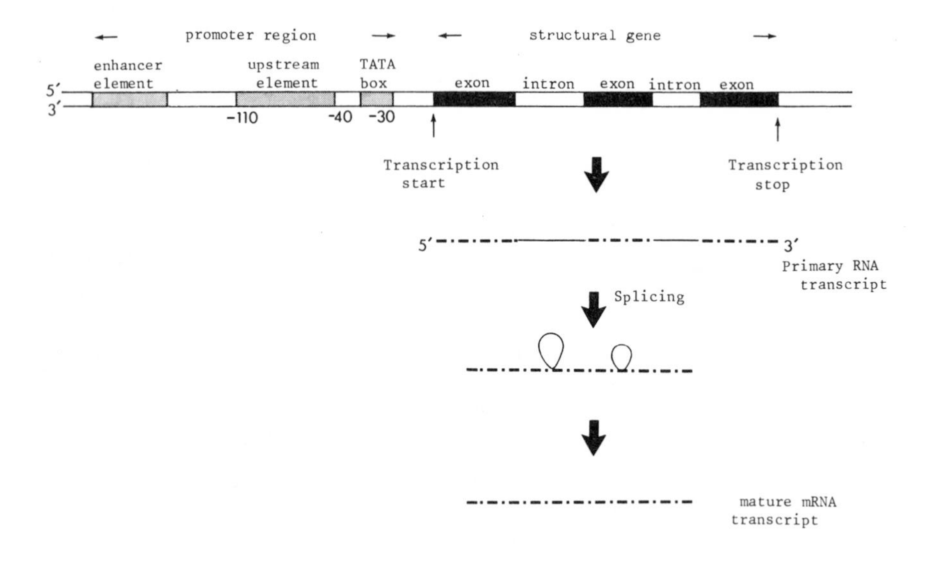

of genomic DNA from a eukaryote required to encode a protein of given size will generally be greater than that from a prokaryote encoding an equivalent-sized protein. Many of the problems associated with the expression of eukaryotic genes in prokaryotes can be overcome by the use of cDNA (copy DNA) (section 3.1.2) or chemically synthesised genes (section 3.1.3).

3.1.2 cDNA

A DNA coding sequence is generally required at some stage for molecular cloning procedures. However, it is often easier or necessary to purify a specific mRNA species and obtain a DNA copy (cDNA) of it, rather than to isolate the gene(s) itself. Production of cDNA enables the cloning and expression in prokaryotes of eukaryotic coding sequences from split genes, which would otherwise not be expressed correctly. Furthermore, cDNA can be used to clone RNA virus genomes (for example, those of influenza, polio and most plant viruses) that do not have an intermediate DNA form. Most eukaryotic mRNA transcripts contain a tract of about 100 A residues at their 3′ termini. This permits purification of mRNA by affinity chromatography using oligo T immobilised on a matrix. In some cases it is possible to enrich for particular mRNA species assuming that they can be identified by *in vitro* translation techniques (section 3.9.3).

A DNA copy is made by hybridising oligo T primers, 10 to 20 nucleotides in length, to the 3′ end of purified mRNA. Avian myeloblastosis virus (AMV) reverse transcriptase[1] is used to synthesise a cDNA copy of the primed RNA molecule (Figure 3.3). In classical cDNA cloning the RNA part of the resultant DNA-RNA hybrid is then removed by alkali treatment. A region of self-complementarity at the 3′ end of the cDNA molecule results in the formation of a hairpin or snap-back structure. This acts as a primer for reverse transcriptase, *E. coli* DNA polymerase I or phage T4 DNA polymerase to synthesise the complementary DNA strand. The 'closed' hairpin end of the duplex cDNA can then be opened by cleavage with the single-strand specific endonuclease S1 from *Aspergillus oryzae.* This results in the loss of about 13 nucleotides from that part of the DNA sequence derived originally from the 5′ end of the mRNA. The termini produced by S1 endonuclease treatment often contain short single-stranded extensions, which must be repaired (section 3.2.3) before further stages in cloning can proceed.

Hairpin-loop cleavage by S1 nuclease has been circumvented by annealing mRNA to a specialised oligo-T-tailed plasmid cloning vector (Okayama and Berg, 1982) (see section 3.5.2.b). cDNA is synthesised with reverse transcriptase and the RNA part of the resulting RNA-DNA hybrid plasmid is replaced with full-length DNA by the joint action of ribonuclease H (RNaseH)[2], DNA polymerase I (section 3.2.3) and *E. coli* DNA ligase (see section 3.4.1). However, the use of such a vector can be avoided by combining a classical oligo-T-primed cDNA synthesis step with an

---

[1] Reverse transcriptase has a DNA polymerase activity on primed RNA or DNA templates (see section 4.9.3).

[2] RNaseH specifically digests the RNA strand of a base-paired RNA-DNA hybrid duplex.

RNaseH-DNA polymerase I-DNA ligase second strand synthesis (Figure 3.3) (Gubler and Hoffman, 1983). This method greatly simplifies the efficient and faithful synthesis of full-length cDNA.

Figure 3.3: Preparation of cDNA

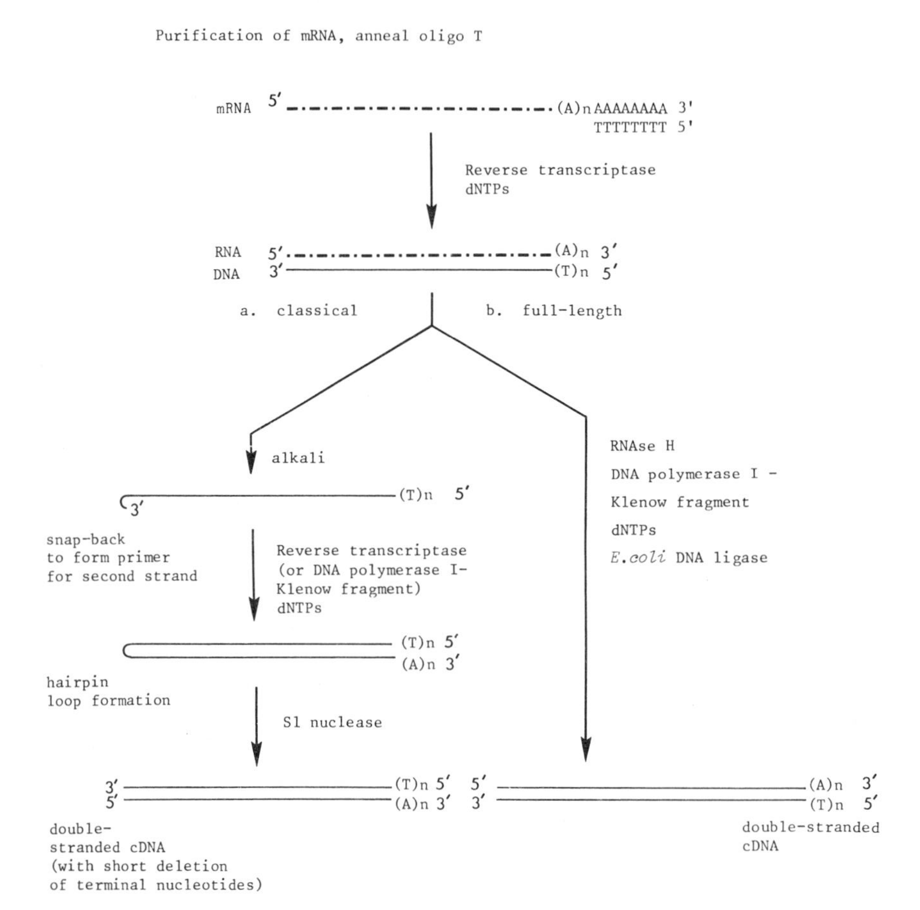

3.1.3 Synthetic genes

The chemical synthesis of specific genes, regulatory sequences, oligonucleotide probes (section 3.7.2), primers (section 3.8.5) and linkers (section 3.4.3) is a technique of increasing importance in genetic manipulation. Synthetic oligonucleotides of the desired sequence can be made readily by automated or semi-automated solid-phase synthesis using either phosphotriester or phosphoramidite chemistries (Gait, 1984; Kaplan,

1985). Techniques are also available for the simultaneous synthesis of large numbers of unique oligonucleotides on a microscale (Frank *et al.*, 1983; Matthes *et al.*, 1984).

Oligonucleotides of >150 bases are not readily synthesised. Long sequences are normally constructed by linking together a number of different oligonucleotide tracts that have been synthesised separately. Determination of the entire coding sequence of the gene of interest, or its deduction from the sequence of mRNA or from the amino acid sequence of the mature polypeptide is an essential prerequisite for gene synthesis. Synthetic genes have been used to direct the synthesis of a number of proteins including human growth hormone and human insulin (see Chapter 5). In these cases the properties of the protein products are essentially the same as those of their 'natural' counterparts.

3.2 CUTTING TECHNIQUES

cDNA and synthetic genes are generally discrete DNA fragments of manageable length for cloning. However, genomic DNA must normally be cut, either randomly or specifically, into smaller tracts prior to cloning.

3.2.1 Mechanical shearing

Duplex DNA can be broken fairly easily if subject to shear forces in solution. Shearing can be achieved by sonication or by high-speed blending. Conditions can be controlled to produce approximately equal-sized fragments of a desired size class. Breakage by these means is essentially at random points along the DNA molecule. The termini produced by shearing tend to have short single-stranded extensions, which must be repaired (section 3.2.3) prior to further stages in the cloning procedure.

3.2.2 Restriction endonucleases

Restriction endonucleases are enzymes that recognise specific sequences within duplex DNA molecules and cut the DNA at or near these sites. About 500 different restriction endonucleases have been discovered, mainly in prokaryotes (Roberts, 1985). In most cases, evidence that such enzymes are actually involved in the process of host-controlled restriction is lacking, but for convenience all the site-specific endonucleases are referred to as restriction enzymes.

Restriction endonucleases have been classified into three groups (Table 3.1). Types I and III have an ATP-dependent restriction (cleavage) activity and a modification (methylation) activity resident in the same multimeric

Table 3.1: Types of restriction endonuclease and their properties

Type	Cofactors	Enzyme structure	Recognition sequence in DNA	Cleavage properties	Example
I	ATP Mg^{2+} S-adenosyl methionine	Enzyme complex of 500 to 600 kdal composed of three separate subunits	13 to 15 base pairs containing interruption of 6 to 8 base pairs	Non-specific cleavage away from recognition site	*Eco*K
II	Mg^{2+}	Normally homodimers of 20 to 70 kdal	4 to 8 base pairs normally with 180° rotational symmetry	Precise cleavage at or near recognition site	*Eco*RI
III	ATP Mg^{2+} (S-adenosyl methionine is stimulatory)	Heterodimers with subunits of 70 and 100 kdal	5 to 6 base pairs with no rotational symmetry	Precise cleavage at a fixed distance 3′ to the recognition site	*Eco*P1

protein. Both these types recognise unmethylated recognition sequences in DNA, but Type I enzymes cleave the DNA at a random site, whereas Type III cleave at a specific site. Type II restriction/modification systems possess separate enzymes for endonuclease and methylase activity and are the most widely used for genetic manipulation. These endonucleases cut within or close to their unmodified recognition sequences, which typically consist of 4 to 8 nucleotide pairs with a two-fold axis of symmetry (Table 3.2). Some restriction enzymes, for example *Hpa*I (recognition sequence GTT↓AAC,[3] cleave DNA at the axis of symmetry to produce blunt (flush)-ended fragments. Other enzymes produce staggered cuts generating fragments with protruding 5′ (for example *Eco*RI, G↓AA*TTC) or 3′ (for example *Pst*I, CTGCA↓G) cohesive (sticky) termini of 1 to 5 bases (*indicates the site of methylation). Such termini can form base pairs with any other cohesive end of the same type, hence permitting the joining of different DNA fragments to produce recombinant molecules.

The majority of recognition sequences for restriction endonucleases are palindromic, that is, the sequence is the same if read 5′ to 3′ from both complementary strands. Some enzymes, for example *Eco*RI, recognise a unique palindromic sequence, whereas others recognise multiple sites (for

[3] By convention restriction sites are written with the sequence 5′ to 3′ as a single strand with the complementary sequence implied. Arrowhead (↓) indicates the point of cleavage.

Table 3.2: Some type II restriction endonucleases and their recognition sequences

Enzyme	Recognition sequence and cleavage site	Source organism
*Acc*I	GT↓$\binom{A}{C}$ and $\binom{G}{T}$AC	*Acinetobacter calcoaceticus*
*Alu*I	AG↓$\overset{*}{C}$T	*Arthrobacter luteus*
*Ava*I	C↓PyCGPuG	*Anabaena variabilis*
*Bam*HI	G↓GAT$\overset{*}{C}$C	*Bacillus amyloliquefaciens* H
*Bcl*I	T↓GATCA	*Bacillus caldolyticus*
*Bgl*I	GCCNNNN↓NGGC	*Bacillus globigii*
*Bgl*II	A↓GATCT	*Bacillus globigii*
*Bst*NI	CC↓$\binom{A}{T}$GG	*Bacillus stearothermophilus*
*Cla*I	AT↓CGAT	*Caryophanon latum* L
Eco RI	G↓A$\overset{*}{A}$TTC	*Escherichia coli* RY13
*Eco*RII	↓C$\overset{*}{C}$$\binom{A}{T}$GG	*Escherichia coli* R245
*Eco*RV	GAT↓ATC	*Escherichia coli* J62 (pLG74)
*Fok*I	GGATG	*Flavobacterium okeanokoites*
*Hae*III	GG↓$\overset{*}{C}$C	*Haemophilus aegyptius*
*Hpa*I	GTT↓A$\overset{*}{A}$C	*Haemophilus parainfluenzae*
*Hpa*II	C↓$\overset{*}{C}$GG	*Haemophilus parainfluenzae*
*Kpn*I	GGTAC↓C	*Klebsiella pneumoniae* OK8
*Not*I	GC↓GGCCGC	*Nocardia otitidis-caviarum*
*Pma*CI	CAC↓GTG	*Pseudomonas maltophila* CB50P
*Pst*I	CTGCA↓G	*Providencia stuartii* 164
*Pvu*I	CGAT↓CG	*Proteus vulgaris*
*Pvu*II	CAG↓CTG	*Proteus vulgaris*
*Rsr*II	CG↓G$\binom{A}{T}$CCG	*Rhodopseudomonas sphaeroides* 630
*Sal*I	G↓TCGAC	*Streptomyces albus* G
*Sau*3A	↓GATC	*Staphylococcus aureus* 3A
*Sma*I	CCC↓GGG	*Serratia marcescens* Sb
*Sst*I	GAGCT↓C	*Streptomyces stanford*
*Taq*I	T↓CG$\overset{*}{A}$	*Thermus aquaticus* YTI
*Xba*I	T↓CTAGA	*Xanthomonas badrii*
*Xho*I	C↓TCGAG	*Xanthomonas holcicola*

*, Site of methylation; Pu, purine; Py, pyrimidine; N, any base.

example *Xho*II recognises A↓GATCC, A↓GATCT, G↓GATCC and G↓GATCT). Enzymes such as *Bgl*I (GCCNNNN↓NGGC) recognise interrupted palindromes, and others, such as *Fok*I
(5′GGATGNNNNNNNNN↓
3′CCTACNNNNNNNNNNNNN↓) recognise non-palindromic sequences [N = any base].

Restriction enzymes that have the same recognition sequence can be isolated from different bacterial species. Such enzymes are called isoschizomers. An example is provided by *Mbo*I (from *Moraxella bovis*) and *Sau*3A (from *Staphylococcus aureus*), both of which recognise the sequence ↓GATC. Furthermore, some restriction enzymes generate cohesive ends that can reanneal with identical termini produced by other enzymes. For instance, DNA cleaved with *Bam*HI (G↓GATCC) has compatible ends with DNA cleaved with *Bcl*I, *Bgl*II, *Mbo*I, *Sau*3A or *Xho*II (Table 3.2). Sometimes the resulting hybrid site generated by reannealing such complementary ends cannot be recut with either enzyme. For example:

5′ A	annealed to	GATCC 3′	AGATCC
3′ TCTAG		G 5′ =	TCTAGG
*Bgl*II cohesive end		*Bam*HI cohesive end	Hybrid site cut by neither enzyme

However, in this case an enzyme, such as *Mbo*I, that recognises only the tetranucleotide GATC core of the hybrid site could be used for cleavage.

Restriction enzymes may be used either singly or in combination to cleave DNA molecules into defined fragments. Assuming that a DNA molecule contains the four deoxyribonucleotides in equal amounts, any given tetranucleotide sequence should occur randomly every 256(4^4) bp and a hexanucleotide sequence every 4096(4^6) bp. In practice, however, the overall and local variation in nucleotide distribution will determine the precise location of particular sequences. Where appropriate sites are found in DNA, they may not be cut due to the secondary structure of the DNA or due to methylation of the sequence. Most strains of *E. coli* contain the *dam* and *dcm* methylases. The *dam* methylase methylates the N^6 position of adenine in the sequence GATC (Hattman *et al.*, 1978) and may interfere with cleavage by endonucleases recognising this site. For example, *Mbo*I (↓GATC) and *Bcl*I (T↓GATCA) will not cut DNA from a *dam*$^+$ *E. coli*, whereas *Sau*3A (↓GATC) will do so. Where it is essential to cleave DNA from *E. coli* with enzymes whose activity is affected by such methylation, *dam*$^-$ mutants must be used. Similarly *dcm*$^-$ mutants are employed if it is necessary to cut DNA from *E. coli* with *Eco*RII (↓CC$\binom{A}{T}$GG) because the *dcm* methylase introduces methyl groups at the C^5 position of cytosine in the sequences CCAGG or CCTGG (Marinus, 1973). Alternatively *Bst*NI can be used instead of *Eco*RII. *Bst*NI recognises the same sequence as *Eco*RII but cuts at a different location (CC↓$\binom{A}{T}$GG) within the sequence. In addition to the normal bases, mammalian DNA contains a proportion of 5-methylcytosine residues, mainly on the 5′ side of G residues (Bird and Southern, 1978). This inhibits cleavage by a number of endonucleases, including *Sal*I(G↓TCGAC) and *Hpa*II(C↓CGG). The effects of methylation on DNA cleavage by various enzymes are summarised by Maniatis and co-workers (1982). The occurrence of such methylation patterns in genomic DNA from different organisms may prevent complete cleavage by certain restriction enzymes, thereby limiting the use of particular cloning strategies.

In some cases the cleavage specificity of restriction endonucleases can be altered by DNA methylation *in vitro*. A DNA methylase is selected whose recognition sequence overlaps the cleavage site(s) of a restriction endonuclease. For example, if DNA is treated with *Taq*I methylase (TCG$\overset{*}{\mathrm{A}}$) it can only subsequently be cut with *Acc*I endonuclease (GT$\left(\begin{smallmatrix}\mathrm{AG}\\ \mathrm{CT}\end{smallmatrix}\right)$AC) at *Acc*I sites with the sequence GT$\left(\begin{smallmatrix}\mathrm{A}\\ \mathrm{C}\end{smallmatrix}\right)$TAC. Similarly, *Msp*I methylation ($\overset{*}{\mathrm{C}}$CGG) blocks cleavage by *Bam*HI endonuclease at the sequence $\overset{*}{\mathrm{C}}$CGGATCC.

Methylation of restriction sites *in vitro* will generally reduce the number of restriction fragments obtained by digesting DNA with a particular endonuclease. In some cases, however, the specificity of DNA recognition can be relaxed so that a reduced number of nucleotides is recognised by altering the conditions of incubation. For example, the relaxed specificity of *Eco*RI is ↓AATT and is referred to as *Eco*RI star (*Eco*RI*) activity. Star activities are found in other endonucleases such as *Bam*HI and *Sal*I and can be used to generate smaller restriction fragments than are generated by the normal enzyme activity.

3.2.3 Repair of termini

It is sometimes necessary to join two DNA molecules that bear incompatible cohesive termini or damaged termini. This can be achieved by blunt-ended ligation (section 3.4.1) provided that the termini have been repaired. Several different methods are available for rendering DNA molecules blunt ended.

(a) Filling in recessed 3′ termini

It is possible to use DNA polymerases to fill in recessed 3′ termini produced by restriction endonuclease cleavage or by physical or enzymatic damage. DNA polymerase I of *E. coli* is a protein of about 112000 daltons, which has 5′→3′ polymerising activity together with 3′→5′ (proofreading) and 5′→3′ exonucleolytic activities. Polymerase I can be cleaved with subtilisin into two active enzyme moieties of 36000 and 76000 daltons. The larger C-terminal or **Klenow fragment** contains both the 5′→3′ polymerase and 3′→5′ exonuclease activities and is used in applications where 5′→3′ exonuclease activity of the entire enzyme would be undesirable. In the presence of the appropriate dNTPs and Mg^{2+}, the Klenow fragment will synthesise a strand complementary to a 5′ protrusion, using the recessed 3′OH end as a primer (Figure 3.4). Phage T4 DNA polymerase can also be used for this filling reaction provided that all four dNTPs are present. In the absence of one or more of the dNTPs this enzyme acts as a 3′→5′ exonuclease.

(b) Removal of protruding 5′ or 3′ termini

Protruding 3′ termini may be removed by using the 3′→5′ exonucleolytic activity of either Klenow fragment or, more commonly, T4 DNA poly-

Figure 3.4: Examples of the repair of DNA termini. (a) End-filling with Klenow fragment of DNA polymerase I. (b) Exonucleolytic removal of 5′ protrusions

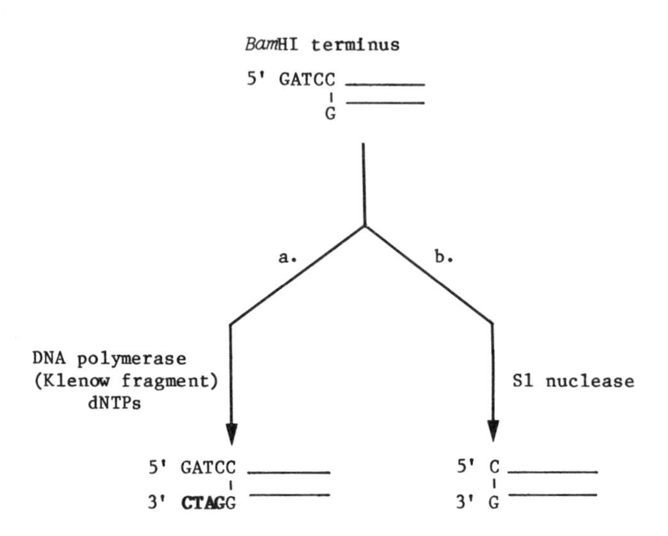

merase. Single-strand protrusions of either polarity may be removed by digestion with single-stranded nucleases, such as S1 nuclease or mung bean nuclease (Figure 3.4). Since these nucleases may have some activity on double-stranded DNA, the termini are normally repaired with Klenow fragment after nuclease digestion in order to render the DNA molecules flush-ended. Nuclease BAL31 (from *Alteromonas espejiana* BAL31) possesses a highly specific endodeoxyribonuclease and exonuclease activity, and catalyses the removal of mononucleotides or short oligonucleotides from both 5′ and 3′ termini of double-stranded DNA. BAL31 also has a single-strand specific endonuclease activity similar to that of S1 nuclease. Controlled digestion of DNA with BAL31 nuclease, followed by repair with Klenow fragment, provides a convenient means of creating blunt-ended molecules.

(c) Partial filling of 5′ protruding termini

Termini with different 5′ single-stranded protrusions can sometimes be joined by partial filling in of the single strands using a controlled reaction with reverse transcriptase (Hung and Wensink, 1984). This can create new single-stranded termini that can be ligated to different termini (Figure 3.5). Ligation can occur where partial filling of these different termini generates suitable protrusions of the same length. These must be one nucleotide long and complementary, or two nucleotides long and complementary or have a mismatch (for example dA : dC) at one position, or three nucleotides long

Figure 3.5: Joining of partially filled restriction termini. Partial filling of termini ensures that molecules do not self-ligate

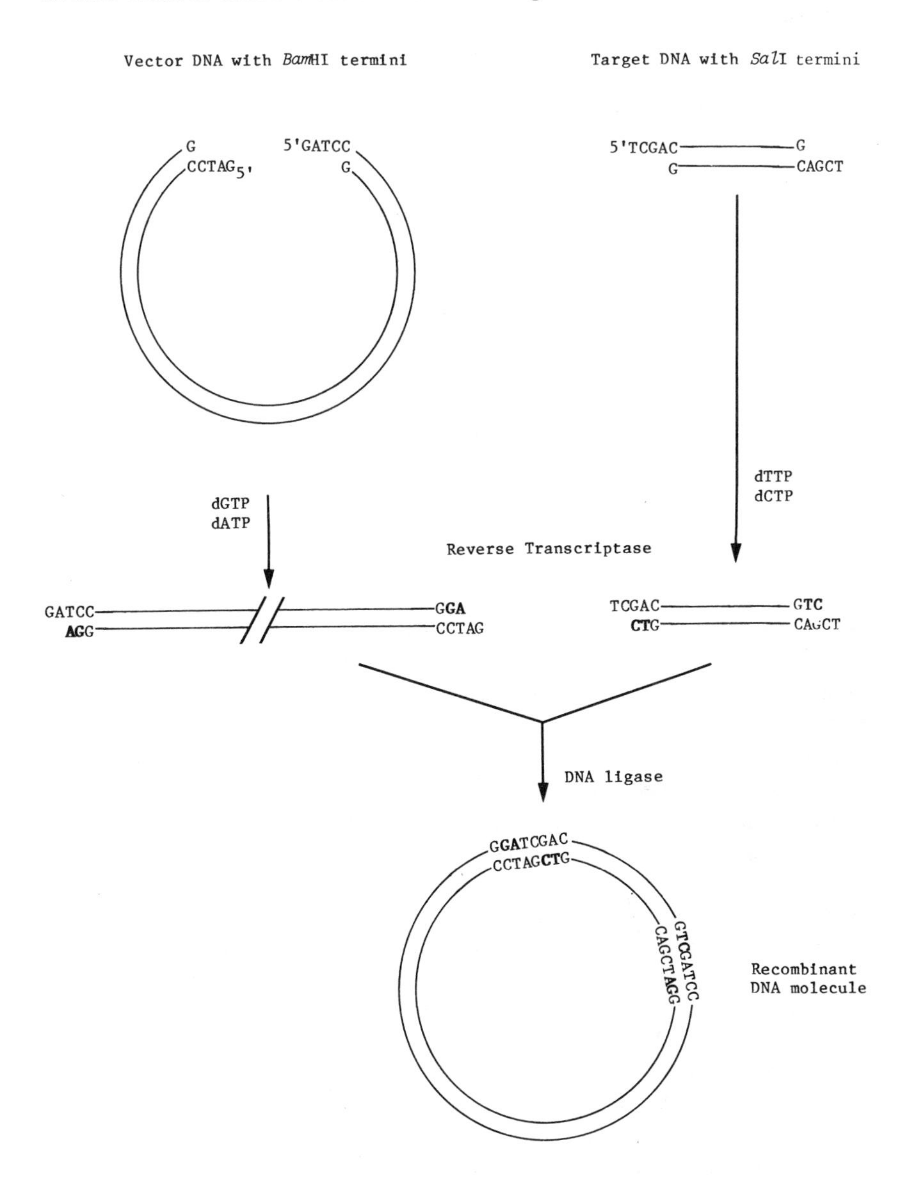

and have a mismatch at the central position. An advantage of this method is that the filling reaction usually prevents self-ligation of DNA fragments.

3.3 GENE ENRICHMENT

Cloning specific genes is facilitated if the DNA can be enriched for the desired sequence. Genomic DNA fragments can be fractionated on the basis of size by sucrose gradient centrifugation, or, more commonly, by agarose gel electrophoresis. Fractionation of DNA by column chromatography using the matrix RPC-5 can enrich for specific sequences on the basis of size and other factors, such as AT-richness (Thompson *et al.*, 1983). In order to perform such enrichments it is helpful if a DNA or RNA hybridisation probe, specific for the desired gene, is available so that elution of the required sequence can be monitored. Monitoring can also be performed by detecting the presence of specific proteins using fractionated DNA fragments to direct a coupled transcription/translation system (section 3.9.4). Alternatively genes that bind specific proteins can be detected by a binding assay.

3.4 JOINING TECHNIQUES

3.4.1 Ligation

E. coli and bacteriophage T4 both produce DNA ligases which are capable of sealing single-stranded nicks in duplex DNA molecules by forming a phosphodiester bond between adjacent 5′ phosphate and 3′ hydroxyl groups. DNA ligase from *E. coli* requires NAD as a cofactor and is capable of ligating DNA molecules with cohesive but not with blunt ends *in vitro*. T4 DNA ligase requires ATP and will join cohesive ends and blunt-ended DNA molecules. For this reason T4 ligase is more widely used for cloning. However, *E. coli* DNA ligase will join blunt-ended molecules in the presence of polyethylene glycol (PEG) (Zimmerman and Pheiffer, 1983). Unlike T4 DNA ligase, which will join RNA to DNA, *E. coli* DNA ligase will only join DNA molecules. By virtue of its specificity *E. coli* DNA ligase is the enzyme of choice in some strategies for cloning cDNA (see sections 3.1.2 and 3.5.2.b). The activity of T4 DNA ligase on blunt-ended molecules is stimulated by T4 RNA ligase (Sugino *et al.*, 1977) and by PEG (Hayashi *et al.*, 1985).

Blunt-ended ligations require high DNA concentrations and much larger amounts of DNA ligase than are necessary for sticky-ended ligations. Ligation of blunt-ended molecules is less efficient since the reaction is solely dependent on random collisions of termini. Addition of polyethylene glycol to ligation mixtures produces macromolecular crowd-

adv.

ing which favours such collisions. Conditions for either cohesive or blunt-ended ligation must be altered to suit the type of recombinant molecule to be constructed. DNA cloning experiments using vectors with circular genomes, such as plasmids, normally require that two types of molecule, linearised vector and passenger fragment, are ligated to produce a circular hybrid (Figure 3.6). (Vector or hybrid molecules that fail to circularise transform cells with much lower efficiency than circular molecules; see section 2.2.1.) Using linear vectors such as phage λ, three separate DNA molecules in the order left vector arm→passenger→right vector arm need to be ligated to produce a linear hybrid molecule (Figure 3.6).

The outcome of a ligation depends largely upon the ratio of vector to potential inserts and the absolute concentration of the participating DNA molecules (Dugaiczyk *et al.*, 1975; Maniatis *et al.*, 1982; Legerski and Robberson, 1985). The ratio of individual cyclised vector or insert molecules to concatameric (end-to-end linked molecules including vector-vector, insert-insert and hybrids) molecules depends on j, the effective concentration of one end of a DNA molecule in the neighbourhood of the other end of the same molecule, and i, the concentration of all complementary termini in the ligation mixture. These concentrations determine whether the end of a DNA molecule will make contact with its opposite end or with the end of a different molecule. When $j > i$, the formation of circular vector or insert molecules should be favoured, and when $i > j$, concatamer formation should be favoured. In practice, the values of i and j will alter as ligation progresses, and various combinations of molecules, joined together as circular or linear structures, will result (Legerski and Robberson, 1985). Intermolecular ligations are favoured at 150 mM to 300 mM NaCl when using T4 DNA ligase in the presence of 10% w/v PEG. Below 150 mM NaCl a mixture of intramolecular and intermolecular products is formed (Hayashi *et al.*, 1985).

3.4.2 Homopolymer tailing

Often passenger DNA is blunt-ended or does not have appropriate cohesive termini for cloning into a specific site on a vector. It may therefore be necessary, as, for example, in cDNA cloning, to create artificial cohesive termini either by the use of linkers (section 3.4.3) or by homopolymer tailing.

Calf-thymus terminal deoxynucleotidyl transferase is capable of adding deoxyribonucleotides to the 3′ ends of single- or double-stranded DNA. This enzyme will add nucleotides to protruding 3′ termini (or to recessed 3′ termini in the presence of cobalt ions) of double-stranded DNA (Lobban and Kaiser, 1973). Terminal transferase permits the addition of complementary homopolymer tails (50 to 150 dA or dT long, for dA.dT tailing

Figure 3.6: Formation of desired ligation products

a. Cloning in vectors with circular genomes

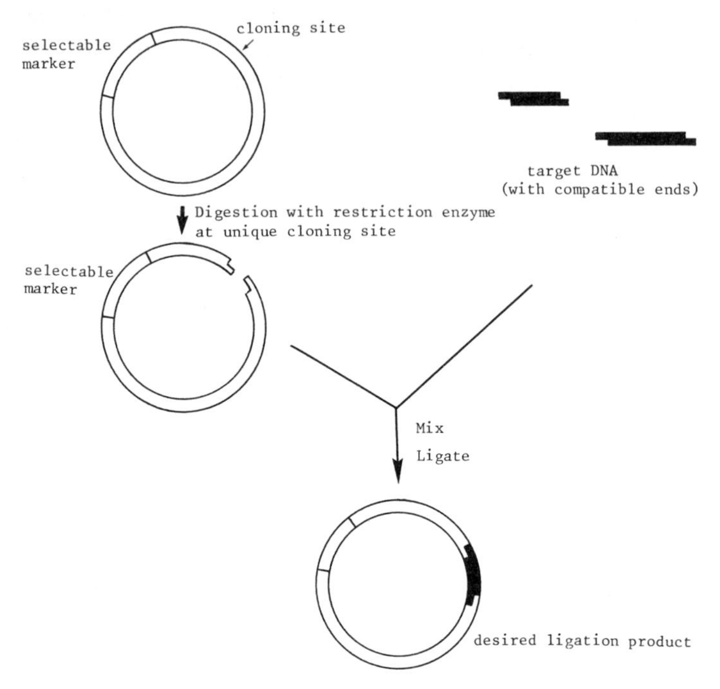

b. Cloning in vectors with linear genomes

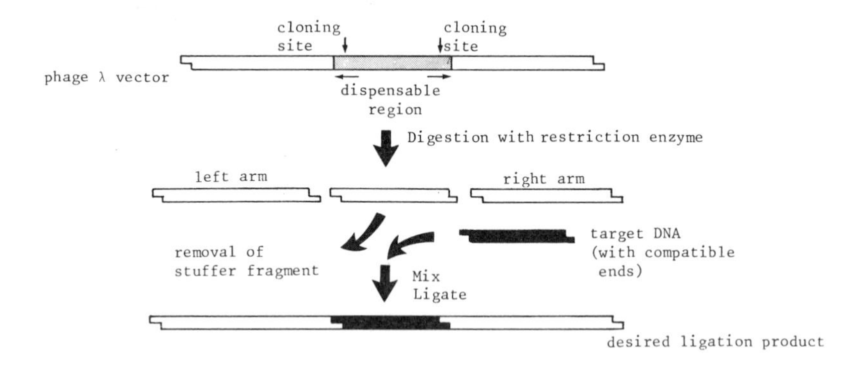

and about 20 dG or dC, for dG.dC tailing) to plasmid vector and passenger DNA. These tails can reanneal to form open circular hybrid molecules (Figure 3.7), which can be ligated *in vitro* or more commonly *in vivo* following transformation, to produce functional recombinant molecules. A major advantage of homopolymer tailing is that self-annealing of donor or passenger molecules *in vitro* is precluded and only hybrid molecules can form by annealing. After cloning, the inserted DNA may be excised from the hybrid molecule by cutting with restriction endonucleases that have sites close to, but outside, the insert. Addition of dG residues to the 3′ protruding ends produced by *Pst*I digestion allows reannealing with dC-tailed insert DNA and generally regenerates a *Pst*I site at each end of the insert. These sites may be used for excision of the insert after cloning. With dA.dT tailing the passenger DNA can be excised by selectively denaturing the A-T rich flanking tails (Hofstetter *et al.*, 1976). The resulting single strands derived from the tails can then be preferentially cleaved by S1 or mung bean endonuclease (section 3.2.3).

Figure 3.7: Homopolymer tailing of DNA molecules

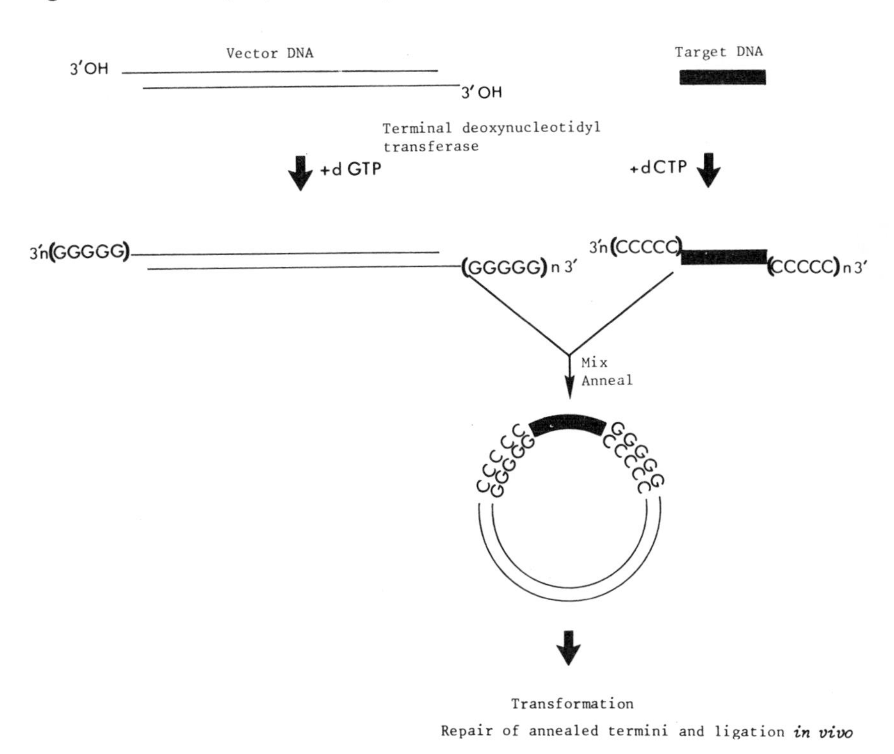

3.4.3 Linkers and adaptors

A more convenient approach to joining DNA molecules is to utilise synthetic oligodeoxynucleotides to form a defined sequence linking vector and passenger. **Linkers** are single-stranded synthetic oligodeoxynucleotides that have a self-complementary base sequence (Bahl *et al.*, 1976). On self-annealing the single strands form blunt-ended duplexes containing one or more restriction sites (Figure 3.8). Such linkers provide a means of introducing any blunt-ended DNA duplex molecule into a particular restriction site on a cloning vector. The duplex linkers are phosphorylated with polynucleotide kinase and ligated to the passenger DNA under conditions for blunt-ended ligation (Figure 3.8). Because of the small size and high concentration, multiple copies of the linker generally ligate to the ends of passenger DNA. In order to expose the cohesive ends of that restriction site adjacent to the insert, the ligated DNA must, therefore, be digested to completion with the appropriate restriction endonuclease. However, the

Figure 3.8: Joining of blunt-ended DNA to a vector using linkers

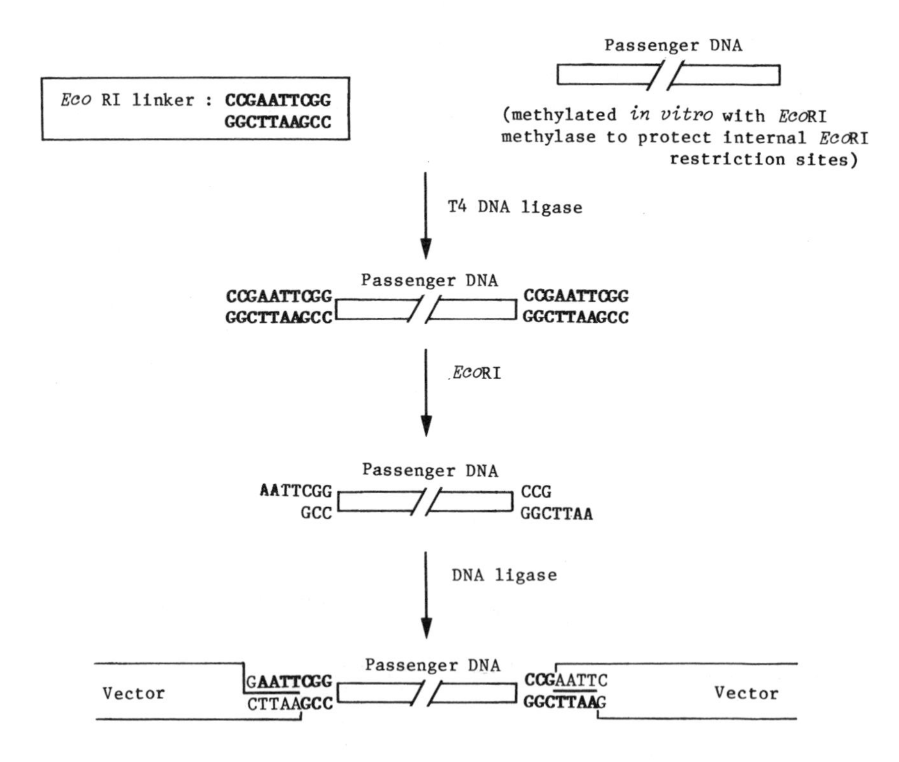

passenger DNA may contain internal sites for the enzyme in question. In such cases an alternative linker and restriction endonuclease might be used. If this is not feasible, the internal restriction sites could be protected by *in vitro* methylation using the appropriate modification methylase, where this is available (Maniatis *et al.*, 1978) or by *in vivo* methylation in an appropriate modifying host (O'Connor and Humphreys, 1982). Linkers ensure that restriction sites are regenerated at both ends of the passenger segment in the hybrid molecule and hence permit easy recovery of the cloned fragment. By judicious use of linkers it is possible to insert DNA into a vector and recover it from any desired restriction site.

A more general solution to the problem of internal restriction sites is the use of **adaptors** (Bahl *et al.*, 1976). These are synthetic oligodeoxynucleotides that can be used to join two incompatible cohesive ends, two blunt ends or a combination of both. Such adaptors are of several types, preformed, conversion and single-stranded.

Preformed adaptors are short DNA duplexes with at least one cohesive end. The problem of internal cleavage of the insert DNA can be overcome by using a preformed adaptor that will introduce a new restriction site. Figure 3.9 illustrates the use of an adaptor to provide *Bam*HI cohesive ends and sites for *Hpa*II and *Sma*I attached to passenger DNA. DNA molecules ligated to the *Bam*HI adaptor can be inserted into a *Bam*HI site in a vector. After cloning, passenger DNA can be excised from the hybrid by using any one of the enzymes that recognise the restriction sites within

Figure 3.9: Use of preformed adaptors

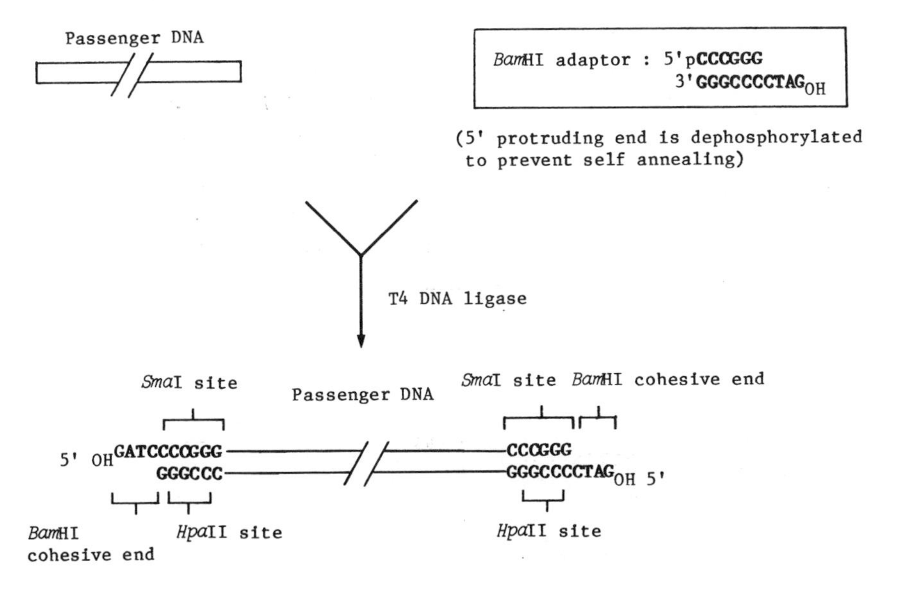

the adaptor region. The choice of enzyme would depend upon the presence or absence of relevant sites within the passenger DNA.

Conversion adaptors are synthetic oligonucleotides bearing different cohesive restriction termini. Such adaptors enable vector molecules that have been cleaved with one endonuclease to be joined to passenger fragments that have been cleaved with another. Often these adaptors contain internal restriction sites that permit recovery of the passenger fragment; for example, the *Eco*RI-*Bam*HI adaptor (Figure 3.10) contains a site for *Xho*I.

Single-stranded adaptors can be used to make 3′-protruding cohesive ends (generated, for example; by *Pst*I cleavage) compatible with 5′-protruding ends (produced, for example, by cleavage with *Eco*RI) (Figure 3.11). Such adaptors permit the insertion of passenger fragments into sites on vectors from which they would otherwise be precluded because of incompatible cohesive ends (Lathe *et al.*, 1982). These adaptors can be designed such that, after annealing, a two-nucleotide single-stranded gap remains between vector and passenger molecule. This gap can be repaired *in vitro* or *in vivo* to regenerate the cut sites for both the restriction endonucleases involved.

Figure 3.10: Use of an *Eco*RI to *Bam*HI conversion adaptor

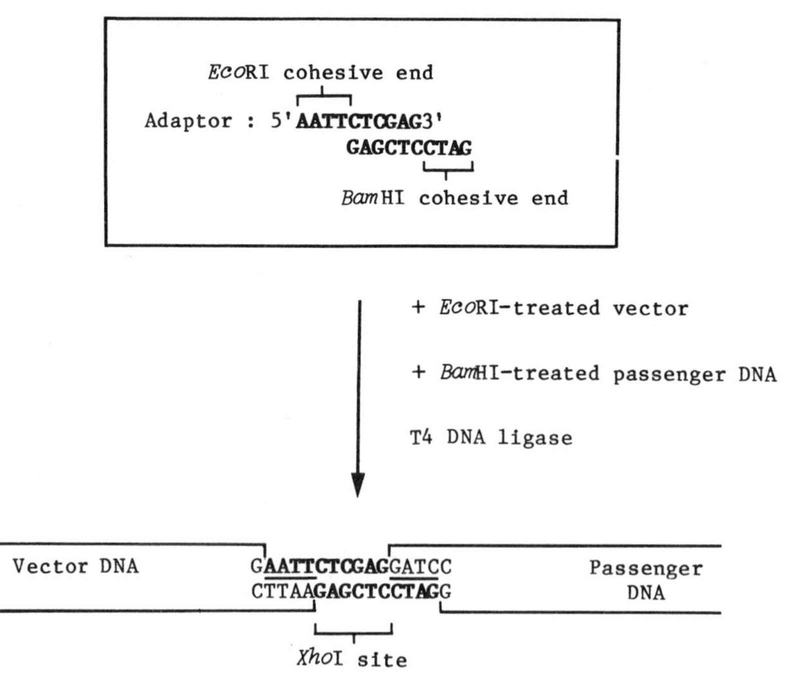

Figure 3.11: Use of a single-stranded adaptor

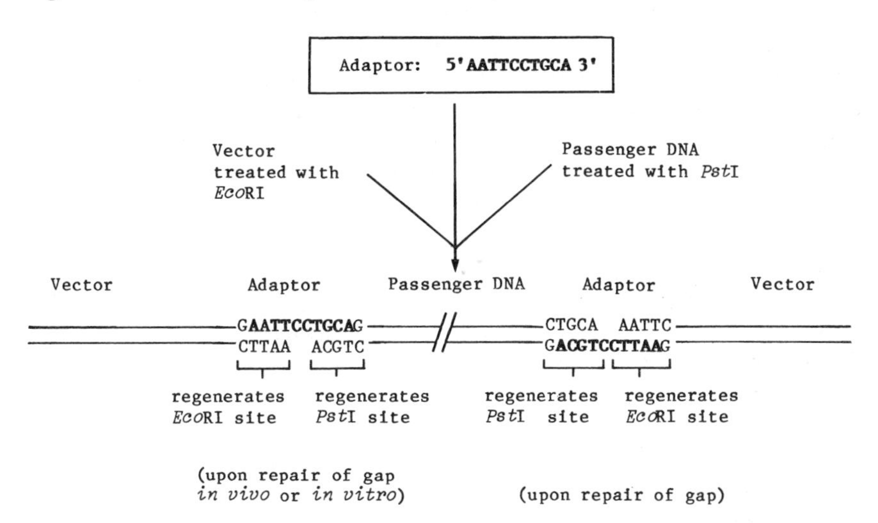

3.4.4 Prevention of self-ligation of vector or passenger DNA

In a ligation mixture some vector molecules will inevitably circularise or join with other vector molecules rather than with passenger molecules. Passenger DNA can also self-ligate. The presence of self-ligated molecules reduces the probability of recovering desired recombinant clones. Self-ligation can be reduced by optimising ligation conditions (section 3.4.1). Alternatively, homopolymer tailing (section 3.4.2) can be used to eliminate the problem. Where homopolymer tailing is undesirable other strategies may be adopted:

(i) *Directional (forced) cloning.* Directional cloning is possible where a vector possesses two (or more) unique restriction sites within a nonessential region. Cleavage at these two sites will remove non-essential DNA and generate vector molecules with different cohesive termini. If these termini are noncomplementary, individual vector molecules cannot circularise. Single passenger DNA fragments generated by the same combination of restriction enzymes will also be unable to circularise, but will be able to anneal with the appropriate termini of the vector (Figure 3.12). A high percentage of the ligated molecules will consequently be the desired hybrids. This approach, of necessity, directs the insertion of the passenger molecule into the vector in one orientation only. Directional cloning may be exploited in DNA sequencing strategies (section 3.8.5) and in directing expression of cloned genes (see Chapter 5).

Figure 3.12: Directional (forced) cloning

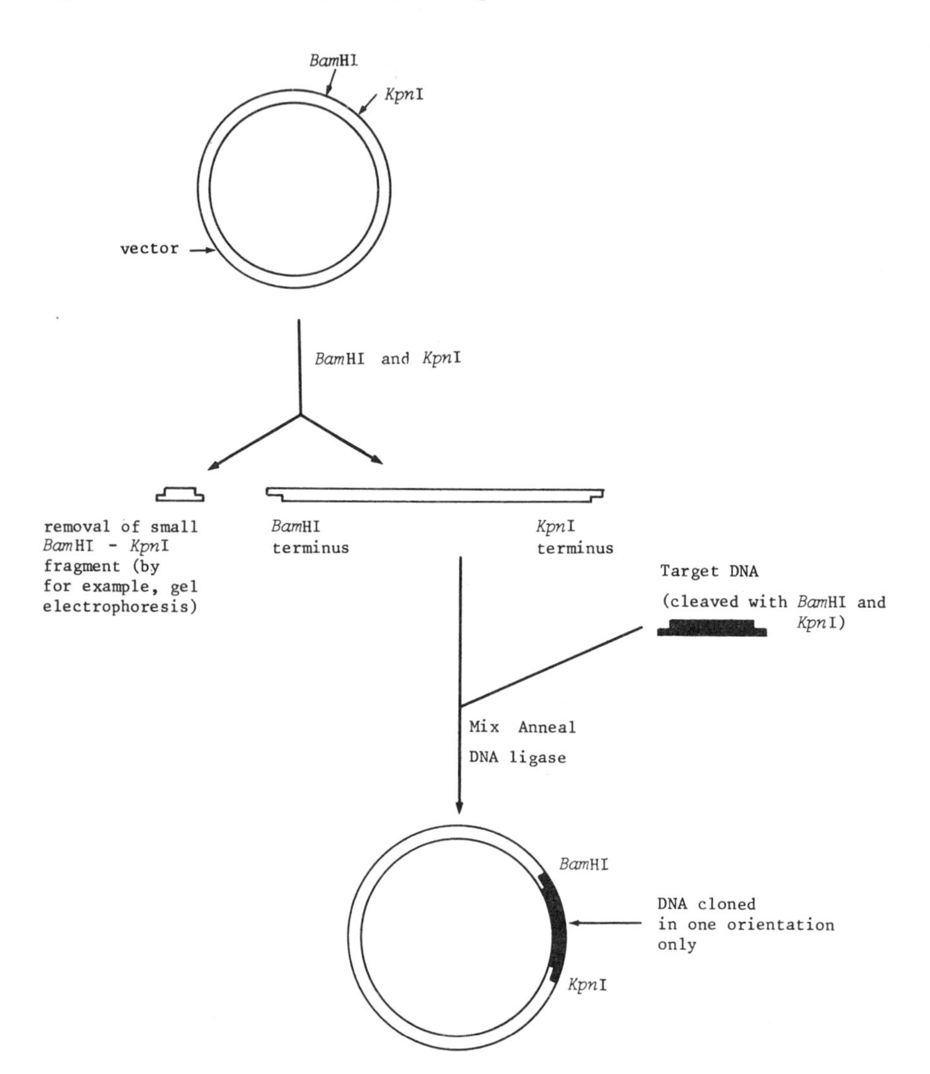

(ii) *Dephosphorylation of termini.* DNA ligase will catalyse the formation of a phosphodiester bond between adjacent nucleotides if one contains a 5′-phosphate group and the other a 3′-hydroxyl group. Thus, removal of terminal 5′-phosphate groups from the cleaved vector (or less usually passenger DNA) will prevent self-ligation. Dephosphorylation of termini can be carried out by treating linearised DNA with bacterial alkaline phosphatase or calf intestinal phosphatase. Vector molecules with dephosphorylated termini can then only religate with phosphorylated passenger DNA fragments to produce functional recombinant molecules (Figure 3.13).

3.5 HOST-VECTOR SYSTEMS

Vectors and their hosts form integrated systems for constructing and maintaining recombinant DNA molecules. The choice of a particular host-vector system depends on a variety of factors, including ease and safety of manipulations and the likelihood of expression of cloned genes. Many vectors have been designed for use in *E. coli* K12, where knowledge of genetics and physiology is generally in advance of that of other micro-

Figure 3.13: Prevention of self-ligation of DNA molecules by dephosphorylation of termini

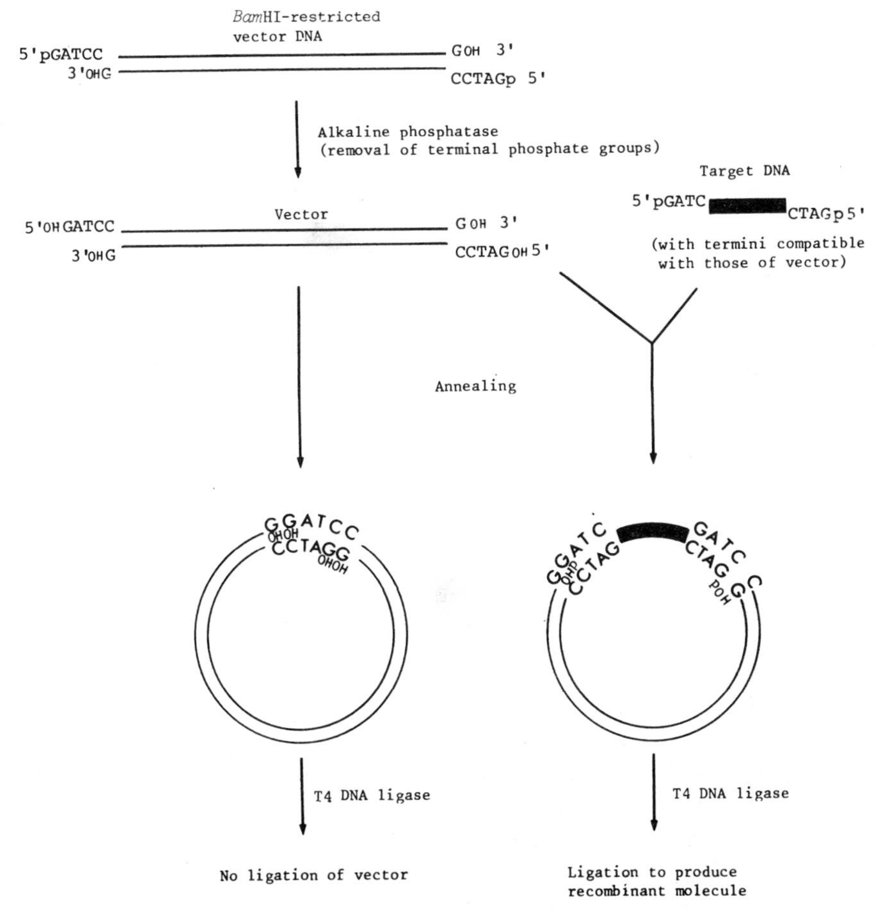

organisms. However, it has become necessary, due to the inapplicability of *E. coli* for many purposes, to develop vector systems for alternative organisms. For a comprehensive description of cloning vectors the reader is referred to Pouwels *et al.* (1985).

3.5.1 Properties of cloning vectors

Cloning vectors are replicons based on viruses, plasmids (or other extrachromosomal elements) or autonomously replicating sequences derived, for example, from chromosomes. Desirable properties for cloning vectors are:

(i) They should be capable of autonomous replication in at least one host organism.
(ii) They should be of small size, since this aids the preparation of vector DNA and reduces the complexity of analysing recombinant molecules.
(iii) Normally they should be capable of amplifying the cloned sequence. For this reason most vectors are present in host cells as multiple copies either stably, in the case of plasmid vectors, or transiently during virus infection, in the case of viral vectors. High copy number facilitates purification of large amounts of vector and/or passenger DNA and generally assists in maximising expression of cloned genes (see Chapter 5).
(iv) There should be a unique cleavage site (or in some cases two sites) for a range of restriction endonucleases, in nonessential regions of the vector genome. Cleavage of the vector at multiple sites reduces the likelihood that functional recombinant molecules will be formed during ligation.
(v) They should normally possess one or more genetic markers located away from the cloning site enabling selection for clones carrying the vector (with or without inserts).
(vi) Ideally they should permit detection, by simple genetic tests, of the presence of passenger DNA inserted at a cloning site. This enables clones containing hybrid molecules to be distinguished from those containing regenerated vector molecules.
(vii) Where expression of the cloned DNA is desired, the vector should have appropriate transcriptional and translational signals located adjacent to the cloning site(s).
(viii) Where biological containment is particularly important, for example in cloning the genes of known toxins, it is essential that the vector be prevented from transferring itself or its passenger DNA to other organisms.

3.5.2 *Escherichia coli* systems

(a) Host strains

Strains of *E. coli* commonly used in cloning experiments are described in Table 3.3. Choice of host is guided largely by an ability to accept and maintain particular vectors. Several strains, such as χ1776, have been partly 'disabled' for use as 'safe' hosts in potentially hazardous cloning experiments. Most cloning experiments can, however, be carried out with strains that are considerably less disabled and hence more easily handled than these hosts. Many commonly used host strains, such as HB101 and DH1, are *recA*$^-$. This limits loss or rearrangement of passenger DNA by homologous recombination events. It also reduces the possibility that cloned sequences might be transferred to other replicons and hence provides a degree of biological containment. On the other hand, the *recA*$^-$ mutation considerably decreases the viability of host strains (by up to 50%). This may be disadvantageous particularly in large-scale culture of recombinant bacteria. Furthermore, some phage λ vectors cannot plaque on *recA*$^-$ hosts (Brammar, 1982, and section 3.5.2.d), and recoveries of homopolymer-tailed recombinant molecules are lower in recombination-deficient strains (Peacock *et al.*, 1981). Most hosts used are defective (*hsdR*$^-$ or *hsdS*$^-$) in the

Table 3.3: Some commonly used *E. coli* host strains

Strain	Genotype*	Comments	Reference
C600	F$^-$, *thr, leuB, tonA, lacY, supE, thi*, λ$^-$	General-purpose host	Appleyard, 1954
DH1	F$^-$, *supE, gyrA, recA, endA, hsdR* (r_k^-, m_k^+) λ$^-$	Transforms with high efficiency	Hanahan, 1983
HB101	F$^-$, *ara, leu, proA, lacY, supE, galK, recA, endA, rpsL, xyl, mtl, thi, hsdS* (r_B^-, m_B^-), λ$^-$	General-purpose host for heterospecific cloning with plasmids	Boyer and Roulland-Dussoix, 1969
RR1	as for HB101 but *recA*$^+$	Used for efficient transformation by recombinant molecules formed by homopolymer tailing	Peacock *et al.*, 1981
JM105	F′ (*traD, proAB, lacI*^{q}*, lacZ*ΔM15) *thi, rpsL, endA, sbcB, hsdR*, Δ(*lac-proAB*)	Restriction-less host for M13 sequencing vectors	Yanisch-Perron *et al.*, 1985
χ1776	F$^-$, *tonA, dapD, minA, glnV, supE*, Δ_{40} (*gal-uvrB*), *minB, rfb, gyrA, thyA, metC, oms*-1, *oms*-2, Δ_{29} (*bioH-asd*), *cycA, cycB, hsdR*, λ$^-$	High-containment host	Curtiss *et al.*, 1977
Q359	*supF, hsdR*$_k^-$, *hsdM*$_k^+$ φ80, P2	Suppressor host to select Spi$^-$ recombinants in λ vectors	Karn *et al.*, 1980

*For genotype designations see Bachmann (1984); Δ, deletion.

*Eco*K system, in order to facilitate the cloning of heterospecific DNA in *E. coli.* For reasons that are not fully understood, certain genes can be cloned in some strains of *E. coli* but not in others. For example, recombinant plasmids carrying the *Bsu*RI restriction modification genes can be used to transform *E. coli* RR1 or HB101 but not DH1 (Kiss *et al.*, 1985). DNA containing methylated cytosines also produces much lower frequencies of transformation than equivalent nonmethylated DNA (Blumenthal *et al.*, 1985). Some strains such as RR1, (Table 3.3), are relatively tolerant of C-methylated DNA and are defective in the *mcrA* and/or *mcrB* genes which encode restriction systems specific for such DNA.

(b) Plasmid vectors

Plasmid replicons have formed the basis of the majority of commonly used cloning vectors in *E. coli* and in other microorganisms. Many of the plasmid cloning vectors for *E. coli* derive from the small multicopy (10-20 copies per chromosome equivalent) plasmid ColE1, its close relative pMB1 or the unrelated plasmid p15A (for reviews see Maniatis *et al.*, 1982; Thompson, 1982). All these plasmids require DNA polymerase I for their replication and continue to replicate in the absence of protein synthesis. The copy number of ColE1 replicons can be amplified up to several thousand copies per cell in the presence of inhibitors of protein synthesis, such as chloramphenicol or spectinomycin. This occurs because the initiation of chromosomal replication is inhibited, whereas that of the plasmid continues. Further amplification of 2- to 3-fold can be obtained by adding high concentrations of uridine to the culture medium (Norgard *et al.*, 1979). However, amplification of p15A replicons by inhibition of protein synthesis results in only 2- to 10-fold increase in number of copies. Some plasmids, for example pKT235, a derivative of the ColD multicopy plasmid, conveniently undergo amplification when the host culture enters stationary phase (Timmis, 1981). Amplification, by whatever means, facilitates purification of the vector and recombinant molecules derived from it. Numerous methods have been devised for the purification of plasmid DNA (see Grinsted and Bennett, 1984, and Thomas, 1984, for examples).

1. General purpose vectors. The most widely used plasmid cloning vector is pBR322 (or its derivatives), a multicopy plasmid specifying resistance to ampicillin and tetracycline and based on a pMB1 (ColE1) replicon (Bolivar *et al.*, 1977) (Figure 3.14). The complete 4363 nucleotide-pair sequence of pBR322 has been determined (Sutcliffe, 1979; Peden, 1983). This information permits the accurate location of restriction sites within the vector molecule, in turn facilitating the design of cloning strategies and the subsequent analysis of recombinants. Several derivatives of pBR322 have been developed. These have elevated copy number and/or additional or

Figure 3.14: Genetic map of pBR322. *bla*, β-lactamase gene conferring ampicillin resistance; *tet*, tetracycline-resistance gene. The *bla* gene is efficiently expressed due to anticlockwise transcription originating at promoters P1 and P3. The *tet* gene is less efficiently transcribed clockwise from P2. The P1 and P2 promoters overlap in the region of the *Hin*dIII site. ⇨, Direction of DNA replication from origin of vegetative replication (*oriV*)

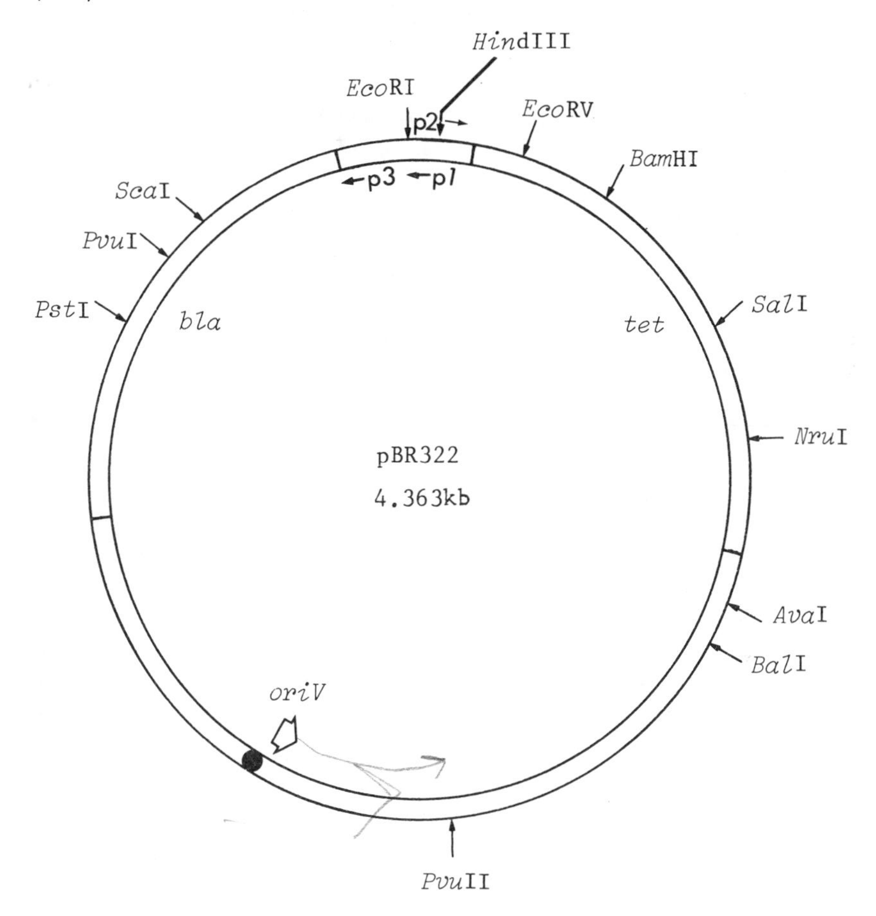

alternative unique restriction sites together with further specialised properties required for cloning (see below). pBR322, which is *mob*⁻, *nic*⁺, can be mobilised by a conjugative plasmid if a suitable (*mob*⁺) nonconjugative plasmid (providing mobility functions *in trans*), such as ColK, is present in the same cell (see section 2.3.1.c). It is theoretically possible for recombinant plasmids to escape from the laboratory by mobilisation to wild-type bacteria. The potential for mobilisation is therefore considered undesirable, particularly when cloning potentially dangerous genes. Consequently 'safe' containment vectors that have lost mobilisation functions have been designed. An example is pAT153 in which the *Hae*IIB and *Hae*IIG fragments of pBR322 have been deleted to produce a plasmid that is not only *nic*$^-$ (and hence cannot be mobilised), but also has a copy number 1.5 to 3 times higher than pBR322 (Twigg and Sherratt, 1980; see also section 5.4.1).

Sometimes it is neither possible nor desirable to use a ColE1-type replicon, such as pBR322. For example, it may be necessary to analyse the function of a cloned sequence by complementation with a second sequence already cloned in a ColE1 replicon. In such a case the first sequence would have to be cloned in a vector compatible with ColE1. Suitable alternatives include vectors based on p15A, such as pACYC184 (Figure 3.15), or on wide host range IncQ plasmids (section 3.5.3).

Cloning in plasmid vectors normally involves linearisation of the vector by cleavage at a unique restriction site located within a structural gene that specifies an identifiable phenotype (usually and most conveniently a resistance character). The vector can then be ligated to target DNA, bearing appropriate termini, to re-form a circular molecule (Figure 3.6). The ligation mixture is used to transform *E. coli*, selecting for a genetic marker carried by the vector at a position distant from the cloning site. Transformants containing recombinant molecules can be identified by **insertional inactivation** whereby the inserted target DNA disrupts the function of the gene containing the cloning site. Where the cloning site is within an antibiotic resistance gene, such insertional inactivation results in transformants sensitive to the appropriate antibiotic. Insertional inactivation thus allows the distinction, after replica plating, of transformants containing recombinant plasmids from those that contain reconstituted vector molecules. (It should be noted that although cells will take up ligation products comprising passenger DNA alone, in almost all cases such molecules carry no easily selectable marker and will also be unable to replicate.)

pBR322 contains unique *Pvu*I and *Pst*I sites in the ampicillin-resistance (*bla*) gene and *Bam*HI, *Eco*RV, *Nru*I and *Sal*I sites in the tetracycline-resistance (*tet*) gene, which may be used for the detection of recombinant formation by insertional inactivation. This plasmid also contains a unique *Eco*RI site, but not within a structural gene. Therefore several pBR322 derivatives have been constructed. An example is pBR325 which carries a chloramphenicol-resistance gene with an approximately central *Eco*RI site.

pBR322 contains a single *Hin*dIII site, which lies in the P2 promoter region of the *tet* determinant (Figure 3.14) (Stuber and Bujard, 1981). Insertions of passenger DNA at this site separate the *tet* structural gene from its promoter. However, insertions do not necessarily produce a Tc^s phenotype, particularly if the insert contains a promoter that is in the correct orientation and is capable of reading through to the *tet* gene

Figure 3.15: Genetic map of pACYC184. *cat*, chloramphenicol acetyl transferase gene conferring resistance to chloramphenicol. This gene is transcribed from the P5 promoter. The *tet* gene is transcribed from the P2 promoter, which (as in pBR322) overlaps with the divergently transcribed promoter P1 in the region of the *Hin*dIII site

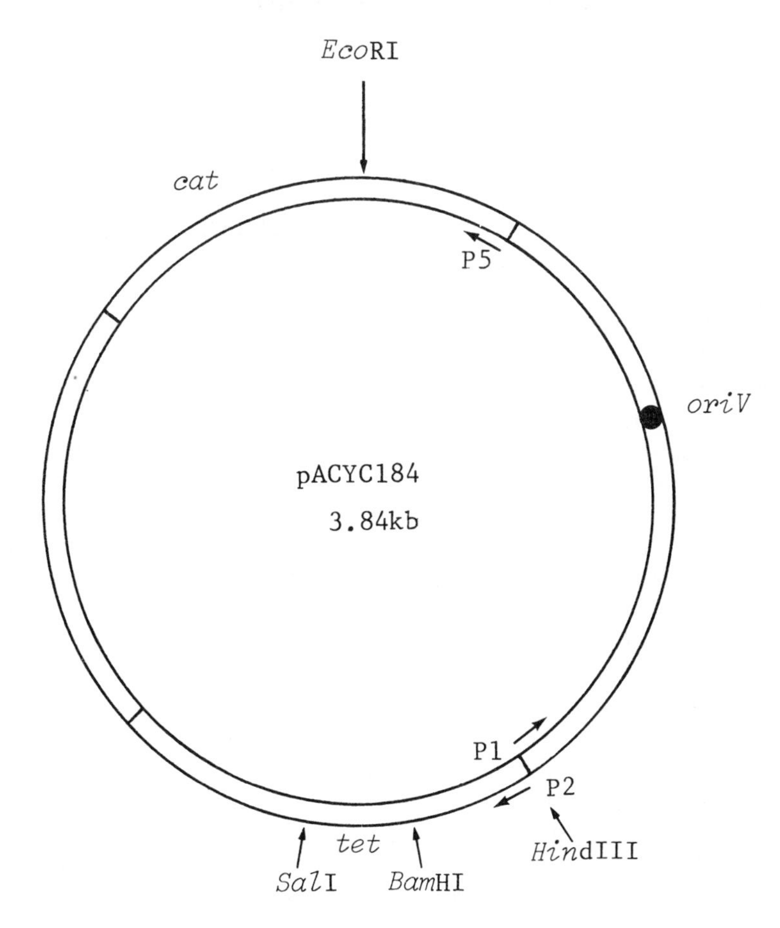

(Widera *et al.*, 1978). Even where Tc^s derivatives of pBR322 containing an insertion at the *Hin*dIII site are isolated, they may give rise spontaneously to Tc^r clones (typically at a frequency of 10^{-6} per cell). Such resistant clones arise as a consequence of deletions within the cloned region that bring a promoter, previously masked by all or part of the insertion, into a position that allows read-through transcription of the *tet* gene (Primrose and Ehrlich, 1981). Thus the use of insertional inactivation as a primary screen for recombinant molecules must be viewed with caution in those instances where insertion at a cloning site leaves the coding sequence of the 'inactivated' structural gene intact.

2. Low copy number vectors. Some cloned genes, particularly those encoding membrane-associated proteins, are expressed poorly or are lethal to *E. coli* when carried on multicopy, but not low copy, plasmids. Hence there is a requirement for low copy number vectors. Vectors of this type have been developed from plasmids such as pSC101, for example, pLG338 (Stoker *et al.*, 1982), F or RK2 (see, for example, sections 3.5.3 and 3.5.8). Plasmid vectors that can be reversibly switched between a low copy and high copy number mode have particular advantages in directing the expression of foreign genes (see section 5.4).

3. cDNA cloning vectors. Okayama and Berg (1982) have developed vectors (for example pSVT186) based on pBR322 for the efficient cloning of full-length cDNA copies (section 3.5.7). Use of these vectors involves a combination of homopolymer tailing and adaptor techniques. The poly-T sequence on the vector is used to prime synthesis of an RNA-DNA hybrid. The RNA component of the hybrid is subsequently removed with RNaseH and replaced by DNA using DNA polymerase I and *E. coli* DNA ligase (section 3.1.2). Analogous procedures have been applied to the cloning of cDNA in plasmid vectors such as pUC9 (Heidecker and Messing, 1983).

4. Direct (positive) selection procedures for recombinant formation. Screening transformants for the presence of recombinant clones by insertional inactivation is a negative selection process which is extremely laborious when large numbers of clones have to be examined. Methods that permit the direct selection of transformants carrying recombinant plasmids are therefore favoured since they reduce considerably the screening necessary to identify a particular clone. Enrichment for recombinants may be carried out where insertional inactivation of either tetracycline or chloramphenicol-resistance genes has occurred. Tetracycline and chloramphenicol are both bacteriostatic drugs. Cells that have become Tc^s or Cm^s as a result of insertional inactivation will therefore stop growing in the presence of the appropriate drug. Non-growing cells are refractory to the effects of bacteriocidal antibiotics (for example penicillins) that kill actively growing cells. Thus in the presence of tetracycline and D-cycloserine, for example, Tc^r cells are killed whereas Tc^s cells survive (Bolivar *et al.*, 1977). Tc^s cells that contain recombinant plasmids can then be recovered after removing the antibiotics.

A simpler system for positive selection of insertional inactivation of the tetracycline-resistance determinant has been devised (Bochner *et al.*, 1980; Maloy and Nunn, 1981). This method relies on the fact that Tc^r cells are hypersensitive to lipophilic chelating agents, such as fusaric acid. Thus if the cloning strategy involves insertional inactivation of the *tet* gene, transformants containing recombinant plasmids may be selected directly by plating on medium containing an appropriate antibiotic (for example ampicillin) to select for the vector replicon, and fusaric acid to kill all the Tc^r cells.

A more reliable approach to positive selection has been the construction of direct selection plasmid vectors. Such vectors typically use insertional inactivation of a gene on the vector that would otherwise be lethal to the host under particular cultural conditions (Table 3.4). Approaching 100% of transformants obtained with such vectors should theoretically contain recombinant plasmids. In practice, however, some transformants contain plasmids bearing deletions of the 'lethal' gene.

(c) Phage M13 vectors

The filamentous single-stranded DNA coliphage M13 (and its relatives) absorbs specifically to F pili. Inside an infected cell the circular single-stranded M13 genome of 6.4 kb is converted to the duplex replicative form (RF), which subsequently replicates to produce a pool of about 100 RF molecules per cell. Unlike most phages, M13 does not kill its host, but merely reduces its growth rate. The duplex RF molecules are replicated to produce single-stranded (+) M13 genomes, which are constantly packaged into filaments and extruded through the cell wall. The phages can be easily recovered from the medium and the DNA purified by phenol extraction and ethanol precipitation.

RF molecules of M13 can be obtained by lysing infected cells and used like plasmids for cloning DNA. Recombinant molecules can be introduced into *E. coli* by transfection (section 3.5.2.g). Since progeny filaments merely increase in length to incorporate a larger genome, the phage can be used to clone a variety of different-sized DNA fragments. However, in practice, inserts of greater than 5 kb tend to be unstable in some hosts due to restriction. The problem of restriction-mediated deletion of insert DNA can however, be overcome by the use of restriction-less host strains. Such strains as JM105 allow the cloning of unmodified DNA fragments of 10 kb or more (Yanisch-Perron *et al.*, 1985).

Cloning of DNA fragments has been facilitated by inserting the promoter and N-terminus of the *E. coli lacZ* gene into the intercistronic region between genes II and IV on the RF of M13 (Barnes, 1980) (Figure 3.16). The first 145 codons of the *lacZ* gene are translated to produce the α-peptide of β-galactosidase. This is enzymatically inactive, but will complement deletions in *lacZ* (for example *lacZ*ΔM15) that remove the amino terminus of β-galactosidase. For this reason host strains for M13 vectors normally contain a complete deletion of *lacZ*, but carry an F plasmid bearing the *lacZ*ΔM15 deletion (Messing *et al.*, 1981). Synthetic DNA fragments containing an array of cloning sites can be inserted into the M13 genome without affecting either infectivity or the synthesis of α-peptide. However, cloning foreign DNA into these restriction sites will insertionally inactivate synthesis of α-peptide and hence lead to loss of β-galactosidase

Table 3.4: Some direct (positive) selection plasmid vectors

Vector	Selectable target gene on plasmid	Positive selection indicator	Cloning sites in selectable gene	Reference
pKN80	Mu *kil* gene	Survival to Kil (killing) function of phage Mu	*Hin*dIII, *Hpa*I	Schumann, 1979
pKY2289	colicin E1 (*cea*) gene	Survival to mitomycin C induction of colicin E1	*Eco*RI, *Xma*I	Ozaki *et al.*, 1980
pTR262	λcI regulating *tet* gene	Resistance to tetracycline caused by release from *c*I-mediated repression	*Bcl*I, *Hin*dIII	Roberts *et al.*, 1980
pNO7523	*rpsL*$^+$ in *rpsL*$^-$ host	Resistance to streptomycin (streptomycin-sensitivity encoded by *rpsL*$^+$ is dominant to streptomycin resistance of *rpsL*$^-$ allele)	*Hpa*I, *Sma*I	Dean, 1981
pGJ53	2394 bp palindromic sequence formed from IRs of Tn*5*	Survival to lethal effects of large palindrome on plasmid	*Hin*dIII	Hagan and Warren, 1982
pHE3	*pheS*$^+$ (phenylalanyl-tRNA synthetase α-subunit) in *pheS*$^-$ host	Resistance to *p*-fluorophenylalanine	*Pst*I	Hennecke *et al.*, 1982
pLV57/pLV59	*Eco*RI endonuclease gene in host carrying temperature-sensitive *Eco*RI methylase gene	Survival to lethal effects of *Eco*RI endonuclease at >37°C when *Eco*RI modification methylase is non-functional	*Bgl*II, *Hin*dIII, *Pst*I	O'Connor and Humphreys, 1982
pLA7	*ush*$^+$ (UDP-glucose hydrolase) gene in *ush*$^-$, *upp*$^-$ (uracil phosphoribosyltransferase) host	Resistance to 5-fluorouracil and 5′AMP	*Bcl*I	Burns and Beacham, 1984

Figure 3.16: Genetic map of M13mp series vectors. *ori*, Origin of DNA replication; ⇨, direction of DNA replication for the (+) and (−) strands; roman numerals indicate M13 genes; gene III encodes the minor coat (pilot) protein, gene VIII encodes the major coat protein, gene II product is required for DNA replication and gene V protein binds to newly synthesised (+) strands to prevent formation of complementary (−) strands prior to the incorporation of the M13 genome into progeny virions. The remaining genes are required for virus assembly and maturation. *lacI*, Gene for *lac* repressor; op, *lac* operator-promoter; *lacZ'*, N-terminal sequence of *lacZ* gene encoding the α-peptide of β-galactosidase

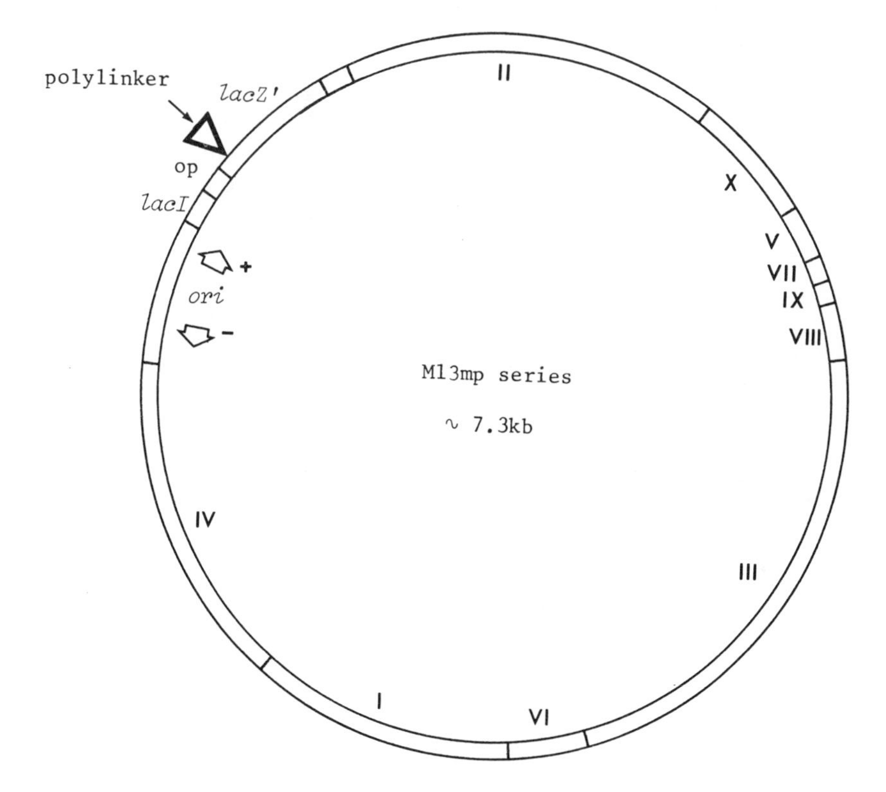

activity. The expression of functional β-galactosidase in clones may be monitored readily on agar plates containing the *lac* inducer isopropylthio-galactoside (IPTG) and chromogenic substrates, such as 5-bromo-4-chloro-3-indolyl-β-D-galactoside (Xgal). Lac^+ clones give blue plaques,[4] whereas recombinant clones are Lac^- and give white plaques.

With some vectors, such as M13mp7, the cloning region contains two

[4] M13 does not produce plaques in the strict sense because it does not lyse host cells. However, the reduced growth rate of infected cells is manifested as a point of poor growth with the appearance of a turbid plaque on a lawn of sensitive bacteria.

sites for each of a number of restriction enzymes, such as *Eco*RI or *Bam*HI (positioned symmetrically about a central unique *Pst*I site) (Messing *et al.*, 1981). When DNA is cloned into such sites, either strand of a cloned restriction fragment may become part of the viral (+) strand depending on its orientation relative to the M13 genome. More convenient vectors have been developed containing single sites for several enzymes to permit directional cloning of foreign DNA (Messing and Vieira, 1982). Pairs of these vectors, for example M13mp18 and M13mp19, which have the same multiple restriction site region but in opposite orientations (Figure 3.17), can be used to guarantee that both strands of a DNA fragment can be cloned as a (+) strand. M13 vectors are particularly useful in applications where large amounts of single-stranded DNA are required. They are therefore of great importance in rapid DNA sequencing methods (section 3.8.5), in directed mutagenesis (section 4.9) and in preparing DNA as a hybridisation probe. Sequencing of DNA cloned in M13 vectors is greatly simplified by the availability of single-stranded master (universal) M13 primers, which can be hybridised to the region adjacent to the multiple cloning sites (section 3.8.5). Some of the principles and techniques used in M13 cloning systems have been employed in the design of plasmid vectors for use in DNA sequencing. For example, the plasmid vectors pUC8 and pUC9 contain the M13 cloning region in opposite orientations

Figure 3.17: The polylinker regions of some M13mp and pUC vectors

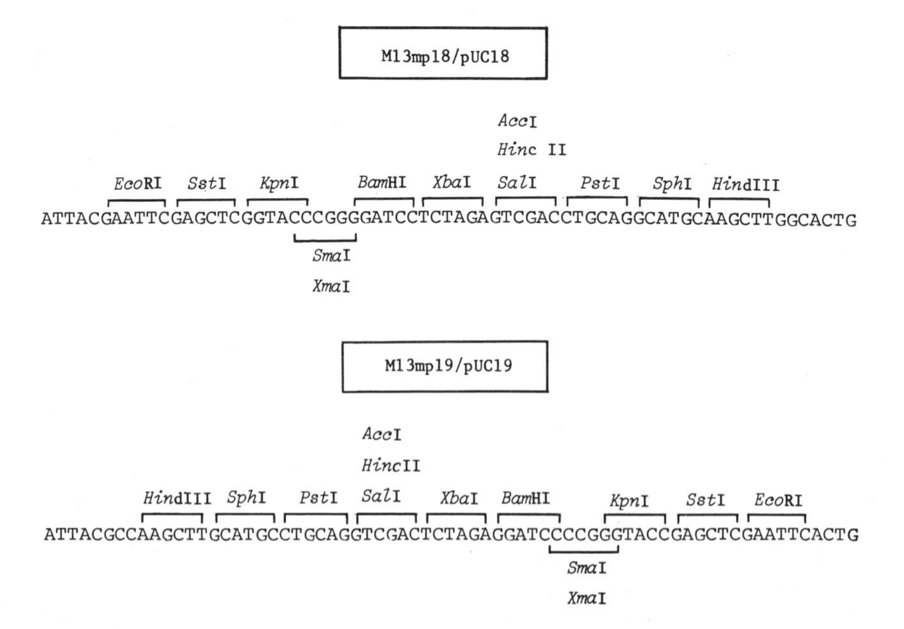

(from M13mp8 and M13mp9 respectively) (Vieira and Messing, 1982). This permits the directional cloning of doubly digested DNA in either orientation with respect to the *lac* promoter and the use of universal M13 primers for sequencing inserts. It is also possible to use the pUC vectors for the construction of cDNA libraries (section 3.6.2). The presence of inserted DNA within the polylinker region is indicated by the production of white (as opposed to blue without the insert) colonies on Xgal plates.

The ability to convert DNA cloned in a plasmid into a single-stranded form is of great advantage in DNA sequencing and site-directed mutagenesis procedures. A segment of the genome of the single-stranded DNA phage f1, containing all the *cis*-acting elements necessary for DNA replication and morphogenesis, can be cloned as a double-stranded fragment in plasmids, such as pBR322. Superinfection of strains carrying such recombinant plasmids with f1 phages results in the production of virions containing either f1 single-stranded or hybrid plasmid single-stranded DNA at about equal frequency (Dotto *et al.*, 1981). Insertion of the appropriate f1 DNA segment into pUC plasmids has enabled the construction of the pEMBL family of vectors, which combine a plasmid origin of replication, polylinker and *lacZ* indicator region with the ability to package single-stranded DNA into virions (Dente *et al.*, 1983). The orientation of the f1 fragment in these plasmids determines which of the two strands is encapsidated into f1 capsids. The β-galactosidase gene is packaged in f1 virions as antisense (non-coding) in pEMBL(+) and sense (coding) in pEMBL(−) vectors.

(d) Phage λ vectors

The wealth of knowledge concerning the genetic and physical organisation of bacteriophage λ (see, for example, Hendrix *et al.*, 1983) has greatly facilitated its use as a cloning vector. The genome of λ is double-stranded linear DNA of 48.6 kb with single-stranded 5′ complementary protrusions of 12 nucleotides (Sanger *et al.*, 1982). These cohesive termini anneal to permit recircularisation soon after the λ DNA molecule enters a host cell. Upon infection the phage may either enter the lytic cycle or lysogenise the host (see section 2.4.2). Establishment and maintenance of lysogeny require the synthesis of λ*c*I repressor. Inactivation of this repressor either by treatment with agents that damage DNA or by using conditional mutants, for example $\lambda c\mathrm{I}_{857}$, that produce a temperature-sensitive repressor will initiate lytic development.

Packaging of DNA into λ capsids is constrained by a stringent requirement for size. Only molecules of between 78 and 105% of the wild-type genome length are packaged to produce viable virus particles. This limits the amount of DNA that can be cloned in λ vectors. However, only about 50% of the genome of λ is essential for growth and plaque formation. Space available for cloning foreign DNA can therefore be increased by removing some or all of the nonessential regions. The wild-type λ genome contains multiple sites for many of the restriction enzymes commonly used for cloning. However, it is possible to remove one or more of these sites by a variety of *in vivo* and *in vitro* manipulations (Brammar, 1982; Murray, 1983). For example, sites may be added or removed by replacing sequences, such as those encoding the immunity (*c*I) region, with similar DNA sequences from other lamboid phages, such as ϕ80 or ϕ434. Such manipulations, coupled with deletions to increase cloning capacity and prevent lysogenisation, have been used to construct a wide variety of λ vectors for cloning DNA fragments with various cohesive termini (Table 3.5). Some of these vectors contain amber (*am*) mutations in essential genes in order to make the vectors dependent on an amber-suppressing host for purposes of biological containment and/or to assist in recombinational mapping techniques (see section 3.7.2). A useful development has been the introduction of polylinkers into the cloning regions of λ to facilitate re-excision of passenger DNA from recombinant phages (see for example Loenen and Blattner, 1983). (For reviews of λ vectors see Brammar, 1982; Maniatis *et al.*, 1982; Murray, 1983.)

1. Types of λ vector

(i) *Insertion vectors.* These have single targets for one or more restriction

Table 3.5: Some bacteriophage lambda vectors and their characteristics

Vector	Restriction enzyme used for cloning	Insert size (kb) Minimum	Insert size (kb) Maximum	Recognition of recombinants
λNM641	*Eco*RI	0	11.6	Clear plaques
λNM781	*Eco*RI	2.2	15.2	Lac⁻ due to insertional inactivation of *supE* gene
Charon 4A	*Eco*RI	7.1	20.1	Lac⁻ due to removal of *lac* DNA
	*Xba*I	0	5.64	—
λgt11	*Eco*RI	0	8.3	Lac⁻ due to formation of β-galactosidase fusion protein
L47.1	*Bam*HI	4.7	19.6	Spi⁻
	*Eco*RI	8.6	24.1	Spi⁻
	*Hin*dIII	7.1	21.6	Spi⁻
	*Xho*I	0	13.3	—
λgtWES.T5622	*Eco*RI	2.1	15.1	Propagation in ColIb containing host

Lac⁻, inability to ferment lactose due to absence of functional β-galactosidase; Spi⁻, forms plaques on a P2 lysogen.

enzyme(s) in a nonessential region of the λ genome (Figure 3.18). The sites are preferably in genes with readily identifiable phenotypes that permit detection of recombinant phages by insertional inactivation (Table 3.5). Only DNA fragments up to about 10 kb may be cloned in insertion vectors due to the upper size limit imposed by packaging.

(ii) *Replacement (substitution) vectors.* Such vectors have a pair of widely spaced restriction sites (either two sites for the same or one site for each of two different enzymes) spanning a nonessential region of the λ genome (Figure 3.18). Replacement vectors possess a lower limit of about 2 kb and an upper limit of about 20 kb to the size of DNA fragment that can be inserted. Cloning with such vectors normally involves cleavage of vector DNA with an appropriate enzyme(s). The essential left and right arms of the vector are annealed by their cohesive termini and separated from the central nonessential segment by sucrose gradient centrifugation or agarose gel electrophoresis. The vectors arms are heated to dissociate their annealed cohesive ends and ligated to foreign DNA fragments bearing appropriate cohesive termini. The resulting recombinants can be introduced into *E. coli* by *in vitro* packaging or, less efficiently, by transfection (section 3.5.2.g). Recombinant DNA molecules may be selected during packaging, since molecules formed by ligating the left vector arm directly to the right arm are generally too small to be packaged.

2. Identification of recombinant phages by size selection. Sometimes it is possible to distinguish between parental phages and recombinants simply on the basis of genome size. Alterations in genome size affect buoyant density of λ phage particles. CsCl density gradient centrifugation can therefore be used to separate recombinant phages exhibiting decreased or increased genome size relative to the parental phage (Philippsen *et al.*, 1978). Increased genome size enhances the sensitivity of phage particles to disruption by certain chelating agents, such as pyrophosphate or EDTA. It is thus possible to distinguish recombinant from parental phages by using plates containing predetermined inhibitory concentrations of pyrophosphate (Parkinson and Huskey, 1971). Size selection may also be imposed by using a *pel*$^-$(*pe*netration of *l*ambda) host (Scandella and Arber, 1974), which prevents growth of phages of reduced genome size.

3. Genetic screens for recombinants. Recombinant phages may be distinguished from parental phages by a number of genetic tests. The *c*I gene encodes the λ repressor, which is responsible for maintaining λ as a prophage and for rendering lysogens immune to infection by phages of the same immunity type. Phages with a functional *c*I gene produce turbid plaques on a lawn of sensitive bacteria due to the presence of lysogenised

Figure 3.18: Types of cloning vectors derived from bacteriophage λ. Only certain λ genes are shown. Functions of those genes not described in the text can be found in Hendrix *et al.* (1983). The *nin5* deletion removes a region between the *P* and *Q* genes (including the rightward transcription terminator tR_2) rendering delayed early transcription independent of the *N*-gene antiterminator product

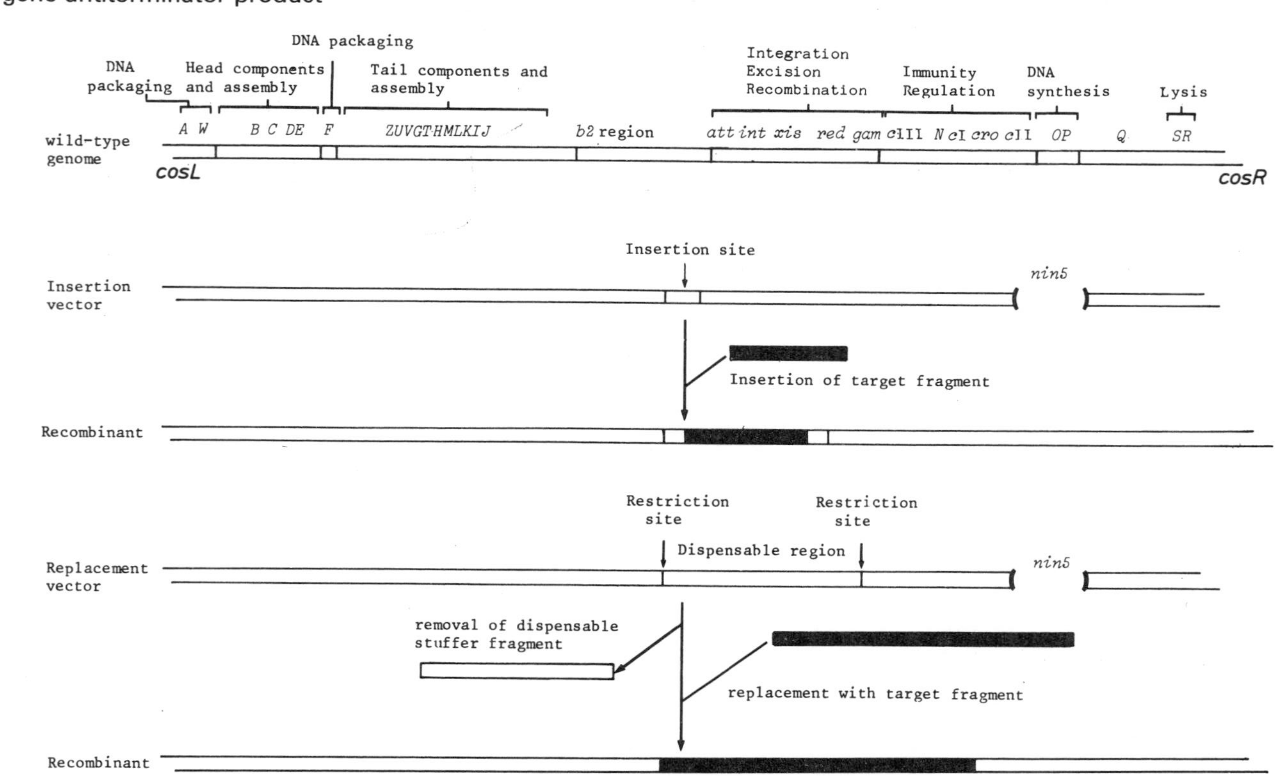

cells within each plaque. Inactivation of the *c*I gene, by, for example, the insertion of foreign DNA at a unique site within the gene, leads to the production of clear plaques.

The *E. coli lacZ* gene carried by certain λ vectors can be used in conjunction with the Xgal indicator system to detect recombinants. Vectors carrying either an intact *lacZ* gene or that part of *lacZ* necessary to produce α-complementation in an appropriate host will give blue plaques on Xgal plates. The presence of multiple copies of the *lac* operator, produced during phage replication, is sufficient to titrate out the *lac* repressor, encoded by the chromosomal *lac* gene in a Lac^+ host, leading to induction of β-galactosidase. Insertion into the *lac* region, or its replacement by foreign DNA, produces the white plaque phenotype, which is indicative of recombinants. Host strains carrying amber-suppressible *lacZ* mutations may also be used in conjunction with replacement vectors carrying either the *E. coli supE* (glutamine tRNA amber suppressor) gene (which can be used for cloning *Eco*RI fragments), or the *supF* (tyrosine tRNA amber suppressor) gene (for cloning *Hin*dIII fragments), for detecting recombinants. In either case replacement of the central *sup* region of the vector with foreign DNA will produce white plaques.

Wild-type λ does not form plaques on strains of *E. coli* that carry the P2 prophage and is thus referred to as having a Spi^+ (*s*ensitive to *P*2 *i*nterference) phenotype. However, λ derivatives that have lost the *gam* and *red* genes (both of which are required for vegetative phage recombination and lie in the nonessential region of the genome)[5] are Spi^- and will grow on P2 lysogens (Zissler *et al.*, 1971). Several replacement vectors such as λ L47.1 and λ Charon 30 have the *red* and *gam* genes in their central filler regions. Accordingly these can be used as direct selection vectors since only recombinant phages will be Spi^-. Complete expression of the Spi^- phenotype depends upon the presence of a *chi* (*c*ross-over *h*ot-spot *i*ndicator) site within the phage genome. *Chi* sites are sequences (5′ GCTGGTGG 3′) that are hot-spots for *recA*-mediated recombination and are required for the efficient maturation of the circular dimeric replicative forms of *red*$^-$, *gam*$^-$ vectors. Such closed circular molecules produced by θ-form (Cairns) DNA replication are resistant to degradation by *recBC* exonuclease. In contrast, the catenated linear λ DNA molecules produced by rolling circle replication are destroyed in *gam*$^-$ vectors by the *recBC* gene product. In the absence of a *chi* site in the phage DNA, development of the virus is poor and the plaques produced on P2 lysogens are small and of variable size. The necessary *chi* site(s) for maturation of Spi^- phages may be

[5] The *gam* gene product is an inhibitor of the host *recBC* gene product, exonuclease V, which degrades catenated linear λ DNA molecules produced by rolling circle replication. The *red* gene product is required for vegetative recombination of λ DNA molecules.

provided fortuitously within the cloned insert. However, in order to obtain plaques of relatively uniform diameter with *red*⁻, *gam*⁻ vectors, it is preferable that one or more of the vector arms contains a *chi* site. For example, λL47.1 contains a *chi* site in its left arm and λ1059 in its right arm (Brammar, 1982; Maniatis *et al.*, 1982). Spi⁻ phages cannot be grown on *recA*⁻ hosts (Lam *et al.*, 1974). This may be disadvantageous since *recA*⁻ strains are often used to prevent recombinational rearrangement of cloned sequences. However, Spi⁻ recombinant phages can, if necessary, be propagated in *recBC*⁻ hosts without apparent rearrangement (Jeffreys *et al.*, 1982). Alternatively λ Charon vectors that are *red*⁻, *gam*⁺ (and therefore inhibit the activity of the *recBC* product) have been developed that can be used for propagating large DNA fragments in *rec*A⁻ strains without rearrangement (Loenen and Blattner, 1983).

An alternative direct selection strategy is employed with the vector λgt WES.T5622, which carries a replaceable *Eco*RI fragment bearing the A3 gene of bacteriophage T5. The A3 gene prevents the vector from infecting hosts carrying the ColIb plasmid. Thus only recombinant phages can propagate in such hosts (Davison *et al.*, 1979).

4. Screening using antibody probes — λgt11. The phage vector λgt11 (Young and Davis, 1983) can be used to isolate specific genes, provided that an appropriate antibody is available to identify clones expressing the relevant antigens (see section 3.7.1.c). Foreign genomic or cDNA is inserted at the unique *Eco*RI site, which lies in the β-galactosidase gene carried by this vector (Figure 3.19). Phages containing inserts of up to 8.3 kb produce proteins that are fused to the N terminal portion of β-

Figure 3.19: λgt11, an expression vector for immunological screening of cloned DNA. *lacZ*, β-galactosidase gene; p, *lac* promoter; cI_{857}, temperature-sensitive λ repressor gene; *Sam*, amber-suppressible mutation in the λ *S* lysis gene; ▼, antigenic site

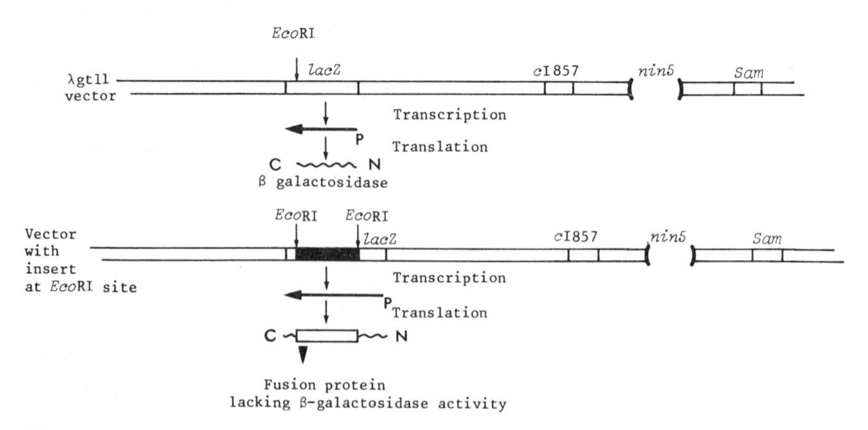

galactosidase. The resulting hybrid proteins lack β-galactosidase activity and recombinant phages therefore do not form blue plaques on a lawn of *lac*$^-$ bacteria growing on Xgal plates. The correct expression of foreign DNA in the phage *lacZ* gene depends on the orientation and reading frame of the insert with respect to the β-galactosidase coding sequence. Only about one-sixth of recombinants containing a specific cDNA produce β-galactosidase fused to the protein of interest (Young and Davis, 1983). In order to achieve efficient lysogeny the *E. coli* host used for cloning with λgt11 is an *hflA*$^-$ (high-frequency lysogeny) mutant. The λgt11 vector itself carries a cI_{857} temperature-sensitive repressor gene and an amber-suppressible mutation in the λ *S* gene rendering the phage lysis-defective. Consequently λgt11 recombinants form lysogens very efficiently at 32°C, but may be induced by raising the temperature to 42°C. In nonsuppressing *E. coli* (particularly those defective in the Lon protease (see section 5.5.1)) large quantities of phage encoded proteins are produced upon induction without cellular lysis. Specific antigens produced by single recombinants may be detected against a background of up to 10^6 colonies on a single agar plate using the immunological screening techniques outlined in section 3.7.1.c.

(e) Cosmids

Cosmids are vectors that comprise a plasmid origin of replication, a selectable marker, one or more unique cloning sites and the cohesive end (*cos*) site of λ. Such vectors were developed initially by Collins and Hohn (1978). Cosmids can be used to clone large fragments of DNA by exploiting the λ *in vitro* packaging system (see section 3.5.2.g). The only ligation products in a cosmid cloning experiment that can be packaged are those that consist of two cosmid molecules in the same orientation flanking a DNA insert that separates the two *cos* sites by ~37-51 kb (Figure 3.20). A cosmid vector is normally about 5 kb in length, which means that up to 45 kb of foreign DNA can be incorporated into each recombinant.

Problems can be encountered with cosmid cloning because vector multimers of the correct size may be packaged. It is also possible that passenger DNA sequences that were not contiguous in the original genome may be ligated together and subsequently packaged. These problems may, however, be overcome by a method developed by Ish-Horowicz and Burke (1981) for the cosmid pJB8, a derivative of pAT153. This procedure involves digesting a sample of vector DNA with *Hin*dIII which cleaves at a unique site on one side of the *cos* site and another sample with *Sal*I which cleaves at a site on the other side. This enables the isolation of 'left-' and 'right'-hand cosmid vector molecules, which are then dephosphorylated by treatment with alkaline phosphatase (section 3.4.4). Both vector preparations are subsequently cleaved with *Bam*HI to generate 'left' and 'right' *cos* fragments. The phosphorylated *Bam*HI termini of these fragments may

Figure 3.20: Cloning in a cosmid vector. *ori*, Origin of replication; *cos*, cohesive end site

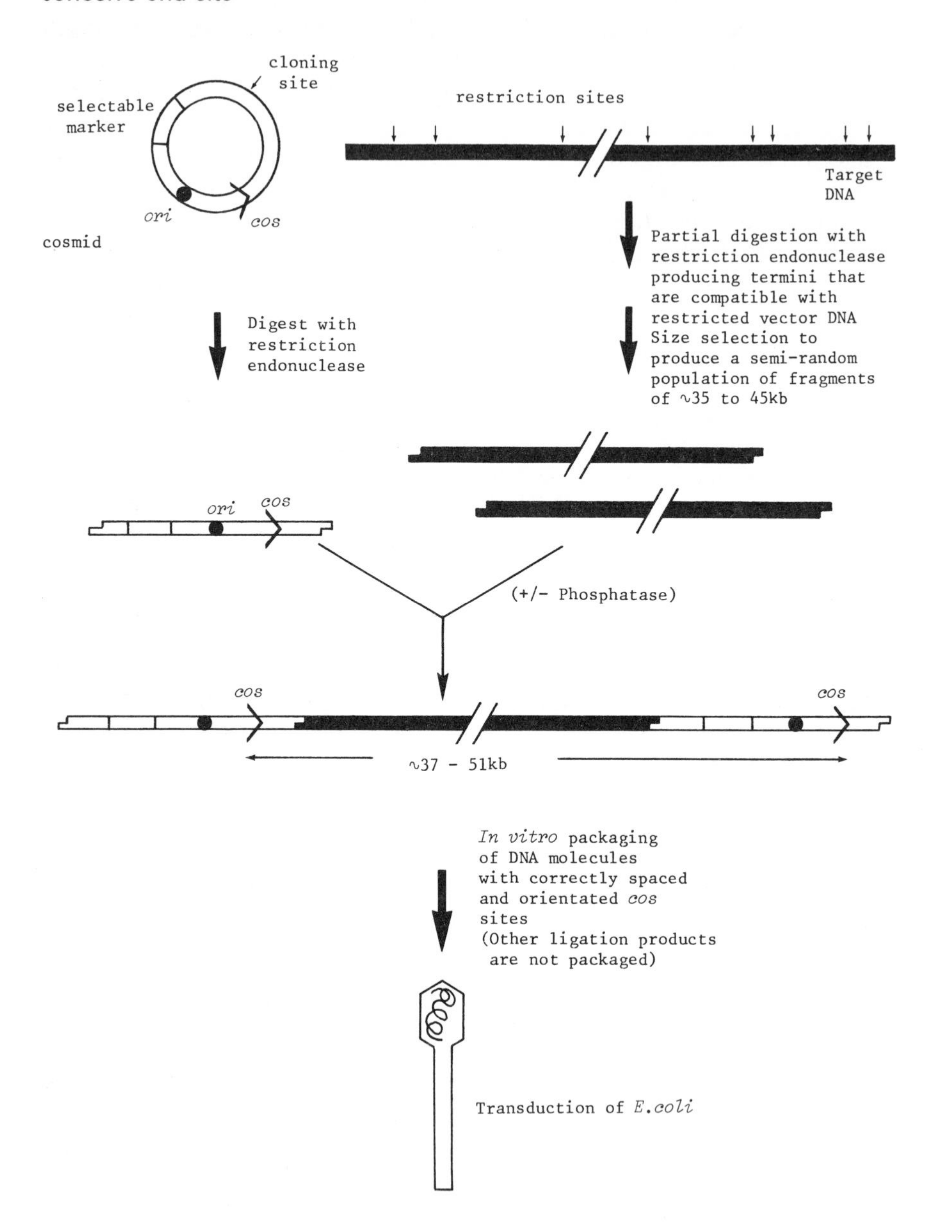

then be ligated to dephosphorylated passenger DNA fragments of ~32-46 kb generated by partial *Sau*3A or *Mbo*I digestion and treatment with alkaline phosphatase. Of the potential ligation products generated, only recombinants with the structure: 'left' *cos* fragment–passenger DNA (32-46 kb)–'right' *cos* fragment, are packaged. The experimental design effectively ensures that noncontiguous target sequences do not become linked into recombinant molecules, and overcomes the problem of vector-vector association.

Cosmid cloning has been simplified by the construction of plasmids, such as c2XB, that carry two *cos* sites separated by a unique blunt-ended (*Sma*I) restriction site (Bates and Swift, 1983). Use of this cosmid eliminates the need for preparing separate right and left *cos* fragments.

Cosmids combine many of the advantages of plasmid and phage vectors and are preferred for certain purposes, such as the formation of gene libraries. The size selection imposed by *in vitro* packaging ensures that large DNA fragments are cloned. Cosmids also have a higher cloning capacity than either plasmid or phage vectors. Recombinant cosmids can generally be maintained stably by using the plasmid origin of replication. They may be transferred to other *E. coli* strains by transformation or mobilisation. However, it is more efficient to use transduction following infection of clones harbouring cosmids with λcI_{857} phage. At 37°C or higher this phage will enter the lytic cycle and package cosmid molecules *in vivo.* The resulting lysate can be used to infect a sensitive host and selection imposed for the cosmid. Growth at 37°C avoids the formation of λ lysogens.

(f) Phasmids

Phasmids are hybrids formed between small multicopy plasmids and bacteriophages such as P4 (Kahn and Helinski, 1978) or λ (Brenner *et al.*, 1982). A phasmid can be propagated as a plasmid or lytically as a phage. Lytic functions of a phasmid can be switched off by propagation in the appropriate lysogen where the plasmid origin of replication is used for maintenance. The phasmid may replicate as a phage if propagated in a nonlysogenic strain. In the case of phasmids based on λ, such as λ1130, the temperature-sensitive gene cI_{857} carried by the vector may be used to effect the switch between replication modes, simply by growing the host at the permissive (plasmid mode) or restrictive (phage mode) temperature (Brenner *et al.*, 1982). The phasmids constructed by Brenner and co-workers (1982) contain multiple *att*λ sites at the junction between the plasmid and phage modules. Recombination between *att*λ sites can thus effect release of the plasmid component as desired. This can be achieved by propagating the phasmid in a host strain that carries a prophage providing both λ integrase for recombination between *att* sites and λ repressor to prevent phage replication.

Phasmids are particularly useful in the generation and analysis of mutations exhibiting nonselectable or lethal phenotypes, such as those affecting the replication of plasmids (Cesareni *et al.*, 1982). Phasmids may also be used as phage replacement vectors (Brenner *et al.*, 1982) and for directing the high-level expression of proteins from cloned sequences by replication in the phage mode.

(g) Reintroduction of vectors into hosts

1. Transformation and transfection. Recombinant molecules may be readily introduced into $CaCl_2$-treated *E. coli* by transformation (with plasmid vectors) or transfection (with phage vectors) (see section 2.2). Both processes are relatively inefficient. At best only 1 to 10% of cells are transformed or transfected (Saunders *et al.*, 1984). Dose-response experiments suggest that a single plasmid DNA molecule is sufficient to produce a transformant in *E. coli* (Hanahan, 1983; Saunders *et al.*, 1984). When the DNA is in excess of about 10^2 plasmid molecules per viable cell, competent *E. coli* become saturated in their ability to be transformed. For this reason subsaturating concentrations of DNA are used in order to obtain the maximum number of transformants per μg of recombinant DNA. Between 10^6 and 10^9 transformants per μg DNA can be obtained with a variety of transformation protocols. Although each transformant can arise from the uptake of a single plasmid molecule, it is evident that at saturating DNA concentrations a proportion of transformed cells (ranging from about 20% to 80%) acquire two or more plasmid molecules. This may be undesirable since individual transformants may then acquire two different recombinant molecules. The problem of multiple transformation events can be lessened by substantially reducing the concentration of transforming DNA.

A major limitation of plasmid transformation is that large DNA molecules ($>$20 kb) are taken up much less efficiently than small ones. This limits the utility of plasmid vectors for cloning large DNA fragments and leads to bias towards the incorporation of small DNA fragments in shotgun cloning experiments.

2. In vitro *packaging of λ DNA.* Linear λ or cosmid DNA may be packaged into mature phage particles *in vitro* by using extracts of bacteria that have been infected with different λ mutants that are defective in phage assembly. Between 10^7 and 10^8 plaque-forming units per μg DNA (equivalent to up to 0.5% of the total λ DNA molecules present in the packaging reaction) may be obtained (Enquist and Sternberg, 1979; Hohn, 1979). The packaging mixture is generally composed of complementary extracts from pairs of A^- and E^- or D^- and E^- mutants of λ (Figure 3.21). The *A* gene protein is involved in the insertion of DNA into bacteriophage heads and cleavage of concatenated precursor DNA at the λ *cos* sites. The *D* gene

Figure 3.21: *In vitro* packaging of DNA by phage λ

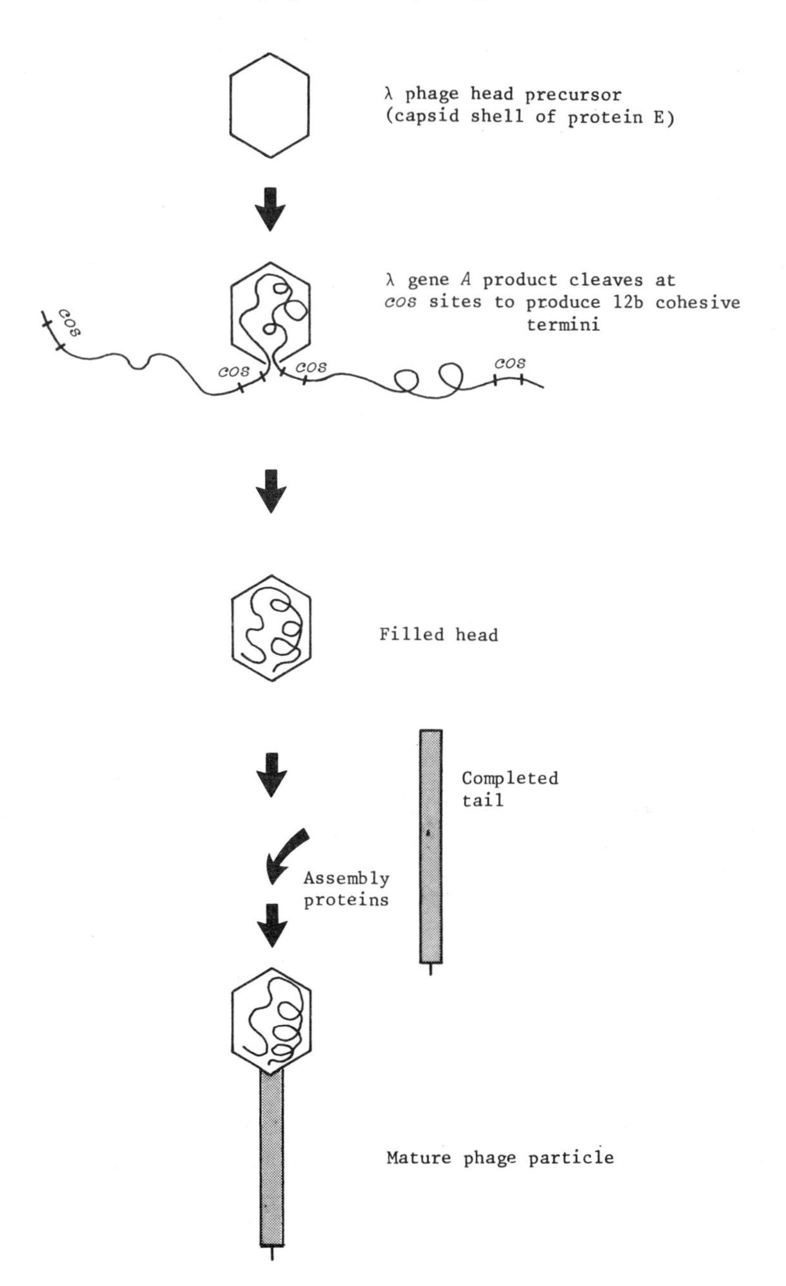

protein is involved in the insertion of λ DNA into the head precursor and subsequent maturation of the head. The *E* gene protein is the major component of the phage head. Lysogens used for making the packaging extracts carry the cI_{857} temperature-sensitive mutation in λ repressor, permitting induction of the lysogens as desired. The prophages also carry the *Sam7* mutation (an amber mutation in gene *S* which is required for cell lysis), which causes the accumulation of capsid components in non-suppressing hosts; the *b2* deletion, which inactivates the *att* site and reduces the packaging of endogenous λ DNA; and a $redB^-$ mutation, which eliminates the generalised recombination system of the phage. In a $recA^-$ host, the $redB^-$ mutation minimises recombination between endogenous λ DNA present in the extract and added recombinant genomes. The problem of packaging the endogenous phage DNA present in the extract can be largely overcome by using a single extract prepared from a host carrying a λ phage that expresses all the genes necessary for packaging but is deleted for the *cos* site and is hence unpackageable (Rosenberg *et al.*, 1985).

3.5.3 Cloning in Gram-negative bacteria other than *E. coli*

Escherichia coli may be inappropriate as a host for many cloning experiments because the genes of interest are not expressed in this background. Furthermore, it may be necessary to clone DNA directly in Gram-negative bacteria other than *E. coli* for purposes such as strain improvement (see Chapter 6). Most of the vectors for *E. coli* have a host range that is limited to this organism and to closely related enterobacteria. Vectors for use in other Gram-negative bacteria, for example pseudomonads (Bagdasarian and Timmis, 1982), may be developed by manipulation of indigenous phage or plasmid replicons. However, this is likely to be time-consuming and of limited application. A more efficient approach may be to utilise exogenous broad host range plasmids as vectors. Such vectors may be used in order to clone DNA sequences directly in the particular species from which they derive (self-cloning). Alternatively, primary cloning and manipulation of such DNA sequences may be carried out in genetically well characterised species, such as *E. coli* or *Pseudomonas aeruginosa.* The cloned DNA may subsequently be returned to the species of interest by utilising the broad host range capability of the vector. The vectors may be transferred by transformation or mobilisation.

(a) Broad host range vectors based on IncP and IncW plasmids

Conjugative plasmids of Inc groups P and W possess the ability to transfer to and to replicate stably in most Gram-negative bacteria. The utility of these plasmids as cloning vectors is limited by their low copy number, high molecular weight and lack of useful cloning sites. However, such plasmids may be manipulated *in vitro* to reduce their size, to alter copy number or to add resistance genes to create more useful cloning vectors. Several 'mini' derivatives of the IncP plasmid RP1 (alias RP4 alias RK2) and of the IncW plasmid pSa have been constructed in this way (Table 3.6). Such derivative plasmids have lost their conjugative ability, but retain broad host range when introduced into bacteria by mobilisation or transformation. IncP- and IncW-based cosmids have also been constructed in order to exploit the *in vitro* packaging system for primary cloning in *E. coli.* A useful strategy can be adopted using the cloning vector pRK290 (Ditta *et al.*, 1980). This is a 20 kb derivative of RK2 that is Tc^r and nonconjugative but mobilisation-proficient. Recombinants formed by cloning foreign DNA in this plasmid can be mobilised from *E. coli* to a wide range of bacteria by the helper plasmid, pRK2013, which is a ColE1-like plasmid carrying the *tra* genes of RK2. pRK2013 is self-transmissible but cannot replicate in bacteria that are not closely related to *E. coli.* This system therefore enables the transfer and maintenance of recombinants based on pRK290 in organisms other than *E. coli,* without inheritance of the helper plasmid.

(b) Vectors based on IncQ plasmids

Small, nonconjugative plasmids belonging to the IncQ group and typically specifying resistance to streptomycin and sulphonamides can replicate in most Gram-negative bacteria. Such plasmids are maintained as multiple (15-20) copies, but unlike ColE1 are not amplifiable. Wild-type IncQ plasmids, such as NTP2, RSF1010 and R300B, contain few useful restric-

Table 3.6: Some broad-host range plasmid vectors for cloning in Gram-negative bacteria

Vector	Replicon	Size (kb)	Copy number	Phenotype*	Cloning sites	Reference
pSa747	IncW	15	2-4	Km, Sp, Mob^+ cosmid	*Bam*HI, *Bgl*II, *Eco*RI, *Hin*dIII *Pvu*II, *Sst*I	Tait *et al.*, 1983
pR0164	IncP	15.8	Multiple	Cb, Tc, Mob^+	*Bam*HI, *Hin*dIII	Olsen *et al.*, 1982
pRK290	IncP	20	2-4	Tc, Mob^+	*Bgl*II, *Eco*RI	Ditta *et al.*, 1980
pTB70	IncQ	17.6	15-20	Km (Tn*5*), Sm, Su, Mob^+	*Bam*HI, *Eco*RI, *Sal*I	Barth *et al.*, 1981
pFG7	IncQ	13.3	15-20	Ap, Tc, Mob^+	*Cla*I, *Bam*HI, *Hin*dIII, *Sal*I	Gautier and Bonewald 1980
pKT258	IncQ	11.5	15-20	Ap, Sm, Mob^-, cosmid	*Eco*RI, *Sst*I	Bagdasarian and Timmis, 1982

*Ap, Cb, Km, Sm, Sp, Su, Tc, resistance to ampicillin, carbenicillin, kanamycin, streptomycin, spectinomycin, sulphonamides, tetracycline, respectively.

tion sites. Furthermore these plasmids are large (8 to 9 kb) in comparison with pBR322 (about 4.3 kb) and its relatives. Attempts to reduce the size of IncQ plasmids by deletion have not been particularly successful, largely because the essential maintenance genes are scattered around the genome (Bagdasarian *et al.*, 1981).

Some improvement in the capabilities of these plasmids as cloning vectors has been made by forming cointegrates *in vitro* with pBR322 to provide a range of cloning sites (Wood *et al.*, 1981). However, such hybrid vectors are rather unwieldy due to their size and are unstable in some species, notably *Pseudomonas aeruginosa.* A more promising approach has been to insert alternative antibiotic resistance genes into IncQ replicons either in addition to or in place of the streptomycin and sulphonamide-resistance determinants (Barth *et al.*, 1981; Bagdasarian and Timmis, 1982) (Table 3.6). Grinter (1983) has devised a system for inserting cloned DNA into the genomes of many different Gram-negative bacteria by transposition. This consists of a carrier IncP plasmid derived from RP4 and containing a sequence into which foreign DNA can be cloned. The sequence is transposable when transposition functions are provided in *trans* by an IncQ helper plasmid based on R300B.

3.5.4 Gene cloning in Gram-positive bacteria

(a) Bacillus

Bacillus subtilis is the best characterised of all Gram-positive bacteria. It has a well defined genetic map (Piggot and Hoch, 1985) and efficient systems for transformation and transduction (see sections 2.2 and 2.4). In addition *B. subtilis* is a commercially important producer of peptide antibiotics and extracellular enzymes, such as proteases. Furthermore, the organism is non-pathogenic, which makes it a 'safe' host for cloning potentially hazardous genes. However, *B. subtilis* does sporulate readily, thus increasing the probability that cloned genes would survive outside the laboratory or fermenter. Asporogenous mutants with increased autolytic activity may however, be used as high-containment host strains.

The indigenous plasmids of bacilli do not bear any easily selectable markers. However, small multicopy resistance plasmids from *Staphylococcus aureus* can replicate stably in *B. subtilis.* Several such plasmids have been used as vectors in *Bacillus,* notably the chloramphenicol-resistant plasmid pC194 with unique *Bgl*I, *Hae*III, *Hin*dIII and *Hpa*II sites (Ehrlich *et al.*, 1982), pUB110 with unique *Ava*I, *Bam*HI, *Bgl*II (within the kanamycin-resistance gene), *Eco*RI, *Pvu*II and *Xba*I sites (Gryczan *et al.*, 1978) and pE194 with *Bcl*I, *Hae*III, *Hpa*I (within the erythromycin-resistance gene), *Pst*I and *Xba*I sites (Horinouchi and Weisblum, 1982). The copy number of pC194 and pUB110 can be increased by amplification during growth in the presence of hydroxyurea. These plasmids have been modified to produce expression vectors (see Chapter 5), dual-replicon shuttle vectors (see section 3.5.8) and low copy number vectors. A direct selection vector, pBD214, based on a hybrid between pUB110 and pC194, has been constructed by Gryczan and Dubnau (1982). pBD214 carries a *thy* gene (encoding the ability to manufacture thymine) containing cloning sites for insertional inactivation and a chloramphenicol-resistance gene. Thy^+ bacteria are sensitive to the drug trimethoprim whereas Thy^- strains are resistant. The thy^+ gene of pBD214 therefore renders a thy^- *B. subtilis* host sensitive to trimethoprim. However, inactivation of the plasmid thy^+ gene by insertion of foreign DNA produces Thy^-, trimethoprim-resistant, chloramphenicol-resistant transformants.

Plasmid vectors are becoming available for other *Bacillus* species. For example, vectors have been derived from the *Bacillus cereus* tetracycline-resistance plasmid pBC16 and from cryptic plasmids of *Bacillus megaterium* and thermophilic bacilli (see Dubnau, 1982, 1985).

Several bacteriophage vectors, including derivatives of the temperate phages p11, ϕ105 and NP02 (Dubnau, 1982) have been developed for *B. subtilis.* The virulent phage SPP1 has been used to construct a direct selection *Bam*HI replacement vector, called SPP1v (Heilmann and Reeve, 1982). Flock (1983) has constructed a series of pCOS plasmids (based on pC194) carrying the cohesive end region of ϕ105. DNA cloned in COS plasmids and subsequently introduced into *B. subtilis* by transformation may be transduced between strains of *B. subtilis* using ϕ105.

Various problems have been encountered when cloning in *B. subtilis.* Shotgun cloning with plasmid vectors is very inefficient as a consequence of the requirement for circular oligomeric transforming DNA and because of damage inflicted on DNA during uptake (section 2.2.1). This can be overcome by using recombinational rescue with a host containing a plasmid that is homologous to the vector. Alternatively, protoplast transformation may be employed. However, high molecular weight DNA transforms *B. subtilis* protoplasts with low efficiency. Unmodified plasmid or phage DNA is also severely restricted in *B. subtilis* (Dubnau, 1982). Instability, particularly of Gram-negative DNA cloned in *B. subtilis* vectors, has been noted even in rec^- backgrounds. *recE*-dependent recombination between plasmids carrying homologous sequences or between plasmids and the host chromosome may be responsible for rearrangements in cloned DNA. Homologous recombination may, however, be deliberately exploited to insert cloned DNA into the chromosome of *B. subtilis* (see section 5.4.3).

(b) Streptococci

Many streptococci are commercially important, particularly in the dairy industry, and others are significant human or animal pathogens. Relatively little is known about the genetics of these organisms. However, vectors have been derived from nonconjugative resistance plasmids found in pathogenic or commensal streptococci for cloning in *Streptococcus sanguis* and *Streptococcus pneumoniae.* Such plasmids and the lactose-fermenting plasmids of lactobacilli should form the basis of gene cloning systems for lactic acid bacteria of industrial relevance.

A high proportion of recombinant plasmids obtained by cloning in streptococci contain extensive deletions or other rearrangements, which presumably occur during transformation (Macrina *et al.*, 1982). Transformation of intact streptococcal cells, like *B. subtilis,* involves uptake of DNA as a single strand. In contrast to *B. subtilis,* monomeric plasmid DNA is active in transformation. However, two monomeric plasmid molecules can be taken up to produce a single streptococcal transformant (Saunders and Guild, 1981). This means that direct cloning in streptococci is very inefficient. Helper cloning using a recipient cell carrying a plasmid that is homologous to the plasmid cloning vector, or use of a shuttle-vector strategy (see section 3.5.8) may overcome this limitation.

(c) Other Gram-positive eubacteria

Although resistance plasmids from *S. aureus* have been essential for the development of gene cloning in *B. subtilis,* much less progress has been made in developing cloning systems for staphylococci. This is due largely to the relatively inefficient transformation system, the strong restriction barrier against foreign DNA and the pathogenicity of *S. aureus.* The non-pathogenic species, *Staphylococcus carnosus,* which is of importance in food technology and has an efficient protoplast transformation system, has, however, been developed as an alternative Gram-positive host for cloning (Thudt *et al.*, 1985). Cloning systems are also under development for commercially important clostridia, such as *Clostridium thermocellum* and *C. butyricum.*

(d) Streptomyces

Streptomycetes are of considerable industrial importance, particularly in the production of antibiotics, other secondary metabolites and extracellular enzymes. *Streptomyces coelicolor* and *Streptomyces lividans* may be transformed efficiently as protoplasts and have been sufficiently well characterised genetically to serve as effective hosts for gene cloning (Chater *et al.*, 1982).

The conjugative plasmid, SCP2, from *S. coelicolor* and the high-fertility variant SCP2* (probably a mutant derepressed for transfer), have been used for the cloning of biosynthetic and antibiotic-resistance genes (Thompson *et al.*, 1982a,b). However, the use of SCP2* as a cloning vector is limited by its high molecular weight (31 kb), low copy number (1-5 per chromosome equivalent) and lack of restriction sites in nonessential DNA. Smaller derivatives have been constructed (Bibb *et al.*, 1983). A DNA sequence on the chromosome of *S. coelicolor* A3(2), together with variable regions that flank this sequence, can give rise to a series of plasmids that are capable of autonomous replication in *S. lividans* (Bibb *et al.*, 1981). One of these plasmids, SLP1.2, is low copy number (4 to 5 per chromosome equivalent) and about 4.5 kb. It also contains a dispensable region with restriction sites suitable for cloning. This plasmid has been used to clone a number of antibiotic resistance genes, for example the viomycin-phosphotransferase (*vph*) gene from *Streptomyces vinaceus* (which is particularly useful because the gene can be expressed in *E. coli*), in order to generate derivatives with different cloning sites (Thompson *et al.*, 1982a,b). One such derivative is the versatile vector pIJ61, which carries a thiostrepton-resistance gene and a neomycin-resistance gene containing a unique *Pst*I site. Cloning into pIJ61 is also possible with several other enzymes including *Bam*HI, *Bcl*I, *Eco*RI, *Kpn*I and *Sst*I. Specialised promoter-cloning vectors have also been derived from SLP1.2 (Bibb and Cohen, 1982). pIJ101, a multicopy conjugative plasmid of 8.9 kb (Figure 3.22) isolated originally from *Streptomyces violaceoruber*, has proved useful as the basis for broad host range vectors capable of being maintained in almost all *Streptomyces* species (Kieser *et al.*, 1982). pFJ105, a 4.7 kb thiostrepton-resistance plasmid (originally derived from a larger *Streptomyces granuloruber* plasmid) may also be used as a wide host range vector. It can be transferred to a number of species, including a tylosin-producing strain of *Streptomyces fradiae* and a narasin-producing strain of *Streptomyces aureofaciens* (Richardson *et al.*, 1982). Bifunctional plasmid shuttle vectors capable of replicating in both *Streptomyces* and *E. coli* have been constructed from this and several other *Streptomyces* plasmids (see section 3.5.8).

Streptomycetes contain many phages, of which the temperate phage ϕC31 from *S. coelicolor* has proved the most popular as a cloning vector. ϕC31 has a fairly wide host range and has a double-stranded DNA genome with cohesive ends. Deletions may be introduced into the phage in order to increase cloning capacity, without affecting viability (Chater *et al.*, 1982). Hybrid derivatives of ϕC31 and pBR322, capable of replicating as a phage in streptomycetes and as a plasmid in *E. coli*, have also proved useful as cloning vectors (Chater *et al.*, 1982; Bibb *et al.*, 1983). Such hybrids may be used as shuttle vectors (see section 3.5.8), replacement vectors, or, where the phage has an intact attachment (*attP*) site, may provide an opportunity to insert cloned genes into the *Streptomyces* chromosome during lysogenisation (Bibb *et al.*, 1983). Deletion of the *attP* site together with other regions of the ϕC31 genome precludes normal lysogenisation. Vectors containing such deletions, for example ϕC31.KC400 (Chater *et al.*, 1982, and section 3.7.1) can therefore only form stable lysogens by homologous recombinantion with the *Streptomyces* genome (selection of lysogens is by expression of viomycin resistance). Insertion into the genome

Figure 3.22: The broad host range *Streptomyces* plasmid pIJ101. Various derivatives of pIJ101 have been constructed. For *in vitro* genetic manipulation the conjugative transfer functions are generally inactivated to decrease vector size and to prevent conjugation following transformation. Conjugative derivatives of pIJ101 can be used for *in vivo* manipulations, such as chromosome mobilisation. The sites of insertion of *Streptomyces* DNA fragments bearing a thiostrepton-resistance (*tsr*) gene and a viomycin-resistance (*vph*) gene are indicated for the conjugative derivative pIJ643. Insertion of the *vph* fragment into the spread function region results in a small pock phenotype but the plasmid retains all other properties of the parental plasmid

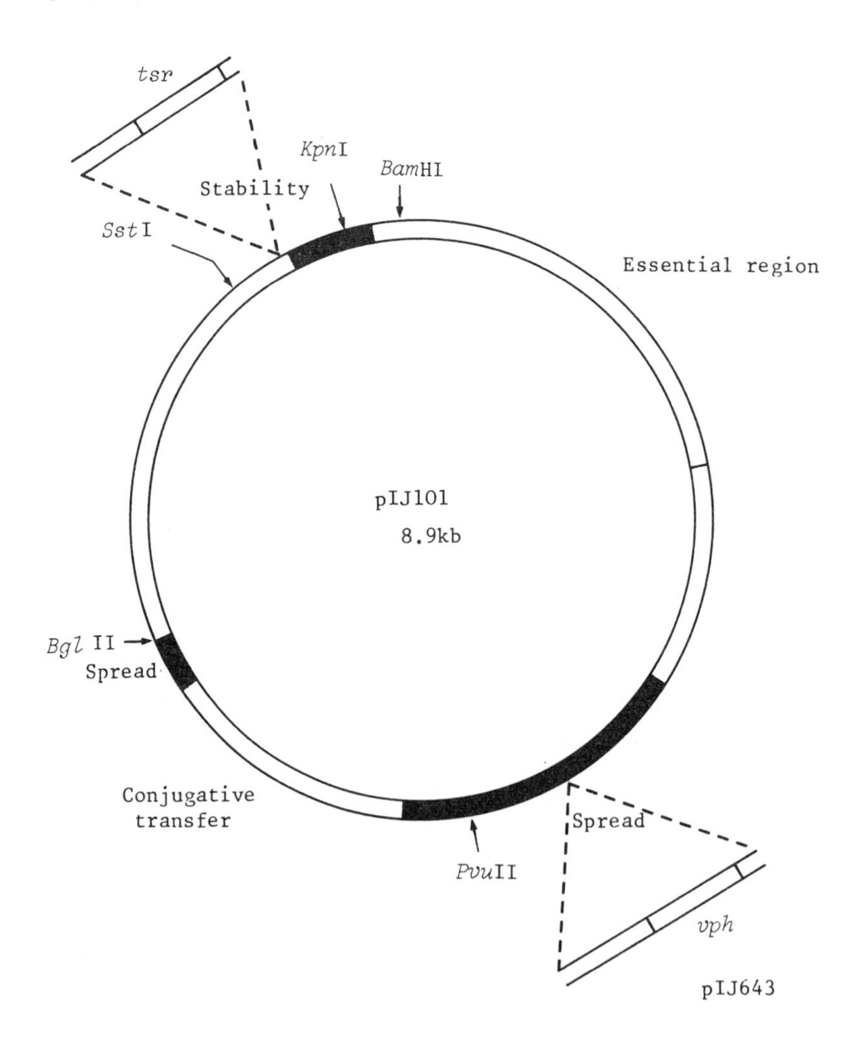

may occur by host-mediated recombination of the vector with the resident ϕC31 prophage. In nonlysogenic host strains recombination occurs between DNA cloned in the vector and homologous regions of the host genome. This permits the use of 'mutational cloning' for the isolation of *Streptomyces* genes with detectable phenotypes (see section 3.7.1).

Streptomyces cloning systems have a number of advantages over those developed for other Gram-positive organisms. Protoplast transformation systems are highly efficient (see section 2.2.1), and available vectors permit efficient shotgun-cloning of DNA directly in *S. coelicolor* or *S. lividans.* In addition, expression of heterologous bacterial genes is generally efficient in these species. Some problems have, however, been encountered with the accumulation of repetitious DNA within the DNA segments cloned in *Streptomyces.* This presumably results from transposition of DNA sequences from the *Streptomyces* genome.

3.5.5 Cloning in yeasts

Yeasts are not only of great importance in the brewing, food and chemical industries, but also possess properties that are advantageous for expressing proteins from cloned DNA (see section 5.7). Genetic studies on yeast have largely focused on *Saccharomyces cerevisiae* (Strathern *et al.*, 1982) and a variety of vectors have been developed for cloning in this organism. Many of these vectors are hybrids of yeast DNA and *E. coli* plasmids, and serve as shuttle vectors (section 3.5.8). Yeast vectors can also be modified by incorporation of λ *cos* sites to take advantage of the λ *in vitro* packaging system.

(a) Integrating vectors (YIp)

Several vectors have been constructed that are based on ColE1 replicons and contain cloned yeast chromosomal genes, for example *LEU* or *HIS.* Selection for the vector in yeast is by complementation of leu^- or his^- hosts. YIp vectors replicate as plasmids in *E. coli,* but cannot be maintained in *S. cerevisiae* unless they can integrate into the chromosome by host-mediated homologous recombination. Transformation of yeast with YIp vectors therefore occurs at low frequency (1-10 transformants per μg DNA). Integration into the yeast chromosome generally results in a tandem duplication of the cloned yeast sequences (Figure 3.23). Such insertions of DNA are unstable due to recombination between the repeated sequences. The result of this recombination may be the loss of one of the repeats plus the *E. coli* part of the vector. As a consequence it is possible to use integrating vectors for placing DNA segments within the yeast genome.

Figure 3.23: Some recombination events in the integration of YIp vectors. Some of the possible recombination events that give rise to transformants when a YIp vector carrying a yeast *LEU2*$^+$ gene is introduced into a *leu2*$^-$ strain of *S. cerevisiae* are shown. The positions of cross-overs affect the precise nature of insertion. Only general classes of recombination event are shown. In (a) (the most frequently observed event) and (b) loss of the *E. coli* plasmid sequences can occur by secondary recombination events within the chromosome. Alternatively plasmid sequences may be lost by unequal sister chromatid exchange. This results in loss of plasmid sequences from one chromatid and duplication in the other

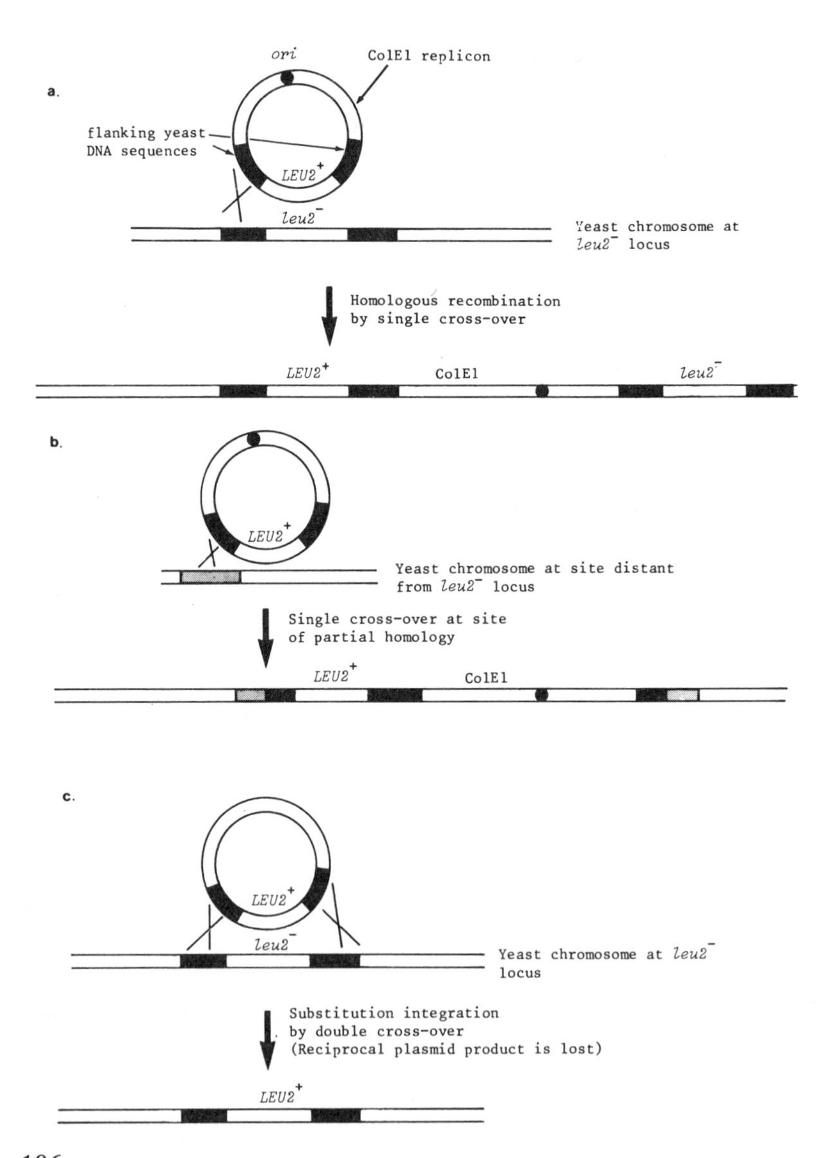

(b) Two-micron plasmid vectors (YEp)

Many *Saccharomyces* species contain circular DNA of 2 μm in contour length. In yeast, 2 μm DNA is maintained stably at about 50-100 copies per cell and behaves in a similar way to bacterial plasmids (Hinnen and Meyhack, 1982; Hollenberg, 1982). Vectors have been constructed by attaching all or part of the 2 μm plasmid *in vitro* to selectable yeast markers (for example *LEU*) and usually also to *E. coli* plasmid DNA (Beggs, 1982; Hollenberg, 1982) (Figure 3.24). Derivatives lacking the 2 μm replication origin may still function as vectors because most of the yeast strains used contain endogenous 2 μm sequences. The incoming vector molecule can therefore be rescued by homologous recombination. Vectors based on 2 μm plasmids combine high transformation frequencies (10^3 to 10^4 times higher than YIp vectors) with the ability to propagate cloned sequences as multiple copies.

Recombinants based on YEp vectors tend to exhibit structural instability. This may result from recombination between either vector and resident 2 μm sequences or the inverted repeat sequences present in the 2 μm plasmid molecule (Hinnen and Meyhack, 1982). Segregational instability (ranging from 1 to 40% per cell division) can also occur and probably results from incompatibility between the vector and endogenous 2 μm plasmids. A solution to segregational loss, applicable to vectors such as pJDB219, is to use cir^0 strains, which lack the 2 μm plasmid (Beggs, 1982). However, other vectors, for example pMP78, which are rep^0 and do not contain the 2 μm origin, are more stable in cir^+ than in cir^0 yeasts (Hollenberg, 1982).

(c) Vectors with chromosomal replication origins (YRp)

Vectors composed of the yeast *TRP* or *ARG* regions cloned into bacterial plasmids have been found to give high frequencies of transformation (10^3 to 10^4 transformants per μg DNA) (Kingsman *et al.*, 1979). These regions contain autonomously replicating sequences (ARS). Further ARS sequences can be obtained from yeasts and other eukaryotes by cloning randomly cleaved chromosomal DNA into YIp vectors (Stinchcomb *et al.*, 1980).

(d) Vectors carrying a yeast centromere (YCp)

Clarke and Carbon (1980) have constructed yeast vectors containing a chromosomal replication origin (as in YRp vectors) plus the centromere of *S. cerevisiae* chromosome III. These plasmids behave as stable circular yeast minichromosomes and may be used as single copy number vectors.

(e) Development of other yeast vectors

Additional yeast vectors may be developed from mitochondrial (mt) DNA or from the linear killer toxin plasmids found, for example, in the cheese

Figure 3.24: The 2 μm plasmid yeast vector pJDB219. The YEp vector pJDB219 consists of the entire 2 μm plasmid cloned into the unique *Eco*RI site on the *E. coli* plasmid pMB9. The 2 μm sequences contain a fragment of yeast chromosomal DNA including the *LEU2* gene inserted at the *Pst*I site. IR, Inverted repeat sequence of 599 bp. Recombination between these sequences reverses the orientation of the remaining unique sequences of the 2 μm plasmid. Preparations of 2 μm-based plasmid vectors consist of a mixed population of A and B forms that differ in sequence orientation. Lightly stippled area indicates pMB9 sequences; P, promoter for tetracycline resistance (*tet*) gene; →, direction of transcription

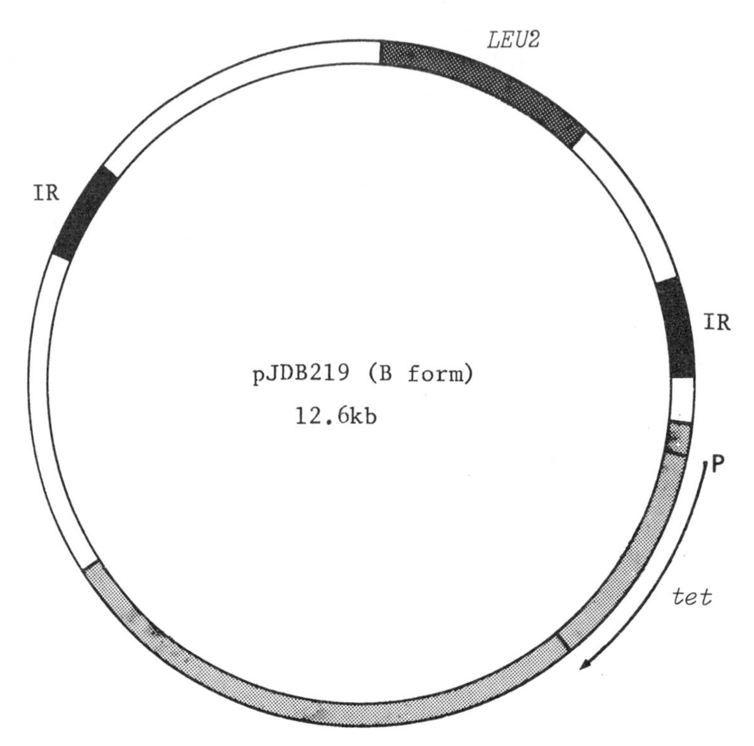

yeast, *Kluyveromyces lactis*. Prospects for the development of yeast vectors (and those for other eukaryotes) have been enhanced by the discovery that the resistance phenotype of bacterial transposon Tn *903* (*601*) is expressed in yeast (Jimenez and Davies, 1980). Tn*903* mediates resistance to aminoglycoside antibiotics including the drug G418, which is an inhibitor of eukaryotic protein synthesis. Resistance to G418 can therefore be used as a selective phenotype. YIp and YEp vectors carrying Tn *903* have been successfully employed in cloning various yeast genes (see, for example, Webster and Dickson, 1983).

3.5.6 Cloning in filamentous fungi

Cloning systems for filamentous fungi are currently less well developed than those available for yeasts. *Neurospora crassa* and *Aspergillus nidulans* have been well characterised genetically and may prove suitable as host organisms. Vectors analogous to YIp plasmids have been developed for *N. crassa* by cloning *Neurospora* nuclear genes, for example *qa-2*$^+$, into pBR322 or similar *E. coli* replicons (Case, 1982). These vectors may be maintained in *E. coli* and used to transform *N. crassa* by selecting for complementation of stable *qa-2*$^-$ chromosomal mutations. Transformants contain the cloned *N. crassa qa-2*$^+$ gene integrated into chromosomal DNA, in some cases flanked by pBR322 sequences.

Hybrids formed between pBR325 and defective mitochondrial (mt) DNA have been shown to replicate autonomously in *Podospora anserina* (Stahl *et al.*, 1982). Some strains of *N. crassa* and *P. anserina* contain mitochondrial plasmids which are distinct from mt DNA. Hybrids between mitochondrial plasmids from *N. crassa*, the *qa-2*$^+$ gene and pBR325 transform *Neurospora* with 5- to 10-fold higher efficiency than integrating vectors. Such hybrids, for example pALS-1 (Stohl and Lambowitz, 1983) and pDV1001 (Hughes *et al.*, 1983), should prove useful cloning vectors because they replicate autonomously in the nucleus and/or cytosol of *N. crassa.* A high-frequency transforming vector for *A. nidulans* carrying the *pyr4* (orotidine-5′-phosphate decarboxylase) gene of *N. crassa* and the self-replicating *ans1* sequence from *A. nidulans* has been constructed by Ballance and Turner (1985).

3.5.7 Cloning in higher eukaryotes

(a) Animal cells

As yet no suitable replicons analogous to bacterial plasmids have been found in animal cells. For this reason genomes of animal viruses, in particular Simian Virus 40 (SV40), have been developed as vectors for such cells. SV40 has a circular double-stranded genome, whose complete 5243 bp nucleotide sequence is known (Tooze, 1980). The virus can enter one of two life cycles, depending upon the host cell line employed. In permissive cells (generally monkey cells) a productive lytic cycle occurs, whereas in nonpermissive rat or mouse cells viral replication is blocked and the cells become transformed.[6] The SV40 genome is transcribed as early and late regions, with transcription proceeding anticlockwise and clockwise respect-

[6] In this context transformation refers to the heritable alteration of an animal cell line such that it is no longer confined to growing as a monolayer and can proliferate in a semi-solid medium without being attached to a solid substratum.

ively from the origin of DNA replication. The early transcripts encode small and large T antigen (a multifunctional protein required for viral replication and transformation). Late transcription is required for synthesis of viral proteins and the production of mature virions. Since DNA molecules of between 70 and 100% of the SV40 genomic length can be encapsidated (Rigby, 1982) deletants of the late region may be used as cloning vectors. Late replacement vectors, such as SVGT-5, have been constructed, which retain the SV40 late promoter, mRNA splice junction and polyadenylation site (Mulligan *et al.*, 1979) permitting the expression of cloned cDNA in animal cells. Replacement vectors that are defective due to deletion of part of the early or late regions may be propagated and encapsidated by co-infection of a permissive cell line with a helper virus. For late replacement vectors the helper is usually an SV40 *tsA* mutant, which is conditionally defective for early functions. This has the advantage that the helper virus can be eliminated as desired by growing infected cells at the restrictive temperature.

Other animal viruses that may form the basis of useful vectors include adenoviruses or natural hybrids between adenoviruses and SV40 (Sambrook and Grodzicker, 1980), papilloma viruses (Law *et al.*, 1981), vaccinia virus (see section 7.2) and retroviruses (Bernstein *et al.*, 1985). Bovine papilloma virus is of particular interest because its genome is maintained extrachromosomally (at 20-200 copies per cell) in transformed or tumour cells.

The utility of animal viruses as vectors has been enhanced by addition of selectable markers. The thymidine kinase (*tk*) gene of Herpes Simplex Virus (HSV) has been used as a marker in conjunction with tk^- mutant cell lines. Unlike wild-type cells Tk^- cells are unable to grow in medium containing hypoxanthine, aminopterin and thymine (HAT medium). Hence, acquisition of the HSV *tk* gene permits the growth of transfectants in HAT medium. However, tk^- mutants can only be obtained for a few cell lines, which limits the effectiveness of the marker. A further system involves the mouse dihydrofolate reductase (*dhfr*) gene which permits the growth of $DHFR^-$ cell lines in the absence of the supplements thymine, glycine and purines (Subramani *et al.*, 1981). High-level expression of the mouse *dhfr* gene or the expression of the *E. coli dhfr* gene in mouse cells confers resistance to the folate inhibitor methotrexate, which can therefore be used as a selective marker. The aminoglycoside phosphotransferase of Tn *903* (see section 3.5.5) may also be used as a selective marker in animal cells, since it confers resistance to the aminoglycoside antibiotic G418 (Colbère-Garapin *et al.*, 1981).

The use of such selective markers coupled to a viral origin of replication, in the absence of genes necessary for the viral life cycle, allows the construction of vectors that are not subject to packaging constraints since they can be introduced into animal cells by methods that do not require encapsidation. DNA may be introduced into mammalian cells, albeit inefficiently, by a calcium precipitation transfection technique (Graham and van der Eb, 1973). It is possible to introduce DNA that lacks any readily identifiable phenotype into animal cells by the technique of cotransfection (see section 2.2.3). This involves exposing cells to DNA containing a selectable marker, for example the *tk* gene, in the presence of a vast excess of unmarked DNA. Alternatively DNA may be introduced by microinjection or in liposomes (Gruss and Khoury, 1982).

(b) Plant cells

The development of plant vectors based on viruses has been hindered by the fact that most plant viruses have RNA genomes. The two DNA virus groups likely to be useful for cloning in plants are caulimoviruses and geminiviruses. Of the caulimovirus group, Cauliflower Mosaic Virus (CaMV) has received most attention as a potential vector. Ti (tumour inducing) plasmids of the plant pathogen *Agrobacterium tumefaciens* are often regarded as vectors for natural genetic engineering since they can transfer part of their DNA (T-DNA) to plant cells. The properties and uses of Ti plasmids and other vectors for plants are considered in Chapter 8.

3.5.8 Shuttle vectors

Many applications in molecular genetics require the transfer of genes between unrelated species. Broad host range vectors exist for Gram-negative bacteria (section 3.5.3) and *Streptomyces* (section 3.5.4). However, replicons capable of maintenance in all possible combinations of unrelated hosts are unlikely to exist naturally. A shuttle or bifunctional replicon vector may be used in these cases. Such vectors normally comprise an *E. coli* plasmid (or part of such a plasmid), such as pBR322, ligated *in vitro* to a plasmid or virus replicon from another species. Shuttle vectors can be made, for example, for *E. coli*/*B. subtilis*, *E. coli*/yeast or *E. coli*/mammalian cells (Table 3.7).

The shuttle-vector strategy permits the exploitation of the many manipulative procedures, such as amplification, available in *E. coli* (or other genetically well characterised species, such as *B. subtilis* or *S. cerevisiae*), while retaining the ability to maintain genes in less well defined genetic backgrounds. The ability to transfer cloned genes across species boundaries is of potential value in the genetic manipulation of industrially important species and is an essential requirement for many experiments in reversed genetics (see section 4.10)

Table 3.7: Some shuttle vectors and their properties

		Mode 1		Mode 2		
Vector	Size (kb)	Host	Replicon	Host	Replicon	Reference
pHV23	9.4	*E. coli*	pBR322 (Ap, Cm)	*B. subtilis*	pC194 (Cm, Tc)	Ehrlich *et al.*, 1982
pLV21	13.0	Most Gram-negative bacteria	NTP2 (Km, Su)	*B. subtilis*	pUB110 (Km, Sm)	Kinghorn *et al.*, 1981
pIJ28	22.2	*E. coli*	pBR322 (Ap)	*Streptomyces lividans*	SLP1.2 (Nm)	Bibb *et al.*, 1983
ϕ31.KC400	38.3	*E. coli*	pBR322 (Ap)	*Streptomyces lividans*	ϕ31 (Vi)	Chater and Bruton, 1983
pJDB219	12.6	*E. coli*	pMB9 (Tc)	*S. cerevisiae*	2 μm plasmid (*LEU2*)	Beggs, 1982
pYR12GR	8.3	*E. coli*	pBR322 (Ap, Km, Tc)	*S. cerevisiae*	yeast *ARS1* (G418, *TRP1, URA3*)	Webster and Dickson, 1983
pBPV-β1	17.2	*E. coli*	pBR322 (Ap)	Mouse cells	Bovine papillomavirus	Di Maio *et al.*, 1982
pSV2-*gpt*	4.8	*E. coli*	pBR322 (Ap)	Monkey cells	SV40 (*gpt*)	Mulligan and Berg, 1981

ARS1, yeast autonomous replication sequence 1; *LEU2*, *TRP1*, *URA3*, yeast biosynthetic genes for leucine, tryptophan and uracil respectively; *gpt*, *E. coli* xanthine-guanine phosphoribosyltransferase gene. Cm, resistance to chloramphenicol; G418, resistance to aminoglycoside G418; Nm, resistance to neomycin; Vi, resistance to viomycin; remaining resistance phenotypes as in Table 3.6.

3.6 FORMATION OF GENE LIBRARIES

3.6.1 Genomic libraries

The construction of a gene library (bank) of an organism is often an important preliminary to screening for particular cloned sequences. Such banks provide a starting point for the physical and genetical characterisation of entire genomes (see section 3.10). Gene libraries are constructed (normally in *E. coli*) such that sequences representative of a complete genome (or of individual chromosomes of higher organisms where these can be separated) are cloned in a common vector molecule. The aim is to clone separately each of a series of random, overlapping fragments representing the entire target genome. Often the fragmented genome is fractionated on the basis of size to ensure that the cloned fragments are of a length that approaches the maximum cloning capacity of the vector. This in turn ensures efficient packaging in λ or cosmid vectors and minimises the risk of linking into the vector several smaller fragments that may not have been contiguous in the intact genome. Controlled shearing of DNA results in a series of fragments with random end-points. The sampling variability therefore follows a Poisson distribution and the number of clones required to make a representative bank can be calculated from the formula:

$$N = \frac{\ln(1 - P)}{\ln(1 - f)}$$

where P is the required probability, f is the fractional proportion of the genome in a single recombinant and N is the necessary number of recombinant clones (Clarke and Carbon, 1976). For example, 7.0×10^5 separate clones would be required to produce a probability of 99% that all possible sequences were included in a library of 20 kb fragments of a mammalian genome of 3×10^6 kb. For bacterial genomes (about 4×10^3 kb) the number of clones required for a similar bank would be about three orders of magnitude lower. The larger the bank the greater the chance of incorporating all possible sequences. The library can be smaller if the average size of cloned insert is large. For this reason phage or cosmid vectors, which have a high capacity for passenger DNA, are normally the vehicles of choice for constructing libraries.

The incorporation of randomly sheared fragments into libraries is relatively inefficient due, for example, to the necessity to add linkers to the blunt-ended DNA molecules. Therefore, gene libraries are more commonly constructed using target DNA that has been subjected to partial digestion with a frequently cutting restriction endonuclease, such as *Sau*3A or *Mbo*I. The resulting fragments can then be inserted into compatible unique cloning sites (for example *Bam*HI or *Bgl*II) on a vector. This procedure can be used to generate a population of DNA molecules that constitute an overlapping set of sequences from a particular genome (see, for example, Figure 3.20). The fragments generated in this way are, however, not totally random since their termini are of necessity dictated by the positions of the restriction sites. Furthermore, the distribution of restriction cleavage sites for a particular enzyme may be such that not all regions of the genome can be cleaved to generate fragments of suitable size for *in vitro* packaging. Use of the Clarke and Carbon formula for constructing libraries with partial restriction digests can, therefore, result in the omission of some sequences. Accordingly, more complex formulae have been devised to determine the degree of sequence representation in gene libraries with greater precision (Seed *et al.*, 1982).

3.6.2 cDNA libraries

Clone banks can also be made from cDNA. cDNA libraries from eukaryotes are usually easier to screen because there are considerably fewer different mRNA sequences (1 to 3×10^4) in a typical eukaryotic cell than the number of potentially clonable DNA restriction fragments (about 10^5 to 10^6). Furthermore, a cloned cDNA sequence may be used subsequently as a hybridisation or recombination probe to locate the gene itself in a genomic library (see section 3.7.2). Estimating the number of cDNA clones necessary for complete representation in a bank is complicated by the fact that individual mRNA species are produced in different amounts. Furthermore, cells from particular tissue types or at particular stages in development will produce varying amounts of specific mRNA species. The abundance of a particular mRNA class can be determined by hybridising its cDNA copy to an excess of total cytoplasmic mRNA and plotting the extent of hybridisation against the Rot value (the product of RNA concentration and time) (Bishop *et al.*, 1974). Three discrete abundance classes of mRNA can be found in typical eukaryotic cells: high, medium and low. In a representative cDNA bank the number of clones containing a particular sequence will therefore be proportional to the abundance of that sequence in the mRNA population. If the bank is of sufficient size to include the low abundance sequences, the high- and medium-abundance sequences should automatically be included. The formula of Clarke and Carbon (1976) can be used to determine the number of clones that will be required. For a 99% probability of including all the low-abundance class a bank of the order of 1.7×10^5 clones is required (Williams, 1981).

3.6.3 Amplification and stability of gene libraries

DNA that has been packaged *in vitro* (see section 3.5.2) from a recombinant ligation mixture remains stable in phage particles for long periods. This provides a means of maintaining reference libraries of cloned sequences. However, it is often necessary to amplify libraries in order to improve the detection of particular clones. This involves infecting (or transforming) sensitive cells and maintaining a bank either as a phage lysate or as bacterial clones carrying recombinant cosmids or plasmids. Some foreign DNA sequences are, however, detrimental to the vector or host and hence may be lost or underrepresented during amplification of the library. Furthermore, if tandemly repetitious sequences are cloned in λ, deletion of DNA may occur by recombination during propagation of the phage.

3.7 IDENTIFICATION OF PARTICULAR CLONED SEQUENCES

3.7.1 Strategies dependent upon expression of the cloned sequence

The identification of specific clones is facilitated where the cloned sequence is expressed as a protein. (Factors that determine the expression of genes in different hosts are discussed in Chapter 5.)

(a) Direct selection

Occasionally cloned genes confer phenotypes on host cells that can be selected for directly, at the reintroduction stage. For example, cDNA for mouse DHFR has been cloned in *E. coli* by exploiting the fact that mouse DHFR is more resistant to the antibiotic trimethoprim than the analogous bacterial enzyme (Chang *et al.*, 1978). Transformants expressing the gene for mouse DHFR can therefore be selected directly on agar containing trimethoprim. Clones with specific recombinant molecules can also be identified by their ability to complement mutational defects (such as those of auxotrophy) in the host chromosome. This is more likely to succeed when cloning DNA into a homospecific or closely related host, where the expressed product is liable to be compatible with the physiological infrastructure. However, complementation of *E. coli* auxotrophs has been used to clone amino acid biosynthetic genes from a number of yeasts and other fungi such as *Neurospora crassa.*

(b) Mutational cloning

Mutational cloning (Chater and Bruton, 1983) was originally devised for the isolation of *Streptomyces* genes with identifiable phenotypes. However, there is no reason why the principles involved should not be applicable to other organisms. The technique depends upon homologous recombination between DNA cloned in ϕC31-pBR322 vectors and the *Streptomyces* genome in order to form viomycin-resistant ϕC31.KC400 lysogens (see section 3.5.4). If the cloned sequence lacks both a promoter and sequences encoding essential regions of the carboxy terminus of the protein, recombination with homologous genomic sequences will cause gene disruption and produce a mutant phenotype, hence the term 'mutational cloning' (see Figure 3.25). On the other hand, if the cloned fragment contains the appropriate transcriptional and translational signals, homologous recombination will result in synthesis of a functional mRNA transcript, and no mutant phenotype will be observed. DNA fragments identified by mutational cloning may subsequently be used as probes for Southern transfers (see section 3.8.2) or to screen gene libraries and isolate any gene intact with its control signals.

(c) Immunological screening

Immunological methods can be used to detect the synthesis of a particular protein by recombinants, provided that specific antibodies, either as polyclonal antibodies raised, for example, in rabbits, or monoclonal antibodies[7] are available. An immunological approach has the advantage that the protein does not have to possess a selectable phenotype. It is also possible to detect a particular clone from among 10^6 or more plaques or colonies. Furthermore, the expressed protein may lack its normal structure and/or activity (for example, it might be a fusion protein) while still retaining the ability to react with specific antibody. Recombinant phage plaques or lysed colonies producing soluble protein antigens can be detected by the formation of precipitin lines. However, this is less reliable and less widely applicable than solid-phase immunoassays, which are capable of detecting as little as one antigen molecule per bacterium. The most widely used solid-phase radioimmunoassays are based on the method of Broome and Gilbert (1978). Such methods rely on the ability of antigens present in a replica plate of plaques or colonies to bind to a polyvinyl, nitrocellulose or equivalent membrane. (Normally, intact colonies are lysed by exposure to chloroform vapour, by thermoinduction of a lysogen (such as λ cI_{857}) or by spraying with a virulent phage (such as coliphage T6).) The method originally employed by Broome and Gilbert (1978) depended on the ability of the immunoglobulin fraction IgG in antiserum raised against the target antigen to bind strongly to the surface of polyvinyl discs. However, simpler methods such as fixing the lysed clones directly on to nitrocellulose membranes are now available.

---

[7] Monoclonal antibodies are monospecific immunoglobulin molecules produced by hybrid cells formed by fusing the immunoglobulin-producing lymphocytes to a tumour cell line (for a review see Kohler, 1985).

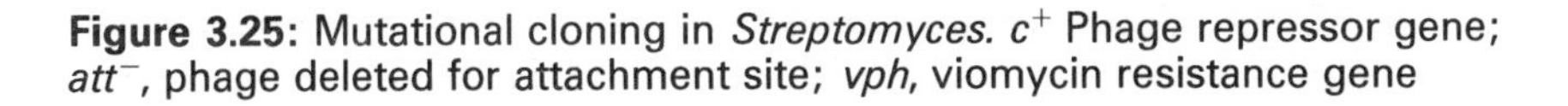

Figure 3.25: Mutational cloning in *Streptomyces*. c^+ Phage repressor gene; att^-, phage deleted for attachment site; *vph*, viomycin resistance gene

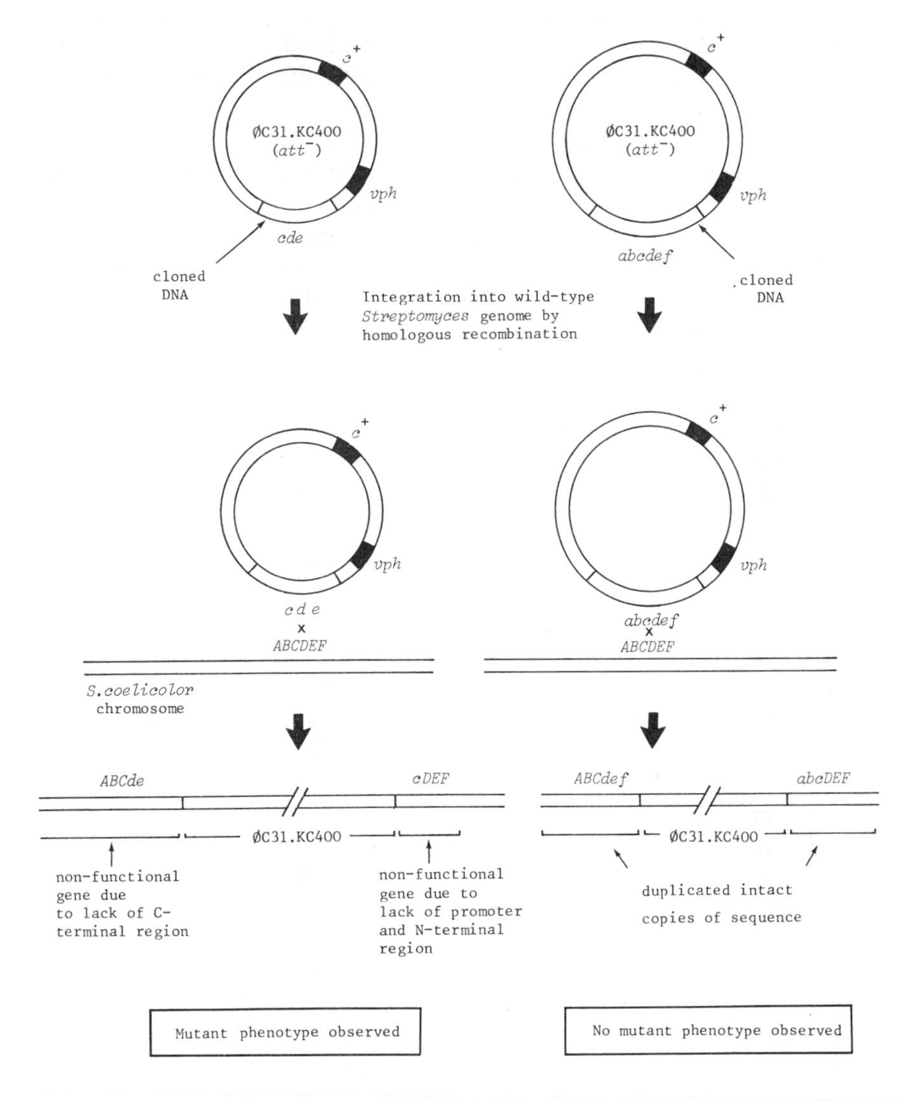

The presence of specific antigens on membrane replicas can be detected in number of ways:

(i) The disc can be incubated with specific IgG that has been iodinated with ^{125}I. Binding of the labelled antibody can be detected by autoradiography using X-ray photographic film exposed to the membrane. The position of the blackened spots caused by radiation-induced silver deposition in the developed film correspond to the positions of colonies or plaques producing a specific antigen on the master plate. Enzyme-linked immunosorbent assay (ELISA) can also be used. An ELISA system that is frequently employed depends upon the use of a horseradish peroxidase/antibody conjugate, which is precipitated by complexing with specific antigen. A chromogenic assay for peroxidase (which involves the formation of a red-brown pigment) permits the localisation of antigen-antibody complexes and hence of the clone producing the desired protein (Kaplan *et al.*, 1983).

(ii) Antigens bound to the disc can be reacted with unlabelled specific antibody whose presence can in turn be detected using labelled *Staphylococcus aureus* protein A. This protein binds specifically to the Fc portion of IgG (Erlich *et al.*, 1979) (Figure 3.26). Protein A binds weakly to some monoclonal antibodies. Therefore a labelled second antibody that binds specifically to the first mono- or polyclonal antibody must be used in place of protein A in these cases.

(d) Open reading frame (ORF) cloning

Some target genes may be poorly expressed in *E. coli* or difficult to detect for other reasons. If the DNA sequence is known, it is often possible to manipulate a gene *in vitro* to obtain efficient expression (see Chapter 5). However, it may be simpler to express any translational open reading frame (ORF) in the target DNA as a tripartite gene fusion (see section 3.9.5). Open reading frame or ORF vectors, which have been designed for this purpose, are usually derivatives of pBR322. They contain an efficient promoter, ribosome-binding site, an initiation codon for translation, and sequences encoding the N terminus of a protein such as β-galactosidase or λ*c*I repressor (Weinstock, 1984). Downstream of this so-called initiator region (and separated from it by suitable restriction sites for cloning) is a *lacZ* sequence that lacks transcriptional and translational signals and several of its N terminal codons (Figure 3.27). The *lacZ* sequence is also out of frame with respect to the initiator sequence so that these vectors do not encode functional β-galactosidase and confer a Lac^- phenotype on hosts. Insertion of a DNA sequence of appropriate length encoding an ORF into the cloning site will align the reading frame of the initiator sequence with that of the *lacZ* gene. This will result in synthesis of a tripartite (tribrid) polypeptide with β-galactosidase activity. Tribrid proteins with ORFs contributing as few as 5 to more than 500 amino acids to the N terminus retain β-galactosidase activity (Silhavy and Beckwith, 1985). The Lac^+ phenotype of such fusions can therefore be used to identify recombinant clones in which an ORF has been correctly expressed. Furthermore, the resulting fusion proteins are much larger ($>$100 000 daltons) than most *E. coli* proteins, which assists purification.

ORF vectors are particularly useful because they can be used to express

Figure 3.26: Detection of specific clones using antibodies as probes. (a) Examples of the detection methods for solid-phase immunoassays. The detection system may be a specific labelled antibody (as in (i)), or an unlabelled specific antibody plus labelled *S. aureus* protein A (as in (ii)) (with unlabelled primary antibody a second labelled antibody directed against the primary antibody may be used in place of protein A). Either technique may involve any of these detection systems. If protein A is used for the sandwich technique, it is usually necessary to coat the membrane with the $F(ab)'_2$ fragment of the IgG molecule. This fragment lacks the Fc portion which contains the receptor for protein A. (b) An autoradiograph of a membrane labelled using the direct attachment technique. The membrane contained a replica of about 150 recombinant colonies from a gene library of *Neisseria gonorrhoeae* DNA that had been cloned in *E. coli* on pBR322. The detection system was a monoclonal antibody specific for the pilin subunit of the pilus of *N. gonorrhoeae*. Specific antigen-antibody complexes were detected using ^{125}I-labelled protein A. (Courtesy of I. Nicolson)

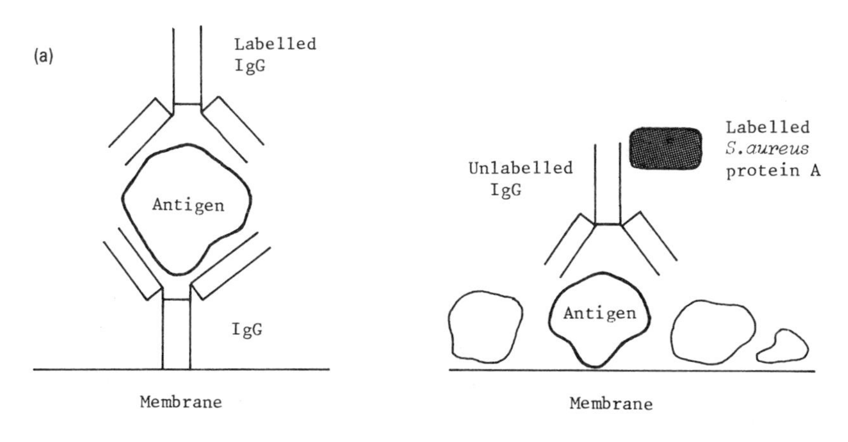

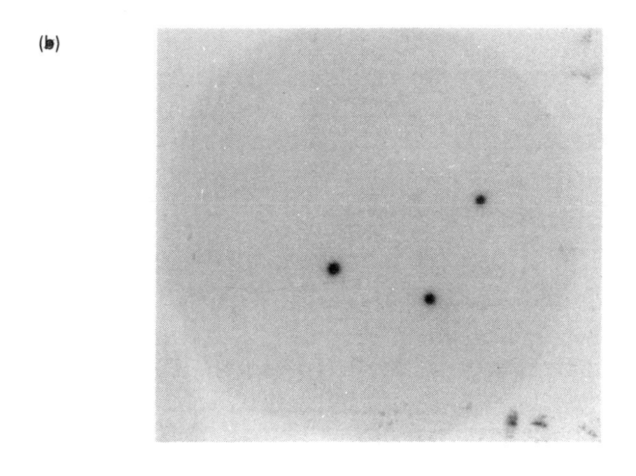

Figure 3.27: Open reading frame (ORF) vector cloning. ORF, DNA fragment bearing open reading frame; p, promoter; S-D, Shine-Dalgarno sequence; ATG, translational initiation codon. The initiator gene may be derived from N-terminal *lacZ* sequences or from other genes such as the *cea* gene of ColE1 or the *ompF* gene of *E. coli*

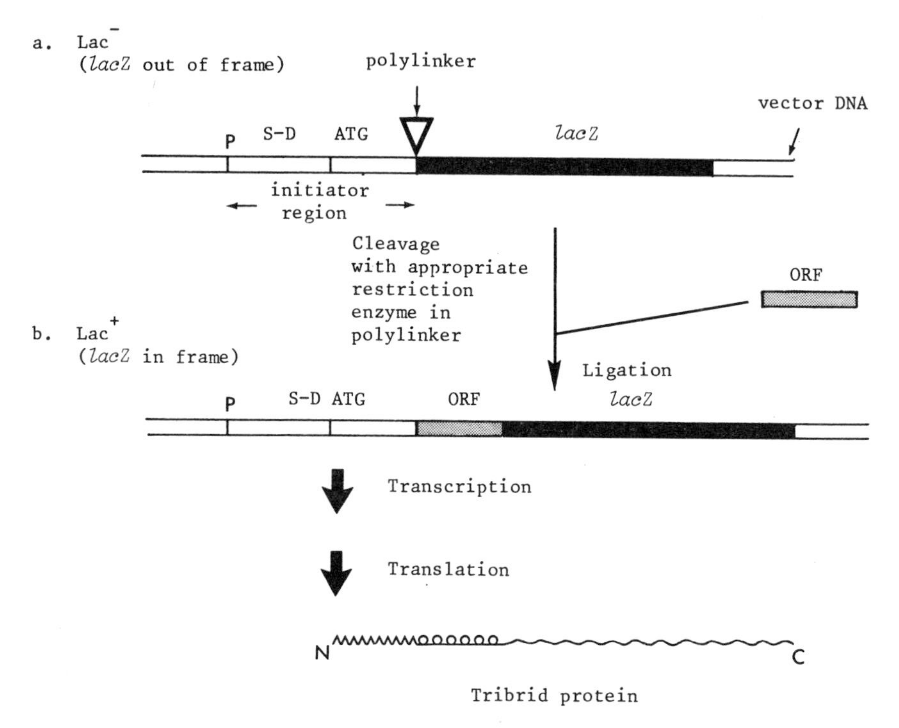

a foreign DNA sequence without prior knowledge of the properties of that sequence. These vectors may also be used for cloning in the absence of a specific assay for the gene product concerned. Where an appropriate antibody is available, Lac^+ clones expressing ORF sequences can be identified by the immunological methods described in section 3.7.1.c. DNA from clones expressing a specific antigen (which may consist of a peptide sequence of as little as ten amino acids) can subsequently be utilised as a probe to identify clones containing the complete gene sequence from a gene library.

ORF vectors may additionally be used to identify gene products by raising antibodies to the tribrid proteins produced when a particular DNA sequence is cloned. Some of the antibodies raised against the resulting tripartite protein will be specific for the polypeptide encoded by the cloned sequence. These antibodies may be used to identify the native protein product of the target gene by Western blotting (see section 3.9.1).

3.7.2 Identification techniques not dependent upon expression of the cloned sequence

(a) Nucleic acid hybridisation

1. Labelling nucleic acid molecules. Nucleic acid hybridisation may be used to detect specific cloned DNA sequences in recombinants provided that a suitable labelled single-stranded DNA or RNA probe is available. The probe may be a previously cloned homologous DNA sequence, specific mRNA (or other RNA), cDNA or synthetic oligonucleotide homologous to part of the gene sought. Probe DNA can be labelled *in vitro* by the incorporation of radiolabelled (usually ^{32}P or ^{35}S) nucleoside triphosphates, and the presence of the radioactive DNA can then be detected by autoradiography. An alternative to radiolabelling is the use of biotinylated nucleotides. The presence of biotinylated DNA can be detected by exploiting the affinity of the egg white glycoprotein, avidin, or a similar protein, streptavidin from *Streptomyces avidini*, for biotin. Avidin or streptavidin can be linked to fluorescent-labelled antibodies that can be quantified by fluorescence measurements, or to enzymes, such as horseradish peroxidase or alkaline phosphatase, whose activity can be detected by colorimetric assays (Langer *et al.*, 1981). Such tests claim to be able to detect as little as 10^{-14} g of specific DNA.

Probe DNA molecules can be end-labelled at their 5′ termini using T4 polynucleotide kinase and [γ-^{32}P] ATP. Terminal deoxynucleotidyl transferase can be used to add the labelled phosphate group from [α-^{32}P]-deoxyribonucleotides or [α-^{32}P]-cordycephin triphosphates to 3′ termini (Tu and Cohen, 1980; Deng and Wu, 1983). For double-stranded DNA fragments bearing 5′ terminal protrusions, the 3′ termini may also be labelled by end-filling with Klenow fragment and [α-^{32}P] deoxyribonucleoside triphosphates (dNTPs). End-labelling methods are generally restricted to the labelling of short oligonucleotide probes. For longer probes alternative methods are generally more efficient.

High specific activity probes of defined length can be generated by using the replacement synthesis reaction of phage T4 DNA polymerase. This enzyme acts as a 3′→5′ exonuclease in the absence of one or more of the four required dNTPs (Figure 3.28). If the missing dNTP is added back to the reaction together with radiolabelled (or biotinylated) dNTPs, the degraded strand is resynthesised and hence labelled (O'Farrell *et al.*, 1980). The **nick-translation** reaction of *E. coli* DNA polymerase I has been widely exploited for labelling DNA probes (Rigby *et al.*, 1977). This enzyme will add deoxyribonucleotides in a 5′→3′ direction to a DNA or RNA primer with a 3′-OH terminus (Kornberg, 1980). The substrate for nick translation is duplex DNA that has been nicked at random (usually by limited digestion with pancreatic DNase I). The 5′→3′ exonuclease activity of polymerase I removes nucleotides from the exposed 5′ terminus generated at a

Figure 3.28: The replacement synthesis reaction of phage T4 DNA polymerase

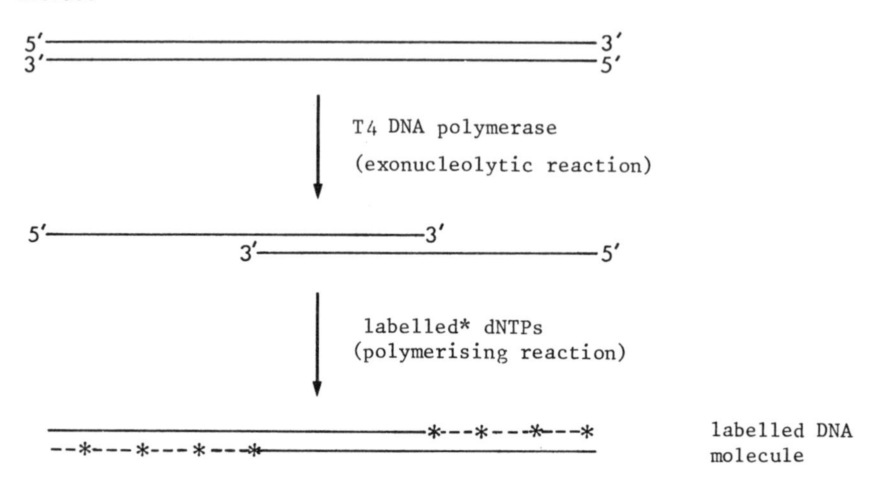

nick. This in turn exposes the template strand (Figure 3.29). The degraded strand can be resynthesised by the polymerising activity of polymerase I. Incorporation of one or more labelled dNTPs ensures that the newly synthesised DNA is labelled uniformly to high specific activity. **Random priming** (Feinberg and Vogelstein, 1983) provides an effective alternative to nick translation, producing high specific activity DNA probes of 100-2000 bases in length from very small amounts (as little as 20 ng) of template. The template is single-stranged DNA obtained from recombinant M13 viral DNA or by denaturing restriction fragments. Template DNA is primed by annealing it to random hexanucleotide primers obtained by nucleolytic degradation of calf thymus or salmon sperm DNA. Klenow fragment in the presence of labelled dNTPs is used to extend the primed template (Figure 3.30).

RNA probes can be labelled *in vitro* by addition of radioactive RNA precursors to the growth medium. It is also possible to synthesise RNA (cRNA) *in vitro* using *E. coli* RNA polymerase, labelled ribonucleoside triphosphates and appropriate template DNA. Specific RNA probes of very high specific activity can be produced from DNA that has been cloned in special plasmid cloning vectors, such as SP62-PL (Butler and Chamberlain, 1982) (Figure 3.31). The gene of interest is first cloned into the polylinker region utilising appropriate restriction sites. This vector contains the extremely active *Salmonella typhimurium* bacteriophage SP6 promoter, which is located upstream of and reads through the polylinker region. Purified recombinant plasmid DNA that has been linearised can be used as a template for an *in vitro* transcription reaction using SP6 RNA polymerase. Initiation of transcription is stringently dependent on the SP6 promoter

Figure 3.29: The nick-translation reaction of *E. coli* DNA polymerase I

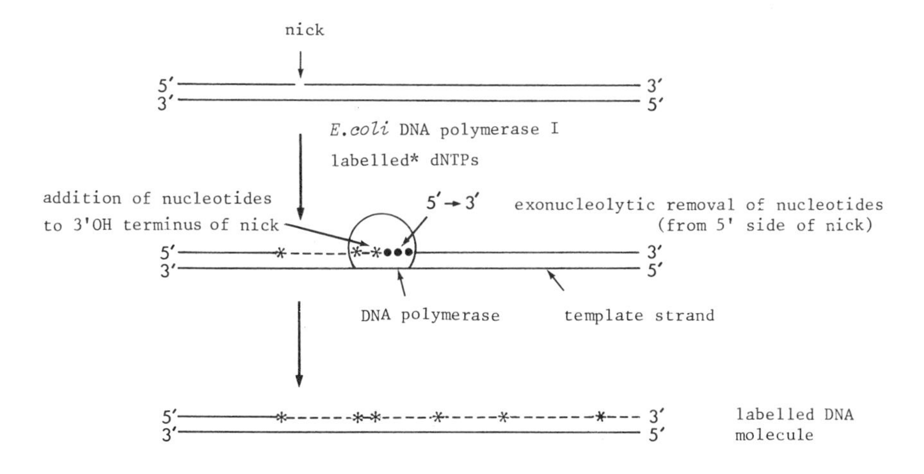

Figure 3.30: Random hexanucleotide primer labelling of DNA

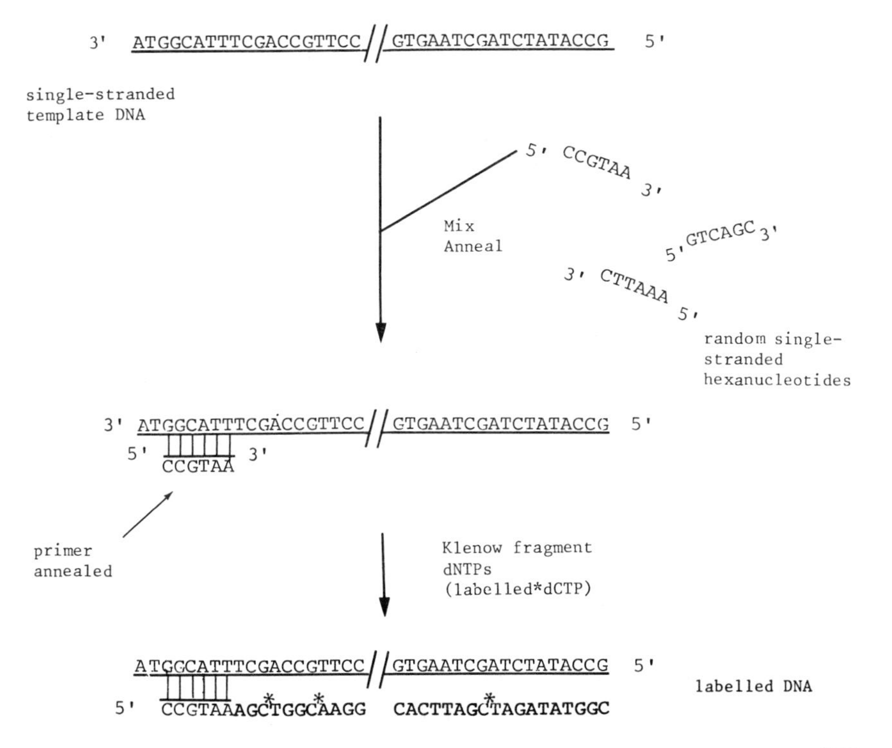

Figure 3.31: Use of SP6 RNA polymerase for *in vitro* transcription. pSP62 was constructed by cloning a DNA fragment containing the *Salmonella* phage SP6 promoter into the *Bam*HI site on pBR322. The promoter is located about 42 bases upstream of a polylinker region taken from M13mp10. →, Direction of transcription; *bla*, β-lactamase gene; *oriV*, origin of vegetative replication

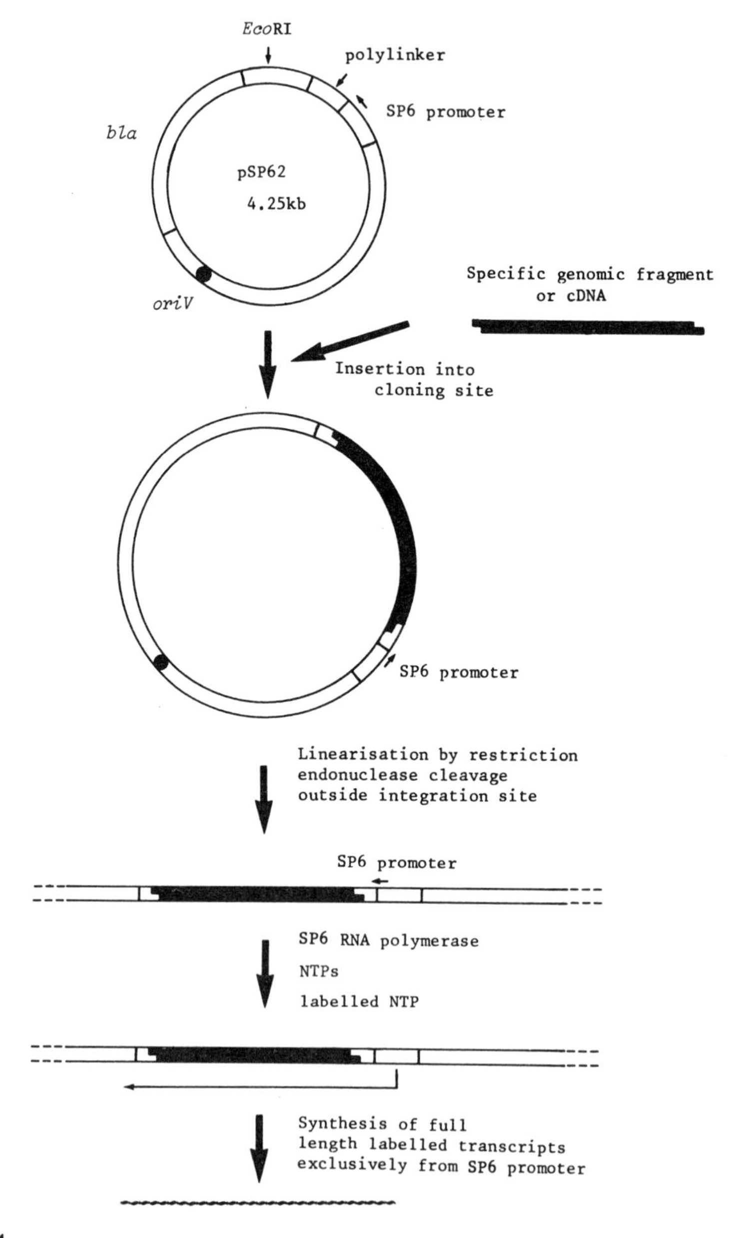

and therefore all transcripts made will read through the cloned gene(s). This technique permits the production of single RNA species that can be labelled to high specific activities, without the need to perform strand separations and other manipulations often required with DNA probes. Similar systems are available that employ promoters and RNA polymerases from *E. coli* phage T3 or T7.

2. In situ *colony (or plaque) RNA-DNA or DNA-DNA hybridisation.* These hybridisation techniques, based largely on the method of Grunstein and Hogness (1975), have greatly facilitated the screening of recombinant clones (for a review see Meinkoth and Wahl, 1984). The method depends upon the transfer of replicas of colonies (or plaques) from agar plates to nitrocellulose or nylon fibre discs. Although originally devised for colonies of *E. coli*, the Grunstein and Hogness technique has been adapted for yeast colonies, phage plaques and SV40 induced plaques in animal cell lines. Colonies are first lysed on the filters, and the DNA is denatured *in situ* (with alkali) and then fixed to the filters. Labelled probe DNA or RNA is hybridised to DNA on the filters and, after washing away unbound probe, clones that have hybridised can be detected by autoradiography (Figure 3.32).

(b) Screening by recombination: the πVX *system*

The homologous recombination system of *E. coli* can be exploited in order to detect a specific cloned sequence (Maniatis *et al.*, 1982). The technique involves insertion of probe sequences into the 902 bp plasmid, πVX, which contains a ColE1 origin of replication, a synthetic *supF* gene and a polylinker containing multiple cloning sites. The plasmid is maintained in *E. coli* by selecting for the ability of the *supF* gene to suppress amber mutations in the *bla* and *tet* genes of a second resident plasmid p3 (a derivative of RP1) and hence confer resistance to ampicillin and tetracycline on the host.

Genomic libraries constructed in a phage λ vector (bearing amber mutations in genes *A* and *B*) are used to infect rec^+ *E. coli* carrying the πVX : probe plasmid. If the probe sequence cloned in πVX is homologous with a sequence present in an infecting phage, reciprocal recombination will result in the insertion of the πVX : probe plasmid into the phage genome. The resulting recombinant phages can be identified by their ability to form plaques on a nonsuppressing host, since the *supF* gene on the πVX portion of the recombinant will suppress the *Aam* and *Bam* mutations. This method permits the rapid isolation of particular cloned sequences and can be used to retrieve underrepresented recombinant bacteriophages from gene libraries (Maniatis *et al.*, 1982; Seed *et al.*, 1982). A major drawback of this system is that the reciprocal recombination event necessary for detection of the cloned sequence may cause disruption of that sequence.

Figure 3.32: Colony hybridisation for the detection of specific recombinant clones

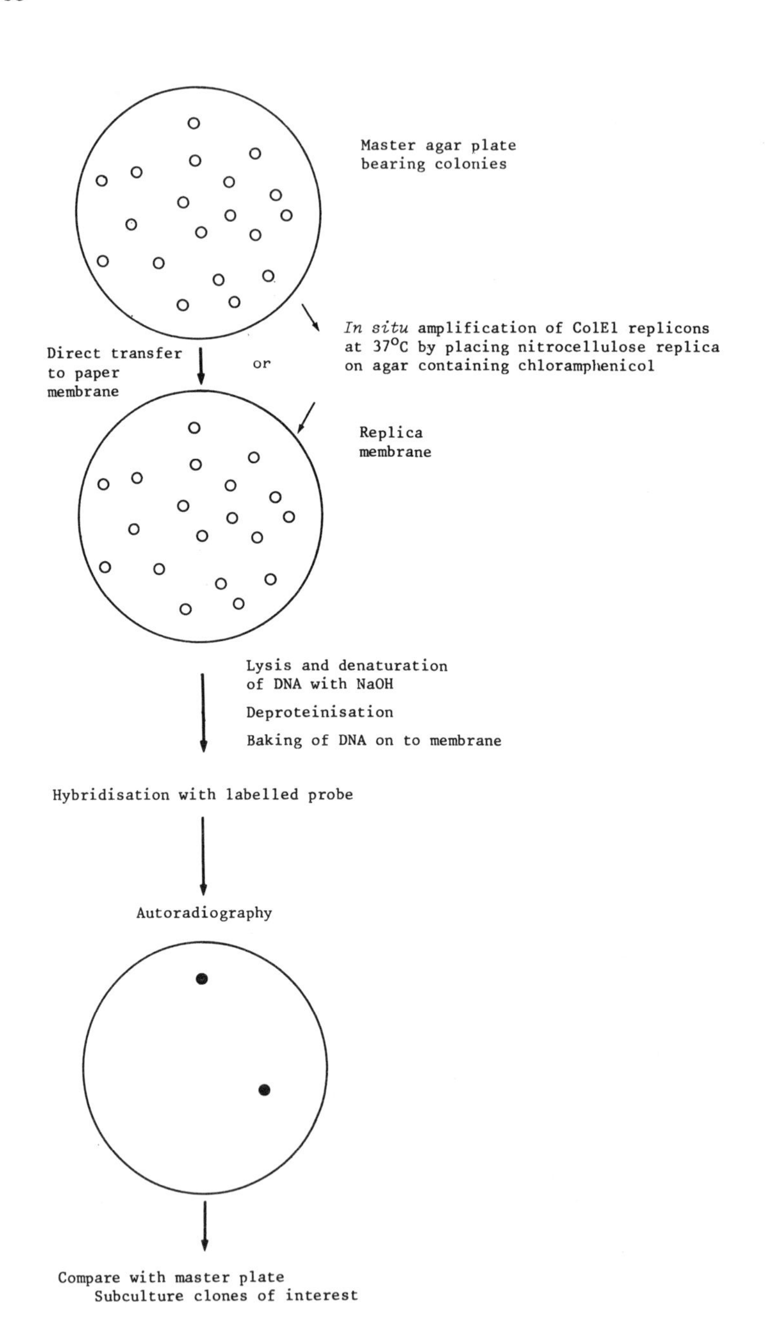

This can, however, be overcome by cloning sequences that lie immediately adjacent to the gene of interest into πVX and using the resultant recombinant as a probe in a second round of screening.

(c) Identification of clones by molecular size

In some cases probes for the detection of cloned sequences are not available and specific clones can only be detected by the characteristic size and/or restriction endonuclease cleavage pattern of the passenger DNA. Numerous microscale methods are available for preparing plasmid or phage molecules rapidly from small-scale cultures, colonies or plaques (see Grinsted and Bennett, 1984; Thomas, 1984, and Maniatis *et al.*, 1982).

3.8 STRUCTURAL ANALYSIS OF CLONED DNA SEQUENCES

As a preliminary to detailed structural analysis, recombinant molecules are isolated from primary clones and the molecular weight of the passenger DNA is determined. Passenger DNA, either attached to the original vector or subcloned in an alternative vector, is then reintroduced into the host in order to confirm that a particular sequence is responsible for encoding the desired character.

3.8.1 Restriction mapping

A detailed restriction endonuclease cleavage map is normally a prerequisite for further structural and functional studies of a cloned sequence. The map is necessary for subcloning strategies designed to locate the minimum part of the cloned sequence that actually codes for the desired characteristic. It is also required for obtaining fragments of DNA that are suitable for sequencing (section 3.8.5).

A cloned DNA fragment can normally be excised from the vector by cleavage with the restriction enzyme(s) used in constructing the recombinant. In some cases the ligated termini of vector and passenger DNA do not recreate the original cleavage site(s) (section 3.2.2). Fragments may therefore be excised by cutting with endonucleases with sites in the vector adjacent to the cloning sites used and that are absent from the cloned sequence. Passenger DNA can be separated from the vector molecule by electrophoresis in agarose or acrylamide (for fragments $<$0.5 kb) gels containing ethidium bromide. Ethidium bromide binds to nucleic acids and fluoresces when illuminated with UV light. This permits visualisation of bands of DNA in gels. DNA bands can be isolated from gels by various techniques that include dissolving the agarose in potassium iodide followed by adsorption of the DNA to a glass powder, recovery of DNA from low melting temperature agarose, or electroelution of DNA fragments into a trough cut in the gel ahead of the band.

Restriction maps are constructed by analysis of the sizes of subfragments (determined by gel electrophoresis) produced by complete and/or partial cleavage of the cloned DNA fragment with single and pairs of restriction endonucleases (Southern, 1979). Frequently cutting endonucleases, with tetranucleotide recognition sequences, normally generate too many subfragments to produce sensible cleavage patterns. Sites for such enzymes can be mapped by digestion of small subfragments of the original cloned sequence. Such sites may be mapped more conveniently by measuring the sizes of partial digestion products of the original fragment that has been radiolabelled at one end only (Smith and Birnstiel, 1976). In this case the fragments form a simple overlapping series with a common labelled terminus. The number of radiolabelled partial digestion products is equivalent to the number of restriction sites in the DNA.

3.8.2 Southern transfer (blotting)

Particular DNA sequences in cloned and genomic DNA fragments may be localised and identified by the hybridisation technique of Southern (1975). This involves alkali denaturation *in situ* of DNA fragments that have been separated on an agarose gel and their subsequent transfer, either by buffer diffusion or electrolytically, to a nitrocellulose filter (or similar matrix such as nylon (Reed and Mann, 1985)) (Figure 3.33). (Transfer of fragments of >10 kb is relatively inefficient, but can be increased by UV-induced photodynamic nicking of DNA in the presence of ethidium bromide, or by acid depurination immediately prior to transfer. Small DNA fragments (<200 bp) are poorly resolved in agarose gels and bind inefficiently to nitrocellulose. These problems may be overcome by electrophoresis in agarose gels containing 50% formamide and transferring the fragments to diazobenzyloxymethylated (DBM) paper (Sun *et al.*, 1982).) The DNA is immobilised on the filter (by baking at 80°C *in vacuo* in the case of nitrocellulose or by UV irradiation for nylon membranes) and then hybridised to radiolabelled (or biotinylated) probe DNA or RNA. Autoradiography is used to locate the position(s) of DNA bands that are complementary to the probe.

Nucleic acids that have been separated on agarose gels can also be detected by direct hybridisation of the dried gel with an appropriate probe (Tsao *et al.*, 1983). This does not involve transfer to a support and is therefore cheaper and more rapid than conventional Southern transfers.

Figure 3.33: Southern transfer hybridisation. (a) Agarose gel electrophoresis of restriction fragments of a plasmid carrying part of the *Neisseria meningitidis* genome. Lane 1, digestion with *Cla*I and *Hin*dIII; lane 2, digestion with *Cla*I and *Pvu*II; lane 3, digestion with *Cla*I and *Eco*RV; lane 4, molecular weight standards (*Bacillus subtilis* phage SPP1 DNA digested with *Eco*RI); lane 5, digestion with *Hin*dIII and *Sal*I; lane 6, digestion with *Eco*RV and *Pvu*II; lane 7, digestion with *Eco*RV and *Sal*I; lane 8, digestion with *Sal*I alone. (b) Autoradiograph of a nitrocellulose membrane blot of the agarose gel in (a) that had been hybridised to probe DNA consisting of the pilin structural gene from *Neisseria gonorrhoeae*. The probe was labelled with [α-^{32}P]dCTP by the random hexanucleotide priming technique. (Courtesy of A. Perry)

(a)

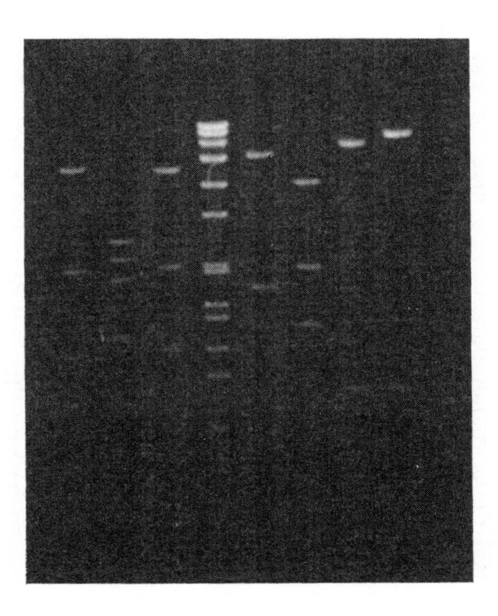

(b)

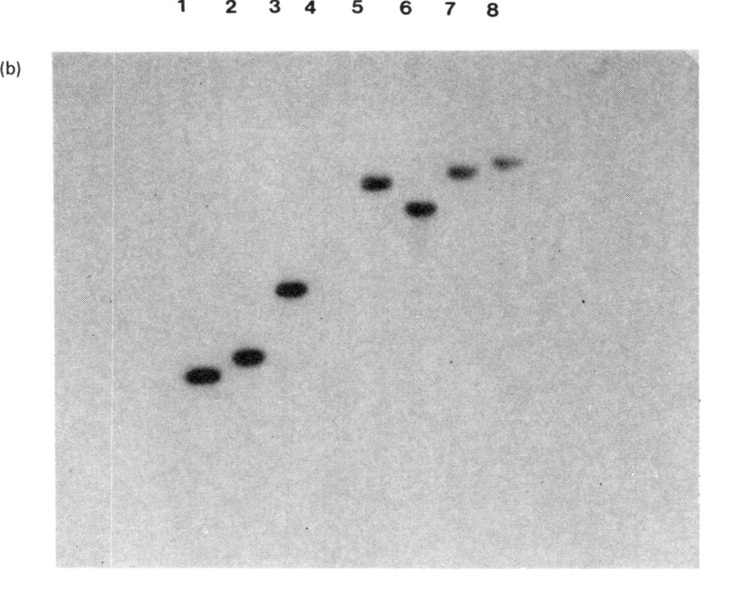

3.8.3 Electron microscopy

Electron microscopy enables single- and double-stranded regions of DNA molecules to be distinguished. DNA is normally prepared for electron microscopy by spreading on to a film of cytochrome *c* and shadowing with platinum or other metals (Brack, 1981). By using selective denaturation conditions it is possible to determine regions of DNA that are AT rich, since they melt out more readily (Inman and Schnos, 1970). Such **denaturation mapping** may be useful, for example, in locating promoter regions, which are typically AT rich. Controlled denaturation and renaturation may also be used to permit the formation of intrastrand base pairing. The creation of stem-loop structures, which may be visualised by electron microscopy, is indicative of inverted repeat sequences within the cloned DNA. Regions of single- and double-stranded DNA present in heteroduplex DNA molecules formed by annealing single-stranded DNA from two different DNA molecules can be mapped (**heteroduplex mapping**) by electron microscopy (Brack, 1981). This permits the degree of sequence relatedness between DNA molecules to be determined. Mapping of heteroduplexes formed between cDNA and appropriate genomic DNA enables the position of introns and exons to be localised. Electron microscopic techniques can also be used to locate regions within cloned DNA molecules that bind specific proteins, such as RNA polymerase or repressors.

3.8.4 Mapping regions on cloned DNA molecules that are transcribed

RNA species, homologous to cloned DNA sequences, may be identified by **Northern transfer** (blotting). RNA species are separated on the basis of size by electrophoresis through agarose gels containing formaldehyde or glyoxal and dimethylsulphoxide. The separated RNA species are then blotted on to nitrocellulose filters by a procedure similar to that used for Southern transfer (section 3.8.2) and hybridised to labelled probe DNA. Autoradiography is used to locate the position(s) of RNA bands that are complementary to the probe.

RNA can hybridise to homologous double-stranded DNA and displace the identical strand to form an R-loop (Figure 3.34). Under appropriate conditions of denaturation the resulting loops can be visualised and measured in the electron microscope (Brack, 1981). **R-loop mapping** permits the location of transcribed regions relative to a known restriction site and of exons and introns in eukaryotic DNA. The ends of RNA molecules and their splice points relative to the template DNA sequence can be determined by **S1 endonuclease mapping** (Berk and Sharp, 1977; Favaloro *et al.*, 1980). S1 endonuclease (or alternatively exonuclease VII)

Figure 3.34: R-loop mapping

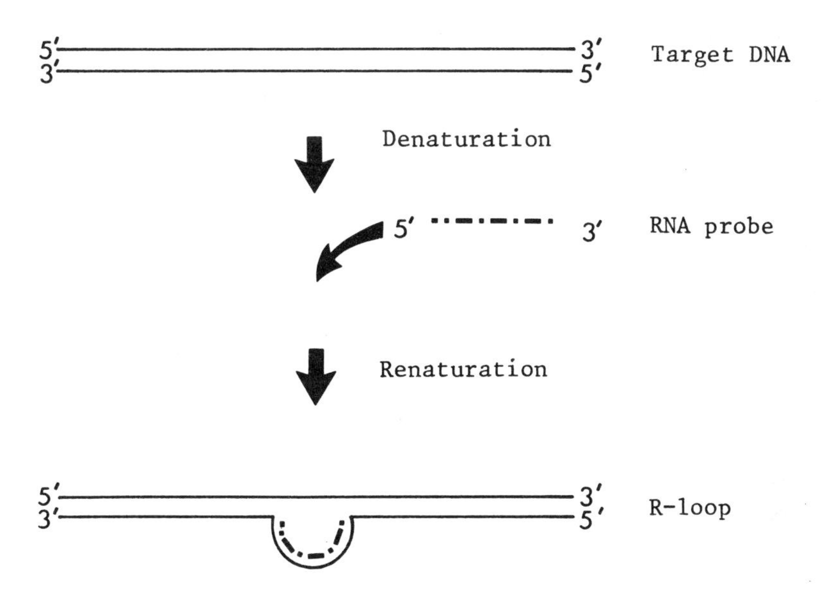

can be used to digest the single-stranded DNA regions of hybrids formed between template (coding) DNA and mRNA (Figure 3.35). The size of the resulting S1-resistant fragment can then be determined by electrophoresis under denaturing conditions. If the DNA is terminally labelled at either the 3′ or 5′ end, and if the labelled end is base paired in the hybrid, the precise termini of the RNA transcript can be mapped by Maxam and Gilbert sequencing (section 3.8.5).

3.8.5 DNA sequence analysis

The ultimate structural characterisation of a cloned fragment depends upon determining its primary DNA sequence. Methods for sequencing DNA are now so rapid (for a review see Brown, 1984) that fine structure restriction mapping is often unnecessary; the positions of all restriction sites may be determined directly from the sequence by a computer search. It is also possible by analysis of the sequence to identify specific regions, such as promoters, ribosome binding sites, transcription terminators, introns and open reading frames (a protein coding region lying between a translational start codon (ATG or GTG) and an in-phase termination codon (TAA or TGA)). Furthermore, sequence analysis of protein coding regions permits determination of the frequency with which a particular species uses the available codons (**codon usage**). Sequence handling computer programs

Figure 3.35: S1 nuclease mapping. (a) Determination of length of transcript. (b) Determination of 5′ end of transcript

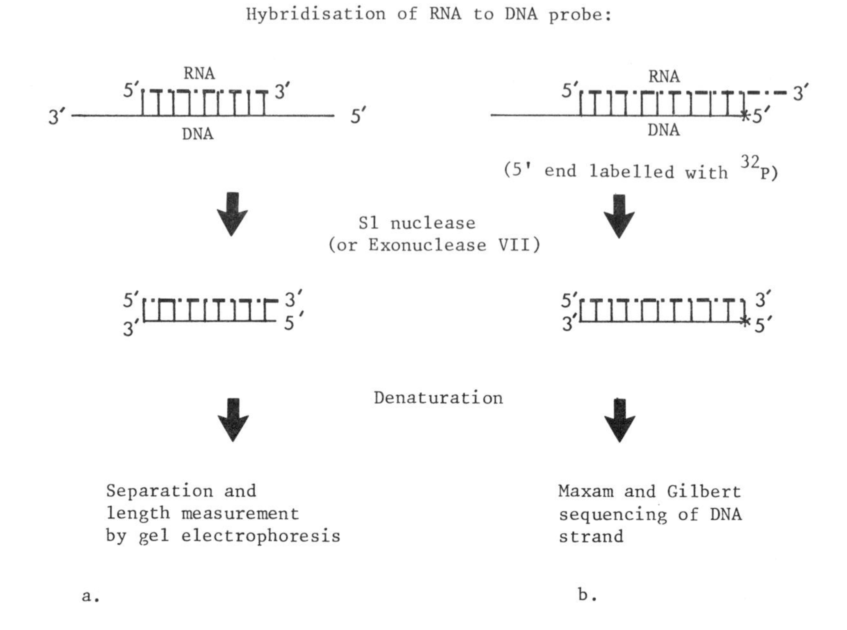

are normally necessary to assemble and process the large amount of information obtained from sequencing even relatively small genes (Staden, 1986).

The two procedures commonly used for DNA sequencing are the chemical method of Maxam and Gilbert (1977) and the chain termination method of Sanger and co-workers (1980).

(a) The Maxam and Gilbert technique

This depends upon a series of specific chemical reactions that break a terminally labelled[8] DNA molecule at specific nucleotides. Conditions are

[8] Fragments to be sequenced may be labelled at their 5′ termini using [γ-^{32}P]-ATP and T4 polynucleotide kinase. Alternatively 3′ termini can be labelled using an [α-^{32}P]-labelled deoxyribonucleotide and terminal deoxynucleotidyl transferase or by repairing the 5′ protruding termini produced by restriction endonuclease cleavage, with labelled dNTPs and T4 or *E. coli* DNA polymerase. Both ends of a double-stranded DNA molecule will normally be labelled by these procedures. In order to simplify sequence analysis it is necessary to obtain molecules labelled at one end only. This can be done by digestion with a restriction enzyme that cleaves the labelled fragment at a unique site near one end. Alternatively, a specific single strand can be separated and purified.

adjusted so that on average only one nucleotide is reacted per DNA molecule. This produces a family of radioactive DNA fragments extending from the same labelled terminus to the various positions of the reacted nucleotide. The products of four (or sometimes more) separate reactions that cleave DNA at adenine or guanine (A + G), at guanine (G), at cytosine (C) and at cytosine or thymine (C + T) (and sometimes additionally A > C and T alone) are subjected to electrophoresis on a denaturing polyacrylamide gel. The sequence can then be read from the pattern of labelled bands on an autoradiogram (Figure 3.36). This method is applicable to both double- and single-stranded DNA and can be used for sequencing DNA fragments isolated directly from gels.

(b) The Sanger chain termination method

The dideoxy chain-terminator procedure of Sanger and co-workers (1980) has become the basis of the most widely used rapid DNA sequencing methods. Unlike the Maxam and Gilbert method this technique has an absolute requirement for single-stranded DNA. For this reason the procedure is generally employed in conjunction with phage M13 or similar cloning vectors (section 3.5.2). Sequencing by the dideoxy method exploits the ability of the Klenow fragment of DNA polymerase I to synthesise a complementary copy of the single-stranded target sequence. Polymerisation is directed from a primer, which may be single-stranded DNA derived from a short restriction fragment or a synthetic single-stranded oligonucleotide, annealed to the template adjacent to the target sequence. Synthetic universal primers are available that anneal specifically to regions of the M13mp series genome immediately adjacent to the multiple cloning site (polylinker) region (Heidecker *et al.*, 1980) (Figure 3.37). Three types of M13 primer can be used:

(i) **sequencing primers**, such as 5′GTAAAACGACGGCCAGT 3′ or 5′ GTTTTCCCAGTCACGAC 3′, which hybridise to the viral (+) strand of the *lac* DNA in M13 vectors. The 3′ OH end of the primer is localised about 30 nucleotides downstream of the *Hin*dIII site in the polylinker region (Figure 3.37). This primer therefore allows sequencing of the 5′ region of the (−) strand of any DNA that has been cloned in M13.

(ii) **reverse sequencing primers**, for example 5′ AACAGCTATGACCATG 3′, which anneal only to the (−) strand with the 3′ OH end localised just upstream of the polylinker region (Figure 3.37). Such primers allow sequencing of the 5′ region of the (+) strand of DNA that has been cloned in M13 phage or pUC *lac* plasmid vectors (see section 3.5.2). In order to sequence with these primers the plasmid DNA or double-stranded RF form of M13 must first be linearised by restriction endonuclease cleavage. This allows the single strands

Figure 3.36: Maxam and Gilbert DNA sequencing. Sequence determination is in the 5′ to 3′ direction of the left-hand end of the upper strand of the original target fragment. The complementary strand can be sequenced in the 3′ to 5′ direction if the corresponding (left-hand) 3′ end is labelled, for example by end-filling with Klenow fragment and ^{32}P-labelled dNTPs. P*, ^{32}P-labelled phosphate group; A + G, chemical cleavage at adenine or guanine; G, cleavage at guanine alone; A > C cleavage at adenine or less efficiently at cytosine; C + T, cleavage at cytosine or thymine; C at cytosine alone. Conditions are adjusted to give cleavage at only one base in each labelled molecule in the region to be sequenced

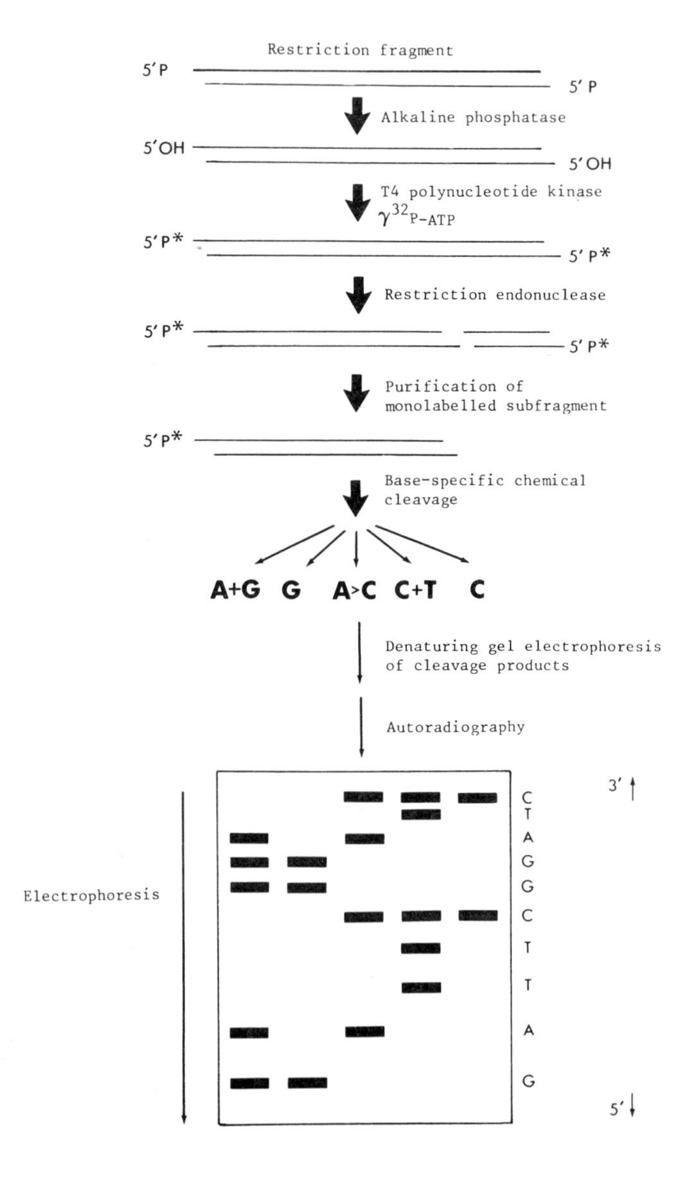

Figure 3.37: Location of the primer regions on M13 vectors

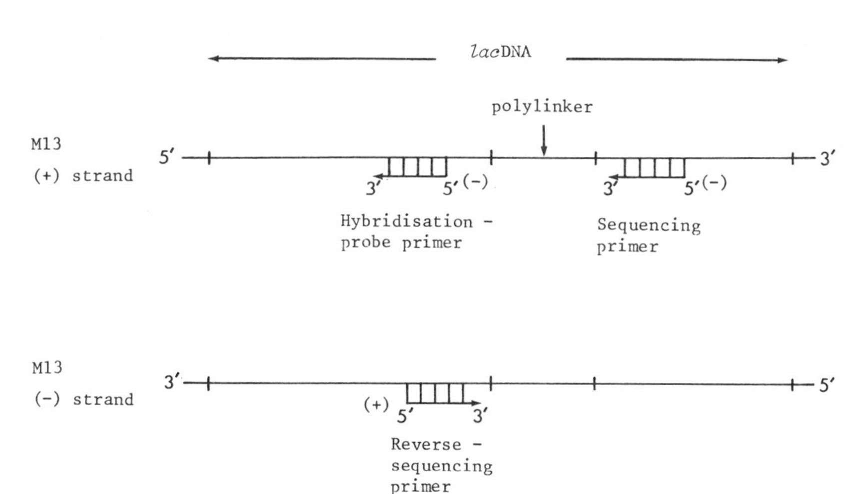

to be separated on boiling and hence available for annealing to the primer. Alternatively a (−) strand template for DNA cloned in M13 phage vectors can be generated from the viral (+) strand by using the sequencing primer, Klenow fragment and the four dNTPs.

(iii) **hybridisation-probe primers**, for example 5′ CACAATTCCA-CACAAC 3′, which anneal to the (+) strand of *lac* in M13 vectors at a position upstream of the polylinker (Figure 3.37). Such primers do not, therefore, prime sequencing through the cloned region. A hybridisation primer is used to generate highly labelled DNA molecules that are partially or completely double-stranded in viral sequences and single-stranded in the cloned sequence (Figure 3.38). The single-stranded region can then be used as a hybridisation probe, with the aid of the radiolabelled double-stranded viral sequence (Hu and Messing, 1983).

Suitable primed molecules for dideoxy chain termination sequencing can be derived from double-stranded DNA (for example, DNA cloned in a plasmid vector) by controlled exonuclease III[9] digestion (Guo and Wu, 1983). DNA cloned in pBR322 can be sequenced by annealing specific pBR322 primers to linearised and denatured plasmid DNA.

[9] Exonuclease III catalyses the stepwise 3′→5′ removal of 5′ mononucleotides from double-stranded DNA. The enzyme is also an endonuclease specific for apurinic DNA and contains both 3′-phosphatase and RNase H activities. Exonuclease III acts at nicks in DNA as well as at termini.

Figure 3.38: Use of the M13-hybridisation primer to generate specific hybridisation probes

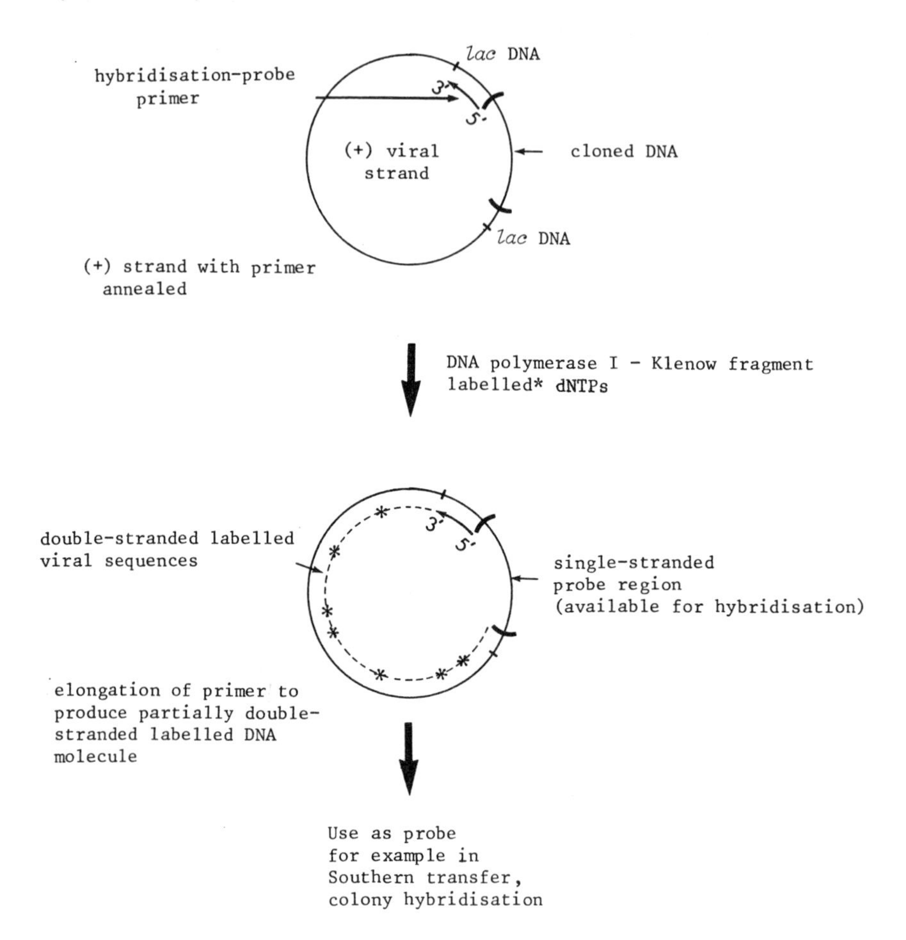

Sequencing is carried out in four separate reaction mixtures, each containing primed template, Klenow fragment, the four dNTPs (one or more of which is labelled with ^{32}P or ^{32}S) and one of the four 2′3′-dideoxy-nucleoside triphosphates (ddATP, ddCTP, ddGTP or ddTTP). The dideoxynucleotide analogue can be incorporated into the nascent DNA chain opposite the appropriate complementary base in the template strand. However, the 3′ end so created is not a substrate for chain elongation. Consequently, growing chains are terminated wherever a dideoxynucleotide is incorporated. It is possible to ensure that chain termination occurs at random locations of a particular nucleotide along the nascent DNA strand by carefully adjusting the ratio of appropriate dNTP to ddNTP in the mixture. This produces a family of radiolabelled strands each having a common 5′ end and a base-specific 3′ end for each of the four reaction mixtures. Products of each reaction are separated by denaturing polyacrylamide gel electrophoresis and the positions of the terminated chains are revealed as radioactive bands on an autoradiograph of the gel. The sequence can then be read 5′ to 3′ in ascending order from the bands (from fragments of smallest to largest molecular weight) (Figure 3.39). Sequence determination of the complementary strand can be achieved by inverting the original cloned sequence (clone turn around) with respect to the primer region. Alternatively the same DNA fragment may be inserted by directional cloning into both members of a pair of M13 vectors, such as M13mp18 and M13mp19. Since the polylinker regions of these vectors are inverted with respect to each other (Figure 3.17), both strands of the target sequence will be cloned as the (+) (viral) strand in one or the other vector.

The chain terminator method allows sequences of about 200 to 300 bases to be read reliably from one gel. Longer chain termination products are insufficiently resolved by gel electrophoresis. Thus the ability to determine a nucleotide sequence is limited by the distance of that sequence from a suitable primer. This problem may be overcome by shotgun cloning small, randomly cleaved fragments of the target sequence into M13 vectors and sequencing individual fragments. The complete sequence of the target molecule is then deduced by integrating the sequences of the smaller fragments. However, much of the data obtained from sequencing large DNA molecules by this approach is redundant since it relates to discontinuous and overlapping segments scattered randomly over both strands.

In order to overcome the problem of redundant information, **kilo-sequencing** strategies have been developed (see for example Barnes and Bevan, 1983). For the technique of Barnes and Bevan to be effective, the target sequence, which can be up to 14 kb, must lack a site for one of the following endonucleases, *Bgl*II, *Eco*RI, *Hin*dIII, *Pst*I or *Xba*I. Unique sites for all of these enzymes lie adjacent to the primer region of the specialised M13 vectors that have been constructed for kilo-sequencing (Barnes and Bevan, 1983). RF DNA of recombinant phages carrying the region to be sequenced is first nicked randomly with DNase I in the presence of ethidium bromide.[10] (Conditions are adjusted so that on average only one nick occurs per molecule.) The nicks are then extended with exonuclease III and the resulting single-strand gaps are cut across with BAL31 nuclease (Figure 3.40). The ends of the resulting linear DNA molecules are repaired with Klenow fragment and ligated to *Eco*RI linkers (or other appropriate linkers depending upon which sites are absent from the target sequence). If the DNA is digested with *Eco*RI (or other appropriate enzyme) and recircularised with T4 DNA ligase, a population of

[10] At appropriate concentrations ethidium bromide inhibits endonuclease activity sufficiently to ensure that only single-stranded scissions occur.

Figure 3.39: Dideoxynucleotide chain termination sequencing. (a) Protocol for M13 sequencing. dd, Dideoxy-; d, deoxy-; N, any nucleotide. (b) Autoradiograph of part of a sequencing gel. (Courtesy of A. Perry)

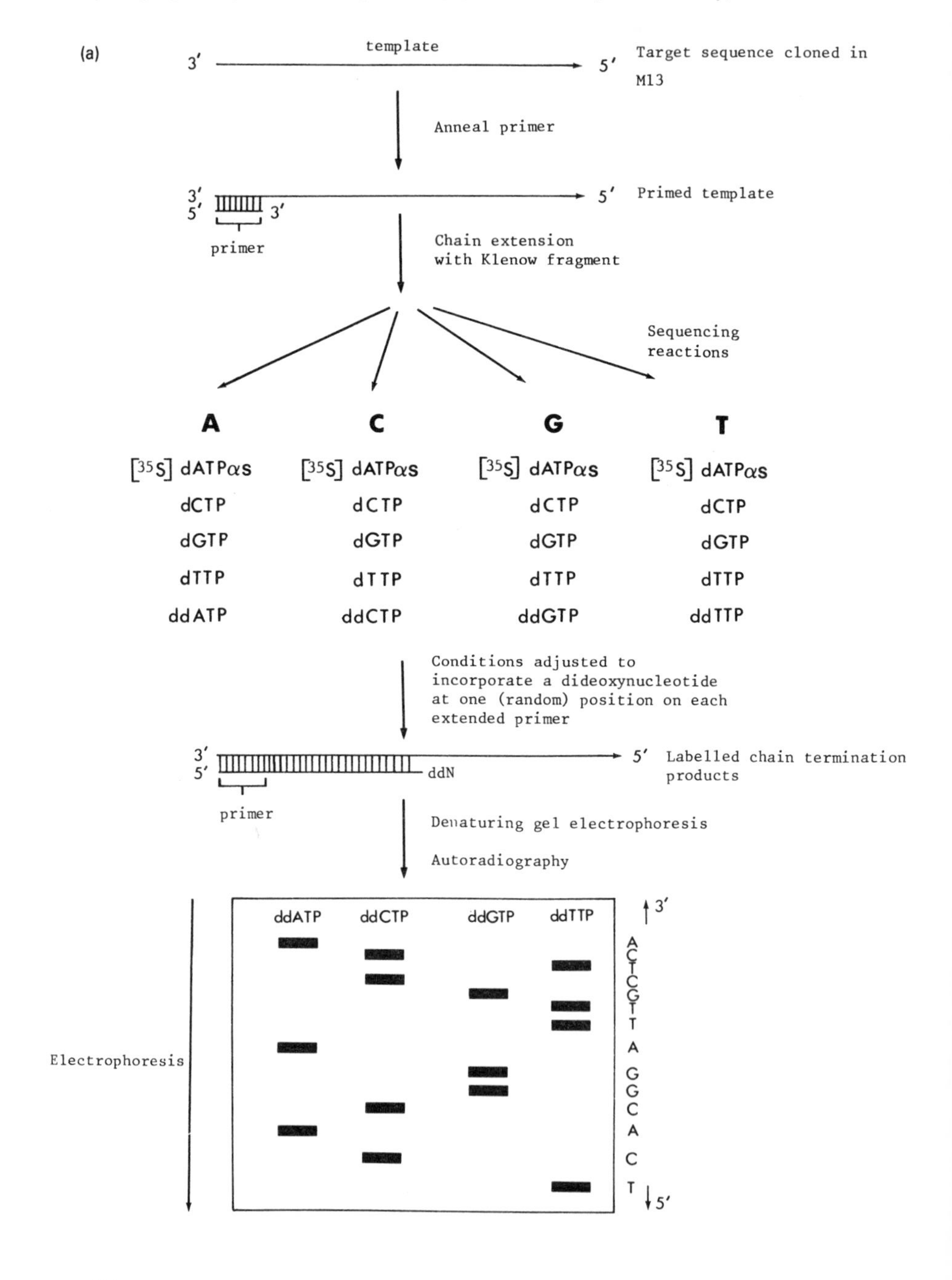

(b)

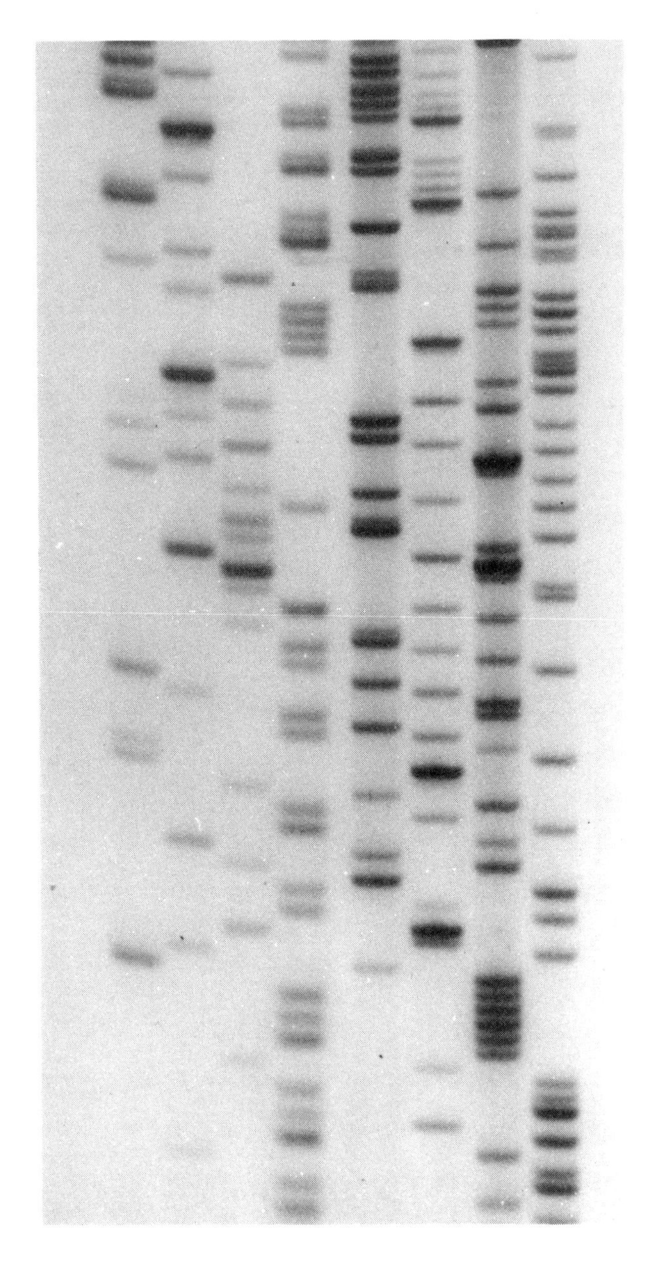

Figure 3.40: The kilo-sequencing strategy. (a) Specialised kilosequencing vector mWB2341. (b) Sequencing strategy. DNase, pancreatic deoxyribonuclease I. After Barnes and Bevan (1983)

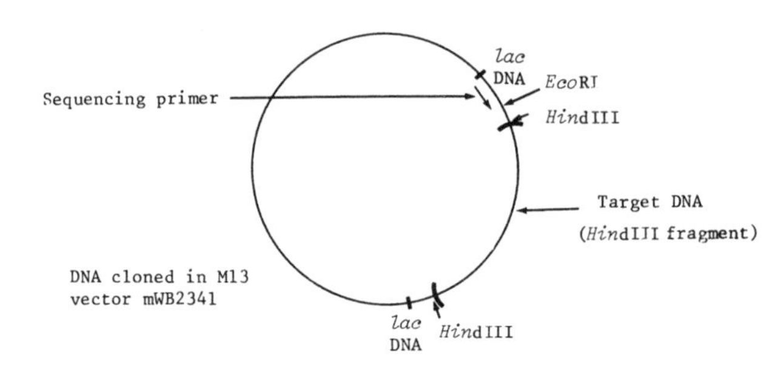

deleted molecules is created. The deletions extend varying distances across the target DNA from the unique *Eco*RI site in the vector. Deleted molecules are used to transfect *E. coli* and progeny phages are pooled and purified from the transfected cells. Viable phage particles can then be separated on the basis of size by electrophoresis in agarose gels. Since M13 filament length alters in accordance with the amount of DNA packaged into the virion, this method allows the isolation of deletant recombinant phages of any size. The deletions are anchored at the unique restriction site adjacent to the sequencing primer on the vector. This enables the isolation of deletant phages whose DNA can act as a template for dideoxy-sequencing beginning at any desired location along the target DNA. By using a spectrum of recombinant phages of decreasing size (and hence containing deletions of increasing size) it is possible to obtain sequence information in an ordered manner from the 5′ to the 3′ end of a large DNA molecule.

A simpler method for producing an ordered set of subclones for sequencing has been devised by Henikoff (1985). M13 recombinant DNA is treated with two restriction endonucleases that cut between the position of insertion of the cloned fragment and the sequencing primer annealing site (Figure 3.41). The enzyme that cuts closer to the primer annealing site must produce a four-base 3′-protruding terminus, whereas the other enzyme must produce a blunt or 5′-protruding terminus on the insert side. Digestion of the resulting cleaved M13 DNA with exonuclease III for varying time periods will produce a family of molecules with large single-stranded regions that extend only into the insert region. Exonuclease III does not digest in the direction of the primer region because the 3′ protrusion produced by restriction cleavage is not a suitable substrate. A subsequent digestion of the DNA with the single-strand specific nuclease S1 produces blunt-ended molecules that can be ligated together to circularise the DNA prior to transformation of *E. coli*.

A second strategy for the ordered sequencing of large DNA fragments has been made possible by the development of methods for the microscale synthesis of specific oligonucleotide primers (section 3.1.3). Large DNA fragments are cloned in M13 vectors and the sequence of the first 200 or so bases adjacent to the universal sequencing primer region is determined. A second, specific primer corresponding to 15-20 bases at the 3′ end of that newly determined sequence is then custom synthesised. This new primer is subsequently used to prime a further series of dideoxy-sequencing reactions (Figure 3.42). This process is repeated with a third, specific primer and so on until the entire sequence of the cloned insert is determined. This ordered stepwise procession along the target sequence allows rapid DNA sequencing without the need for repeated subcloning.

(c) Genomic sequencing

Church and Gilbert (1984) have devised a method for sequencing DNA

Figure 3.41: Preparation of ordered deletions using exonuclease III for M13 sequencing

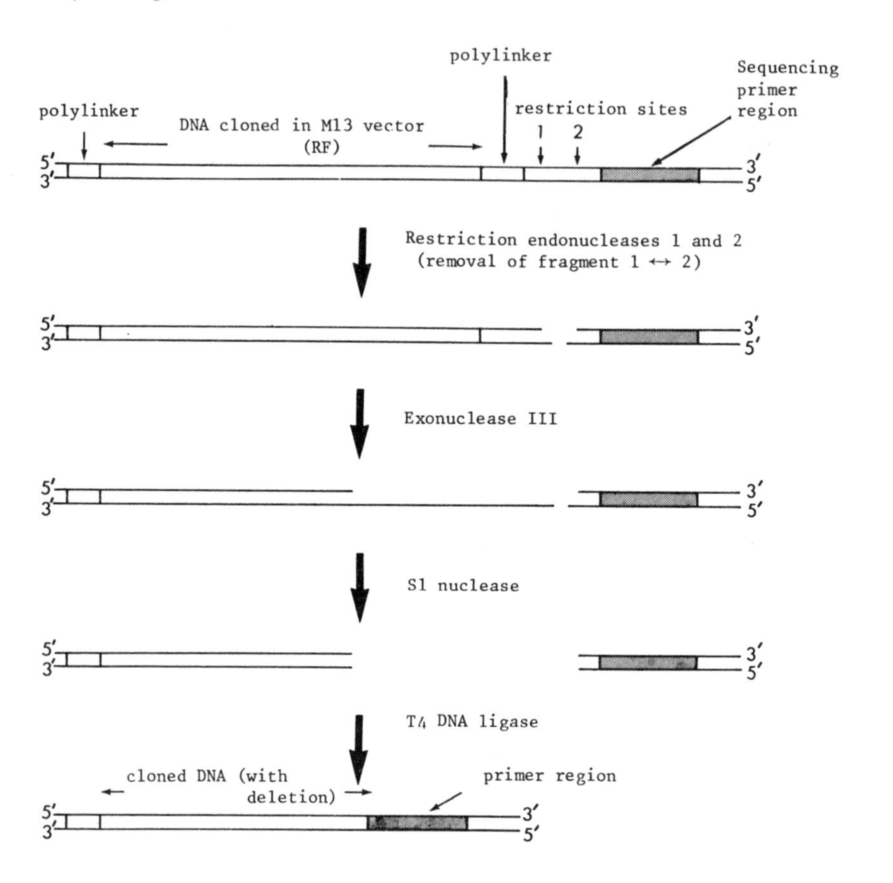

Figure 3.42: Ordered M13 sequencing using sequential custom synthesis of specific primers. 1-4, Custom primers synthesised as appropriate depending on the 3′ end of the sequence determined previously

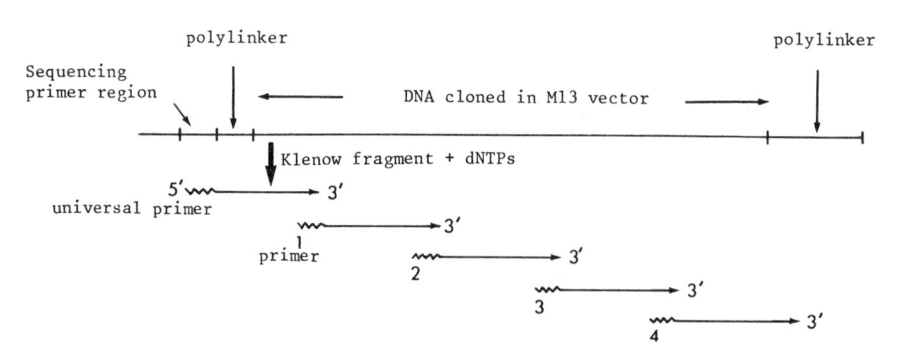

that can be used directly on uncloned DNA to study, for example, methylation patterns (Nick *et al.*, 1986) and to analyse DNA-protein interactions. The method combines the chemical sequencing procedure of Maxam and Gilbert (1977) with detection of specific DNA sequences by Southern transfer hybridisation (section 3.8.2). Total cellular DNA is first digested to completion with a restriction endonuclease. Samples of the digested DNA are subsequently subjected separately to partial cleavage at A, C, G or T residues and the products are separated on denaturing polyacrylamide gels. The DNA is transferred to nylon membranes, fixed by UV irradiation and hybridised to a ^{32}P-labelled nucleic acid probe (generally 100-200 nucleotides long). Since the sequencing cleavage products are unlabelled, this means that only fragments homologous to the probe and thence in the area of interest will be detected (Figure 3.43). The complete sequence of that region can be read directly from an autoradiograph of the membrane. This method is not only extremely sensitive but also allows the rehybridisation of membranes with different probes specific for either strand of the DNA and/or different regions.

3.9 FUNCTIONAL ANALYSIS OF CLONED DNA FRAGMENTS

3.9.1 Identification of proteins

Functional analysis of a cloned sequence serves to confirm the identity of a particular cloned fragment as the producer of a desired protein (or RNA species). Moreover, such analysis provides information concerning the regulation and compartmentalisation of the protein within the host organism. This is of particular importance where high-level expression of that protein is required (see section 5.5.5).

Proteins can be labelled *in vivo* by incorporation of radioactive amino acids that have been added to growth media. Alternatively radiolabelled amino acids can be added to an *in vitro* translation system (section 3.9.3). The resulting radiolabelled proteins can be separated from extracts by one- or two-dimensional sodium dodecyl sulphate-polyacrylamide gel electrophoresis. However, the large number of different proteins encoded by the host genome (about 1000 to 1500 for *E. coli*, depending upon the growth conditions (Neidhardt *et al.*, 1983)) hampers the detection of specific polypeptides encoded by recombinant DNA molecules. Individual polypeptides may be identified *in situ* on polyacrylamide gels if an enzyme-specific staining procedure is available. For example, β-lactamases can be detected by their ability to hydrolyse the chromogenic cephalosporin nitrocefin. (This substrate is yellow and its hydrolysed product is red. Protein bands with β-lactamase activity therefore stain red in the presence of this substrate.) Where specific antibodies are available, individual proteins may be

Figure 3.43: Genomic sequencing

Complete digestion of chromosomal DNA
with a restriction endonuclease

Restriction fragment

Partial chemical cleavage
in separate reactions for each
of the four bases

Example of single-stranded products from C-specific cleavage:

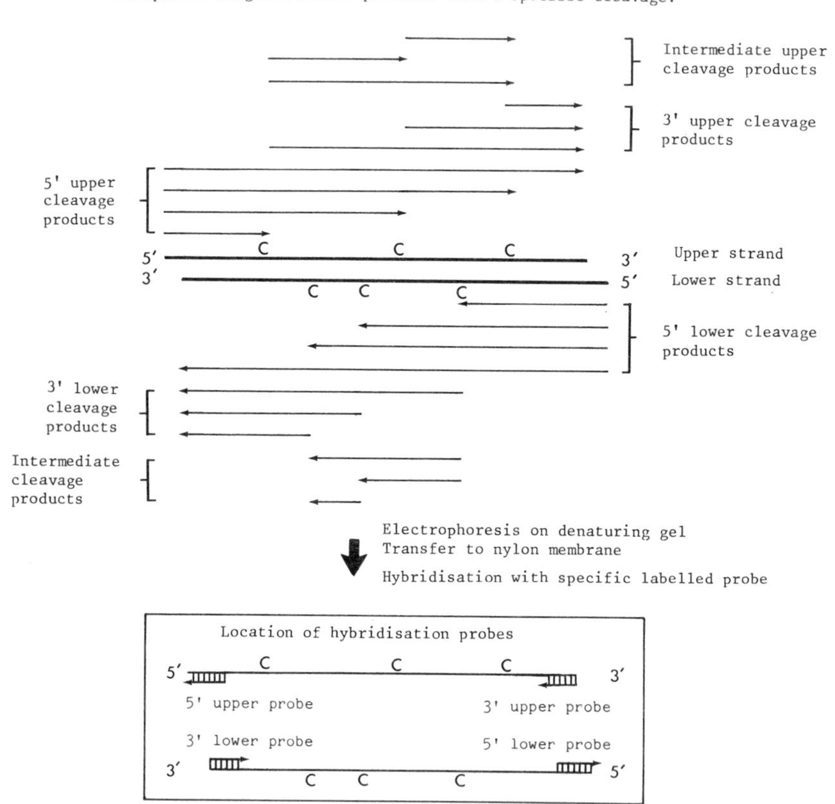

Autoradiography

Example of cleavage products detected with the 5' upper probe

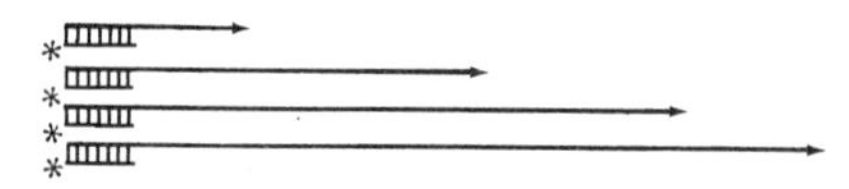

identified by the formation of antigen-antibody precipitin complexes. Screening for specific proteins by more sensitive radioimmunoassay or ELISA techniques is possible if proteins that have been separated by electrophoresis in an acrylamide gel are first transferred to a matrix, such as a nitrocellulose sheet, by the **Western blotting** (transfer) procedure (Burnette, 1981).

3.9.2 *In vivo* transcription and translation of cloned DNA sequences

(a) Minicells

Minicell-producing mutants of *E. coli* (and other bacteria) are defective in cell division such that the septum forms abnormally towards one of the cell poles. The product of this aberrant division is a small chromosome-less cell (the minicell) and a large but otherwise normal cell (Dougan and Kehoe, 1984). Minicells may be readily separated from viable parental cells by sucrose density gradient centrifugation. The minicells are inviable but possess most of the metabolic abilities, including those of transcription and translation, of parental cells. If the parental strain carries a plasmid, some plasmid copies are segregated into the minicells, where they may replicate to a limited extent and plasmid-encoded genes may be expressed. The translation of chromosomal gene products is not totally eliminated in minicells due to long-lived mRNA species acquired at cell division. However, background translation from such mRNA can be decreased if a period is allowed for *in vivo* degradation of endogenous mRNA prior to the labelling of minicells (usually by incubation in medium containing either ^{35}S-methionine for labelling proteins or ^{32}P-labelled RNA precursors for labelling transcription products).

Minicells provide a simple *in vivo* system for identification of plasmid-specified products in the absence of a background of chromosomally specified products. A comparison of the polypeptides specified by a recombinant molecule with those encoded by the vector plasmid is usually sufficient to determine which proteins are encoded by a cloned fragment (Figure 3.44). The synthesis of λ-specific proteins encoded by recombinant λ phages can be selectively depressed by infecting minicells that contain a plasmid that overproduces the *c*I repressor (Reeve, 1978). Passenger DNA cloned in λ will therefore be expressed (provided such DNA possesses its own promoters) in the absence of a background of λ-specific proteins.

(b) Maxicells

Proteins encoded by recombinant plasmids can be specifically labelled with ^{35}S-methionine in cultures of *recA*$^-$ *uvrA*$^-$ *E. coli* that have been heavily irradiated with UV to form maxicells (Sancar *et al.*, 1979). The chromo-

Figure 3.44: Analysis of polypeptides specified by recombinant plasmids in minicells. An autoradiograph of radiolabelled polypeptides from protein extracts of minicells labelled with ^{35}S-methionine. The polypeptides were separated by SDS-polyacrylamide gel electrophoresis. Lane 1, Polypeptides produced by the plasmid vector pLV59; lanes 2 and 3, polypeptides produced by recombinants formed by cloning fragments of *Klebsiella aerogenes* chromosomal DNA into the unique *Bgl*II site on the vector. (Courtesy of P. Allen)

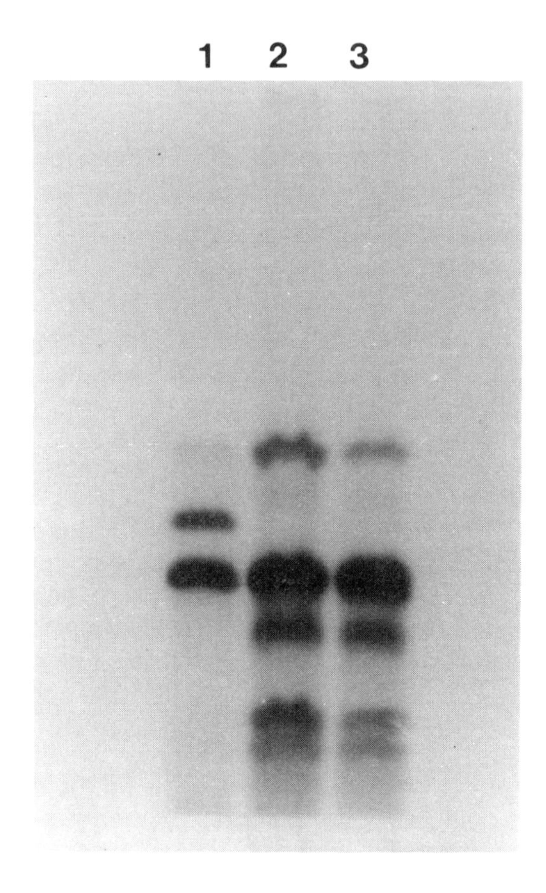

somal DNA in maxicells is extensively degraded due to unrepaired damage to DNA (see section 4.3). If cells contain multicopy plasmids, irradiation conditions can be adjusted so that sufficient plasmid DNA molecules per cell escape a lethal hit with UV and can continue to replicate. Consequently, plasmid-encoded proteins continue to be synthesised, whereas chromosomal proteins are not. The maxicell system avoids the time-consuming steps involved in purifying minicells, but the UV dose must be carefully determined in order to reduce chromosomal expression sufficiently.

(c) Chloramphenicol amplification

Treatment of cultures of *E. coli* with chloramphenicol results in the amplification (up to 2000 copies per cell) of ColE1-type replicons (section 3.5.2). Proteins encoded by amplifiable plasmids are synthesised preferentially during the period of recovery from prolonged amplification when the chloramphenicol is removed from the medium (Neidhardt *et al.*, 1983). This is attributable to the high dosage of plasmid-borne relative to chromosomally-borne genes.

3.9.3 Translation of purified RNA

The ability to translate purified RNA and subsequently identify products provides a means of correlating a cloned DNA sequence with the proteins it encodes.

(a) In vivo *translation*

When immature eggs (oocytes) of the toad (*Xenopus* are microinjected with eukaryotic (but not prokaryotic) mRNA, they can direct the synthesis of appropriate proteins. The injected mRNA is highly stable and is efficiently translated in oocytes. Furthermore, enzymes in the oocytes correctly modify (for example by glycosylation) the encoded proteins.

(b) In vitro *(cell-free) translation*

RNA preparations can be translated *in vitro* either in lysates of rabbit reticulocytes or in extracts of wheat germ. *In vitro* translation systems provide greater experimental control than *in vivo* methods. However, *in vitro* techniques are generally more time-consuming. Furthermore functions such as the proteolytic processing involved in transport of proteins across membranes (see section 5.5.5) may be absent.

(c) Hybrid released (selected) translation (HRT)

mRNA specific for a cloned DNA sequence can be isolated by hybridising cellular RNA from a putative recombinant clone to cloned DNA that has been denatured and bound to a nitrocellulose filter. After washing the filter to remove non-specifically bound RNA, the specific hybridised mRNA can be released from RNA-DNA hybrids on the filter by heating in buffers containing formamide or low salt concentrations. The released (selected) RNA preparation may be translated, for example in a cell-free system, in order to identify the proteins encoded by the cloned DNA.

(d) Hybrid arrested translation (HART)

Total mRNA from a putative recombinant clone can be hybridised to the relevant cloned DNA under conditions that favour the formation of RNA-

DNA hybrids. The mixture of nucleic acid can then be used to direct a cell-free translation system. Since RNA that is hybridised to DNA cannot direct the synthesis of polypeptides in such a system, the presence of an mRNA species specific for the cloned DNA is detected by the absence of appropriate translation products. These products should, however, be synthesised if the specific mRNA-DNA hybrids are dissociated, by a period of heating, prior to their addition to the translation system.

3.9.4 *In vitro* coupled transcription/translation of cloned DNA

It is possible to use DNA cloned in *E. coli*, without recourse to the isolation of RNA, to drive an *in vitro* coupled transcription/translation system (Zubay, 1973). Various modifications of the original Zubay system have been developed. All of these systems utilise cell-free extracts of *E. coli* containing the necessary enzymes and other components for transcription and translation, to which a mixture of cofactors and low molecular weight compounds is added. Nuclease inhibitors, such as diethyl pyrocarbonate, are also included since these systems are extremely sensitive to contamination with RNases. DNA added to the system is transcribed and translated faithfully and the resultant polypeptides are labelled by the incorporation of ^{35}S-methionine. Specific radiolabelled polypeptides may be selectively recovered from the mixture of translation products and identified by immunoprecipitation with appropriate antibodies, prior to analysis on SDS-polyacrylamide gels. It is possible to utilise supercoiled DNA molecules (for example plasmid vector-insert hybrids) or linear DNA fragments (such as a cloned insert alone) as the template[11] (Pratt *et al.*, 1981). Thus polypeptides encoded by a cloned sequence alone can be distinguished from polypeptides resulting from transcriptional read-through from DNA sequences in the vector to sequences in the cloned insert. Restriction sites within a gene of interest can also be identified since cleavage of the template DNA with the appropriate endonuclease should result in elimination of a specific polypeptide. Genes from organisms such as *Staphylococcus aureus* and *Bacillus subtilis* can be expressed in coupled transcription/translation systems derived from *E. coli* or *Streptomyces lividans*. However, the development of cell-free systems utilising extracts from other microorganisms may be necessary in order to extend the range of genes that can be expressed directly from cloned DNA.

[11] There is evidence that certain genes may require supercoiling of the template DNA for efficient expression *in vivo*. However, supercoiling may be relatively unimportant *in vitro* so that linear DNA can be transcribed faithfully, albeit sometimes less efficiently than equivalent supercoiled DNA. The efficiency of transcription/translation of linear DNA fragments can be improved by the use of cell-free extracts from exonuclease V-deficient (*recBC*$^-$) *E. coli*.

3.9.5 Identification of regulatory sequences in cloned DNA

The location of control regions, such as promoters and transcription terminators, within cloned DNA can be deduced from the primary DNA sequence (see section 5.1). However, functional tests can only be carried out if the expression of the genes that such regions normally regulate can be readily quantified. A general solution to this problem is to replace the normal coding sequence of these genes with a coding sequence whose product can be assayed conveniently. The *E. coli* structural gene *lacZ* for β-galactosidase provides such a system. *lacZ* can be readily isolated free of its normal translational initiation site and/or promoter (Silhavy and Beckwith, 1985). Such *lacZ* derivatives can be fused *in vitro* or *in vivo* to DNA sequences of interest to create transcription (operon or promoter) or transcription-translation (gene) fusions (Figure 3.45). Fusions may be created by cloning the sequence of interest upstream of the β-galactosidase coding region in special cloning vectors or by inserting a gene cartridge carrying *lacZ* or some other gene (see below) into genes that have already been cloned. Fusions of both types may also be made by using special fusion vectors derived from defective Mu (Mu d) phages (Casadaban and Cohen, 1980).

Figure 3.45: Operon and gene fusions to *lacZ*. For (a) a hybrid polypeptide from ATG(*X*) to the C terminus of *lacZ* might be produced instead of X′ polypeptide. β-galactosidase activity indicates strength of expression signal (s). p*X*, Promoter of gene *X*, ATG, translational start; *X*′, ′*X*, parts of interrupted *X* gene; *lacZ*′, *lacZ* gene minus translational start

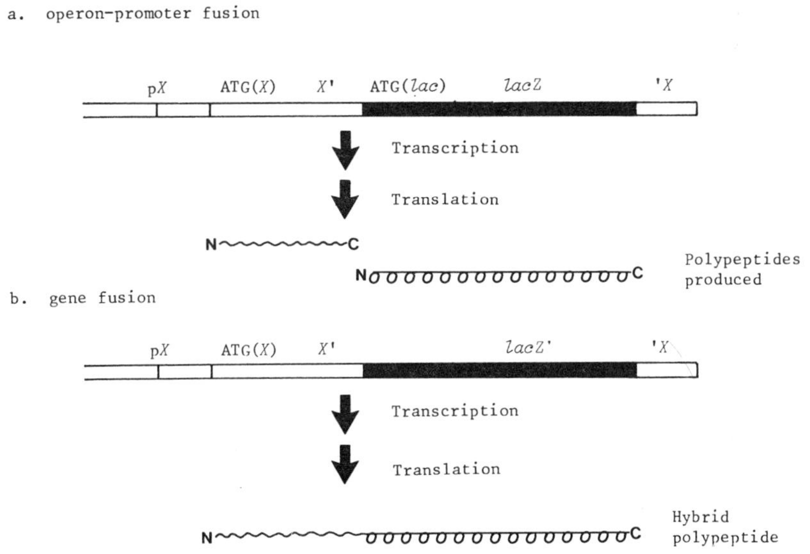

Mu d(Ap^r, *lac*) phages carry the *bla* gene of Tn*3* in place of some Mu genes. These defective phages can insert by transposition into virtually any site on the bacterial chromosome or into DNA cloned on low copy number vectors. Insertions can be selected for by virtue of the ampicillin resistance that they confer on the host. The *lac* gene carried by Mu d phages may lack either transcription or transcription-translation signals, depending on the phage. Expression of *lac* therefore only occurs if the Mu d(Ap^r, *lac*) inserts in the correct orientation into a transcription plus or minus a translation unit (Figure 3.46). Insertion of Mu d(Ap^r, *lac*) into a gene normally leads to inactivation of that gene, and as a consequence of polarity will prevent expression of genes further downstream in the same transcription unit. A problem with Mu d phages is that the resulting fusions are generally unstable due to secondary transposition events. Furthermore, the *cts* gene of the phage renders the recipient strain temperature sensitive for growth. A stratagem that circumvents both problems is to replace the Mu d phage part of the fusion *in situ* with a hybrid λp*lac* Mu phage that lacks Mu transposition functions. This is achieved by reciprocal recombination between common sequences on Mu d and the hybrid phage (Silhavy and Beckwith,

Figure 3.46: Analysis of transcription signals using Mu d*lac*. *pX*, promoter of gene *X* (when Mu d inserts to form a fusion, the *X′* indicates the 5′ end of the gene and ′X and 3′ end); *s, c*, ends of Mu genome; *lacZ, Y, A′*, *lac* operon minus part of the 3′ end of the *A* gene; *bla*, β-lactamase gene; Mu d, defective Mu genome. Insertions are detected by selection for resistance to ampicillin, loss of *X* function and Lac$^+$ phenotype

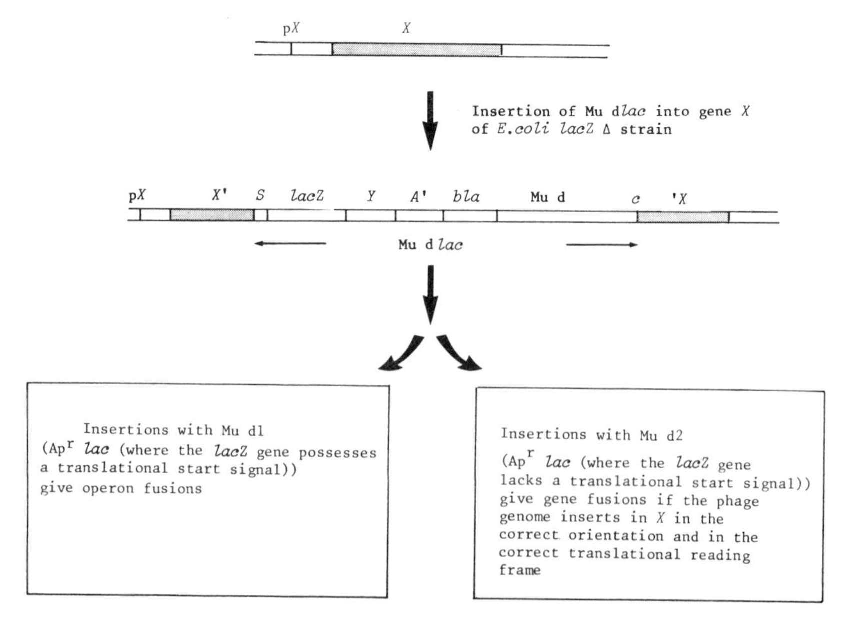

1985). An additional advantage of λp*lac* Mu phages is that, unlike Mu d phages, they can be used to form fusions to DNA cloned in high copy number vectors.

As an alternative to defective phage fusions, a portable test gene (**gene cartridge**), for example the chloramphenicol acetyl transferase (CAT) gene of Tn*9*, lacking its own promoter may be inserted *in vitro* downstream of a putative cloned promoter sequence (Close and Rodriguez, 1982) (Figure 3.47). Cloned sequences containing functional promoters can be identified

Figure 3.47: Use of a chloramphenicol acetyl transferase *Hin*dIII gene cartridge to measure promoter strength. ATG, translational start codon; p, promoter; →, direction of transcription; S-D, Shine-Dalgarno sequence; TAA, translational termination codon; CAT, chloramphenicol acetyl transferase

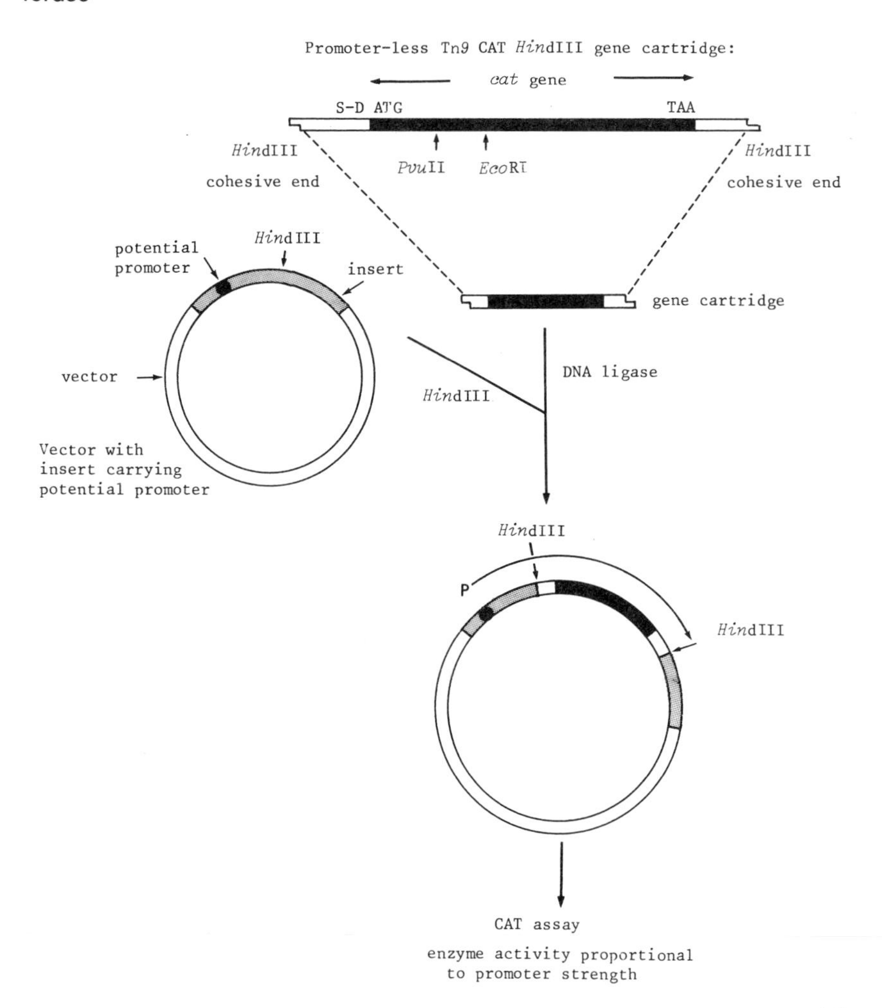

by the ability to promote expression of the gene for CAT and hence confer chloramphenicol resistance on the host. The relative strengths and regulatory mechanisms of different promoters can be determined by assaying the activity of CAT produced by different clones. The insertion of a gene cartridge often results in the interruption of coding regions controlled by the 'promoter' under study. Such insertional inactivation permits investigation of the regulation of genes that might otherwise be lethal to the host. The cartridge system also permits study of the regulation of cloned genes that specify no readily identifiable phenotype.

3.10 GENOME WALKING

The assignment of a gene to a particular cloned fragment provides the starting point for a more comprehensive characterisation of an individual chromosome or entire genome. Each gene in a genomic library should be represented on several different overlapping cloned sequences. A cloned fragment containing a given gene or DNA sequence can therefore be used as a hybridisation or recombination probe (section 3.7.2) to locate other clones in the library containing sequences that were immediately adjacent to that gene in the intact genome. DNA from these clones can be used in turn as probes to locate further flanking sequences. The resulting 'walk' along the genome continues until the ends of a linear genome (or the start point on a circular genome) are reached. **Genome walking** enables the ordering of cloned fragments from a library and the assignment of genes to particular locations on the genome or on a particular chromosome of an organism. It is not necessary to know the function of any of the genes involved in a genome walk. The technique therefore permits a simple physical map of a genome to be constructed from the sum of the individual properties (for example restriction cleavage patterns) of the ordered cloned fragments.

BIBLIOGRAPHY

Appleyard, R.K. (1954) Segregation of new lysogenic types during growth of a doubly lysogenic strain derived from *Escherichia coli* K12. *Genetics* **39**, 440-52

Bachmann, B.J. (1984) Linkage map of *Escherichia coli* K12. In *Genetic maps* (S.J. O'Brien, ed.), pp. 145-61. Cold Spring Harbor Laboratory, Cold Spring Harbor NY

Bagdasarian, M. and Timmis, K.N. (1982) Host:vector systems for gene cloning in *Pseudomonas. Curr. Top. Microbiol. Immunol.* **96**, 47-67

Bagdasarian, M., Lurz, R., Ruckert, B., Franklin, F.C.H., Bagdasarian, M.M., Frey, J. and Timmis, K.N. (1981) Specific purpose plasmid cloning vectors. II Broad host range, high copy number, RSF1010-derived vectors, and a host vector system for gene cloning in *Pseudomonas. Gene* **16**, 237-47

Bahl, C.P., Marians, K.J., Wu, R., Slawinsky, J. and Narang, S.A. (1976) A general method for inserting specific DNA sequences. *Gene*, **1**, 81-92

Ballance, D.J. and Turner, G. (1985) Development of a high-frequency transforming vector for *Aspergillus nidulans*. *Gene* **36**, 321-31

Barnes, W. (1980) DNA cloning with single-stranded phage vectors. In *Genetic engineering*, vol. 2 (J.K. Setlow and A. Hollaender, eds), pp. 185-200. Plenum Press, New York

Barnes, W.M. and Bevan, M. (1983) Kilosequencing: an ordered strategy for rapid DNA sequence data acquisition. *Nucl. Acids Res.* **11**, 349-68

Barth, P.T., Tobin, L. and Sharpe, G.S. (1981) Development of broad host range plasmid vectors. In *Molecular biology, pathogenicity and ecology of bacterial plasmids* (S.B. Levy, R.C. Clowes and E.L. Koenig, eds), pp. 439-48. Plenum Press, New York

Bates, P.F. and Swift, R.A. (1983) Double *cos* site vectors: simplified cosmid cloning. *Gene* **26**, 137-46

Beggs, J.D. (1982) Gene cloning in yeast. In *Genetic engineering 2* (R. Williamson, ed.), pp. 175-203. Academic Press, London and New York

Berk, A.J. and Sharp, P.A. (1977) Sizing and mapping of early adenovirus mRNAs by gel electrophoresis of S1-endonuclease-digested hybrids. *Cell* **12**, 721-32

Bernstein, A., Berger, S., Huszar, D. and Dick, J. (1985) Gene transfer with retrovirus vectors. In *Genetic engineering*, vol. 7 (J.K. Setlow and A. Hollaender, eds), pp. 235-61. Plenum Press, New York and London

Bibb, M.J. and Cohen, S.N. (1982) Gene expression in *Streptomyces*: construction and application of promoter-probe plasmid vectors in *Streptomyces lividans*. *Mol. Gen. Genet.* **187**, 265-77

Bibb, M.J., Ward, J.M., Kieser, T., Cohen, S.N. and Hopwood, D.A. (1981) Excision of chromosomal DNA sequences from *Streptomyces coelicolor* forms a novel family of plasmids detectable in *Streptomyces lividans*. *Mol. Gen. Genet.* **184**, 230-40

Bibb, M.J., Chater, K.F. and Hopwood, D.A. (1983) Developments in *Streptomyces* cloning. In *Experimental manipulation of gene expression* (M. Inouye, ed.), pp. 53-82. Academic Press, New York

Bird, A.P. and Southern, E.M. (1978) The use of restriction enzymes to study eukaryotic DNA methylation. I. The methylation pattern in ribosomal DNA from *Xenopus laevis*. *J. Mol. Biol.* **118**, 27-47

Bishop, J.D., Morton, J.G., Rosbash, M. and Richardson, M. (1974) Three abundance classes in Hela cell messenger RNA. *Nature (London)* **250**, 199-204

Blumenthal, R.M., Gregory, S.A. and Cooperider, J.S. (1985) Cloning of a restriction-modification system from *Proteus vulgaris* and its use in analysing a methylase-sensitive phenotype in *Escherichia coli*. *J. Bacteriol.* **164**, 501-9

Bochner, B.R., Huang, H.C., Schieven, G.L. and Ames, B.N. (1980) Positive selection for loss of tetracycline-resistance. *J. Bacteriol.* **143**, 926-33

Bolivar, F., Rodriguez, R.L., Greene, P.J., Betlach, M.C., Heynecker, H.C., Boyer, H.W., Crosa, J.H. and Falkow, S. (1977) Construction and characterization of new cloning vehicles. II. A multipurpose cloning system. *Gene* **2**, 95-113

Boyer, H. and Roulland-Dussoix, D. (1969) A complementation analysis of the restriction and modification of DNA in *Escherichia coli*. *J. Mol. Biol.* **41**, 459-72

Brack, C. (1981) DNA electron microscopy. *Crit. Rev. Biochem.* **10**, 113-69

Brammar, W.J. (1982) Vectors based on bacteriophage lambda. In *Genetic engineering*, vol. 3 (R. Williamson, ed.), pp. 53-81. Academic Press, London and New York

Brenner, S., Cesareni, G. and Karn, J. (1982) Phasmids: hybrids between ColE1 replicons and bacteriophage lambda. *Gene* **17**, 27-44

Broome, S. and Gilbert, W. (1978) Immunological screening method to detect specific translation products. *Proc. Natl Acad. Sci. USA*, **75**, 2746-9

Brown, N. (1984) DNA sequencing. In *Methods in microbiology*, vol. 17 (P.M. Bennett and J. Grinsted, eds), pp. 259-313. Academic Press, London and New York

Burnette, W.N. (1981) 'Western blotting': electrophoretic transfer of proteins from sodium dodecyl sulfate-polyacrylamide gels to unmodified nitrocellulose and radiographic detection with antibody and radioiodinated protein A. *Anal. Biochem.* **112**, 195-203

Burns, D.M. and Beacham, W.R. (1984) Positive selection vectors: a small plasmid vector useful for the direct selection of *Sau*3A-generated overlapping DNA fragments. *Gene* **27**, 323-5

Butler, E.T. and Chamberlain, M.J. (1982) Bacteriophage SP6-specific RNA polymerase. I. Isolation and characterisation of the enzyme. *J. Biol. Chem.* **257**, 5772-8

Casadaban, M.J. and Cohen, S.N. (1980) Analysis of gene control signals by DNA fusion and cloning in *Escherichia coli. J. Mol. Biol.* **138**, 179-207

Case, M.E. (1982) Transformation of *Neurospora crassa* using recombinant plasmid DNA. In *Basic life sciences vol. 19. Engineering microorganisms for chemicals* (A. Hollaender, R.D. De Moss, J. Konisky, D. Savage and R.S. Wolfe, eds), pp. 87-100. Plenum Press, New York

Cesareni, G., Castagnoli, L. and Brenner, S. (1982) The phasmid as a tool for plasmid genetics. II. Isolation of point mutations that affect replication of a ColE1-related plasmid. *Genet. Res. Camb.* **40**, 233-47

Chang, A.C.Y., Nunberg, J.H., Kaufman, R.J., Erlich, H.A. and Cohen, S.N. (1978) Phenotypic expression in *E. coli* of a DNA sequence coding for mouse dihydrofolate reductase. *Nature (London)*, **257**, 617-24

Chater, K.F. and Bruton, C.J. (1983) Mutational cloning in *Streptomyces* and the isolation of antibiotic production genes. *Gene* **26**, 67-78

Chater, K.F., Hopwood, D.A., Kieser, T. and Thompson, C.J. (1982) Gene cloning in *Streptomyces. Curr. Top. Microbiol. Immunol.* **96**, 69-95

Church, G.M. and Gilbert, W. (1984) Genomic sequencing. *Proc. Natl Acad. Sci. USA* **81**, 1991-5

Clarke, L. and Carbon, J. (1976) A colony bank containing synthetic ColE1 hybrid plasmids representative of the entire *E. coli* genome. *Cell* **9**, 91-9

Clarke, L. and Carbon, J. (1980) Isolation of a yeast centromere and construction of functional small circular chromosomes. *Nature (London)* **287**, 504-9

Close, T.J. and Rodriguez, R.L. (1982) Construction and characterization of the chloramphenicol resistance gene cartridge: a new approach to the transcriptional mapping of extrachromosomal elements. *Gene* **20**, 305-16

Colbère-Garapin, F., Horodniceanu, F., Kourilsky, P. and Garapin, A.C. (1981) A new dominant hybrid selective marker for higher eukaryotic cells. *J. Mol. Biol.* **150**, 1-14

Collins, J. and Hohn, B. (1978) Cosmids: a type of plasmid gene-cloning vector that is packageable *in vitro* in bacteriophage λ heads. *Proc. Natl Acad. Sci. USA* **75**, 4242-6

Curtiss, R., Inoue, M., Pereira, D., Hsu, J.C., Alexander, L. and Rock, L. (1977) Construction and use of safer bacterial strains for recombinant DNA research. In *Molecular cloning of recombinant DNA* (W.A. Scott and R. Werner, eds), pp. 99-114. Academic Press, New York

Davison, J., Brunel, F and Merchez, M. (1979) A new host-vector system allowing selection for foreign DNA inserts in bacteriophage λgt*WES. Gene* **8**, 69-80

Dean, D. (1981) A plasmid cloning vector for the direct selection of strains carrying recombinant plasmids. *Gene* **15**, 99-102

Deng, G-R. and Wu, R. (1983) Terminal transferase: use in the tailing of DNA and for *in vitro* mutagenesis. *Methods in enzymology*, vol. 100 (R. Wu, L. Grossman and K. Moldave, eds), pp. 96-116. Academic Press, London and New York

Dente, L., Cesareni, G. and Cortese, R. (1983) pEMBL: a new family of single-stranded plasmids. *Nucl. Acids Res.* **11**, 1645-55

DiMaio, D., Treisman, R. and Maniatis, T. (1982) A bovine papillomavirus vector which propagates as an episome in both mouse and bacterial cells. *Proc. Natl Acad. Sci. USA* **79**, 4030-4

Ditta, G., Stanfield, S., Corbin, D. and Helsinki, D.R. (1980) Broad host range DNA cloning system for Gram-negative bacteria: construction of a gene bank of *Rhizobium meliloti. Proc. Natl Acad. Sci. USA* **77**, 7347-51

Dotto, G.P., Enea, V. and Zinder, N.D. (1981) Functional analysis of bacteriophage f1 intergenic region. *Virology* **114**, 463-73

Dougan, G. and Kehoe, M. (1984) The minicell system as a method for studying expression from plasmid DNA. In *Methods in microbiology*, vol. 17 (P.M. Bennett and J. Grinsted, eds), pp. 233-58. Academic Press, London and New York

Dubnau, D.A. (1982) *The molecular biology of the bacilli*, vol. I, Academic Press, New York

Dubnau, D.A. (1985) *The molecular biology of the bacilli*, vol. II, Academic Press, New York

Dugaiczyk, A., Boyer, H.W. and Goodman, H.M. (1975) Ligation of *Eco*RI endonuclease-generated DNA fragments into linear and circular structures. *J. Mol. Biol.* **96**, 171-84

Ehrlich, S.D., Niaudet, B. and Michel, B. (1982) Use of plasmids from *Staphylococcus aureus* for cloning of DNA in *Bacillus subtilis. Curr. Top. Microbiol. Immunol.* **96**, 19-29

Enquist, L. and Sternberg, N. (1979) *In vitro* packaging of λ *Dam* vectors and their use in cloning DNA fragments. In *Methods in enzymology*, vol. 68 (R. Wu, ed.), pp. 281-98. Academic Press, New York

Erlich, H.A., Cohen, S.N. and McDevitt, H.O. (1979) Immunological detection and characterization of products translated from cloned DNA fragments. In *Methods in enzymology*, vol. 68 (R. Wu, ed.), pp. 443-53. Academic Press, New York

Favaloro, J., Treisman, R. and Kamen, R. (1980) Transcription maps of polyoma virus-specific RNA: analysis by two-dimensional nuclease S1 gel mapping. In *Methods in enzymology*, vol. 65 (L. Grossman and K. Moldave, eds), pp. 718-32. Academic Press, New York

Feinberg, A.P. and Vogelstein, B. (1983) A technique for radiolabelling DNA restriction endonuclease fragments to high specific activity. *Anal. Biochem.* **132**, 6-13

Flock, J-I. (1983) Cosduction:transduction of *Bacillus subtilis* with phage φ105 using a φ105 cosplasmid. *Mol. Gen. Genet.* **189**, 304-8

Frank, R., Heikens, W., Heisterberg-Moutsis, G. and Blocker, H. (1983) A new general method for the simultaneous chemical synthesis of large oligonucleotides: segmental solid supports. *Nucl. Acids Res.* **11**, 4365-77

Gait, M.J. (1984) *Oligonucleotide synthesis, a practical approach.* IRL Press, Oxford

Gautier, F. and Bonewald, R. (1980) Use of plasmid R1162 and its derivatives for gene cloning in the methanol-utilizing *Pseudomonas* AM-1. *Mol. Gen. Genet.* **182**, 99-105

Graham, F.L. and van der Eb, A.J. (1973) A new technique for the assay of infectivity of human adenovirus 5 DNA. *Virology* **52**, 456-67

Grinsted, J. and Bennett, P.M. (1984) Isolation and purification of plasmid DNA. In *Methods in microbiology*, vol. 17 (P.M. Bennett and J. Grinsted, eds), pp. 123-31. Academic Press, London and New York

Grinter, N.J. (1983) A broad-host-range cloning vector transposable to various replicons. *Gene* **21**, 133-43

Grunstein, M. and Hogness, D. (1975) Colony hybridization: a method for the isolation of cloned DNAs that contain a specific gene. *Proc. Natl Acad. Sci. USA* **72**, 3961-5

Gruss, P. and Khoury, G. (1982) Gene transfer into mammalian cells: use of viral vectors to investigate regulatory signals for the expression of eukaryotic genes. *Curr. Top. Microbiol. Immunol.* **96**, 159-70

Gryczan, T.J. and Dubnau, D. (1982) Direct selection of recombinant plasmids in *Bacillus subtilis. Gene* **20**, 459-69

Gryczan, T.J., Contente, S. and Dubnau, D. (1978) Characterization of *Staphylococcus aureus* plasmids introduced by transformation into *Bacillus subtilis. J. Bacteriol.* **134**, 318-29

Gubler, U. and Hoffman, B.J. (1983) A simple and very efficient method for generating cDNA libraries. *Gene* **25**, 263-9

Guo, L-H. and Wu, R. (1983) Exonuclease III: use for DNA sequence analysis and in specific deletions of nucleotides. In *Methods in enzymology*, vol. 100 (R. Wu, L. Grossman and K. Moldave, eds), pp. 60-96. Academic Press, London and New York

Hagan, C.E. and Warren, G.J. (1982) Lethality of palindromic DNA and its use in selection of recombinant plasmids. *Gene* **19**, 147-51

Hanahan, D. (1983) Studies on transformation of *Escherichia coli* with plasmids. *J. Mol. Biol.* **166**, 557-80

Hattman, S., Brooks, J.E. and Maswekar, M. (1978) Sequence specificity of the P1 modification methylase (M.*Eco*P1) and the DNA methylase (M.*Eco dam*) controlled by the *Escherichia coli dam* gene. *J. Mol. Biol.* **126**, 367-80

Hayashi, K., Nakazawa, M., Ishizaki, Y. and Obayashi, A. (1985) Influence of monovalent cations on the activity of T4 DNA ligase in the presence of polyethylene glycol. *Nucl. Acids Res.* **13**, 3261-71

Heidecker, G. and Messing, J. (1983) Sequence analysis of Zein cDNAs obtained by an efficient mRNA cloning method. *Nucl. Acids Res.* **11** 4891-906

Heidecker, G., Messing, J. and Gronenborn, B. (1980) A versatile primer for DNA sequencing in the M13mp2 cloning system. *Gene* **10**, 69-73

Heilmann, H. and Reeve, J.N. (1982) Construction and use of SPP1v, a viral cloning vector for *Bacillus subtilis. Gene* **17**, 91-100

Hendrix, R.W., Roberts, J.W., Stahl, F.W. and Weisberg R.A. (eds) (1983) *Lambda II.* Cold Spring Harbor Laboratory, Cold Spring Harbor, NY

Henikoff, S. (1985) Undirectional digestion with exonuclease III creates targeted breakpoints for DNA sequencing. *Gene* **28**, 351-9

Hennecke, H., Gunther, I. and Binder, F. (1982) A novel cloning vector for the direct selection of recombinant DNA in *E. coli. Gene* **19**, 231-4

Hinnen, A. and Meyhack, B. (1982) Vectors for cloning in yeast. *Curr. Top. Microbiol.* **96**, 101-17

Hofstetter, H., Schambock, A., Van den Berg, J. and Weissmann, C. (1976) Specific excision of the inserted DNA segment from hybrid plasmids constructed by the poly(dA).poly(dT) method. *Biochim. Biophys. Acta* **454**, 587-91

Hohn, B. (1979) *In vitro* packaging of λ and cosmid DNA. In *Methods in enzymology*, vol. 68 (R. Wu, ed.), pp. 299-309. Academic Press, NY

Hollenberg, C.P. (1982) Cloning with 2μ DNA vectors and the expression of foreign genes in *Saccharomyces cerevisiae. Curr. Top. Microbiol. Immunol.* **96**, 119-44

Horinouchi, S. and Weisblum, B. (1982) Nucleotide sequence and functional map of pE194, a plasmid that specifies inducible resistance to macrolide, lincosamide and streptogramin type B antibiotics. *J. Bacteriol.* **150**, 804-14

Hu, N-T. and Messing, J. (1983) The making of strand-specific M13 probes. *Gene* **17**, 271-7

Hughes, K., Case, M.E., Geever, R., Vapnek, D. and Gills, N.H. (1983) Chimeric plasmid that replicates autonomously in both *Escherichia coli* and *Neurospora crassa. Proc. Natl Acad. Sci. USA* **80**, 1053-7

Hung, M-C. and Wensink, P.C. (1984) Different restriction enzyme-generated sticky ends can be joined *in vitro. Nucl. Acids Res.* **12**, 1863-74

Inman, R.B. and Schnos, M. (1970) Partial denaturation of thymine and 5-bromouracil containing DNA in alkali. *J. Mol. Biol.* **49**, 93-8

Ish-Horowicz, D. and Burke, J.F. (1981) Rapid and efficient cosmid vector cloning. *Nucl. Acids Res.* **9**, 2989-98

Jeffreys, A.J., Barrie, P.A., Harris, S., Fawcett, D.H., Nugent, Z.J. and Boyd, A.C. (1982) Isolation and sequence analysis of a hybrid delta-globin pseudogene from the brown lemur. *J. Mol. Biol.* **156**, 487-503

Jimenez, A. and Davies, J. (1980) Expression of a transposable antibiotic resistance element in *Saccharomyces. Nature (London)* **287**, 869-71

Kahn, M. and Helinski, D.R. (1978) Construction of a novel plasmid-phage hybrid: use of the hybrid to demonstrate ColE1 DNA replication *in vivo* in the absence of a ColE1-specified protein. *Proc. Natl Acad. Sci. USA* **75**, 2200-4

Kaplan, B.E. (1985) The automated synthesis of oligodeoxyribonucleotides. *Trends Biotechnol.* **3**, 253-7

Kaplan, D.A., Greenfield, L. and Collier, R.J. (1983) Chromogenic method to screen very large populations of bacteriophage plaques for the presence of specific antigen. *Methods in enzymology*, vol. 100 (R. Wu, L. Grossman and K. Moldave, eds), pp. 342-68. Academic Press, New York

Karn, J., Brenner, S., Barnett, L. and Cesareni, G. (1980) Novel bacteriophage λ cloning vector. *Proc. Natl Acad. Sci. USA* **77**, 5172-8

Kieser, T., Hopwood, D.A., Wright, H.H. and Thompson, D.A. (1982) pIJ101, a multicopy broad host range *Streptomyces* plasmid: functional analysis and development of DNA cloning vector. *Mol. Gen. Genet.* **185**, 223-8

Kinghorn, E.M., Barth, P.T. and Humphreys, G.O. (1981) Gm+ve/Gm−ve plasmid cloning vector with a wide Gm−ve host range. *Proc. Cetus conference on molecular cloning and gene regulation in bacilli*, pp. 83-90. Stanford University Press, Stanford, Mass.

Kingsman, A., Clarke, L., Mortimer, R.K. and Carbon, J. (1979) Replication in *Saccharomyces cerevisiae* of plasmid pBR313 carrying DNA from the yeast *trp* 1 region. *Gene* **7**, 141-52

Kiss, A., Posfai, G., Keller, C.C., Venetianer, P. and Roberts, R.J. (1985) Nucleotide sequence of the *Bsu* RI restriction-modification system. *Nucl. Acids Res.* **13**, 6403-24

Kohler, G. (1985) Derivation and diversification of monoclonal antibodies. *EMBO J.* **4**, 1359-66

Kornberg, A. (1980) *DNA replication.* W.H. Freeman, San Francisco, CA

Lam, S.T., Stahl, M.M., McMilin, K.D. and Stahl, F.W. (1974) Rec-mediated recombinational hot-spot activity in bacteriophage lambda: II. A mutation which causes hot-spot activity. *Genetics* **77**, 425-33

Langer, P.R., Waldrop, A.A. and Ward, D.C. (1981) Enzymatic synthesis of biotin-labelled polynucleotides: novel nucleic acid affinity probes. *Proc. Natl Acad. Sci.*

USA **78**, 6633-7

Lathe, R., Ballend, A., Kohli, V. and Lecocq, J.P. (1982) Fusion of restriction termini using synthetic adaptor oligonucleotides. *Gene* **20**, 187-95

Law, M-F., Lowy, D.R., Dvoretzky, I. and Howley, P.M. (1981) Mouse cells transformed by bovine papillomavirus contain only extrachromosomal viral DNA sequences. *Proc. Natl Acad. Sci. USA* **78**, 2727-31

Legerski, R.J. and Robberson, D.L. (1985) Analysis and optimization of recombinant DNA joining reactions. *J. Mol. Biol.* **181**, 297-312

Lobban, P.E. and Kaiser, A.D. (1973) Enzymatic end-to-end joining of DNA molecules. *J. Mol. Biol.* **78**, 453-71

Loenen, W.A.M. and Blattner, F.R. (1983) Lambda Charon vectors (Ch32, 33, 34 and 35) adapted for DNA cloning in recombination-deficient hosts. *Gene* **26**, 171-9

Macrina, F., Tobian, J.A., Jones, K.R., Evans, R.P. and Clewell, D.B. (1982) A cloning vector able to replicate in *Escherichia coli* and *Streptococcus sanguis. Gene* **19**, 345-53

Maloy, S.R. and Nunn, W.D. (1981) Selection for loss of tetracycline resistance by *Escherichia coli. J. Bacteriol.* **145**, 1110-12

Maniatis, T., Hardison, R.C., Lacy, E., Laver, J., O'Donnell, C., Quon, D., Sim, D.K. and Efstratiadi, A. (1978) The isolation of structural genes from libraries of eucaryotic DNA. *Cell* **15**, 687-701

Maniatis, T., Fritsch, F. and Sambrook, J. (1982) *Molecular cloning. A laboratory manual.* Cold Spring Harbor Laboratory, Cold Spring Harbor, NY

Marinus, M.G. (1973) Location of DNA methylation genes on the *Escherichia coli* K-12 genetic map. *Mol. Gen. Genet.* **127**, 47-55

Matthes, H.W.D., Zenke, W.M., Grundström, T., Staub, A., Winzerith, M. and Chambon, P. (1984) Simultaneous rapid chemical synthesis of over one hundred oligonucleotides on a microscale. *EMBO J.* **3**, 801-5

Maxam, A.M. and Gilbert, W. (1977) A new method for sequencing DNA. *Proc. Natl Acad. Sci. USA* **74**, 560-4

Meinkoth, J. and Wahl, G. (1984) Hybridization of nucleic acids immobilized on solid supports. *Anal. Biochem.* **138**, 267-84

Messing, J. and Vieira, J. (1982) A new pair of M13 vectors for selecting either DNA strand of double digest restriction fragments. *Gene* **19**, 269-76

Messing, J., Crea, R. and Seeburg, P.H. (1981) A system for shotgun DNA sequencing. *Nucl. Acids Res.* **9**, 309-21

Mulligan, R.C. and Berg, P. (1981) Factors governing the expression of a bacterial gene in mammalian cells. *Mol. Cell. Biol.* **1**, 449-59

Mulligan, R.C., Howard, B.H. and Berg, P. (1979) Synthesis of rabbit β-globin in cultured monkey kidney cells following infection with a SV40-β-globin recombinant genome. *Nature (London)* **277**, 108-14

Murray, N.E. (1983) Lambda vectors. In *Lambda II* (R.W. Hendrix, J.W. Roberts, F.W. Stahl and R.A. Weisberg, eds), pp. 677-84. Cold Spring Harbor Laboratory, Cold Spring Harbor, NY

Neidhardt, F.C., Vaughn, V., Phillips, T.A. and Bloch, P.L. (1983) Gene-protein index of *Escherichia coli* K-12. *Microbiol. Rev.*, **47**, 231-84

Nevins, J.R. (1983) The pathway of eukaryotic mRNA formation. *Ann. Rev. Biochem.* **52**, 441-66

Nick, H., Bowen, B., Ferl, R.J. and Gilbert, W. (1986) Detection of cytosine methylation in the maize alcohol dehydrogenase gene by genomic sequencing. *Nature (London)* **319**, 243-6

Norgard, M.V., Emigholz, K. and Monahan, J.J. (1979) Increased amplification of pBR322 plasmid deoxyribonucleic acid in *Escherichia coli* K-12 strains RR1 and

χ1776 grown in the presence of high concentrations of nucleoside. *J. Bacteriol.* **138**, 270-2
O'Connor, C.D. and Humphreys, G.O. (1982) Expression of the *Eco*RI restriction-modification system and the construction of positive-selection cloning vectors. *Gene* **20**, 219-29
O'Farrell, P.H., Kutler, E. and Nalcaniski, M. (1980) A restriction map of the bacteriophage T4 genome. *Mol. Gen. Genet.* **179**, 421-35
Okayama, H. and Berg, P. (1982) High-efficiency cloning of full length cDNA. *Mol. Cell. Biol.* **2**, 161-70
Olsen, R.H., De Busscher, G. and McCambie, W.R. (1982) Development of broad-host-range vectors and gene banks: self-cloning of the *Pseudomonas aeruginosa* PAO chromosome. *J. Bacteriol.* **150**, 60-9
Ozaki, L.S., Maeda, S., Shimada, K. and Takagi, Y. (1980) A novel ColE1::Tn*3* plasmid vector that allows direct selection of hybrid clones in *E. coli. Gene* **8**, 301-14
Parkinson, J.S. and Huskey, R.J. (1971) Deletion mutants of bacteriophage lambda. *J. Mol. Biol.* **56**, 369-84
Peacock, S.L., McIver, C.M. and Monohan, J.J. (1981) Transformation of *E. coli* using homopolymer-linked plasmid chimeras. *Biochim. Biophys. Acta* **655**, 243-50
Peden, K.W.C. (1983) Revised sequence of the tetracycline resistance gene of pBR322. *Gene* **22**, 227-80
Philippsen, P., Kramer, A. and Davis, R.W. (1978) Cloning of the yeast ribosomal DNA repeat unit in *Sst*I and *Hin*dIII lambda vectors using genetic and physical size selections. *J. Mol. Biol.* **123**, 371-86
Piggot, P.J. and Hoch, J.A. (1985) Revised genetic linkage map of *Bacillus subtilis. Micro. Rev.* **49**, 158-79
Pouwels, P.H., Enger-Valk, B.E. and Brammar, W.J. (1985) *Cloning vectors: a laboratory manual.* Elsevier, Amsterdam
Pratt, J., Boulnois, G., Darby, V., Orr, E., Wahle, E. and Holland, I.B. (1981) Identification of gene products programmed by restriction endonuclease DNA fragments using an *E. coli in vitro* system. *Nucl. Acids Res.* **9**, 4459-74
Primrose, S.B. and Ehrlich, S.D. (1981) Isolation of plasmid deletion mutants and study of their instability. *Plasmid* **6**, 193-201
Reed, K.C. and Mann, D.A. (1985) Rapid transfer of DNA from agarose gels to nylon membranes. *Nucl. Acids Res.* **13**, 7207-21
Reeve, J. (1978) Selective expression of transduced or cloned cDNA in minicells containing plasmid pKB280. *Nature (London)* **276**, 728-9
Richardson, M.A., Mabe, J.A., Beerman, N.E., Nakatsukasa, W.M. and Feyerman, J.T. (1982) Development of cloning vehicles from the *Streptomyces* plasmid pFJ103. *Gene* **20**, 451-7
Rigby, P.W.J. (1982) Expression of cloned genes in eukaryotic cells using vector systems derived from viral replicons. In *Genetic engineering 3* (R. Williamson, ed.), pp. 84-141. Academic Press, London and New York
Rigby, P.W.J., Dieckmann, M., Rhodes, C. and Berg, P. (1977). Labelling deoxyribonucleic acid to high specific activity *in vitro* by nick translation with DNA polymerase I. *J. Mol. Biol.* **113**, 237-51
Roberts, R.J. (1985) Restriction and modification enzymes and their recognition sequences. *Nucl. Acids Res.* **13** (supplement) r165-r200
Roberts, T.M., Swanberg, J.L., Poteete, A., Riedel, G. and Bachman, K. (1980) A plasmid cloning vehicle allowing a positive selection for inserted fragments. *Gene* **12**, 123-7
Rosenberg, S.M., Stahl, M.M., Kobayashi, I. and Stahl, F.W. (1985) Clean and

simple one-strain *in vitro* packaging of bacteriophage λ DNA. *ASM News* **51**, 386-8

Sambrook, J. and Grodzicker, T. (1980) Adenovirus-SV40 hybrids: a model system for expression of foreign sequences in an animal virus vector. In *Genetic engineering vol. 2: Principles and methods* (J.K. Setlow and A. Hollaender, eds), pp. 103-14. Plenum Press, New York

Sancar, A., Hack, A.M. and Rupp, W.D. (1979) Simple method for identification of plasmid-coded proteins. *J. Bacteriol.* **137**, 692-3

Sanger, F., Coulson, A.R., Barrell, B.G., Smith, A.J.H. and Roe, B.A. (1980) Cloning in single-stranded bacteriophage as an aid to rapid λ DNA sequencing. *J. Mol. Biol.* **143**, 161-78

Sanger, F., Coulson, A.R., Hong, G.R., Hill, D.F. and Petersen, G.B. (1982) Nucleotide sequence of bacteriophage λ DNA. *J. Mol. Biol.* **162**, 729-73

Saunders, C.W. and Guild, W.R. (1981) Pathway of plasmid transformation in pneumococcus: open circular and linear forms are active. *J. Bacteriol.* **146**, 517-26

Saunders, J.R., Docherty, A. and Humphreys, G.O. (1984) Transformation of bacteria with plasmid DNA. In *Methods in microbiology*, vol. 17 (P.M. Bennett and J. Grinsted, eds), pp. 61-95. Academic Press, London and New York

Scandella, D. and Arber, W. (1974) Phage infection in *Escherichia coli pel* mutants is restored by mutations in genes *V* or *H*. *Virology* **58**, 504-13

Schumann, W. (1979) Construction of an *Hpa*I and *Hin*dII plasmid vector allowing direct selection of transformants harbouring recombinant plasmids. *Mol. Gen. Genet.* **174**, 221-4

Seed, B., Parker, R.C. and Davidson, N. (1982) Representation of DNA sequences in recombinant DNA libraries prepared by restriction enzyme partial digestion. *Gene* **19**, 201-9

Silhavy, T.J. and Beckwith, J.R. (1985) Uses of *lac* fusions for the study of biological problems. *Microbiol. Rev.* **49**, 398-418

Smith, H.O. and Birnstiel, M.L. (1976) A simple method for DNA restriction site mapping. *Nucl. Acids Res.* **3**, 2387-98

Southern, E. (1975) Detection of specific sequences among DNA fragments separated by gel electrophoresis. *J. Mol. Biol.* **98**, 503-17

Southern, E. (1979) Gel electrophoresis of restriction fragments. In *Methods in enzymology*, vol. 68 (R. Wu, ed.), pp. 152-76. Academic Press, New York

Staden, R. (1986) The current status and portability of our sequence handling software. *Nucl. Acids Res.* **14**, 217-32

Stahl, U., Tudzynski, P., Kuck, U. and Esser, K. (1982) Replication and expression of a bacterial-mitochondrial hybrid plasmid in the fungus *Podospora anserina*. *Proc. Natl Acad. Sci. USA* **79**, 3641-5

Stinchcomb, D.T., Thomas, M., Kelly, J., Selker, E. and Davis, R.W. (1980) Eukaryotic DNA segments capable of autonomous replication in yeast. *Proc. Natl Acad. Sci. USA* **77**, 4559-63

Stohl, L.L. and Lambowitz, A.M. (1983) Construction of a shuttle vector for the filamentous fungus *Neurospora crassa*. *Proc. Natl Acad. Sci. USA* **80**, 1058-62

Stoker, N.G., Fairweather, N. and Spratt, B.G. (1982) Versatile low copy number plasmid vectors for cloning in *Escherichia coli*. *Gene* **18**, 329-35

Strathern, J.N., Jones, E.W. and Broach, J.R. (1982) *The molecular biology of the yeast Saccharomyces*. Cold Spring Harbor Laboratory, Cold Spring Harbor, NY

Stuber, D. and Bujard, H. (1981) Organization of transcriptional signals in plasmids pBR322 and pACYC184. *Proc. Natl Acad. Sci. USA* **78**, 167-71

Subramani, S., Mulligan, R. and Berg, P. (1981) Expression of the mouse dihydrofolate reductase complementary deoxyribonucleic acid in Simian Virus 40 vectors. *Mol. Cell. Biol.* **1**, 854-64

Sugino, A., Goodman, H.M., Heynecker, H.L., Boyer, H.W. and Cozarelli, N.R. (1977) Interaction of bacteriophage T4 RNA and DNA ligases in joining duplex DNA at base paired ends. *J. Biol. Chem.* **252**, 3987-94

Sun, Y.L., Xu, Y.Z. and Chambon, P. (1982) A simple and efficient method for the separation and detection of small DNA fragments by electrophoresis in formamide containing agarose gels and Southern blotting to DBM-paper. *Nucl. Acids Res.* **10**, 5753-63

Sutcliffe, J.G. (1979) The complete nucleotide sequence of *Escherichia coli* plasmid pBR322. *Cold Spring Harbor Symp. Quant. Biol.* **43**, 77-90

Tait, R.C. Close, T.J., Lundquist, R.C., Haguja, M., Rodriguez, R.L. and Kado, C.I. (1983) Construction and characterization of a versatile broad host range DNA cloning system for Gram-negative bacteria, *Biotechnology* **1**, 269-75

Thomas, C.M. (1984) Analysis of clones. In *Methods in microbiology*, vol. 17 (P.M. Bennett and J. Grinsted, eds), pp. 163-95. Academic Press, London and New York

Thompson, C.J., Ward, J.M. and Hopwood, D.A. (1982a) Cloning of antibiotic resistance and nutritional genes in streptomycetes. *J. Bacteriol.* **151**, 668-77

Thompson, C.J., Kieser, T., Ward, J.M. and Hopwood, D.A. (1982b) Physical analysis of antibiotic-resistance genes from *Streptomyces* and their use in vector construction. *Gene* **20**, 51-62

Thompson, J.A., Blakesley, R.W., Doran, K., Hough, C.J. and Wells, R.D. (1983) Purification of nucleic acids by RCP-5 ANALOG chromatography: peristaltic and gravity flow applications. In *Methods in enzymology*, vol. 100 (R. Wu, L. Grossman and K. Moldave, eds), pp. 368-99. Academic Press, London and New York

Thompson, R. (1982) Plasmid and phage M13 cloning vectors. In *Genetic engineering*, vol. 3 (R. Williamson, ed.), pp. 1-52. Academic Press, London and New York

Thudt, K., Schleifer, K.H. and Gotz, F. (1985) Cloning and expression of the α-amylase gene from *Bacillus stearothermophilus* in several staphylococcal species. *Gene* **37**, 163-9

Timmis, K. (1981) Gene manipulation *in vitro. Symp. Soc. Gen. Microbiol.* **31**, 49-109

Tooze, J. (1980) *Molecular biology of tumor viruses. Part 2. DNA tumor viruses*, 2nd edn. Cold Spring Harbor Laboratory, Cold Spring Harbor, NY

Tsao, S.G.S., Brunk, C.F. and Pearlman, R.E. (1983) Hybridization of nucleic acids directly in agarose gels. *Anal. Biochem.* **131**, 365-72

Tu, C.P.D. and Cohen, S.N. (1980) 3′-end labelling of DNA with [α-^{32}P] cordycepin-5′-triphosphate. *Gene* **10**, 177-83

Twigg, A. and Sherratt, D.J. (1980) Trans-complementable copy-number mutants of plasmid ColE1. *Nature (London)* **283**, 216-18

Vieira, J. and Messing, J. (1982) The pUC plasmids, a M13mp7 derived system for insertion mutagenesis and sequencing with synthetic universal primers. *Gene* **19**, 259-68

Webster, T.D. and Dickson, R.C. (1983) Direct selection of *Saccharomyces cerevisiae* resistant to the antibiotic G418 following transformation with a DNA vector carrying the kanamycin-resistance gene Tn*903*. *Gene* **26**, 243-52

Weinstock, G.M. (1984) Vectors for expressing open reading frame DNA in *Escherichia coli* using *lacZ* gene fusions. In *Genetic engineering*, vol. 6 (J.K. Setlow and A. Hollaender, eds), pp. 31-48. Plenum Press, New York.

Widera, G., Gautier, P., Lindenmainer, W. and Collins, J. (1978) The expression of tetracycline resistance after insertion of foreign DNA fragments between the

*Eco*RI and *Hin*dIII sites of the plasmid cloning vector pBR322. *Mol. Gen. Genet.* **163**, 301-5

Williams, J.G. (1981) The preparation and screening of a cDNA clone bank. In *Genetic engineering*, vol. 1 (R. Williamson, ed.), pp. 1-59. Academic Press, London and New York

Wood, D.O., Hollinger, M.F. and Tindoe, M.B. (1981) Versatile cloning vector for *Pseudomonas aeruginosa*. *J. Bacteriol.* **145**, 1448-1451

Yanisch-Perron, C., Vieira, J. and Messing, J. (1985) Improved M13 phage cloning vectors and host strains: nucleotide sequences of the M13mp18 and pUC19 vectors. *Gene* **33**, 103-19

Young, R.A. and Davis, R.W. (1983) Efficient isolation of genes by using antibody probes. *Proc. Natl Acad. Sci. USA* **80**, 1194-8

Zimmerman, S.B. and Pheiffer, B.H. (1983) Macromolecular crowding allows blunt-ended ligation by DNA ligases from rat liver or *Escherichia coli*. *Proc. Natl Acad. Sci. USA* **80**, 5852-6

Zissler, J., Singer, E. and Schaefer, F. (1971) The role of recombination in growth of bacteriophage lambda. I The gamma gene. In *The bacteriophage lambda* (A.D. Hershey, ed.), pp. 469-76. Cold Spring Harbor Laboratory, Cold Spring Harbor, NY

Zubay, G. (1973) *In vitro* synthesis of protein in microbial systems. *Ann. Rev. Genet.* **7**, 267-87

4

In Vivo and *In Vitro* Mutagenesis

Mutations are heritable changes in genetic material and can occur spontaneously or can be induced. They provide the ultimate source of genetic variability for microorganisms. Errors in DNA replication, misrepair of DNA damage, transposable genetic elements, environmental chemicals and radiation and the thermodynamic characteristics of nucleic acid molecules can all influence the occurrence of spontaneous (background) mutation. Induced mutations can be effected by various chemical or physical agents termed **mutagens**. Mutagens may act directly on the nucleic acid, altering bases *in situ* and indirectly, often through error-prone repair.

Mutations can be grouped broadly into a number of classes. Lesions affecting a single nucleotide pair are termed point mutations. These include **base substitutions** (transitions and transversions) and **frameshift** mutations involving the addition or deletion of one base pair. Genetic lesions extending over several base pairs are termed **multisite** mutations. Lesions affecting extensive tracts of the genome include inversions, duplications, insertions, translocations and deletions of genetic material. The type of genetic lesion induced is mutagen-dependent. Some mutagens induce all types of molecular change whereas others are more specific. The spectrum of lesions induced is termed the **mutagen specificity**.

Mutations can affect structural genes or regulatory regions of the genome, in turn altering the function (normally by either inactivation or modification of an existing property, rarely by creating a new property) or amount of the gene product. Mutants are fundamental tools for the genetic analysis of microorganisms. Furthermore, mutation techniques remain central to many industrial strain improvement programmes and have a role in programmes for the development of novel products (see Chapter 6). Until recently, procedures for mutant isolation relied largely upon random mutagenesis *in vivo* (often followed by a directed (rational) selection procedure (Chang and Elander, 1979) for specific phenotypes). However, the advent of recombinant DNA technology has enabled a greater degree of control over the mutation process. *In vitro* mutagenesis techniques permit the construction of mutations at predetermined sites, definition of the precise molecular nature of the lesion(s) involved and determination of the functional effect of the mutation. This chapter considers aspects of *in vivo* and *in vitro* mutagenesis.

4.1 THERMODYNAMIC PROPERTIES OF NUCLEIC ACID MOLECULES AND MUTATION

The inherent thermodynamic properties of nucleic acid molecules themselves can influence the occurrence of mutations. Deamination, depurination and depyrimidination of nucleotides can all occur under physiological conditions. Such reactions may affect the fidelity of DNA replication and mutagenesis. Physicochemical properties of the nucleosides, including tautomerism, ionisation and *syn-anti* rotation of bases, can effect mispairing and lead to mutation (Singer and Kuśmierek, 1982).

4.2 FIDELITY OF DNA REPLICATION

Accurate replication of DNA depends upon a multicomponent process involving sequential enzymatic mechanisms (see Kornberg,1980, 1982; Loeb and Kunkel, 1982). Perturbation of this process can result in decreased fidelity and mutagenesis. There are several ways in which the replication machinery serves to reduce errors in DNA replication:[1]

(i) *Discrimination against base misincorporation.* The precise nature of the discriminatory mechanism that operates at the level of base insertion is not clear. It is probably dictated by differences in free energy (ΔG) between correct and incorrect base pairing. This difference may be due, in part, to the chemical structure of the bases *per se*, but it may be enhanced in some way by DNA polymerase (and other proteins).
(ii) *Excision of incorrectly inserted bases (proof-reading).* When an incorrect base has been incorporated it may be removed by the **proof-reading (copy-editing)** function of DNA polymerase. DNA polymerases from bacteria and phages contain a $3' \rightarrow 5'$ exonuclease activity which proof-reads mistakes during polymerisation. Correction occurs immediately following base misincorporation. The relative efficiencies of exonuclease and associated polymerase probably influence the accuracy of replication. Alterations that reduce or eliminate the editing function of DNA polymerase may increase the frequency of

[1] Errors in DNA replication in wild-type strains of *E. coli* have been estimated as one error per 10^8 to 10^{11} base pairs replicated.

errors. Exonucleases (capable of removing mismatched bases) associated with DNA polymerases have also been found in fungi. However, the contribution of such exonuclease activity to the fidelity of DNA replication is not clear (Loeb and Kunkel, 1982).

(iii) *Post-synthetic correction of mismatched bases (mismatch repair).* Misincorporated bases that have escaped proof-reading may be removed by mismatch specific endonucleases. Excised bases can be resynthesised by gap-filling DNA polymerase activity (see Glickman, 1982). DNA methylation is suggested to have a role in enabling the repair system to discriminate newly synthesised DNA and parental template DNA. The *dam* (DNA adenine methylase) gene product of *E. coli* is responsible for this methylation (see section 3.2.2). Delayed methylation of nascent DNA permits recognition of the daughter strand, so that replication errors can be eliminated from that strand. Bacterial strains defective in mismatch repair (for example, *dam*$^-$ mutants) exhibit increased frequencies of mutation.

4.3 DNA REPAIR

Stability of DNA and its faithful replication are crucial to genetic integrity. Damage to DNA, if not repaired, may thus be lethal to cells. Several repair systems exist which can counteract this damage. Although some of them (error-free repair systems (see Table 4.1)) will correct damage induced by mutagens, others (error-prone systems) apparently enhance mutagenesis. Knowledge of DNA repair processes is, therefore, essential to an understanding of the mechanics of mutagenesis.

The activities of repair enzymes that are normally present in the cell are generally sufficient to correct most DNA damage that occurs. However, sudden exposure to a high dose of a mutagen may transiently saturate this repair capacity, in turn inducing many mutations.

Although repair systems appear to be found universally among micro-organisms, this section is confined to systems that operate in *E. coli*, and includes **photoreactivation**, **excision repair**, **post-replication recombinational repair** and **inducible error-prone repair**. (For a more detailed appraisal of various aspects of repair see, for example, Hanawalt *et al.*, 1979, 1981; Cairns *et al.*, 1981; Grossman, 1981; Hall and Mount, 1981; Haynes and Kunz, 1981; Lindahl, 1982; Friedberg, 1985; Hurst and Nasim, 1985; Walker, 1985.)

4.3.1 Photoreactivation

The photoreactivating enzyme, photolyase (*phr* gene product), can catalyse

Table 4.1: Mechanisms involved in repair (error-free) of damaged DNA

Mechanism	Comments
Direct reversal:	
(i) *Photoreactivation*	
N-T^T-N —photolyase→ N-T-T-N	Photolyase operates on pyrimidine dimers, for example thymine dimers (T^T), and requires visible light
(ii) *Transalkylation of* 0^6*-alkylguanine*	
for example:	
O^6-CH_3-G —methyl transferase→ G + CH_3-Enzyme	Methyl transferase is inducible ('adaptive response'). The enzyme transfers the methyl group from alkylated DNA to the enzyme resulting in enzyme inactivation. Unsubstituted guanine is left in the DNA
Damaged base removal:	
(i) *Glycosylation*	Damaged base (R) recognised and removed directly by glycosylase. Hydrolysis of N-glycosidic bond occurs, liberating free modified base and generating an apurinic (AP) (or apyrimidinic (APy)) site
N-R-N —glycosylase→ N-AP-N	
Incision mechanisms:	
Endonucleolytic cleavage	
(a) N↓AP↓N (AP endonuclease)	Incision (↓) of DNA containing an AP or APy site (arising from endogenous depurination/depyrimidination or as a consequence of DNA glycosylase activity): hydrolysis of phosphodiester bond at the 5′ side or at the 3′side of AP site is catalysed by an AP endonuclease
(b) N↓N-N-N-N-N-N-N-T^T-N-N-N↓N (uvr^+ endonuclease)	Incision (↓) of DNA containing bulky adducts, for example pyrimidine dimers, intra- or inter-strand crosslinks, by uvr^+ endonuclease: cleavage of phosphodiester bonds occurs in vicinity of damaged residues
Excision mechanisms:	
Exonucleolytic activity	Excision of DNA by excision exonucleases, DNA polymerase-associated exonucleases
Reinsertion mechanisms:	
DNA polymerisation	Reinsertion of nucleotides is controlled by DNA polymerases of which DNA polymerase I is a likely candidate
Ligation	DNA ligase completes the reinsertion process by the joining of adjacent bases

direct monomerisation of UV-induced pyrimidine dimers. The enzyme binds specifically to UV-irradiated DNA and in the presence of visible light effects breakage of the covalent bond attaching two pyrimidines in a cyclobutane ring.

4.3.2 Excision repair

The most important repair systems in *E. coli* depend upon the excision of an altered nucleotide residue or group. The mechanisms involved in excision repair are not entirely clear. Repair is probably initiated by an incision event involving cleavage of the phosphodiester bond(s) in the proximity of the primary DNA lesion. This may involve an apurinic or apyrimidinic endonuclease (where an apurinic or apyrimidinic site exists) or the UvrABC nuclease (where helix distortions, caused, for example, by pyrimidine dimers or polycyclic hydrocarbon adducts, occur). According to the model for *uvr*$^+$-dependent excision repair, proposed by Sancar and Rupp (1983), the Uvr nuclease incises the DNA strand containing the damaged DNA on both sides and in the vicinity of the lesion (rather than on one side and 5′ to the lesion, as previously suggested: see Hanawalt *et al.*, 1979). The resulting 12- to 13-nucleotide-long single-stranded DNA fragment formed is removed from the DNA producing a gap of that size, which is filled by DNA polymerase I activity and sealed using DNA ligase (Figure 4.1). The Uvr nuclease (excision nuclease or excinuclease) comprises the *uvrA*, *uvrB* and *uvrC* gene products. It has been suggested that the UvrA protein initially binds to the DNA. UvrB protein then associates with UvrA protein and the complex formed moves along the DNA to a site of damage. In the presence of UvrC protein, strand cleavage is catalysed. Displacement of the resultant UvrABC-incised DNA complex may involve proteins, such as the *uvrD* gene product (helicase II). The UvrD protein may mediate a coordinated excision-reinsertion reaction involving DNA polymerase I at the gap occupied by the damaged DNA fragment (Yeung *et al.*, 1983).

Such two-cut activity of the Uvr nuclease may be especially useful where both strands of a DNA molecule are damaged, as in the formation of cross-links. Scissions on both sides of a cross-link would provide a suitable substrate for processing by the recombination mode of repair (see section 4.3.3).

This excision repair process (referred to as **short-patch repair**), which requires the intact complementary strand as template for DNA synthesis, is essentially error-free. Such a process has an important influence over the extent of mutation, since lesions that are excised are much less likely to give rise to mutation than those that are not. Strains deficient in excision repair are more mutable, by various mutagens, than excision-proficient

Figure 4.1: Model for *uvr*$^+$-dependent excision repair (short patch repair). The Uvr nuclease hydrolyses the eighth phosphodiester bond 5′ and the fourth (or sometimes fifth) phosphodiester bond 3′ to the pyrimidine dimer. The resulting oligonucleotide (12 to 13 nucleotides long) carrying the damage is removed. The gap is filled by DNA polymerase I and sealed by DNA ligase (from Sancar and Rupp, 1983)

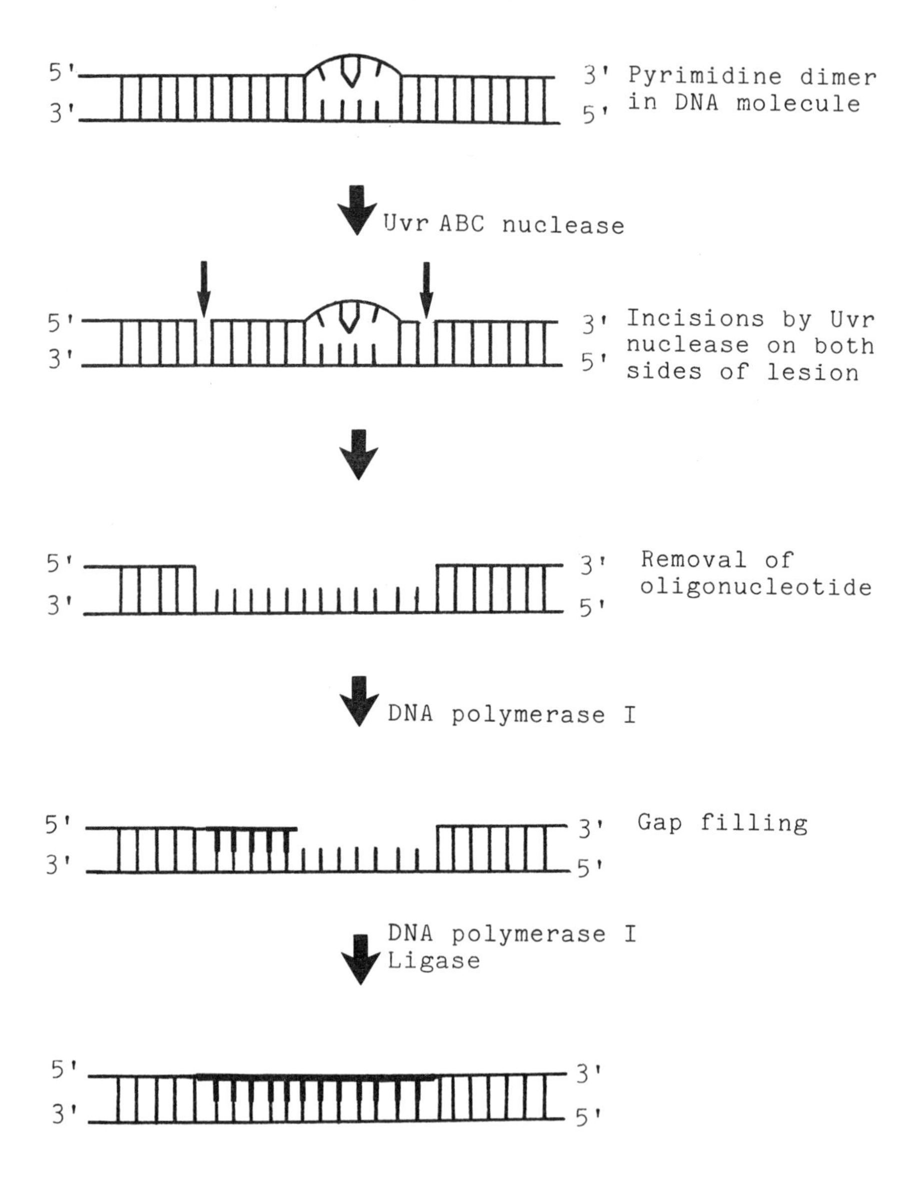

strains. *uvr*$^-$ or *polA*$^-$ mutants exhibit drastically reduced survival after UV treatment, but show high rates of UV-induced mutation among the survivors (Witkin, 1976). There is a second *uvr*$^+$-dependent excision repair process termed **long-patch repair**. This appears to be a form of SOS repair (Cooper, 1981, 1982).

4.3.3 Post-replication recombinational repair (daughter-strand gap repair)

Genetic lesions, such as pyrimidine dimers, can provide blocks to DNA replication. Gaps may be formed in the daughter strand during replication when strand elongation is blocked at such a lesion and subsequently resumes some distance beyond it (see Hall and Mount, 1981). DNA polymerase stalled at the lesion possibly reinitiates at the primer for the next Okazaki fragment downstream (at least for 'lagging' strand synthesis). These gaps may then be substrates for post-replication recombinational repair. Gaps may be filled with homologous DNA from the intact isopolar parental strand. Resultant discontinuities in the parental strand could then be eliminated by repair synthesis using undamaged regions of the newly synthesised daughter strand as template (Figure 4.2). This mechanism, which is apparently *rec*-dependent, effectively promotes recovery rather than repair of the DNA since the primary lesions remain.

4.3.4 Error-prone (mutagenic) repair

Treatments that damage DNA or inhibit DNA replication (for example UV irradiation, alkylating agents, cross-linking agents) activate *recA*-dependent SOS functions (Little and Mount, 1982; Walker, 1984), including enhanced DNA repair (Table 4.2). The precise mechanisms involved in error-prone repair are not known. However, it has been proposed that SOS functions enable replication to bypass lesions (for example pyrimidine dimers) that would otherwise inhibit the replication machinery but with a high frequency of error. The normal 'idling' reaction, which occurs when DNA polymerase stalls at a pyrimidine dimer, seems to be suppressed permitting DNA replication past the dimer.

A model for the regulation of the SOS response involves the *recA* and *lexA* genes (Figure 4.3). The LexA protein represses several unlinked genes (one of these is *recA*) that are believed to have roles in the SOS response. DNA damage or the inhibition of replication triggers reversible activation of a specific protease activity of RecA protein. This protease cleaves and inactivates LexA repressor leading to an increased expression of the *lexA* target genes (including *uvrA*, *uvrB*, *uvrD*, *umuC* and *umuD*)

Figure 4.2: Model for post-replication recombinational repair (daughter-strand gap repair). Lesions, such as pyrimidine dimers, provide blocks to DNA replication. DNA polymerase III will stall at a dimer. Replication apparently resumes at the next available primer (at least for lagging strand synthesis). The gap created in the daughter strand (a) may be filled by sister-strand exchange, in which homologous undamaged DNA is inserted into the gap. The gap formed in the donor molecule is filled in by polymerase I, using undamaged regions of the newly synthesised daughter strand (b) as template and sealed with DNA ligase

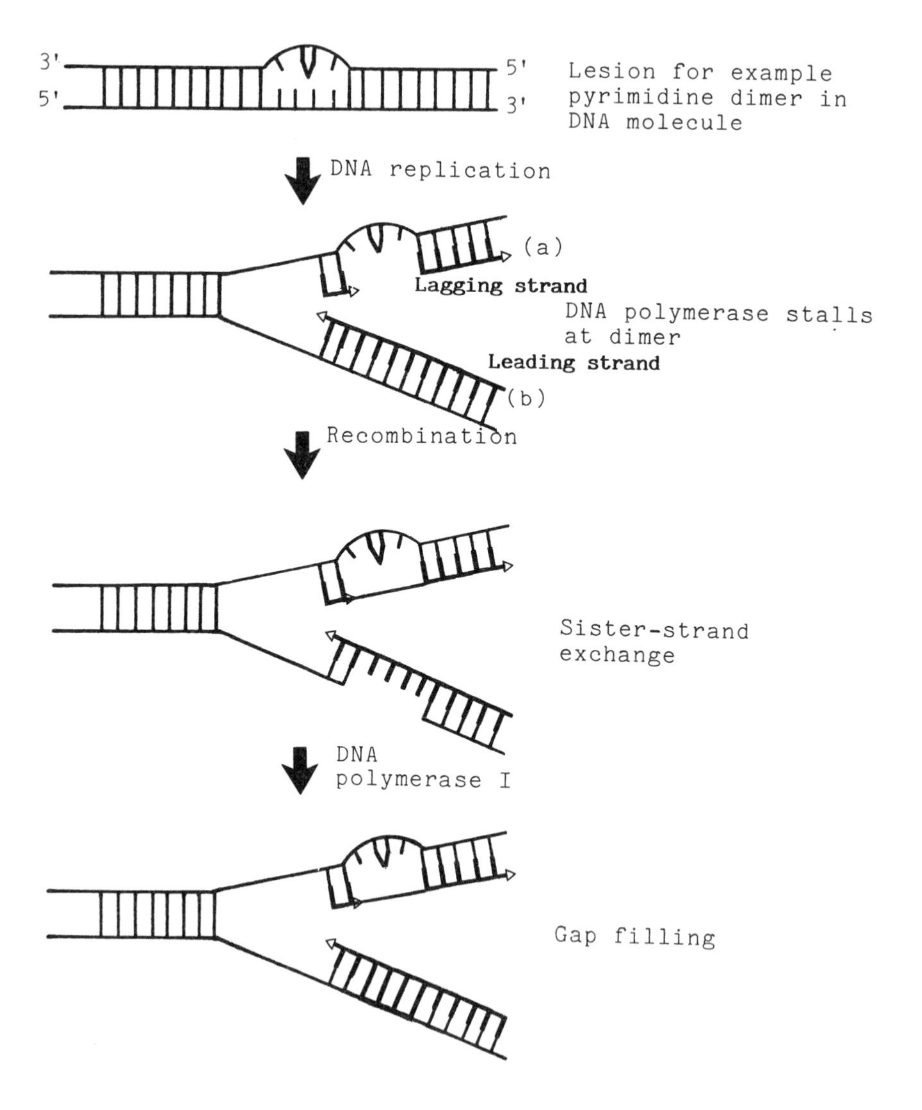

Table 4.2: Some SOS responses

Enhanced DNA repair (post-replication and excision)
Enhanced mutagenesis
Induction of prophage
Inhibition of cell division
Induction of stable DNA replication
Alleviation of host-controlled restriction
Inhibition of respiration

and in turn an enhanced capacity for DNA repair and mutagenesis (Little and Mount, 1982; Walker, 1984). The proteolytic activity of RecA protein might also modify proteins that affect the accuracy of DNA replication (see Hall and Mount, 1981). Alternatively *umuD* and *umuC* gene products may have a role in modifying such proteins (Walker, 1984). Modification of an existing DNA polymerase III, such that the activity of its editing 3′→5′ exonuclease is relaxed; or the formation of a new polymerase (possibly involving *umuC* and *umuD* gene products) with an ability to act on damaged templates could facilitate replication past dimers (transdimer replication) with a high incidence of base misincorporation. Such lack of fidelity of replication may extend beyond the region opposite the lesion, leading to mutation at sites other than the primary lesion (see Figure 4.4).

Different mutagens apparently act via different repair mechanisms. Mutagenic efficiency may be governed, at least in part, by selection of conditions for expression of the specific repair pathway required (see section 4.6.3). Various mutagens, including UV irradiation, ionising radiation and many alkylating agents, appear to rely upon SOS functions. However, ethylmethane sulphonate (EMS) and *N*-methyl-*N*′-nitro-*N*-nitrosoguanidine (N-MNNG) can act independently of *recA*, presumably via non-SOS repair functions (see section 4.4.1).

4.4 CHEMICAL MUTAGENS

The precise mechanism by which a given chemical agent exerts its mutagenic effect *in vivo* is not completely clear. Mutagens can react in various ways with DNA. However, such reactions are not necessarily responsible for the observed mutations. Chemical mutagens may be base modifiers, base analogues or agents that bind to DNA and intercalate between the bases.

Figure 4.3: Model for the regulation of the SOS response. In an uninduced cell the product of *lexA* gene represses a number of unlinked genes, such as *recA*, *uvrA*, *umuDC* and *lexA*, by binding to similar operator sequences in front of each gene. (The amount of RecA protein synthesised in the repressed state is, however, sufficient for homologous recombination.) For induction an SOS-inducing signal reversibly activates a specific protease activity of RecA protein. Proteolytic cleavage of LexA protein occurs and the pools of LexA molecules are reduced. Genes whose operators bind LexA are expressed. Loss of the inducing signal (by, for example, DNA repair) effects the return of RecA to its proteolytically inactive state. The pools of LexA molecules increase, in turn resulting in repression of SOS genes by LexA and a return to the uninduced state. ◇, LexA molecule; △▽, cleaved LexA molecule; ○, RecA molecule (proteolytically inactive); ●, RecA molecule (proteolytically active)

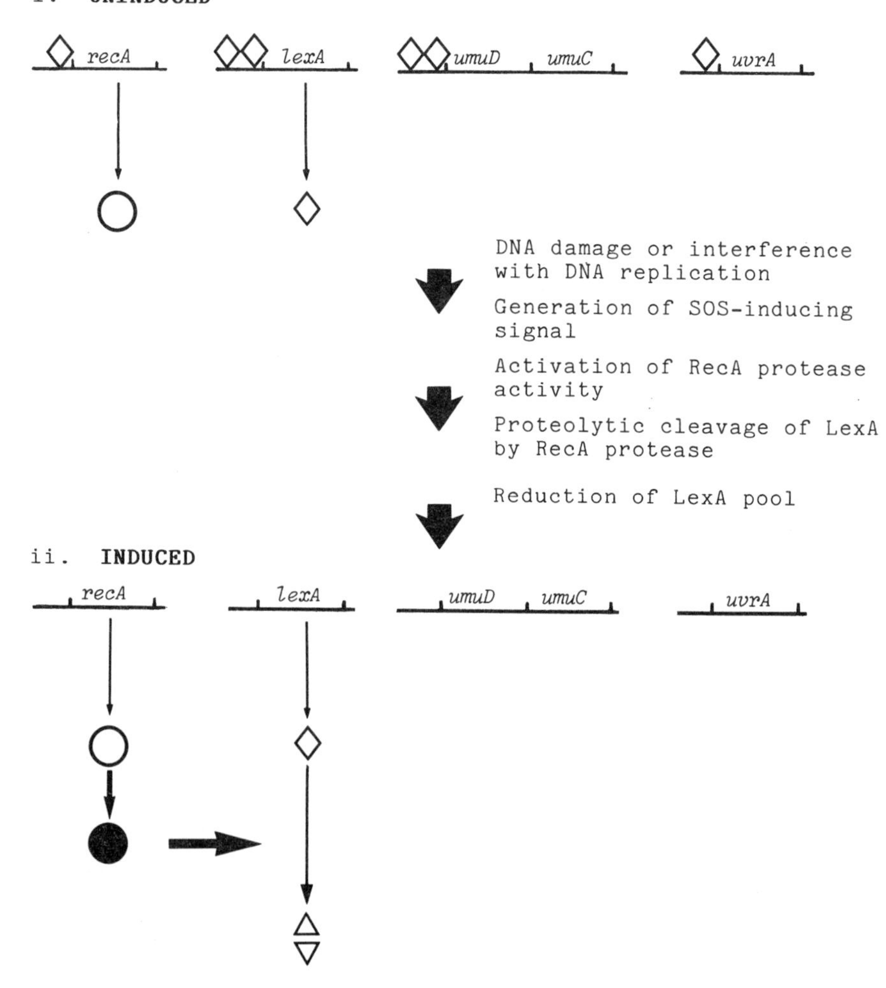

Figure 4.4: Model for error-prone (post-replication) repair. Damaged DNA induces SOS functions that allow DNA replication past dimers. The replication system is error-prone and leads to incorporation of a higher than normal number of mismatched bases (◆). Incorrect bases may be added at sites other than those opposite the lesions. Some of these bases may be removed by the mismatch repair system

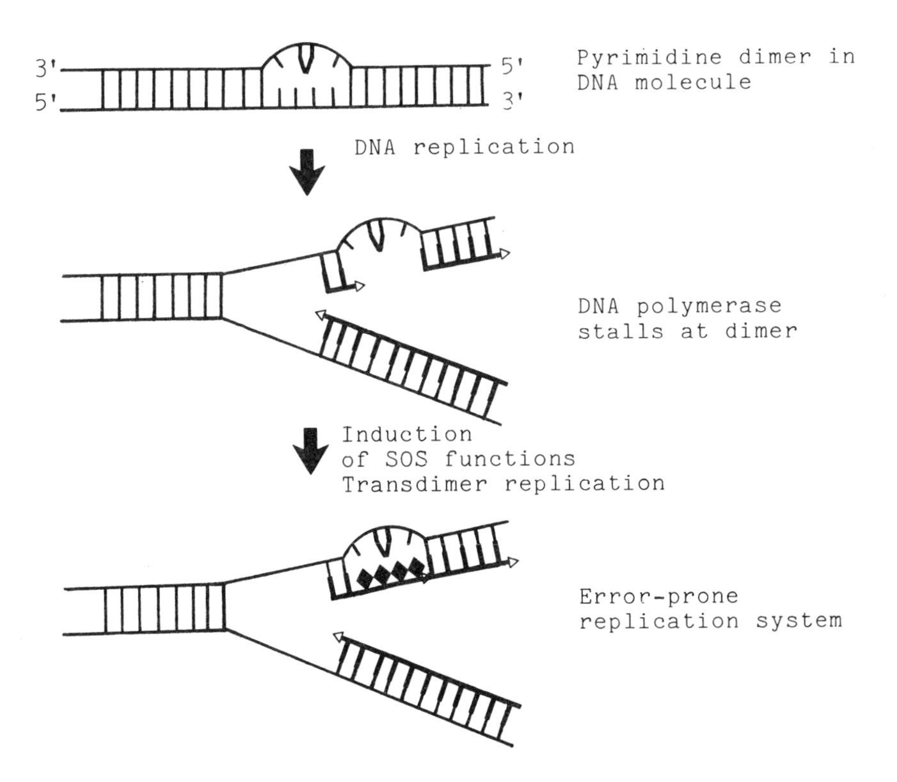

4.4.1 Chemical modification of nucleotides

Various chemical mutagens (Table 4.3) modify the nucleotides of DNA. Details of the chemical reactions involved have been discussed by Singer and Kuśmierek (1982).

(a) Nonalkylating agents

Agents such as hydroxylamine, hydrazine, nitrous acid and bisulphite can effect base alterations and change base pairing. For example, hydroxylamine reacts with pyrimidines (preferentially C) and induces transitions (AT→GC). In addition, it can cause crosslinking of complementary strands.

Table 4.3: Examples of chemical mutagens

1. *Nonalkylating agents*	
Hydroxylamine	$H_2N{-}O{-}H$
Hydrazine	$H_2N{-}NH_2$
Nitrous acid	$O{=}N{-}O{-}H$
Bisulphite	$O{=}S(O^-){-}O{-}H$
2. *Alkylating agents*	
Methylmethane sulphonate (MMS)	$CH_3{-}S({=}O)_2{-}O{-}CH_3$
Ethylmethane sulphonate (EMS)	$CH_3{-}S({=}O)_2{-}O{-}CH_2{-}CH_3$
N-ethyl-*N*-nitrosourea (ENU)	$O{=}N{-}N(CH_2{-}CH_3){-}C({=}O){-}NH_2$
N-methyl-*N*′-nitro-*N*-nitrosoguanidine (N-MNNG)	$O{=}N{-}N(CH_3){-}C({=}NH){-}N(H)(NO_2)$
S-mustard (mustard gas)	$CH_2{-}CH_2Cl$ on ${}^+S$, ring ${}^+S{-}CH_2{-}CH_2$, Cl^-
N-mustard (NM)	$CH_3{-}{}^+N(CH_2{-}CH_2{-}Cl)$, ring ${}^+N{-}CH_2{-}CH_2$, Cl^-

3. *DNA base analogues*

5-Bromouracil (5-BU)
(thymine analogue)

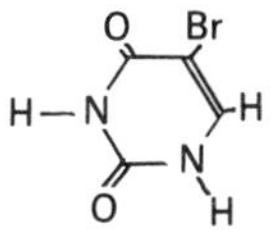

keto form
(pairs with
adenine)

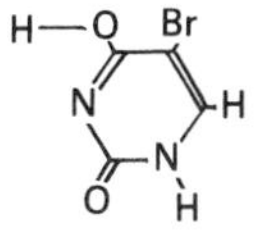

rare enol form
(pairs with
guanine)

2-Aminopurine
(adenine analogue)

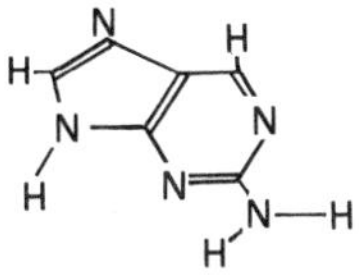

(pairs with both thymine
and cytosine)

N^4-Hydroxycytosine
(thymine analogue)

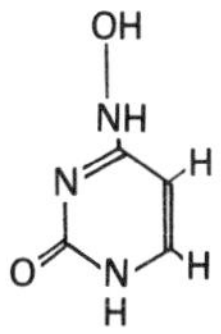

amino form
(pairs with
guanine)

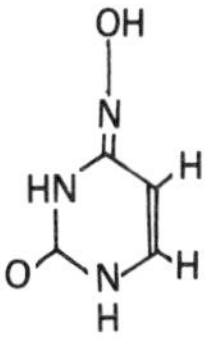

imino form
(pairs with
adenine)

4. *DNA binding agents*

Acridine orange

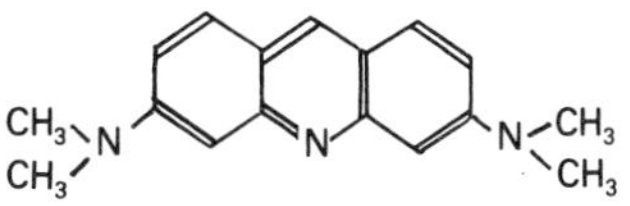

Proflavine

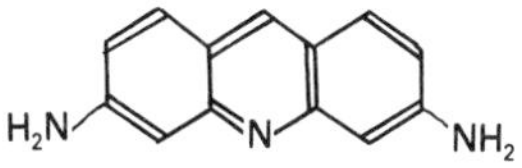

(b) Alkylating agents

Alkylating agents, which constitute the largest and one of the most potent groups of chemical mutagens, have been widely used in industrial programmes. Such agents apparently exert their mutagenic effects through mispairing and misrepair. Most alkylating agents act via SOS functions (see section 4.3.4); however, some, for example N-MNNG and EMS, can act independently of the SOS system. The existence of an N-MNNG immutable strain that can be mutagenised by EMS implicates different mechanisms for the action of these two agents (Cerdá-Olmedo and Ruiz-Vázquez, 1979).

Monofunctional alkylating agents have one alkyl group to donate, whereas bifunctional agents have two reactive groups. Alkylation can occur at various positions in nucleotides. Alkyl sulphates (mono) react almost entirely with nitrogen, and *N*-nitroso compounds (mono) react primarily with oxygen. S-mustard resembles the alkyl sulphates in its reaction, although intra- and inter-strand crosslinking can also ocur. Major alkylation products (3-methyladenine and 7-methylguanine) in DNA treated with simple methylating agents, such as N-MNNG and MMS, can be removed by DNA glycosylases (Lindahl, 1982, and see Table 4.1). However, O^6-methylguanine, which may also be formed, is not normally removed efficiently. Formation of O^6-methylguanine is the most likely cause of mispairing after N-MNNG mutagenesis in *E. coli.* O^6-methylguanine can be effectively removed using a *recA*-independent repair pathway (involving a transmethylase enzyme, see Table 4.1) that is induced if cells are exposed to sublethal concentrations of the alkylating agent. Such inducible repair (the adaptive response) of DNA protects against the mutagenicity of various alkylating agents (Cairns *et al.*, 1981).

Both EMS and N-MNNG preferentially induce GC$\rightarrow$AT transitions. EMS-induced mutations are distributed more or less randomly over the genome (Guerola and Cerdá-Olmedo, 1975). In contrast, N-MNNG acts preferentially at the replicating fork, inducing clusters of mutations in closely linked genes, a phenomenon referred to as (localised) **comutation.** An explanation for such mutagenic events is that N-MNNG can interact with the replication machinery and induce error-prone replication over limited regions of the genome (Cerdá-Olmedo and Ruiz-Vázquez, 1979). Comutation increases the probability that cells carrying mutations in a specific gene will also have mutations in neighbouring genes. However, distant genes that are replicated simultaneously by different replication forks during bidirectional replication may also be affected. Comutation can be exploited in a number of ways, for example:

(i) to isolate mutants that have no readily identifiable phenotype, by selecting for mutants with lesions at nearby loci;
(ii) to alter regulatory elements that are closely linked to structural genes;

(iii) to detect changes in modes of replication;
(iv) to compare genomes of different organisms.

A disadvantage of comutation is that even with low concentrations of mutagen it is difficult to avoid introducing clusters of mutations as opposed to single mutations.

4.4.2 Base analogues

Base analogues (see Table 4.3) can be introduced into DNA in place of their naturally occurring counterparts. Mutagenesis is generally ascribed to mispairing, effected by tautomerisation of the analogue, during DNA replication. Analogues, such as 5-bromouracil (5-BU), induce SOS repair functions (section 4.3.4), which can override the highly specific base pairing that normally occurs during DNA replication. 5-BU and 2-aminopurine (2-AP) both induce AT→GC transitions.

4.4.3 DNA intercalating agents

Certain agents (such as acridine orange and proflavine, Table 4.3) are capable of binding to DNA and causing frameshift mutations, presumably by intercalating between the stacked nucleotide bases (Streisinger *et al.*, 1966).

4.5 PHYSICAL MUTAGENS

UV radiation and other forms of radiation (including X-rays, gamma-rays and neutrons) can induce mutations. Mutagenicity of these agents does not depend directly upon the type of DNA damage caused, but rather upon specific repair functions (SOS functions, section 4.3.4) that act on the damage. This section considers UV-induced mutagenesis in *E. coli*, where repair of UV-damaged DNA is best understood.

Far UV irradiation (254 nm) is normally used for inducing genetic lesions. The main photoproduct induced is the intrastrand pyrimidine dimer (principally the thymine dimer) (Beukers and Berends, 1961). There are two major mechanisms for the removal of these dimers, excision repair (section 4.3.2) and photoreactivation (section 4.3.1). UV-induced mutagenesis apparently results from error-prone repair (SOS repair) of dimers that are not eliminated. Tandem base-pair changes appear to account for a considerable proportion of UV-induced mutations. Presumably such changes occur during transdimer replication (section 4.3.4). Other UV-induced mutagenic events include deletions and frameshifts (Hall and Mount, 1981).

recA$^-$ and *umuC*$^-$ mutations prevent UV-induced mutagenesis, whereas *uvr*$^-$ or *polA*$^-$ mutants exhibit high rates of mutation at low UV doses (Witkin, 1976; Bagg *et al.*, 1981). Organisms that are non-mutable by UV presumably lack the necessary repair functions. In such cases other mutagens that do not rely upon these specific repair functions could be employed.

4.6 MUTAGEN USAGE

In vivo mutagenesis techniques involve treatment of strains with specific mutagens and subsequent screening for desired mutants. There are certain considerations for the use of mutagens.

4.6.1 Choice of mutagen

A number of factors should be considered when choosing between mutagens for a given mutation programme:

(i) *Ease and safety of handling.* Physical mutagens, such as UV, are normally easier to handle than chemical agents. UV can be contained and directed solely at the experimental material. Chemical mutagens require more rigorous safety precautions for their containment (see Kilbey *et al.*, 1977, for procedures for handling mutagens).

(ii) *Mutagen specificity.* Although the molecular basis of mutagen specificity is not completely clear, the various repair systems implicated in mutagenesis presumably influence the spectrum of molecular changes induced. Endogenous repair activities of an organism thus affect not only its mutability by a given mutagen, but also the types of mutations induced. In many cases the precise nature of the molecular change(s) required is not known in advance. It is thus crucial that as wide a spectrum of mutant types as possible is produced. Accordingly, a useful strategy when using mutagens is to alternate between various agents that operate through different repair systems.

4.6.2 Mutagen dose

Mutagenic efficiency (the ratio of mutational to lethal events) depends upon interactions between the mutagen and the specific strain. A dose-response curve for survival and mutagenesis should thus be constructed for each strain used, in order to maximise the probability of producing the desired mutant. The optimum dose of mutagen, in terms of mutants per survivor, should be used. This varies depending upon the type of mutant required (for discussion see Rowlands, 1983). The frequency of mutants (per survivor) at the optimum dose of mutagen may be enhanced either by the use of hypermutable strains (Saunders *et al.*, 1982) or by specific manipulation of the environment in order to inhibit excision repair and/or increase error-prone repair.

4.6.3 Environmental factors

The environment prior to, during and immediately after mutagenesis can influence the frequency and types of mutants. Synergistic effects sometimes occur if material is pretreated with one mutagen before another (see, for example, Talmud, 1977). Such synergism could be due to saturation of the error-free repair system or to induction of enhanced error-prone repair by the first mutagen. Factors such as temperature and nature of the suspending medium during treatment with the mutagen can affect the type of mutational changes. Furthermore, growth on a rich medium after mutagenesis appears to increase the yield of mutants and affects specificity. Presumably, inducible error-prone repair is enhanced under such conditions (Clarke, 1975). Addition of inhibitors of excision repair can also increase the frequency of mutants, but at the expense of survivability (Auerbach, 1976).

4.7 TRANSPOSITION MUTAGENESIS

The ability of transposable elements to insert at various sites within a genome can result in the formation of insertion mutations.[2] Insertion of an element within a gene disrupts the linear continuity of that gene and normally leads to loss of gene function. Insertion mutations thus provide a

---

[2] Insertion mutations are given designations that describe the material inserted. For example, *hisC8691*::Tn*10* designates a specific insertion of Tn*10* within the *hisC* gene. Insertions that are not apparently mutations of a specific gene are given designations that describe their map position. Such designations start with *z*, and the other letters represent map positions in minutes. The second letter designates 10-minute map segments (for example *a* = 0-10; *b* = 10-20 and so on); the third letter designates minutes within any 10-minute sector. Thus a Tn*10* insertion at 44 map minutes near *his* is designated *zee*::Tn*10* (see Campbell *et al.*, 1977; Davis *et al.*, 1980).

means of eliminating gene function without deleting genetic material. When the mutations occur within an operon they are often strongly polar (see section 2.1.3). Each insertion mutant normally carries a single genetic lesion. The probability of an insertion mutation occurring in a given gene is a function of the size of the gene and of the presence of appropriate integration sites for the transposing element. Insertion mutations are generally stable. Revertants can, however, be isolated following precise excision of the transposable element.

In addition to simple insertions, transposable elements can mediate deletions of genetic material (and other rearrangements) (see, for example, Kleckner *et al.*, 1979). Bacteriophage Mu and transposons (particularly resistance transposons and Tn*1000* (γδ)) are widely used in **transposition mutagenesis** (see Kleckner *et al.*, 1977). Such mutagens can often be used where conventional mutagens (physical or chemical agents) prove impracticable.

4.7.1 Use of transposons in mutagenesis

Antibiotic-resistance transposons (and others with readily identifiable phenotypes) are particularly useful in mutagenesis because they provide a means of positive selection. Insertion within a gene not only inactivates that gene, but also provides the mutant with an antibiotic-resistance phenotype. The mutation is, of necessity, linked to the resistance determinant. The choice of Tn used for selection will be determined by the efficiency of its expression in the particular host. For example, the ampicillin-resistance determinant of the Gram-negative transposons Tn*1*/Tn*3* is poorly expressed in some bacterial species (due to the lack of correct processing of the β-lactamase enzyme). There is little information about the functioning of Gram-negative Tns in Gram-positive organisms, although some drug-resistance determinants of Tns do not express in *Bacillus subtilis.* The Gram-positive Tn, Tn*917*, from *Streptococcus faecalis* is, however, functional in *B. subtilis* (Youngman *et al.*, 1983). Moreover, the aminoglycoside-resistance determinants of Tn*5* and Tn*903* (*601*), conferring resistance to the antibiotic G418, are expressed in a wide range of bacteria and in eukaryotes (Jimenez and Davies, 1980; Southern and Berg, 1982).

Induction of mutations is limited by insertional specificity of certain Tns (see section 2.1.1). Some transposons, for example Tn*1* and Tn*3*, insert preferentially into plasmid replicons; whereas others, for example Tn*5*, apparently transpose equally readily into plasmid or chromosome (Sherratt, 1981). Tn*7* is unusual in that it transposes to many sites on plasmids, but to a specific region of the *E. coli* chromosome (between *dnaA* and *ilv*) (Lichtenstein and Brenner, 1982).

Provision of an antibiotic-resistance determinant (or other readily scorable marker) associated with the insertion mutation facilitates genetic manipulation in a number of ways:

(i) Mutants with no easily scorable phenotype can be isolated by selecting for the Tn phenotype.
(ii) Mutations can be generated in genes near to the gene(s) of interest (techniques for isolating these insertion mutations are described below).
(iii) Mutations can be conveniently mapped both physically (by restriction endonuclease cleavage patterns: see section 3.8.1) and genetically. (Insertion mutations generally behave as point mutations in deletion mapping.)
(iv) Provision of genetic markers via Tn insertion either in or near to the gene of interest facilitates movement of the gene to new genetic backgrounds (by *in vivo* or *in vitro* manipulation). This can be very useful in strain development.
(v) Revertants can be recognised due to loss of the Tn phenotype and can be distinguished from pseudorevertants (arising as a result of secondary mutational events at suppressor or bypass loci), which maintain the Tn phenotype. Revertibility can be exploited in strain construction: an insertion mutation near to a gene of interest can be used to transfer that gene to a new host. The insertion mutation can then be eliminated by precise excision of Tn. The new strain will thus be altered only for the gene of interest.
(vi) Tn insertion can be used in localised mutagenesis (see also section 2.4.3) when the element is inserted very near to a given gene. Mutagenesis of transducing phages carrying such insertions, and selection for recipients that exhibit the Tn phenotype after transduction, enable mutations affecting a linked gene to be recovered.

(a) Strategies for introducing Tns into target DNA

Various strategies are available for generating insertion mutations. One approach involves introduction of target DNA (for example, plasmid or phage DNA) into cells carrying the transposon. Following transposition, the target DNA with inserted Tn can be obtained by a number of procedures. Where the target DNA is contained in a conjugative plasmid, conjugal transfer of the plasmid to appropriate recipients and selection for the Tn phenotype enable the insertion mutation to be obtained. In other cases, plasmid DNA may be prepared from the host cells and used to transform suitable recipients selecting for the Tn phenotype. Tn-containing phages may be recovered from phage lysates by transducing suitable recipients and selecting for the Tn phenotype.

Tns can also be introduced into target DNA during mobilisation of a nonconjugative plasmid, where the transfer process involves Tn-mediated cointegrate formation (see section 2.3.1.c). In this case the target DNA may be DNA of either the conjugative or nonconjugative plasmid, depending upon which plasmid is carrying a Tn. Transposition of the element from one plasmid to the other results in the formation of a cointegrate. Following transfer to recipient cells, the cointegrate can be resolved to generate separate conjugative and nonconjugative plasmids each containing a Tn insertion (Figure 4.5). This method is particularly convenient for introducing insertion mutations (commonly utilising Tn *1000* (γδ)) into DNA sequences cloned on small multicopy plasmid vectors (see section 3.5.1). Another approach to the introduction of a Tn into target DNA involves the use of cells that carry the target DNA. The Tn is introduced on a vehicle under conditions that prevent its maintenance in the recipient, and

Figure 4.5: Transposition mutagenesis by mobilisation. The donor cell harbours a conjugative plasmid (Tn donor) and nonconjugative cloning vector. A cointegrate is formed during intermolecular transposition of the Tn. The cointegrate may be transferred to the recipient by conjugation. Resolution of the cointegrate generates the conjugative plasmid and cloning vector each carrying a copy of the Tn. In the population the Tn will be inserted at various sites on the vector and cloned DNA. ⁓,Tn

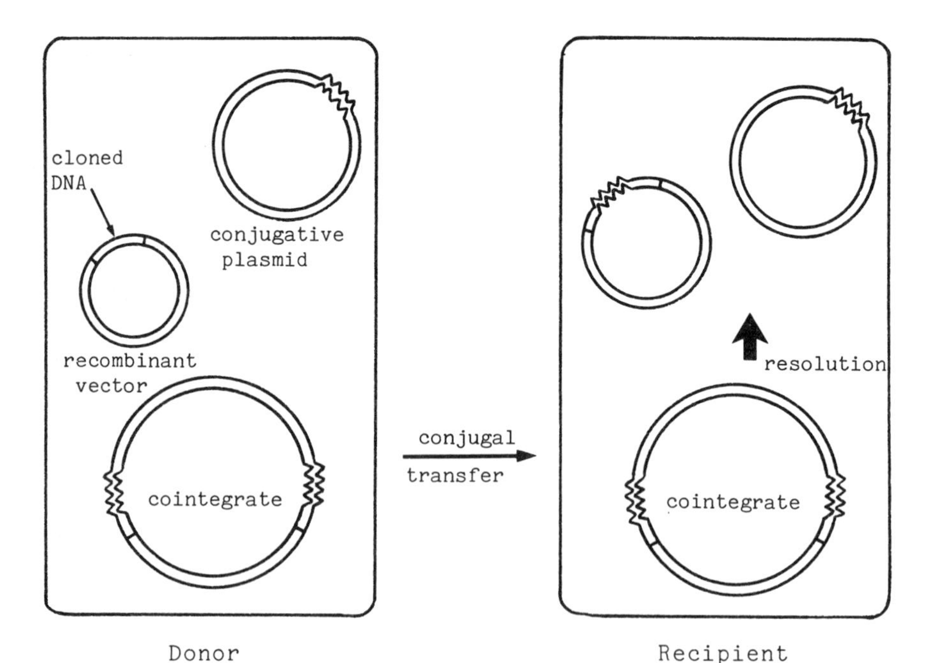

selection is imposed for the Tn phenotype. Recipients exhibiting a stable Tn phenotype will be those in which the Tn has transposed into a functional resident replicon (for example the chromosome). Vehicles based on plasmids or temperate phages are normally used for this purpose.

Plasmid vehicles. There are a number of procedures using plasmids for introducing the Tn into cells and selecting against maintenance of the plasmid, for example:

(i) Plasmid derivatives (**suicide plasmids**) that are incapable of replicating in the recipient may be used. Such plasmids include those carrying a temperature-sensitive mutation affecting replication (for example RP1*ts* (Robinson *et al.*, 1980). At the nonpermissive (restrictive) temperature, replication of the plasmid is blocked (unless the plasmid can integrate into the chromosome) but the transposon can be maintained if it has inserted into the chromosome, or another functional replicon. A second example involves plasmids carrying Mu insertions (for example RP4::Mu::Tn*7* (Van Vliet *et al.*, 1978) and pPH1JI::Mu::Tn*5* (Beringer *et al.*, 1978). The establishment of such plasmids in certain Gram-negative bacteria is reduced relative to that of the parent plasmids. (This is possibly due to Mu functions *per se.* Alternatively/additionally the phage may provide target sites, lacking on the parent plasmid itself, for the host restriction enzymes.) These hybrid plasmids thus provide a means of introducing transposons into particular recipients, without inheritance of the donor plasmid. The use of IncP plasmids in this way enables Tns to be introduced into various Gram-negative bacteria. (It should be noted that by exploiting Mu in this way it is not always clear whether mutagenesis of the recipient is associated with the incoming Tn or Mu or both.)

(ii) Incompatibility can be used to prevent establishment of a donor plasmid that carried a Tn into a recipient cell already harbouring a plasmid of the same Inc group. The Tn can be maintained by transposition to the host chromosome (see, for example, Foster *et al.*, 1975).

(iii) By using specific hosts that are unable to support replication of the vector plasmid and selecting for the Tn phenotype, cells carrying the Tn inserted into recipient replicons can be recovered. For example, *polAts* strains of *E. coli* have been used as recipients for a ColE1-derived vector plasmid (see Kleckner *et al.*, 1977). At the nonpermissive temperature, replication of ColE1 is prevented (since the *polA* gene product, DNA polymerase I, is required for ColE1 replication). The Tn can survive by integrating into a resident replicon. Weiss and Falkow (1983) have used as a vector the plasmid pAS8Rep-1 which contains a ColE1 replicon and RP4 conjugation genes. This system provides a means of introducing Tns into a wide range of bacteria (since RP4 will transfer to various Gram-negative bacteria), while preventing establishment of the plasmid (since ColE1 plasmids replicate in only a limited number of species).

Phage vehicles. Specially constructed defective transducing phages that are unable to lysogenise the host (and are often also reduced in their ability to direct phage growth) can be used as carriers of Tns. If such a phage carrying a Tn is used to infect sensitive host cells and selection imposed for the Tn phenotype after infection, survivors will be those in which the Tn has transposed to the host genome.

Derivatives of phage P22 with mutations that block phage replication (12^-), lysis (13^-), repression (*c2ts*) and integration (int^-) have been used in Tn mutagenesis of *Salmonella* (Davis *et al.*, 1980). Insertion of a Tn into the phage genome can make the composite P22::Tn molecule too large to be packaged *in toto.* The phage particles will contain fragments of the composite genome packaged by the headful mechanism.

Deletion variants of phage λ can be used in *E. coli* mutagenesis. The λ packaging system is site-specific, such that only DNA of length 37 to 51 kb between *cos* sites is packaged. It is therefore normally necessary to delete nonessential DNA from the phage genome in order to accommodate the Tn, without exceeding the maximum packageable length. Deletants such as λ*b221* that lack the attachment site (*att*λ*P*) and the *int* and *xis* genes have been used. Vehicles carrying mutations that prevent the maintenance of lysogeny and phage replication, in addition to the *b221* deletion, may also be used (see Kleckner *et al.*, 1977).

Superinfection immunity can be used to select against the phage vehicle while permitting entry of the Tn. Replication of a phage can be blocked where the recipient cell is a lysogen already carrying an immune prophage.

(b) Isolation of insertion mutations near to specific genes

Generalised transduction (see section 2.4.1) can be exploited to insert a Tn near to a given gene. Random insertions, using, for example, an antibiotic resistance Tn, are generated in a collection of clones (approximately 2000). The clones are pooled and serve as host for the generalised transducing phage. The resultant progeny phages from such an infection are then used to transduce a specific mutant (carrying a mutation in the region of interest) and selection is imposed for restoration of function to that mutant. For example, an auxotrophic recipient may be transduced to prototrophy. Prototrophic transductants are then scored for the Tn phenotype (antibiotic resistance). Resistant clones are likely to have received a single transduced fragment carrying the Tn and the region of interest (although some may be double transductants having acquired two fragments, one carrying the Tn and the other the gene(s) of interest, in which case the Tn is likely to have been inherited at a site remote from the one of interest).

(c) Tn-induced deletions

The ability of Tns to mediate deletions of genetic material can be exploited to generate deletions at particular sites. In some cases, deletion involves loss of Tn and contiguous host DNA. Thus if a strain contains an antibiotic-resistance Tn inserted in a region of interest and selection is made for loss of drug resistance, the clones obtained may carry deletions (of variable size) near the original Tn site. In other cases deletion is not accompanied by loss of Tn, but extends from one end of the inserted element (see Figure 4.6). Such deletions can easily be transferred by virtue of the linked Tn insertion.

4.7.2 Mu-induced mutagenesis

The temperate bacteriophage Mu (mutator)[3] can generate insertion mutations by virtue of its ability to integrate into the host genome to establish lysogeny. Since Mu DNA can insert efficiently at many different sites within a host genome (see, for example, Bukhari and Zipser, 1972; Toussaint and Résibois, 1983) the phage is useful for inducing random insertion mutations. Mu may enter its host by infection of Mu-sensitive strains (such as *E. coli* K12, *Citrobacter freundii* and *Shigella dysenteriae*). Alternatively a conjugative plasmid (notably of the IncP group) carrying a Mu prophage may be employed. The use of broad host range plasmids affords a route for introducing Mu into various Gram-negative bacteria (including *Agrobacterium* spp., *Pseudomonas* spp., *Rhizobium* spp. and *Rhodopseudomonas* spp.) that are resistant to infection by Mu phage. In the case of strains that are sensitive to phage P1 but resistant to Mu, Mu-P1 hybrids that are Mu phages with only the host range of P1 may be used. These facilities extend the range of organisms in which Mu may be used for mutagenesis and other genetic manipulations.

Mu-induced mutations are normally very stable. Revertants can, however, be obtained if the thermoinducible Mu prophage (Mu*cts*) carries mutations in the *B* gene inactivating the killing functions of Mu (Bukhari, 1975). Precise excision, which occurs at a frequency of about 10^{-6} to 10^{-8}, results in restoration of the activity of the gene into which Mu was previously inserted.

Mu derivatives (mini-Mus) carrying internal deletions in the Mu*cts*

[3] Mutator phages have also been identified in *Vibrio cholerae* (Johnson *et al.*, 1981) and *Pseudomonas* (see Kleckner, 1981). These phages are possibly analogous to Mu (and its relative D108) and may prove similarly useful.

Figure 4.6: Tn*10*-promoted deletions. (a) Tetracycline-sensitive deletions in the *his* operon of *Salmonella*. Deletions obtained from tetracycline-sensitive derivatives of *hisG*::Tn*10* insertion. Deletions of variable length extend into chromosomal regions on one side or the other of the Tn*10* insertion. (b) Tetracycline-resistant deletions of prophage P22. The Tn*10* insertion is between gene *9* and the attachment site. Deletions of variable length extend into the prophage genome. ——, Material deleted in individual isolates

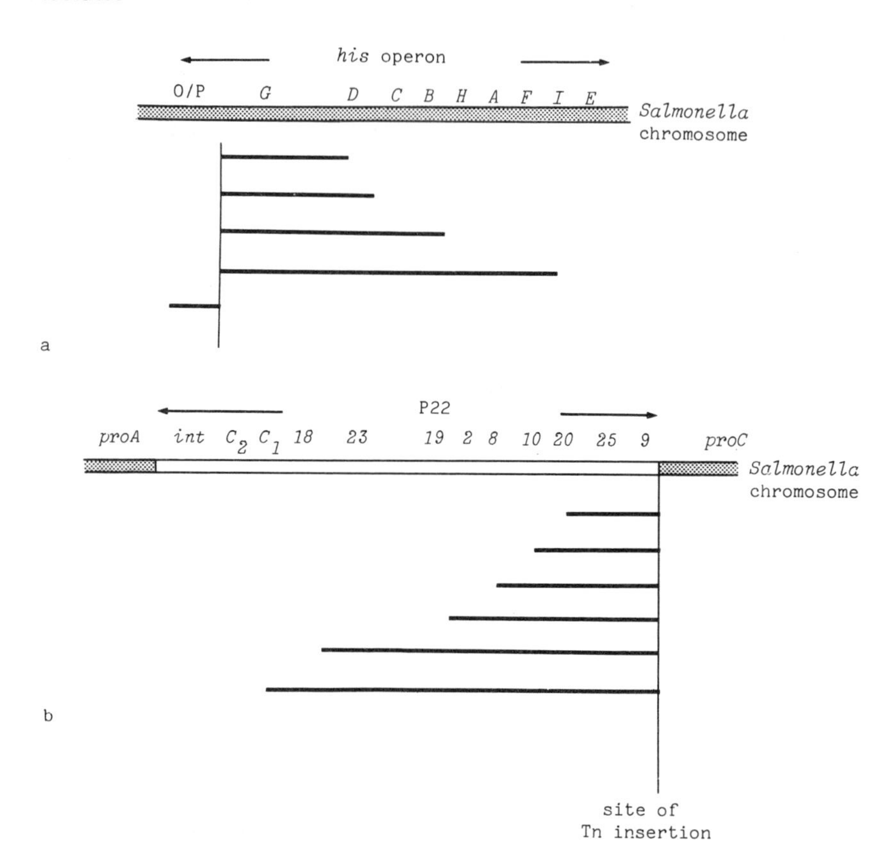

prophage may be used in mutagenesis. Some of these elements retain Mu integrative properties and can transpose to randomly distributed sites on the host genome upon induction, but lack essential functions for lytic development (Faelen *et al.*, 1979). Mini-Mus can thus effectively promote mutations and chromosome rearrangements without killing the host. The availability of derivatives of both Mu*cts* phages and mini-Mus carrying non-transposable antibiotic-resistance markers simplifies genetic analysis. Lysogens of these phages can be readily selected on medium containing the relevant antibiotic (Toussaint, 1985).

(a) Mu-mediated deletions

Mu can promote the deletion of host DNA adjacent to a Mu prophage in a number of ways. Deletions can sometimes arise during the process of lysogenisation. Mu lysogens normally contain simple insertions of the Mu genome. However, some additionally contain deletions adjacent to the site of prophage insertion (Cabezon *et al.*, 1975). Deletions of variable size may be formed by imprecise excision of Mu DNA from a Mu*cts* lysogen following induction at 42°C. Colonies that have lost the *c* end of Mu and therefore have a defective prophage will survive. Host genes adjacent to the prophage site may be lost in such clones. Partial induction of Mu*cts* lysogens (allowing transposition but not completion of the lytic cycle) can stimulate two types of deletion formation (Faelen and Toussaint, 1978). During the lytic cycle Mu transposes to many sites on the host genome. Survivors obtained after exposure to partially inducing conditions (by growing the lysogen at 37°C instead of 42°C) often contain deletions extending from one side of the integrating Mu. Deletions having two variable end points and located far from the integration site of the parental prophage may also be formed. Deletions can also be recovered after fully inducing a mini-Mu A^+B^- prophage, since it does not kill the host upon induction.

4.8 LOCALISED MUTAGENESIS *IN VIVO*

A number of techniques may be used to localise mutagenesis to particular regions of the genome, for example:

(i) *Localised nitrosoguanidine mutagenesis.* In a synchronised culture the maximum frequency of N-MNNG-induced mutations of a gene occurs at the time the gene is replicated. This preference for the replication fork can be exploited to focus mutagenesis on specific tracts of the genome by adding N-MNNG to the culture at the appropriate time (Cerdá-Olmedo and Ruiz-Vázquez, 1979).

(ii) *Mutagenesis during active transcription.* Genes that are being actively transcribed appear to be more mutable by certain mutagens than inactive genes. It is therefore possible to improve the specificity for genes of interest by carrying out mutagenesis during the time of maximum expression of such genes (for example following induction of an operon).

4.9 DIRECTED MUTAGENESIS *IN VITRO*

Classical (*in vivo*) mutation techniques have a number of drawbacks. First, mutagenesis is normally random, which means that the introduction of a desired mutation may be accompanied by undesirable mutations (in adjacent or even unlinked genes). Secondly, the precise location and nature of the lesion can only be determined by time-consuming genetic analysis. Furthermore, it is difficult to obtain mutants with subtle changes in phenotype or which have lethal phenotypes. The advent of recombinant DNA technology has, however, permitted the development of *in vitro* mutagenesis techniques, which permit the direction of mutations to particular segments or even individual nucleotides of a DNA molecule (for a review see Smith, 1985). The use of such techniques is likely to reduce the numbers of organisms that need to be screened before desired mutants are obtained. Directed mutagenesis *in vitro* does, however, require that the gene(s) of interest has been successfully cloned and at least partially characterised (see Chapter 3).

4.9.1 Deletion mutagenesis

Deletion mutagenesis techniques are used to remove segments of DNA from a given gene or as a prelude to the introduction of mutations at a particular site. A variety of methods can be used to delete regions of cloned DNA. Almost all rely on the target DNA being cloned in a circular vector.

(a) Excisional deletion

Precise excision of specific DNA fragments may be effected by complete or partial digestion of the cloned DNA with one or more appropriate restriction endonucleases. Deletants may be obtained after ligation of the DNA *in vitro* and transformation of *E. coli* (Figure 4.7). (Incompatible termini produced by double digestion may be rendered flush by exonucleolytic removal of terminal protrusions or end-filling with DNA polymerase I (see section 3.2.3), prior to blunt-ended ligation.) Excisional deletion requires a detailed knowledge of restriction sites within the target sequence and is limited by the distribution of these sites. Furthermore, the only suitable sites in the target gene may be those for frequently cutting (tetranucleotide-recognising) endonucleases, which would also cut the vector many times. This problem can be overcome by cloning the target DNA into a single-stranded vector, such as M13, and annealing to the target DNA (+) strand a complementary single-stranded DNA fragment that spans the region to be deleted and that contains suitable restriction sites. Since most restriction endonucleases only cleave double-stranded DNA, this procedure will permit the deletion of the specific region that lies within duplex DNA,

Figure 4.7: Excisional deletion with restriction enzymes. See text for details

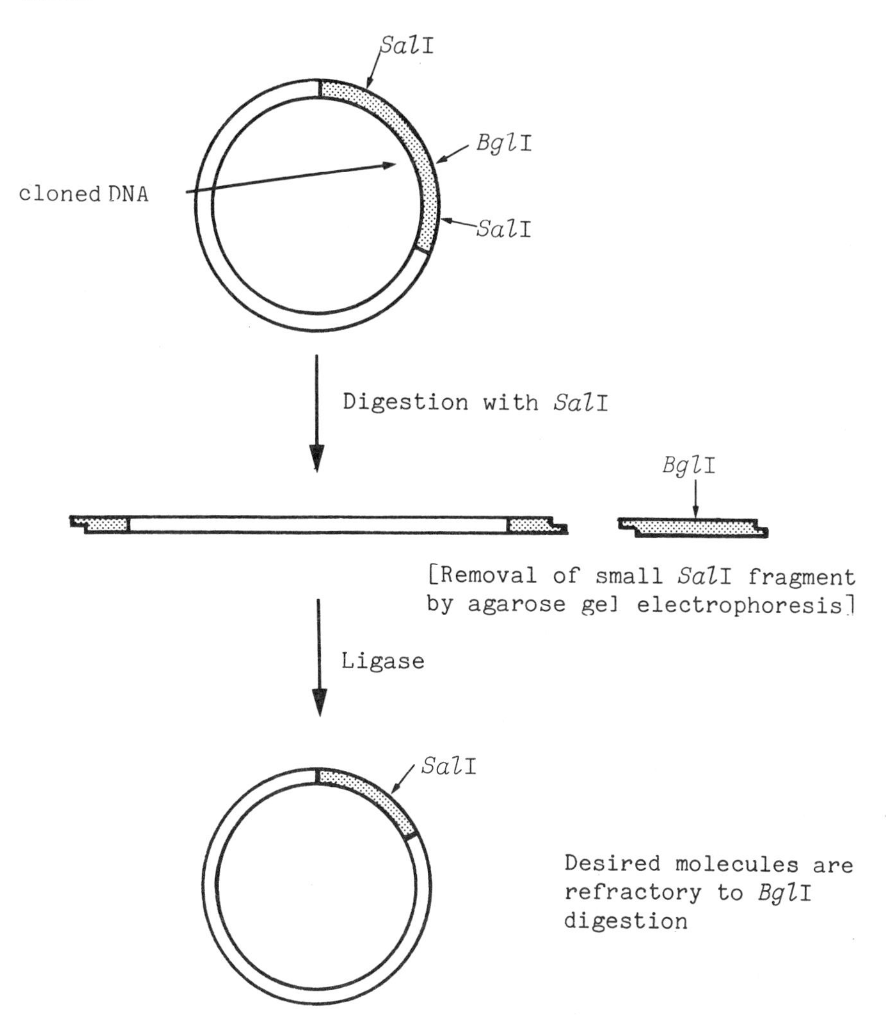

leaving the single-stranded region uncleaved (Figure 4.8). After ligation, the vector carrying the remaining double-stranded DNA may be rendered totally double-stranded by treatment with DNA polymerase, using the remaining duplex region as primer, and introduced into *E. coli* by transfection.

(b) Exonucleolytic deletion in vitro

Deletions may be made by nucleolytic digestion of the two termini produced when a circular recombinant DNA molecule is linearised by a

Figure 4.8: Excisional deletion using a single-stranded vector. *A B C, A′ B′ C′*, complementary sequences, --- new DNA Synthesis. See text for details

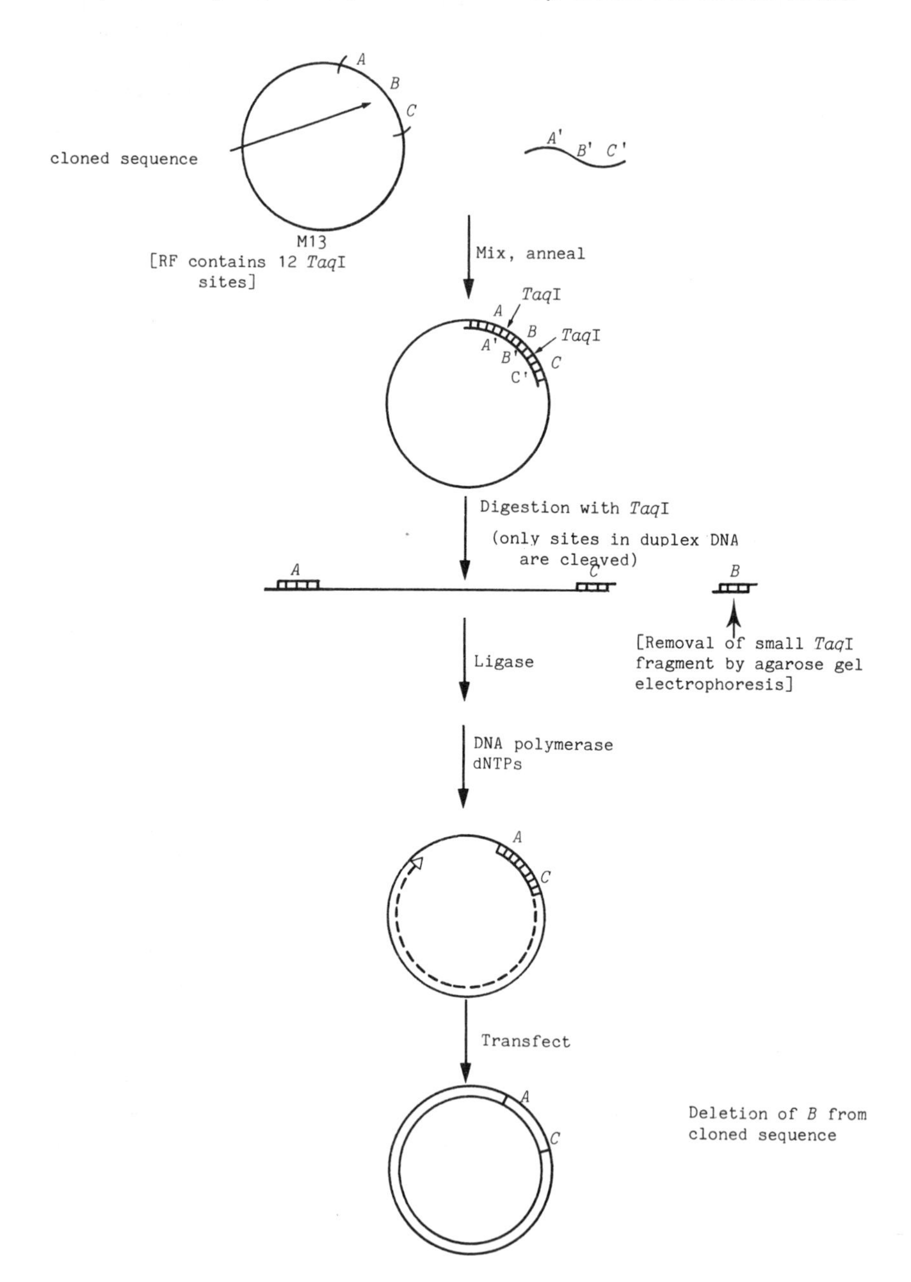

single double-stranded scission within the target region (Figure 4.9). Linearisation may be achieved at a specific site by cleavage of the target molecule with an appropriate restriction endonuclease. Quasi-random cleavages may be made in the target with a restriction endonuclease that normally cuts at multiple sites, but which has been partially inhibited with ethidium bromide so that the enzyme cleaves, on average, at only one site per molecule. Completely random double-stranded breaks in circular DNA can be made by cleavage with pancreatic DNase in the presence of Mn^{2+}. Alternatively, pancreatic DNase in the presence of ethidium bromide can be used to introduce random single-stranded nicks in DNA. These nicks can be extended with exonuclease III and the resulting single-stranded gap cut across with S1 nuclease (Pipas *et al.*, 1980). Varying numbers of nucleotides can be removed from the exposed termini by controlled digestion with one of a number of nucleases, including S1 endonuclease, λ-exonuclease (either alone or in conjunction with S1 nuclease) or BAL31 nuclease. (The properties of these enzymes are described in Chapter 3.) Following *in vitro* deletion, the molecules are ligated and used to transform *E. coli*, where they are subsequently propagated.

Deletions may also be obtained provided the target sequence can be localised as a deletion or substitution loop in a heteroduplex DNA molecule (see section 4.9.3). Treatment of heteroduplexes with S1 nuclease cleaves such single-stranded loops and removes nucleotides from the termini of the exposed strands (Shenk, 1977).

(c) In vivo *deletion*

Recombinant plasmid DNA that has been linearised within the target sequence (either specifically or randomly) *in vitro* may be subjected to exonucleolytic deletion *in vivo*, following transformation of *E. coli*. The proportion of transformants carrying deletant plasmids depends upon the genotype of the recipient and the nature of the termini of the linearised molecule. For example, with plasmid molecules bearing cohesive termini, about 5% of the transformants contain deletants using wild-type *E. coli* as recipient, whereas 45% contain deletants using a *recBC*$^-$, *lop11* mutant[4] (Conley and Saunders, 1984). Higher frequencies of deletion (>90%) can be obtained in wild-type recipients by using molecules with blunt ends. Deletions ranging in size from several base pairs to several kilobase pairs extend from one or both sides of the linearisation site. Such deletions presumably arise as a consequence of exonucleolytic processing of the termini of the plasmid molecules and/or recombination events necessary to recircularise the linear molecules.

---

[4] *recBC* encodes exonuclease V, which is a component of the major recombination pathway of *E. coli*. The *lop11* mutation results in the overproduction of DNA ligase.

Figure 4.9: Exonucleolytic deletion *in vitro*. See text for details

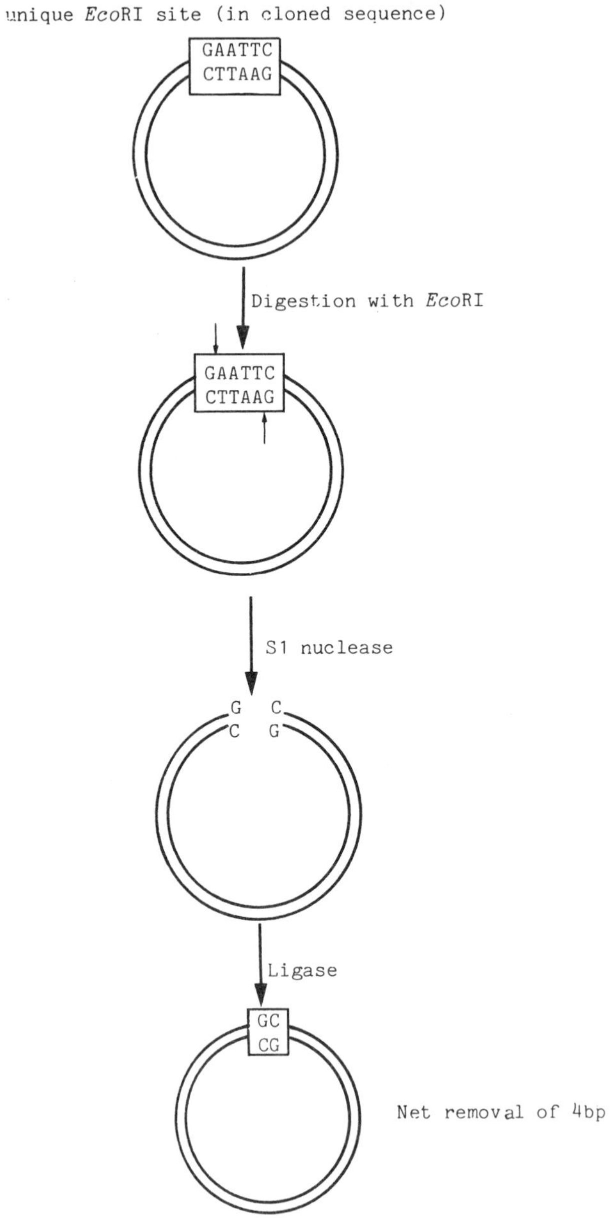

4.9.2 Insertion mutagenesis

Insertions of natural or synthetic DNA at desired points in cloned DNA can be used to interrupt genes, to add new restriction sites, to add regulatory signals, such as promoters (see Chapter 5), or to facilitate second-stage mutations, such as deletions.

(a) Short insertions at restriction sites

Insertions of 1 to 5 bp may be made at specific restriction sites by filling in the 5′ protrusions, produced by appropriate restriction endonucleases, with Klenow fragment or T4 DNA polymerase (section 3.2.3). When the resulting blunt-ended termini are ligated, the resultant molecule contains a duplication of the base sequence represented by the protrusion (Figure 4.10). In most cases this procedure will result in the destruction of the restriction site.

(b) Linker mutagenesis

Synthetic linkers or adaptors (see section 3.4.3) can be inserted, either randomly or specifically, into cloned DNA in order to place restriction sites at novel positions within the molecule (Heffron *et al.*, 1978). The insertion of a linker or adaptor within a gene can cause mutation in several different ways. First, procedures such as the removal or repair of cohesive termini prior to blunt-ended ligation (see section 3.2.3) necessary to insert a linker can be used to generate small deletions or insertions at the site of linker insertion (Figure 4.11). Secondly, the insertion of short synthetic oligonucleotides can alter the coding capacity of the DNA. Such insertions may, for example, produce frameshift mutations or in-frame insertions of two or more codons (Barany, 1985). Thirdly, DNA that lies between two inserted linkers can be excised by digestion with the appropriate restriction endonuclease.

4.9.3 Base substitutions

(a) Localised mutagenesis of DNA fragments

It is possible to localise chemical mutagenesis to purified DNA restriction fragments, DNA fragments cloned in intact plasmid or virus vectors, or DNA fragments contained within transducing particles. Examples of mutagens that have been used for this purpose are hydroxylamine and N-MNNG. Following mutagenesis a target fragment can be reincorporated (in place of the wild-type fragment) into the genome by ligation *in vitro* or by homologous recombination *in vivo* (see section 4.10). The efficiency of fragment mutagenesis is, however, low, because mutations are introduced essentially at random within the target sequence (and also within the vector molecule if this is attached). Some degree of direction has been achieved in

Figure 4.10: Creation of insertion mutations by end-filling. See text for details

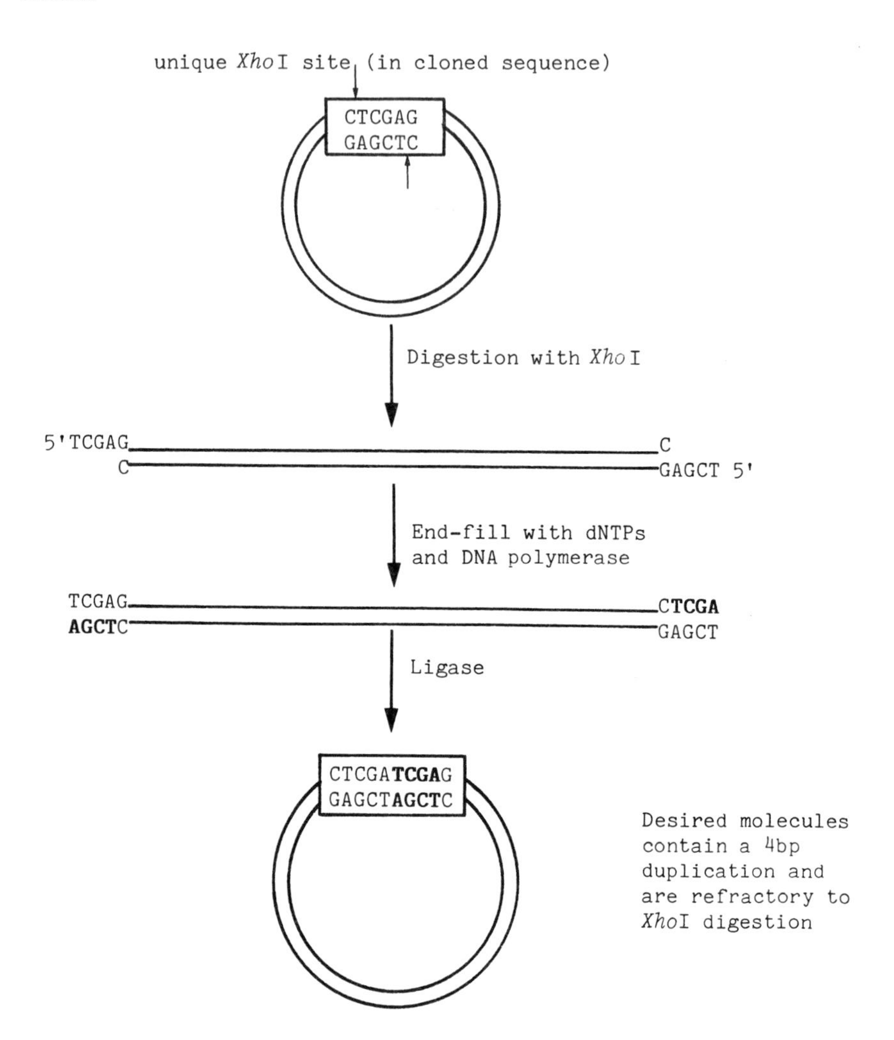

the mutagenesis of phage T7 by covalently linking a polyfunctional nitrogen mustard to purified mRNA transcripts of the target gene (Salganik *et al.*, 1980). The modified RNA species were allowed to form R loops (see section 3.8.4) with the target DNA, and a previously inert alkylating group on the nitrogen mustard was activated chemically to create multiple cross-links between RNA and complementary DNA. When such cross-linked molecules were packaged *in vitro* and used to infect *E. coli*, up to 12% of the progeny phages were found to be mutated in the gene of interest. This procedure may prove to be of general application in targeting

Figure 4.11: Linker mutagenesis. See text for details

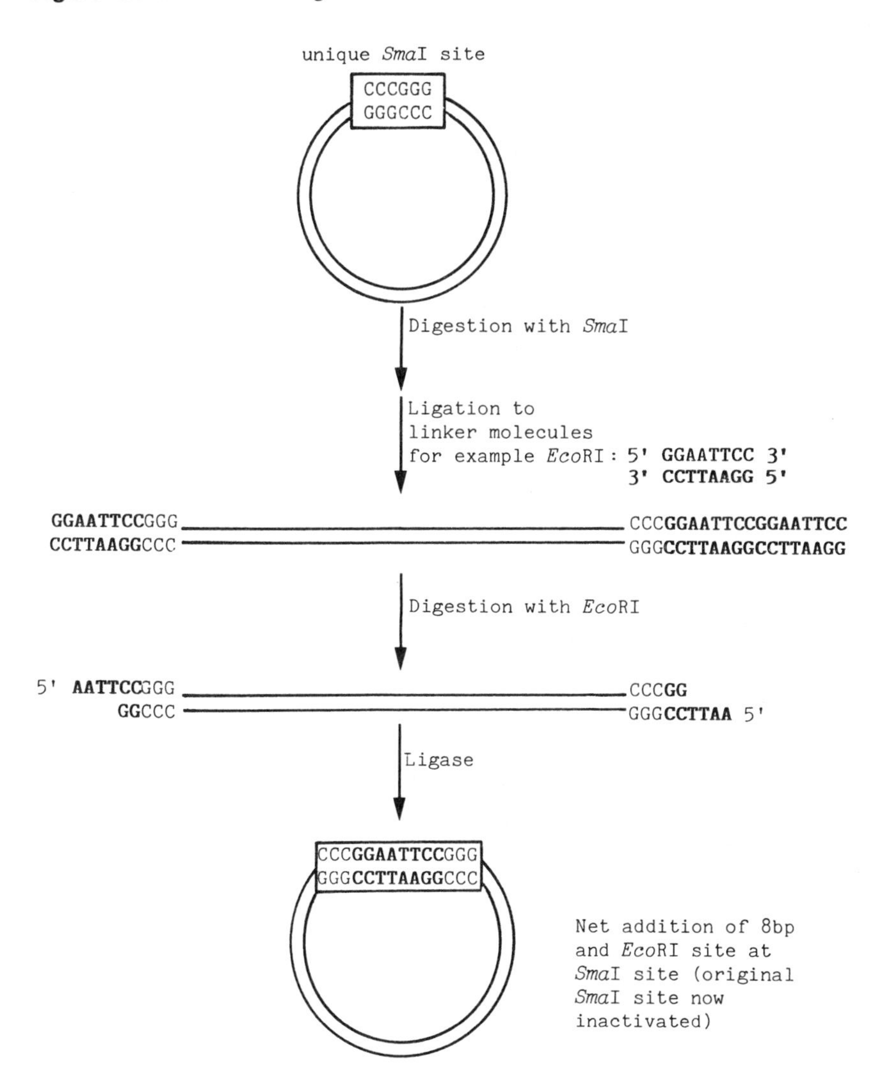

mutation to a specific gene, where the appropriate mRNA species is available.

(b) Segment-directed mutagenesis using bisulphite

Sodium bisulphite catalyses the deamination of cytosine to uracil and causes CG→TA transitions (CG→AT transversions sometimes also occur with this mutagen, but the molecular mechanism involved is not clear). The reaction occurs preferentially in single-stranded DNA. It is thus possible to mutate the C residues selectively in a target sequence by creating single-stranded regions at the desired positions in DNA molecules:

(i) *Mutation at nicks created by endonucleases.* Single-stranded nicks may be made specifically or randomly by treating double-stranded target DNA with restriction or other endonucleases in the presence of ethidium bromide. A nick made at a unique site in the target molecule can be extended by using 5′→3′ or 3′→5′ exonucleases to produce a short single-stranded gap (Figure 4.12). The exposed single-stranded region is treated with sodium bisulphite and the complementary strand is resynthesised using DNA polymerase. When the resultant molecule is introduced into a host cell, a high proportion of the progeny molecules contain transition mutations at one or more positions of C in the original target sequence.

(ii) *Displacement loop mutagenesis.* Stable displacement (D) loops are created when an homologous single-stranded DNA fragment is assimilated into a double-stranded molecule and base-pairs with the complementary strand (Figure 4.13). The single-stranded DNA can be either a natural fragment or a synthetic oligonucleotide. Invasion of double-stranded circular DNA by homologous single-stranded DNA can be catalysed *in vitro* by the RecA protein of *E. coli*, in the presence of Mg^{2+} and ATP (Shortle *et al.*, 1980; Shortle and Botstein, 1983). A nick is then introduced into the D loop with S1 nuclease. The extent of S1 digestion is limited to a nick because the primary scission of the D loop causes relaxation of the partially supercoiled circular molecule resulting in spontaneous displacement of the single-stranded fragment and collapse of the D loop (Figure 4.13). The resulting relaxed molecule is no longer a substrate for S1 nuclease, but contains a segment-specific nick, which can be extended, for example, by using the 5′→3′ exonuclease activity of *Micrococcus luteus* DNA polymerase. This procedure creates a single-stranded region within the target sequence for bisulphite mutagenesis and has the advantage that the siting of the nick is not limited by the availability of suitable restriction sites.

(iii) *Deletion loop mutagenesis.* This technique can be used to direct mutagenesis to a specific region of DNA that has previously been defined by a deletion mutation (see section 4.9.1) (Kalderon *et al.*, 1982; Peden and Nathans, 1982). Wild-type and deletion mutant sequences are cloned separately using the same plasmid vector. DNA from each of the resulting types of recombinant is treated with a different restriction endonuclease that cleaves the molecule once (Figure 4.14). If the two cleaved DNA preparations are mixed, denatured and allowed to anneal, only heteroduplexes will recircularise due to the staggered positions of endonucleolytic cleavage

Figure 4.12: Bisulphite mutagenesis at specific restriction sites. Transformants containing CG→TA transitions at one or more positions in the target region can be isolated. See text for details

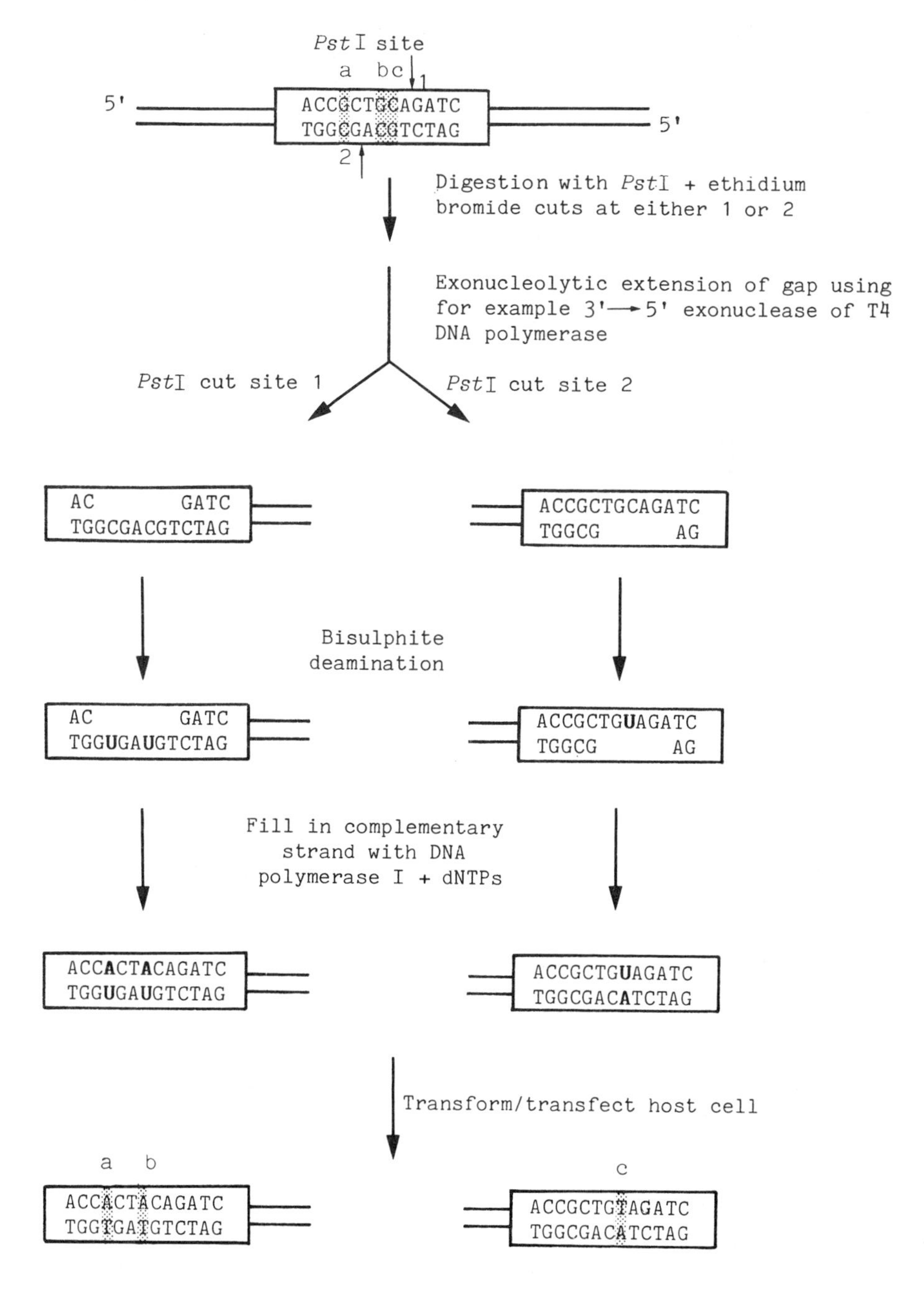

Figure 4.13: Displacement (D) loop mutagenesis. See text for details

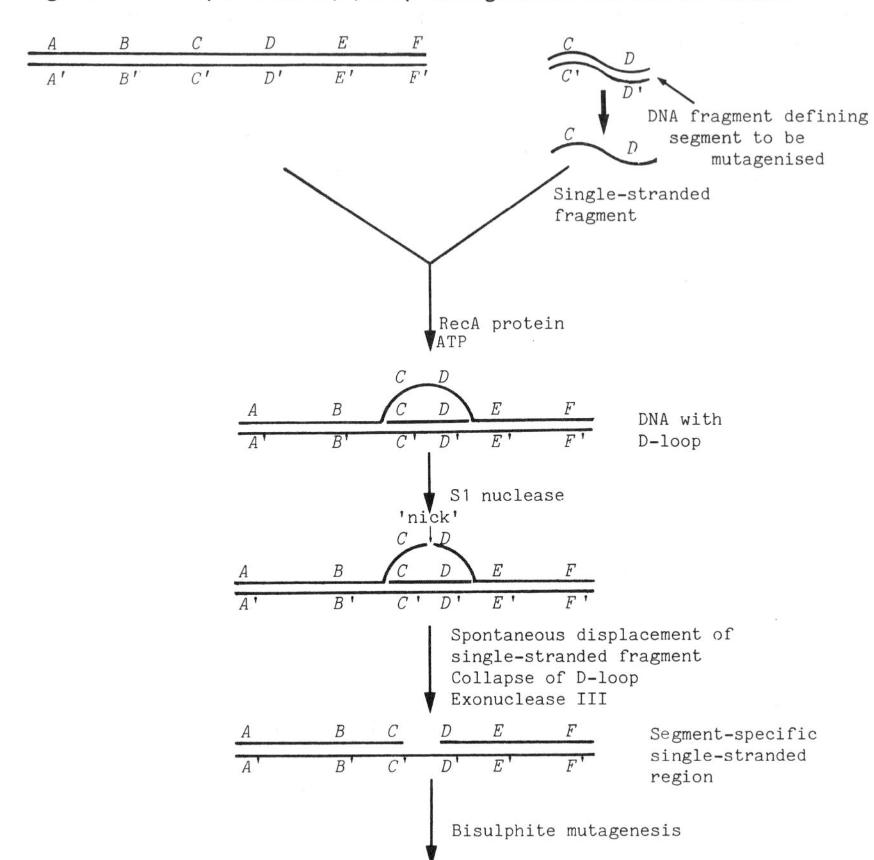

(any homoduplexes that form remain linear and unlike circular molecules transform poorly when the annealed duplexes are reintroduced into *E. coli*). Heteroduplex molecules will contain a single-stranded deletion loop, which is located on the wild-type strand of the duplex at the position opposite the deletion in the mutant strand. This loop is susceptible to sodium bisulphite mutagenesis. After treatment with bisulphite, the duplex mixture may be used to transform an *E. coli ung*$^-$ mutant (defective in the uracil repair enzyme, uracil *N*-glucosidase, which removes U from DNA).

About half[5] of the transformants contain plasmids of wild-type

[5] This would be expected since each heteroduplex molecule contains one parental deletant strand and one parental mutant (or parental wild-type) strand and will replicate semiconservatively to produce 50% of each type of progeny molecule.

Figure 4.14: Deletion loop mutagenesis. About half of the transformants contain plasmids of wild-type length, with one or more CG→TA transitions within the target region. *ung*⁻, Defective in uracil-*N*-glucosidase

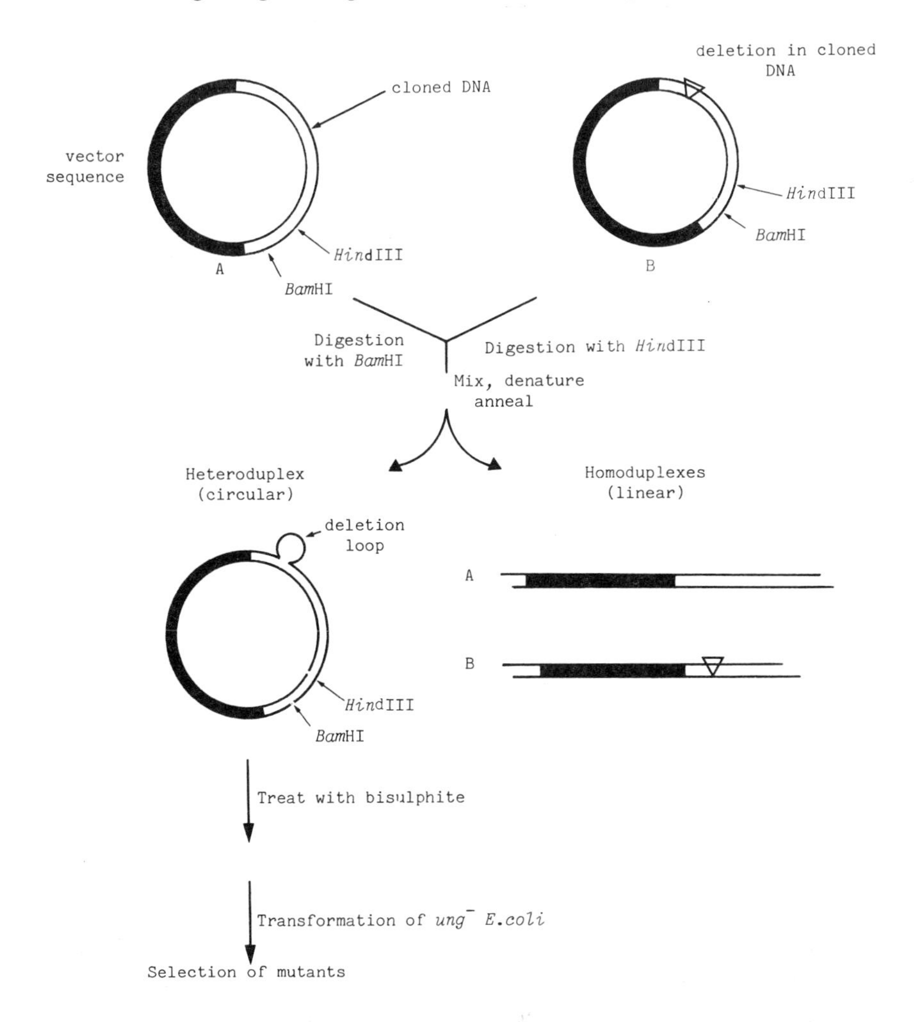

length and most of these contain CG→TA transition mutations at one or more of the C positions within the target region. It is possible to avoid the use of an *ung*⁻ mutant if the mutated deletion loop is treated with Klenow fragment and dNTPs *in vitro* prior to transformation, since the newly synthesised strand should contain the appropriate G→A transitions.

(c) Mutagenesis by nucleotide misincorporation

Mutations can be introduced at nicks in double-stranded DNA by the incorporation of nucleotide analogues during the polymerisation reaction of nick translation (see section 3.7.2). For example N^4-hydroxycytosine can be incorporated in place of dTTP. This analogue pairs with A or G and hence will cause TA→CG transitions at the sites where it substitutes for T.

Misincorporation of normal nucleotides is observed during *in vitro* polymerisation of primed DNA templates by a variety of DNA polymerases. The frequency of mutation caused by such misrepair of double-stranded DNA can be increased by adding beryllium to the reaction, by using Mn^{2+} or Co^{2+} instead of Mg^{2+} as a cofactor for the polymerase, or by omitting one of the four dNTPs required. For example, omitting dCTP from the reaction can lead to substitutions of A, G or T at the normal positions of C (Shortle *et al.*, 1981). The incorporation of noncomplementary bases during polymerisation can be used to direct mutagenesis to specific single-stranded gaps or nicks in DNA. Errors caused by omission of a dNTP during polymerisation of primed DNA templates cannot be corrected by polymerases lacking a 3′→5′ exonuclease activity, such as reverse transcriptase. Once the misincorporation has occurred, the primed DNA now carrying a mismatched nucleotide at its 3′ end can be elongated in the presence of the correct nucleotides (Traboni *et al.*, 1983; Zakour *et al.*, 1984). Thus it is possible to generate any base-pair change in the position adjacent to the 3′OH terminus of the priming strand. This will normally be the only base change introduced, although multiple base-pair substitutions may be made by repeated incubations of the primed-template complex with reverse transcriptase and different combinations of omitted nucleotides. The selection of mutants generated in this way is greatly facilitated if a selectable genetic marker (such as the *lac* α-peptide) can be associated specifically with the DNA strand synthesised *in vitro* and thus linked to any error made during chain elongation.

A convenient way to do this is to clone the wild-type sequence to be mutagenised in a single-stranded phage vector within the α-peptide region such that the reading frame is altered and white plaques are produced on Xgal plates. Single-stranded recombinant phage DNA containing the wild-type target sequence is annealed to a short piece of DNA carrying insertions or deletions of one or two bases that restore the correct translational reading frame[6] for the α-peptide and whose 3′ terminus is complementary to sequences in the locality of the region where base-pair substitutions are to be introduced. After misincorporation has been induced and the primed template extended, phages containing mutated sequences should produce blue plaques when introduced into *E. coli* by transfection.

---

[6] The insertions or deletions for restoration of reading frame may be located outside the gene to be mutagenised, and can conveniently be placed in adjacent vector sequences.

(d) Use of synthetic oligonucleotides

Synthetic oligonucleotides provide a convenient and very precise tool for introducing specific base changes into the target gene. **Oligonucleotide mutagenesis** permits subtle restructuring of genes and may allow the generation of mutants that are phenotypically silent. A short oligonucleotide containing the mutant sequence is normally used as a primer[7] to direct the synthesis *in vitro* of an intact recombinant genome. This can be achieved by annealing the oligonucleotide to a template comprising the target gene cloned into a single-stranded vector, such as M13 (Figure 4.15) (Smith and Gillam, 1981). Alternatively, the target gene can be cloned in a plasmid vector and the recombinant subsequently rendered single-stranded by nicking at a single site, followed by exhaustive digestion with exonuclease III. The complete second strand is then synthesised by using the mutant primer and DNA polymerase and the molecule is sealed with DNA ligase to form a duplex circle comprising one wild-type and one mutant strand. The maximum proportion of mutants that could be obtained when such heteroduplex molecules are used to transform *E. coli* is 50%. In practice the recovery of mutants may be much lower than this. However, it is possible to enrich for mutants by recovering the DNA from the primary transformants and employing it as a template for a further round of DNA synthesis using the mutant primer. By carefully controlling the hybridisation conditions it is possible to ensure that the mutant oligonucleotide primer will anneal only to the mutant strand, which is hence selectively replicated.

The process of recovering mutant genes using single-stranded vectors is generally inefficient partly because of the need to synthesise a complete copy of the mutant strand (second strand synthesis) by polymerisation from the mismatched oligonucleotide primer. Unless complete synthesis and ligation of the (−) strand is achieved, the 5′ end of the mismatched primer may be repaired *in vivo* by the exonucleolytic editing functions of polymerases (Carter *et al.*, 1985). An efficient double priming procedure that avoids the need for complete second strand synthesis has been developed by Norris and colleagues (1983). In this method a mismatched primer and a universal 15-base M13 dideoxy-sequencing primer are used simultaneously on a circular M13 template carrying the target gene (Figure 4.15). After only a brief period of polymerisation with Klenow fragment and treatment with DNA ligase, the extension of the mutant strand is sufficient to permit the excision of a double-stranded restriction fragment

---

[7] The priming oligonucleotide must comprise a minimum of 2 to 3 nucleotides 3′ to and 6 to 7 nucleotides 5′ to the mismatched base in order to achieve site specificity, stability of the heteroduplex and efficient priming (Gillam and Smith, 1979). In practice the oligonucleotides range from about 14 to 40 nucleotides in length.

Figure 4.15: Oligonucleotide mutagenesis. ◆, Mismatch. See text for details

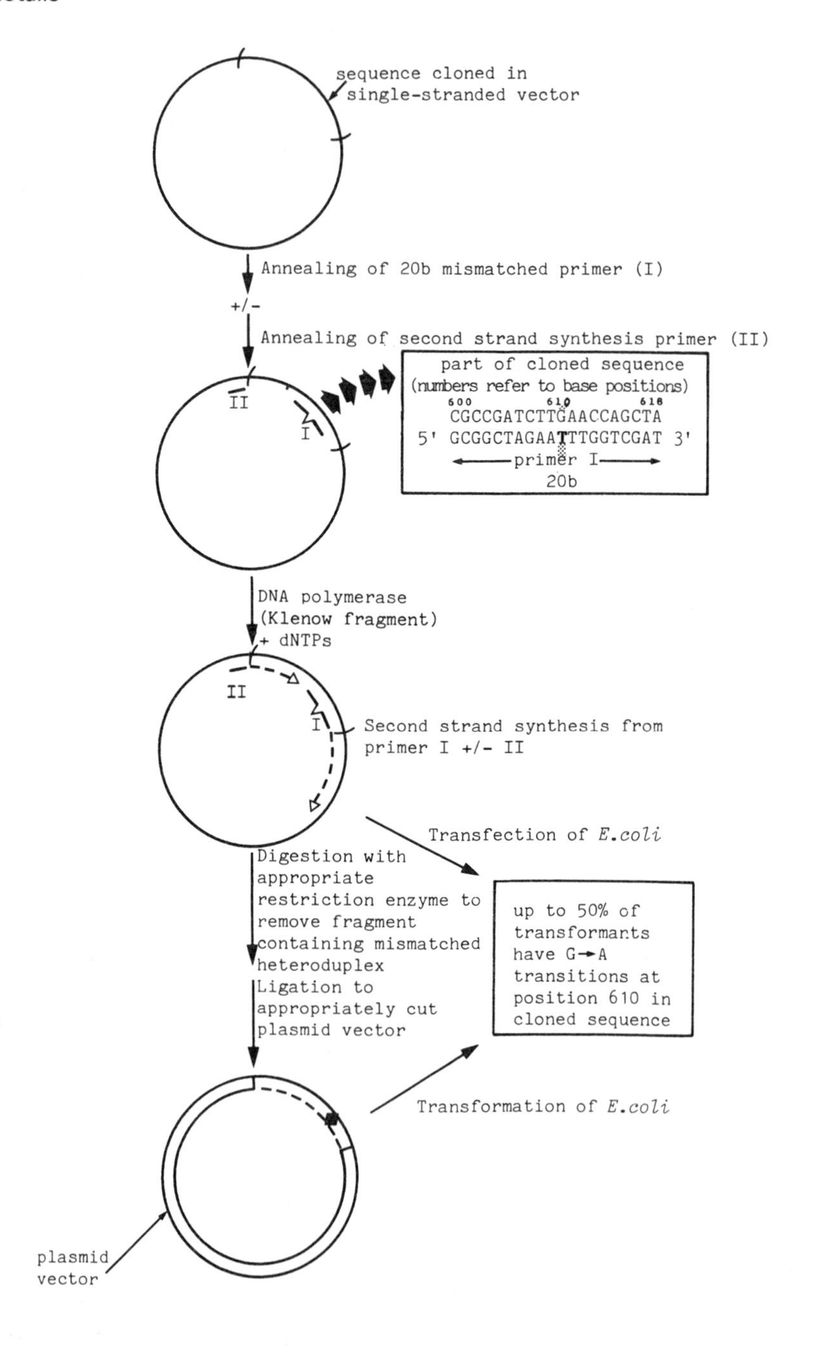

containing the target DNA with mismatched base(s). This double-stranded fragment may be cloned into a plasmid vector, such as pBR322. Over 40% of transformants obtained with such recombinant plasmids contain the desired mutation.

A mismatch in a heteroduplex is likely to be repaired since the mismatched (−) strand, synthesised *in vitro*, will be unmethylated whereas the (+) strand, synthesised *in vivo*, will be methylated. Transfection of a mismatch repair-deficient (for example *mutL*$^-$) strain of *E. coli* (see section 4.12) should therefore assist recovery of mutant progeny whose genomes derived ultimately from (−) strands (Carter *et al.*, 1985). The efficiency of recovery of mutant phages can also be increased by genetic strategies that ensure selection against phages whose genomes derived ultimately from the (+) strand. For example, Bauer and colleagues (1985) have used derivatives of M13mp10 with amber mutations in phage genes I and II. These vectors cannot replicate in a nonsuppressing (*sup*^{0}) host. The target DNA is first cloned in the vector and the (+) strand is annealed to the (−) strand of wild-type M13mp10 to create a gapped heteroduplex (Figure 4.16). An appropriate mutagenic oligonucleotide primer is then annealed to the target and the remaining gaps are filled enzymatically. Transfection of a *sup*^{0} host by the resultant heteroduplex selects for phages whose genomes have derived from the (−) strand.

The problem with selectable amber mutations is that they are lost following a single round of mutagenesis and selection. Multiple rounds of mutagenesis would thus demand that the target gene be recloned into the M13 selection vector following each round. To avoid this, Carter and colleagues (1985) have constructed M13 vectors containing a recognition site for the Type I restriction systems *Eco*B 5′ TGANNNNNNNNTGCT 3′ or *Eco*K 5′ AACNNNNNNGTGC 3′ in the polylinker region. A selection primer that will convert an *Eco*K site in the (+) strand of such a vector into an *Eco*B site is annealed to the template. At the same time a specific mutagenic primer is annealed to the template in the target region. Following repair of gaps the resulting heteroduplex is used to transfect an *Eco*K-restricting host (which selects against progeny phages whose genomes derive from the (+) strand) that is deficient in mismatch repair (Figure 4.17). A second round of mutagenesis can be carried out by annealing a second primer mutagenic for the target and an *Eco*K selection primer to the progeny phage DNA. Progeny phages now containing two mutations in the target sequence can be selected by transfection of an *Eco*B-restricting, repair-deficient host. The process can be repeated as many times as desired by cycling between *Eco*K and *Eco*B selection primers.

Synthetic primers can be used to make deletions between two defined sites in DNA cloned in M13 vectors if one half of the primer is complementary to one site and the other half complementary to a second site, which may be 1000 or more bases away. Insertions can also be made using

Figure 4.16: Genetic enrichment for mutations generated by oligonucleotide mutagenesis in M13 vectors. xI, xII, Amber mutations in M13 genes I and II respectively; *sup*^{0}, non-suppressing host. See text for details

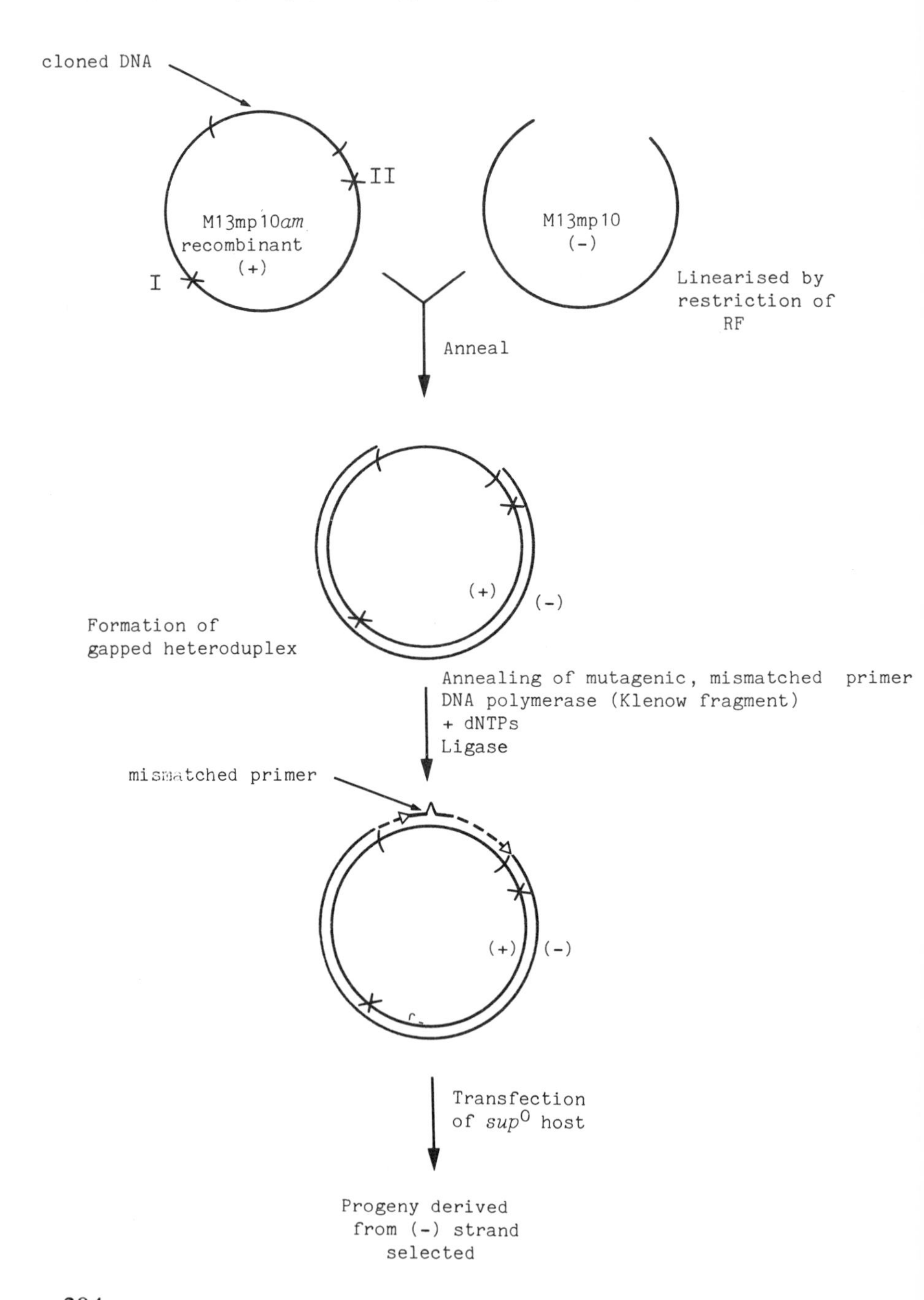

Figure 4.17: Cyclic selection oligonucleotide mutagenesis using coupled primers. Multiple rounds of mutagenesis can be effected by cycling between *Eco*K and *Eco*B selection primers.
*Eco*B selection primer,
5′ CACTAGAATGTCATCGAGG 3′;
*Eco*K selection site on M13K19,
3′ — TCGTGATCTTACAAGAGCTCCGG — 5′;
*Eco*K
*Eco*K selection primer,
5′ CACTAGAATGTTATCGAGG 3′;
*Eco*B selection site,
3′ — TCGTGATCTTACAGTAGCTCCGG — 5′;
*Eco*B
*, Mismatched bases; $r_k^+m_k^+$, *Eco*K, restriction-modification proficient; $r_B^+m_B^+$, *Eco*B, restriction-modification proficient; *mutL*$^-$, mutator (deficient in point mismatch repair). See text for details

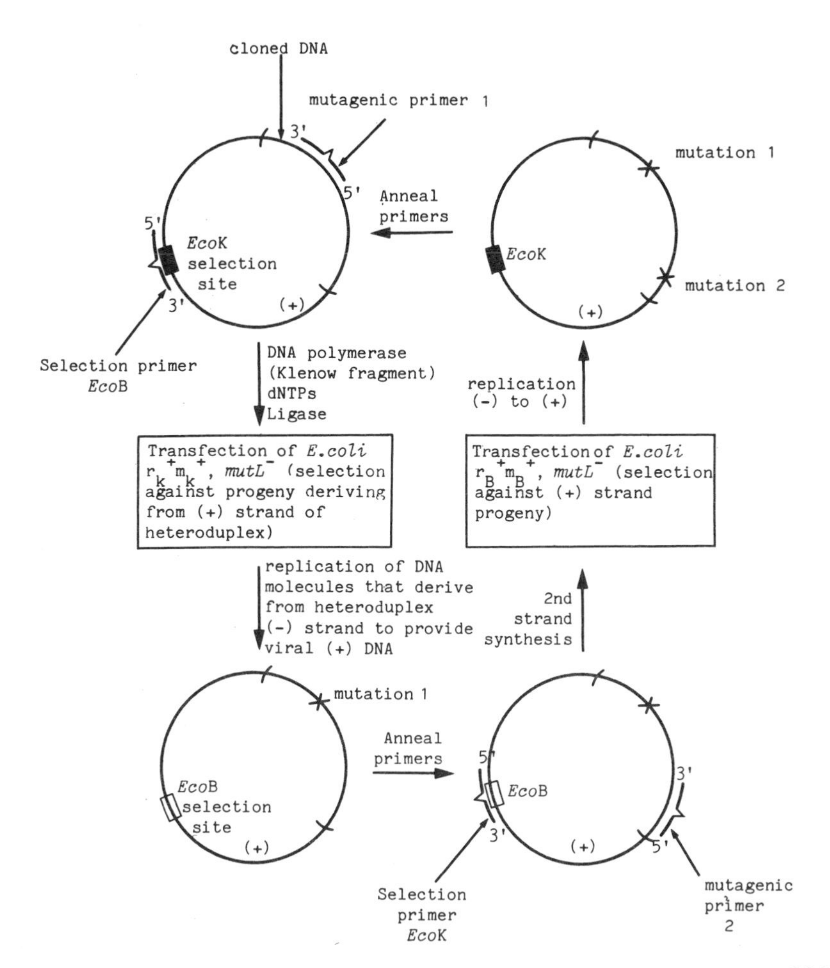

an appropriate primer (that contains the insert) but the length of the primer limits severely the size of insertion that is possible (Figure 4.18) (Osinga *et al.*, 1983).

Mismatched-primer mutagenesis techniques are methods of choice where only a few isolated changes in DNA sequences are sought. However, complicated mutations involving, for example, multiple base changes or inversions of tracts of DNA, require repeated rounds of mutagenesis. Shotgun gene synthesis (Grundström *et al.*, 1985) permits the introduction, in a single step, of complex or multiple mutations. Mutant genes are constructed by shotgun ligation of a specific series of overlapping oligonucleotides, synthesised by rapid, microscale gene synthesis (see section 3.1.3). The synthetic mutant gene is designed to have cohesive ends that allow directional cloning into the polylinker region of an M13 vector (Figure 4.19).

Figure 4.18: Generation of deletions and insertions during oligonucleotide-directed mutagenesis in M13 vectors. See text for details

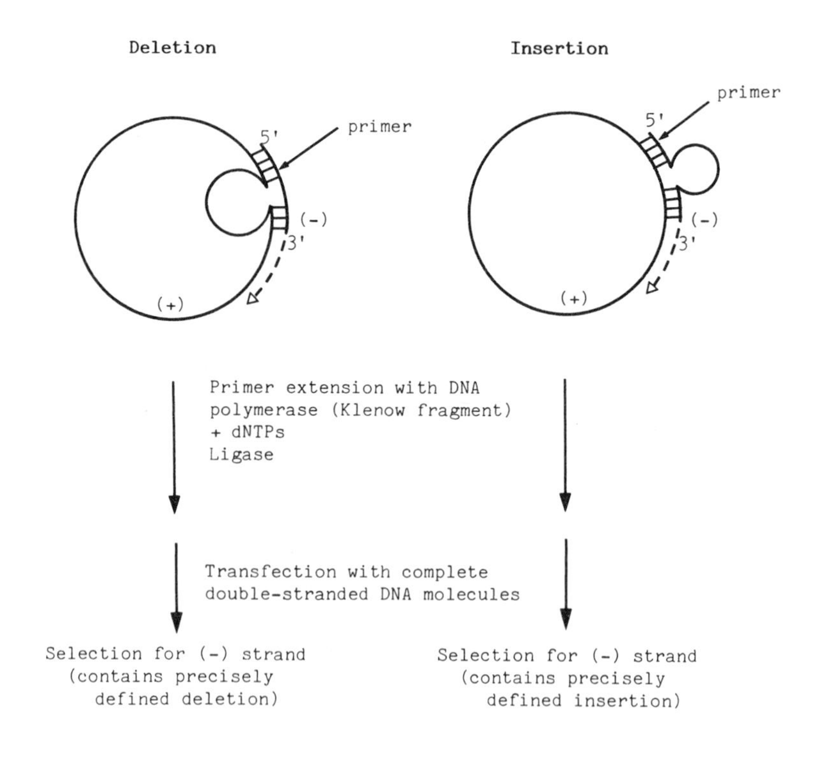

Figure 4.19: Oligonucleotide mutagenesis by shotgun gene synthesis. A to G are overlapping synthetic oligonucleotides. About 50% of ligation products contain the desired mutant gene sequence. See text for details

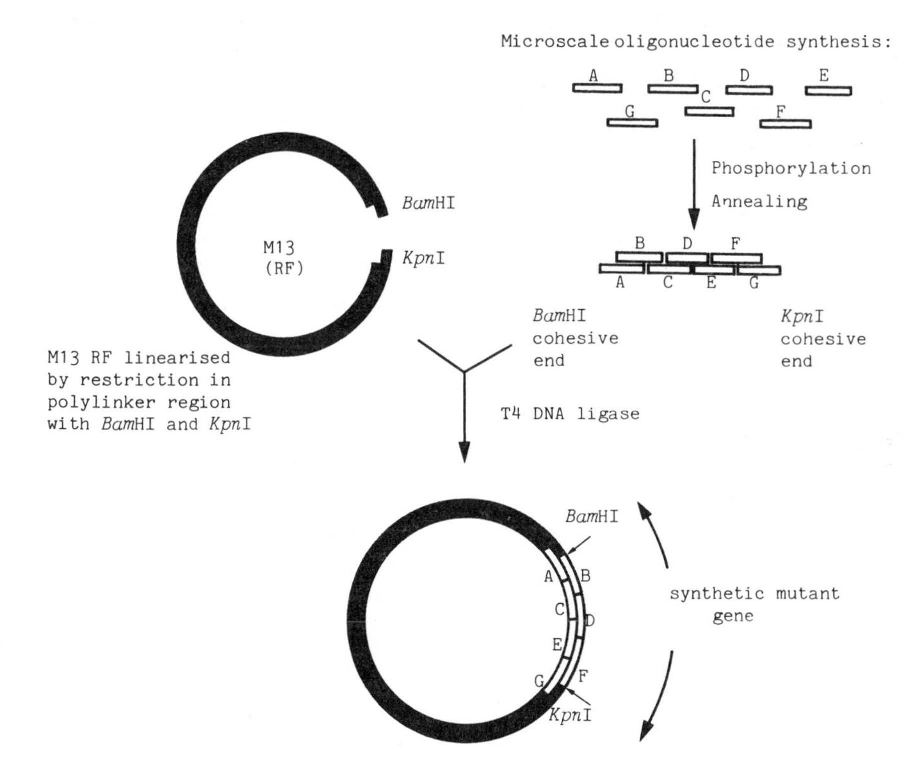

4.10 ALLELE REPLACEMENT (HOMOGENOTISATION)

In vitro techniques permit the cloning and mutagenesis of specific DNA sequences from any organism of interest. Of necessity such procedures are often carried out in *E. coli*. Under these conditions it may not be possible to characterise the phenotype encoded by a mutant gene. The mutated gene must therefore be returned to its normal genetic background in order to study the effect of the mutation(s). In the case of genes from simple genomes, such as plasmids or virus chromosomes, appropriate restriction fragments carrying the mutant allele may be substituted for the wild-type allele by ligation *in vitro*. This approach is not possible when dealing with chromosomal genes. However, if a mutant allele is returned to a strain of the originating organism that is wild-type for the gene in question, homologous recombination *in vivo* may lead to allele replacement (also referred to as transplacement or homogenotisation) (Shortle *et al.*, 1981). This permits the introduction of defined mutations into organisms that are difficult to mutate by conventional methods. The combination of *in vitro* site-directed mutagenesis and allele replacement has been called 'reversed genetics' (Weissmann *et al.*, 1979) since, in contrast to classical genetics, sequences with precisely defined lesions are made first and the phenotypes they encode are determined subsequently.

In order to achieve allele replacement, two sequential cross-overs are required between homologous sequences on the incoming mutated DNA (most conveniently carried on an integrating vector (such as YIp) or a vector conditional for replication) and on the resident wild-type gene(s) (usually on the host chromosome(s)) (Figure 4.20). Replacement can be monitored provided the vector carries a selectable marker. Selection for the marker permits isolation of strains in which the vector has inserted into the host genome flanked by the mutant and wild-type alleles in direct repeat. This insertion occurs as a result of a single cross-over event. A second cross-over between mutant and wild-type sequences can result in looping out of the vector together with a copy of either of these sequences, depending upon the position of the recombinational event. A cross-over at position 2b in Figure 4.20 would result in allele replacement, whereas a cross-over at 2a would lead to retention of the 'wild-type' allele. Provided that the vector cannot replicate autonomously, this second stage can be conveniently monitored by loss of the selectable marker.

Allele replacement is likely to be a successful technique for the mutation of all prokaryotes with functional reciprocal recombination systems. The technique has been demonstrated in yeast, which has a highly efficient system for homologous recombination, but allele replacement in higher eukaryotes, such as cultured mammalian cells, has been less successful due to integration of incoming DNA at multiple sites on the recipient genome.

4.11 FORMATION OF HYBRID GENES BY *IN VIVO* RECOMBINATION

One approach to obtaining a gene product with a preferred combination of properties is to form hybrids between two or more related gene sequences. The formation of hybrid genes by recombinant DNA techniques is limited, to a large extent, by the location of appropriate restriction sites within the sequences that have to be fused. However, it is possible to exploit the homologous recombination system of *E. coli in vivo* in order to create a variety of novel hybrid genes that could not easily be constructed *in vitro*. By using this approach Weber and Weissmann (1983) have constructed novel interferons comprising the C-terminal half of human interferon α1 and the N-terminal half of interferon α2. The hybrid genes were obtained by transforming *E. coli* with linear DNA molecules (constructed *in vitro*) comprising plasmid vector sequences flanked on one side by the α2 inter-

Figure 4.20: Allele replacement. A single recombinational event between homologous sequences on incoming mutated DNA and wild-type resident DNA effects insertion of vector into the host genome. A second recombinational event effects excision of the vector. Depending upon the position of this second cross-over, allele replacement may (b) or may not (a) occur. ◆, Position of mutation

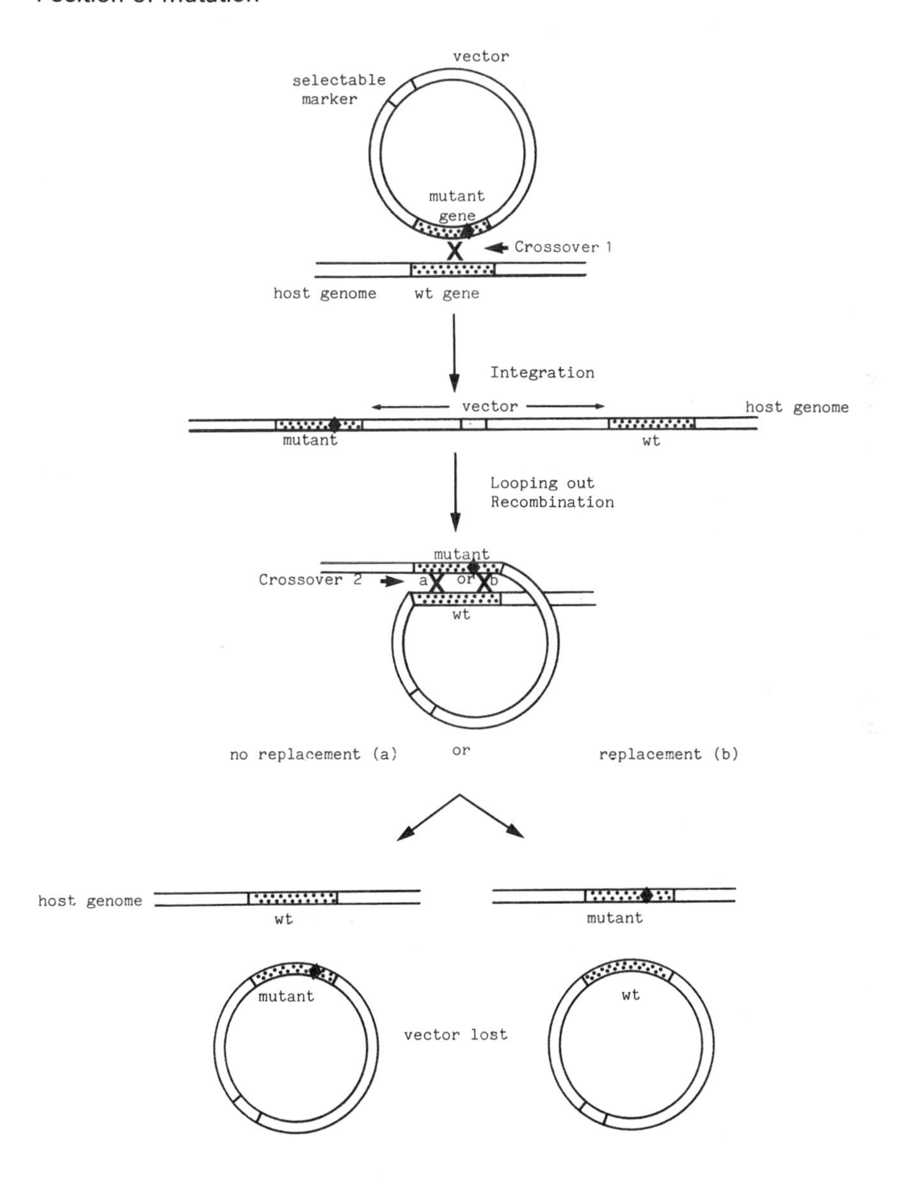

feron gene and on the other by part of the α1 interferon gene. This region of the α1 interferon gene exhibited about 80% sequence homology with the α2 interferon gene. The most likely way for such molecules to recircularise and become established as replicons is by recombination between homologous regions within the flanking gene sequences (Figure 4.21). Various hybrid genes have been generated, depending upon the positions of recombinational events between the interferon sequences. Furthermore, since cross-overs can occur within regions of as little as 3

Figure 4.21: *In vivo* recombination for hybrid gene construction. Various hybrid genes may be generated depending upon the position of the recombinational event(s) involved in recircularisation. Note: the termini of the constructed linear DS molecule are incompatible to prevent recircularisation *in vitro* by end annealing. Km^r, resistance to kanamycin.

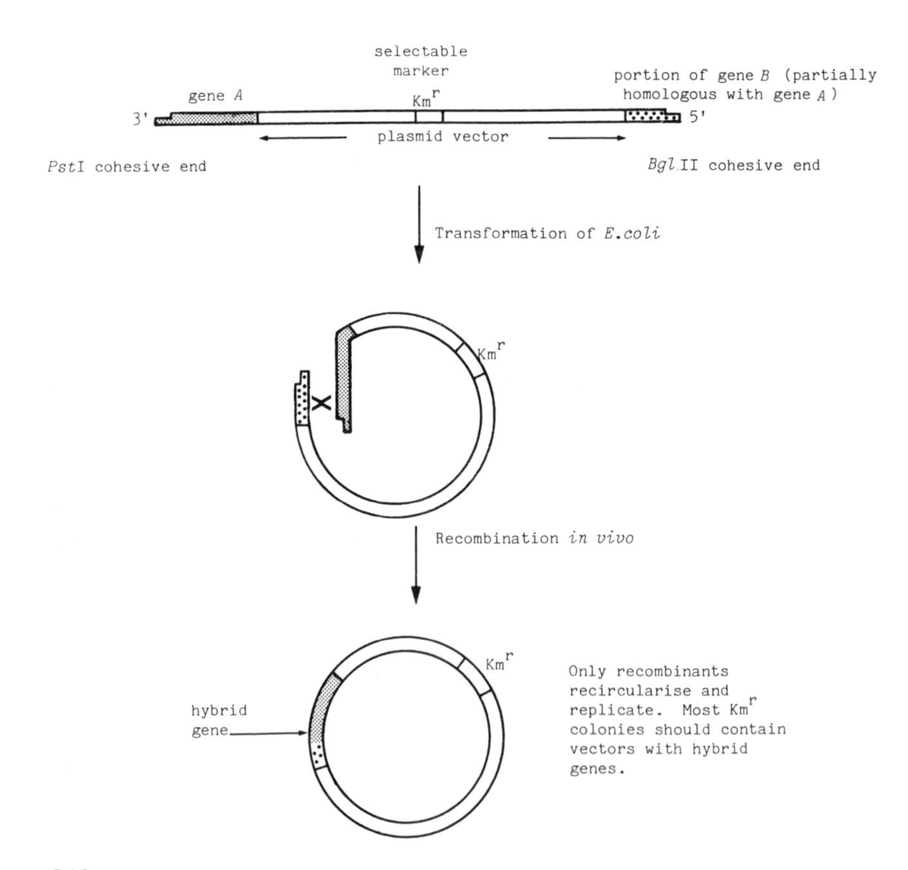

base pairs of homology, a large number of different hybrids between any two partially homologous gene sequences could potentially be generated by this technique.

4.12 MUTATOR AND ANTIMUTATOR ALLELES

Certain genes in a mutant state can alter the frequency of spontaneous and/or induced mutation. **Mutator** alleles (designated *mut*$^-$ in *E. coli*) may encode altered DNA polymerases that are defective in the copy-editing function (see section 4.2.ii) resulting in increased mutation. **Antimutator** alleles may encode mutant polymerases with improved editing efficiency, resulting in reduced frequency of mutation. Mutation of the *dam* gene of *E. coli* can increase the mutation frequency. Methylation of DNA by the methylase (*dam* gene product) appears to have a role in instructing the mismatch repair system to remove incorrect bases from daughter strands following DNA replication (section 4.2.iii). In *dam*$^-$ strains no such methylation occurs. The repair system may, therefore, be unable to discriminate parental from daughter strands and often removes correct parental bases and replaces them with bases complementary to the incorrect bases on the daughter strands. Transposition of transposable elements, such as Tn*5*, Tn*10* and Tn*903*, occurs at higher frequency in *dam*$^-$ hosts. In the case of Tn*10* this is probably due, at least in part, to absence of methylation of a 5′ GATC 3′ sequence in the transposase promoter (pIN) (Roberts *et al.*, 1985). *dam*$^-$ hosts may therefore be utilised to increase the efficiency of transposition mutagenesis.

Certain mutator strains may prove useful in mutation programmes where a higher frequency of mutants per survivor (at the optimum dose of mutagen) is required, or where it would be advantageous to use lower doses of mutagen. Furthermore, mutator strains, in which the frequency of spontaneous mutation is enhanced, can be employed for mutant isolation where treatment with physical or chemical mutagens proves undesirable. Mutator strains may also be used to prevent mismatch repair in oligonucleotide mutagenesis (section 4.9.3.d). The use of mutator strains with increased spontaneous mutability as production strains is, however, likely to lead to enhanced strain degeneracy. Antimutator strains in which the spontaneous but not induced mutability is reduced may be useful for reducing genetic instability.

4.13 PROCEDURES FOR MUTANT ISOLATION

4.13.1 Screens

Rapid identification of the desired mutant(s) following mutagenesis depends upon an efficient screening procedure. It is important that the conditions of the screen are such that the desired genotype is properly expressed. (However, it should be noted that expression under conditions of the screen may be different from that during production.) It is often advisable to include a subcultivation step between mutagenesis and screening in order to allow for expression and segregation of mutations. Screens may be of two types, **random** or **rational**. In the random (nonselective) screen, all (or random) isolates are tested independently for the character(s) of interest. This can be a laborious and painstaking operation. Rational screens are based upon available genetic and biochemical knowledge of the system involved and are generally more efficient than random screens. The rational screen normally serves as a primary or pre-screen, in which the character selected for is one associated with the one of ultimate interest, but one which is more easily scored (Chang and Elander, 1979). The output from such a screen is subsequently analysed to ensure that the desired mutants have been obtained. (Examples of rational selection techniques are described in section 6.2.1.)

Screens may be performed on solid or in liquid medium. Procedures involving liquid culture have normally proved to be more time consuming than plate screens in which large numbers of mutants can be conveniently processed. However, liquid culture conditions more closely mimic production conditions.

The screen may involve some form of bioassay, for example antibiotic-producing organisms may be screened by their ability to inhibit growth of specific indicator organisms in an overlay plate assay. Antibiotic-producing colonies can be detected directly by the appearance of zones of growth inhibition in the overlay. The value of the potency index (PI), defined as the ratio of the diameter of the zone of inhibition to the diameter of the antibiotic-producing colony, provides a means of identifying those colonies producing high yields of antibiotic. However, such an overlay assay does not necessarily distinguish between different antibiotics from the producer strain. Furthermore, application of the method depends upon correlating performance on plates with that in submerged culture.

4.13.2 Selective methods

Selective methods for mutant isolation are generally based upon growth differences between the required mutant and the parental population.

Ideally a positive selection method, using medium that allows growth of the mutant but not of the parent strain, is employed. Many selective media have been designed to enhance the growth of mutant relative to that of the parent strain (see, for example, Clarke, 1975). Plate selection relies upon growth rates or other properties of parental and mutant strains being sufficiently different to distinguish variation in colonial morphology. However, in continuous culture, mutants with relatively small growth-rate differences from the parents can be detected. Cultures can be grown in a chemostat for as many generations as are necessary for a particular mutant to take over. (For a detailed appraisal of selection in chemostats see Dykhuizen and Hartl, 1983.) Where no positive selection procedure is available, various **enrichment techniques** may be applied to ease mutant recovery (see, for example, Hopwood, 1970). These techniques increase the ratio of potentially valuable mutants to the total population. Penicillin or nystatin can be used to enrich for certain types of mutant (for example auxotrophs and conditional mutants) of bacteria or fungi respectively. Addition of the antibiotic to growth medium that permits growth of the parent but not of the mutant strain enriches for the non-growing mutants. **Starvation enrichment** can be applied to enrich for auxotrophs. Omission of a particular growth factor(s) required by the parental strain (and the mutant) from the medium can lead to unbalanced growth of the parent population and lethality. The auxotrophic mutants are unable to grow under these conditions because of the additional lack of their specific nutritional requirements. For example, starvation enrichment has been used in the isolation of auxotrophs of *Schizosaccharomyces pombe* by omission of inositol (required by the starting strains), in addition to specific growth factors required by the potential auxotrophs, from the growth medium (Megnet, 1964).

Filtration or centrifugation techniques may be used in enrichment procedures where differences in shape, size or density of parental and mutant strains exist. Other enrichment methods include the application of **suicide** techniques. These involve the use of substances whose metabolism or assimilation effects cell death. Conditional mutants, temperature sensitive for DNA replication, have been obtained by tritium suicide. Addition of ^{3}H-thymidine to growth medium, under conditions that allow growth of the parental population only, results in incorporation of ^{3}H-thymidine into the DNA of the parents. β-particles emitted upon long-term storage of the culture in the cold are lethal to the parental cells. Such a culture will thus contain an increased proportion of mutants.

BIBLIOGRAPHY

Auerbach, C. (1976) *Mutation research: problems, results and perspectives.* Chapman & Hall, London

Bagg, A., Kenyon, C.J. and Walker, G.C. (1981) Inducibility of a gene product required for UV and chemical mutagenesis in *Escherichia coli. Proc. Natl Acad. Sci. USA* **78**, 5749-53

Barany, F. (1985) Single-stranded hexameric linkers: a system for in-phase insertion mutagenesis and protein engineering. *Gene* **37**, 111-23

Bauer, C.E., Hesse, S.D., Waechter-Brulla, D.A., Lynn, S.P., Gumport, R.I. and Gardner, J.F. (1985). A genetic enrichment for mutations constructed by oligonucleotide-directed mutagenesis. *Gene* **37**, 73-82

Beringer, J.E., Beynon, J.L., Buchanan-Wollaston, A.V. and Johnston, A.W.B. (1978) Transfer of the drug-resistance transposon Tn*5* to *Rhizobium. Nature (London)* **276**, 633-4

Beukers, R. and Berends, W. (1961) The effects of UV irradiation on nucleic acids and their components. *Biochim. Biophys. Acta* **49**, 181-9

Bukhari, A.I. (1975) Reversal of mutator phage Mu integration. *J. Mol. Biol.* **96**, 87-99

Bukhari, A.I. and Zipser, D. (1972) Random insertion of Mu-1 DNA within a single gene. Nature (New Biol.) **236**, 240-3

Cabezon, T., Faelen, M., De Wilde, M., Bollen, A. and Thomas, R. (1975) Expression of ribosomal protein genes in *Escherichia coli. Mol. Gen. Genet.* **137**, 125-9

Cairns, J., Robins, P., Sedgwick, B. and Talmud, P. (1981) The inducible repair of alkylated DNA. *Prog. Nucl. Acids Res. Mol. Biol.*, **26**, 237-44

Campbell, A., Berg, D., Botstein, D., Lederberg, E., Novick, R., Starlinger, P. and Szybalski, W. (1977) Nomenclature of transposable elements in prokaryotes. In *DNA insertion elements, plasmids and episomes* (A.I. Bukhari, J.A. Shapiro and S.L. Adhya, eds), pp. 15-22, Cold Spring Harbor Laboratory, Cold Spring Harbor, NY

Carter, P., Bedouelle, H. and Winter, G. (1985) Improved oligonucleotide site-directed mutagenesis using M13 vectors. *Nucl. Acids Res.* **13**, 4431-43

Cerdá-Olmedo, E. and Ruiz-Vázquez, R. (1979) Nitrosoguanidine mutagenesis. In *Proceedings of the 3rd international symposium on genetics of industrial microorganisms* (O.K. Sebek and A.I. Laskin, eds), pp. 15-20. American Society for Microbiology, Washington, DC

Chang, L.T. and Elander, R.P. (1979) Rational selection for improved cephalosporin C productivity in strains of *Acremonium chrysogenum* Gams. *Devel. Ind. Microbiol.* **20**, 367-79

Clarke, C.H. (1975) Mutagenesis and repair in microorganisms. *Science Progress (Oxford)* **62**, 559-77

Conley, E.C. and Saunders, J.R. (1984) Recombination-dependent recircularization of linearized pBR322 plasmid DNA following transformation of *Escherichia coli. Mol. Gen. Genet.* **194**, 211-18

Cooper, P. (1981) Inducible excision repair in *Escherichia coli.* In *Chromosome damage and repair* (E. Seeberg and K. Kleppe, eds), pp. 139-46. Plenum Press, New York

Cooper, P. (1982) Characterization of long patch excision repair of DNA in ultraviolet-irradiated *Escherichia coli*: an inducible function under *rec-lex* control. *Mol. Gen. Genet.* **185**, 189-97

Davis, R.W., Botstein, D. and Roth, J.R. (1980) *Advanced bacterial genetics. A manual for genetic engineering.* Cold Spring Harbor Laboratory, Cold Spring Harbor, NY

Dykhuizen, D.E. and Hartl, D.L. (1983) Selection in chemostats. *Microbiol. Rev.* **47**, 150-68

Faelen, M. and Toussaint, A. (1978) Stimulation of deletions in the *Escherichia coli* chromosome by partially induced Mu*cts*62 prophages. *J. Bacteriol.* **136**, 477-83

Faelen, M., Résibois, A. and Toussaint, A. (1979) Mini-Mu, an insertion element derived from temperature phage Mu-1. *Cold Spring Harbor Symp. Quant. Biol.* **43**, 1169-77

Foster, T.J., Howe, T.G.B. and Richmond, K.M.V. (1975) Translocation of the tetracycline resistance determinant from R100-1 to the *Escherichia coli* K-12 chromosome. *J. Bacteriol.* **124**, 1153-8

Friedberg, E.C. (1985) *DNA repair.* W.H. Freeman, San Francisco, Ca

Gillam, S. and Smith, M. (1979) Site-specific mutagenesis using synthetic oligodeoxyribonucleotide primers. I. Optimum conditions and minimum oligodeoxyribonucleotide length. *Gene* **8**, 81-97

Glickman, B.W. (1982) Methylation-instructed mismatch correction as a post replication error avoidance mechanism in *Escherichia coli.* In *Molecular and cellular mechanisms of mutagenesis* (J.F. Lemontt and W.M. Generoso, eds), pp. 65-87. Plenum Press, New York

Grossman, L. (1981) Enzymes involved in repair of damaged DNA. *Arch. Biochem. Biophys.* **211**, 511-22

Grundström, T., Zenke, W.M., Wintzerith, M., Matthes, H.W.D., Staub, A. and Chambon, P. (1985) Oligonucleotide-directed mutagenesis by microscale 'shotgun' gene synthesis. *Nucl. Acids Res.* **13**, 3305-16

Guerola, N. and Cerdá-Olmedo, E. (1975) Distribution of mutations induced by ethyl methanesulfonate and ultraviolet radiation in the *Escherichia coli* chromosome. *Mutat. Res.* **29**, 145-7

Hall, J.D. and Mount, D.W. (1981) Mechanisms of DNA replication and mutagenesis in ultraviolet-irradiated bacteria and mammalian cells. *Prog. Nucl. Acids Res. Mol. Biol.* **25**, 53-126

Hanawalt, P.C., Cooper, P.K., Ganesan, A.K. and Smith, C.A. (1979) DNA repair in bacteria and mammalian cells. *Ann. Rev. Biochem.* **48**, 783-836

Hanawalt, P.C., Cooper, P.K. and Smith, C.A. (1981) Repair replication schemes in bacteria and human cells. *Prog. Nucl. Acids Res. Mol. Biol.* **26**, 181-96

Haynes, R.H. and Kunz, B.A. (1981) DNA repair and mutagenesis in yeast. In *The molecular biology of the yeast Saccharomyces. Life cycle and inheritance* (J.N. Strathern, E.W. Jones and J.R. Broach, eds), pp. 371-414. Cold Spring Harbor Laboratory, Cold Spring Harbor, NY

Heffron, F., So, M. and McCarthy, B.J. (1978) *In vitro* mutagenesis of a circular DNA molecule using synthetic restriction sites. *Proc. Natl. Acad. Sci. USA* **75**, 6012-16

Hopwood, D.A. (1970) The isolation of mutants. In *Methods in microbiology,* vol. 3A (J.R. Norris and D.W. Ribbons, eds), pp. 363-433. Academic Press, London and New York

Hurst, A. and Nasim, A. (1985) *Repairable lesions in microorganisms.* Academic Press, Orlando, Fla

Jimenez, A. and Davies, J. (1980) Expression of a transposable antibiotic resistance element in *Saccharomyces. Nature (London)* **287**, 869-71

Johnson, S.R., Liu, B.C.S. and Romig, W.R. (1981) Auxotrophic mutations induced by *Vibrio cholerae* mutator phage VcA1. *FEMS Microbiol. Lett.* **11**, 13-16

Kalderon, D., Oostra, B.A., Ely, B.K. and Smith, A.E. (1982) Deletion loop mutagenesis: a novel method for the construction of point mutations using deletion mutants. *Nucl. Acids Res.* **10**, 5161-71

Kilbey, B.J., Legator, M., Nichols, W. and Ramel, C. (1977) *Handbook for mutagen testing procedures.* Elsevier, Amsterdam, New York and Oxford

Kleckner, N. (1981) Transposable elements in prokaryotes. *Ann. Rev. Genet.* **15**, 341-404

Kleckner, N., Roth, J. and Botstein, D. (1977) Genetic engineering *in vivo* using translocatable drug-resistance elements. New methods in bacterial genetics. *J. Mol. Biol.* **116**, 125-59

Kleckner, N., Reichart, K. and Botstein, D. (1979) Inversions and deletions of the *Salmonella* chromosome generated by the translocatable tetracycline-resistance element. *J. Mol. Biol.* **127**, 89-115

Kornberg, A. (1980) *DNA replication.* W.H. Freeman, New York and San Francisco

Kornberg, A. (1982) *Supplement to DNA replication.* W.H. Freeman, New York and San Francisco

Lichtenstein, C. and Brenner, S. (1982) Unique insertion site of Tn*7* in the *E. coli* chromosome. *Nature (London)* **297**, 601-3

Lindahl, T. (1982) DNA repair enzymes. *Ann. Rev. Biochem.* **51**, 61-87

Little, J.W. and Mount, D.W. (1982) The SOS regulatory system of *Escherichia coli. Cell* **29**, 11-22

Loeb, L.A. and Kunkel, T.A. (1982) Fidelity of DNA synthesis. *Ann. Rev. Biochem.* **51**, 429-57

Megnet, R. (1964) A method for the selection of auxotrophic mutants of the yeast *Schizosaccharomyces pombe. Experientia* **20**, 320-1

Norris, K., Norris, F., Christiensen, L. and Fül, N. (1983) Efficient site-directed mutagenesis by simultaneous use of two primers. *Nucl. Acids Res.* **11**, 5103-12

Osinga, K.A., Van der Blick, A.M., Van der Horst, G., Groot Koerkamp, M.J.A., Tabak, H.F., Veeneman, G.H. and Van Boom, J.H. (1983) *In vitro* site-directed mutagenesis with synthetic DNA oligonucleotides yields unexpected deletions and insertions at high frequency. *Nucl. Acids Res.* **11**, 8595-608

Peden, K.W.C. and Nathans, D. (1982) Local mutagenesis within deletion loops of DNA heteroduplexes. *Proc. Natl Acad. Sci. USA* **79**, 7214-17

Pipas, J.M., Adler, S.P., Peden, K.W.C. and Nathans, D. (1980) Deletion mutants of simian virus 40 that affect the structure of viral tumor antigens. *Cold Spring Harbor Symp. Quant. Biol.* **44**, 285-91

Roberts, D., Hoopes, B.C., McClure, W.R. and Kleckner, N. (1985) IS*10* transposition is regulated by DNA adenine methylation. *Cell* **43**, 117-30

Robinson, M.K., Bennett, P.M., Falkow, S. and Dodd, H.M. (1980) Isolation of a temperature-sensitive derivative of RP1. *Plasmid* **3**, 343-47

Rowlands, R.T. (1983) Industrial fungal genetics and strain selection. In *The filamentous fungi, vol. 4. Fungal technology* (J.E. Smith, D.R. Berry and B. Kristiansen, eds), pp. 346-72. Edward Arnold, London

Salganik, R.I., Diannov, G.L., Ovchinnikova, L.P., Voronina, E.N., Kokoza, E.B. and Mazin, A.V. (1980) Gene-directed mutagenesis in bacteriophage T7 provided by polyalkylating RNAs complementary to selected DNA sites. *Proc. Natl Acad. Sci. USA* **77**, 2796-800

Sancar, A. and Rupp, W.D. (1983) A novel repair enzyme. UvrABC excision nuclease of *Escherichia coli* cuts a DNA strand on both sides of the damaged region. *Cell* **33**, 249-60

Saunders, G., Allsop, A.E. and Holt, G. (1982) Modern developments in mutagenesis. *J. Chem. Tech. Biotechnol.* **32**, 354-64

Shenk, T.E. (1977) A biochemical method for increasing the size of deletion mutations in simian virus 40 DNA. *J. Mol. Biol.* **113**, 503-15

Sherratt, D. (1981) *In vivo* genetic manipulation in bacteria. In *Genetics as a tool in microbiology. Symp. Soc. Gen. Microbiol. vol. 31* (D.A. Hopwood and S.W. Glover, eds), pp. 35-47. Cambridge University Press, Cambridge

Shortle, D. and Botstein, D. (1983) Directed mutagenesis with sodium bisulphite. In *Methods in enzymology, vol. 100. Recombinant DNA part B* (R. Wu., L. Grossman and K. Moldave, eds), pp. 457-68. Academic Press, London and New York

Shortle, D., Koshland, D., Weinstock, G.M. and Botstein, D. (1980) Segment-directed mutagenesis: construction *in vitro* of point mutations limited to a small predetermined region of a circular DNA molecule. *Proc. Natl Acad. Sci. USA* **77**, 5375-9

Shortle, D., DiMaio, D. and Nathans, D. (1981) Directed mutagenesis. *Ann. Rev. Genet.* **15**, 265-94

Singer, B. and Kuśmierek, J.T. (1982) Chemical mutagenesis. *Ann. Rev. Biochem.* **51**, 655-93

Smith, M. (1985) *In vitro* mutagenesis. *Ann. Rev. Genet.* **19**, 423-62

Smith, M. and Gillam, S. (1981) Constructed mutants using synthetic oligodeoxyribonucleotides as site-specific mutagens. In *Genetic engineering*, vol. 3 (J.K. Setlow and A. Hollaender, eds), pp. 1-32. Plenum Press, New York

Southern, P.J. and Berg, P. (1982) Transformation of mammalian cells to antibiotic resistance with a bacterial gene under control of the SV40 early promoter region. *J. Mol. Appl. Genet.* **1**, 327-41

Streisinger, G., Okada, Y., Emrich, J., Newton, J., Tsugita, A., Terzaghi, E. and Inouye, M. (1966) Frameshift mutations and the genetic code. *Cold Spring Harbor Symp. Quant. Biol.* **31**, 77-84

Talmud, P.J. (1977) Mutational synergism between *p*-fluorophenylalanine and UV in *Coprinus lagopus. Mutat. Res.* **43**, 213-22

Toussaint, A. (1985) Bacteriophage Mu and its use as a genetic tool. In *Genetics of bacteria* (J. Scaife, D. Leach and A. Galizzi, eds), pp. 117-46. Academic Press, London and New York

Toussaint, A. and Résibois, A. (1983) Phage Mu: transposition as a life-style. In *Mobile genetic elements* (J.A. Shapiro, ed.), pp. 105-58. Academic Press, New York

Traboni, C., Cortese, R., Ciliberto, G. and Cesareni, G. (1983) A general method to select for M13 clones carrying base pair substitution mutants constructed *in vitro. Nucl. Acids Res.* **11**, 4229-39

Van Vliet, F., Silva, B., van Montagu, M. and Schell, J. (1978) Transfer of RP4::Mu plasmids to *Agrobacterium tumefaciens. Plasmid* **1**, 446-55

Walker, G.C. (1984) Mutagenesis and inducible responses to deoxyribonucleic acid damage in *Escherichia coli. Microbiol. Rev.* **48**, 60-93

Walker, G.C. (1985) Inducible DNA repair systems. *Ann. Rev. Biochem.* **54**, 425-57

Weber, H. and Weissmann, C. (1983) Formation of genes coding for hybrid proteins by recombination between related, cloned genes in *Escherichia coli. Nucl. Acids. Res.*, **11**, 5661-9

Weiss, A.A. and Falkow, S. (1983) Transposon insertion and subsequent donor formation promoted by Tn*501* in *Bordetella pertussis. J. Bacteriol.* **153**, 304-9

Weissmann, C., Nagata, S., Tanaguchi, T., Weber, H. and Meyer, F. (1979) The use of site-directed mutagenesis in reversed genetics. In *Genetic engineering, principles and methods*, vol. 1 (J.K. Setlow and A. Hollaender, eds), pp. 133-50. Plenum Press, New York

Witkin, E.M. (1976) Ultraviolet mutagenesis and inducible DNA repair in *Escherichia coli. Bacteriol. Rev.* **40**, 869-907

Yeung, A.T., Mattes, W.B., Oh, E.Y. and Grossman, L. (1983) Enzymatic properties of purified *Escherichia coli* uvrABC proteins. *Proc. Natl Acad. Sci. USA* **80**, 6157-61

Youngman, P.J., Perkins, J.B. and Losick, R. (1983) Genetic transposition and insertional mutagenesis in *Bacillus subtilis* with *Streptococcus faecalis* transposon Tn*917. Proc. Natl Acad. Sci. USA* **80**, 2305-9

Zakour, R.A., James, E.A. and Loeb, L.A. (1984) Site-specific mutagenesis: insertion of single noncomplementary nucleotides at specific sites by error directed DNA polymerization. *Nucl. Acids Res.* **12**, 6615-28

5

Optimisation of Expression of Cloned Genes

In order to maximise production of proteins from cloned genes it is essential that both transcription and translation are optimised for an appropriate microbial host. This not only improves the yield of such proteins from fermentations, but also assists protein recovery and other aspects of downstream processing. A variety of high-expression vectors has been constructed, principally for *E. coli*, but, increasingly, for those organisms traditionally used in industry. Such expression vectors function by placing the cloned sequence under the control of a strong promoter and a ribosome-binding site that are efficiently recognised by the host machinery for transcription and translation respectively. Gene expression can usually be enhanced by using cloning vectors that alter the dosage of the cloned sequence and ensure that the sequence is maintained stably throughout growth of cultures. This chapter describes improvements in yield of proteins that can be achieved by adopting such strategies. Genetic manipulations necessary to circumvent post-translational barriers to high-level expression of genes in heterologous backgrounds and to assist in the purification of desired proteins from microbial cells are also considered.

5.1 TRANSCRIPTION OF CLONED GENES

5.1.1 Promoters and the regulation of gene expression

Promoters are DNA sequences that direct the binding of RNA polymerase to DNA for the initiation of RNA synthesis. Measurements of the amount of RNA and the protein it encodes indicate that there is considerable variation in the strength of promoters. Sequencing of a number of *E. coli* promoters has revealed structural similarities that are now regarded as characteristic features (Figure 5.1). There are two highly conserved regions, one at about 35 bases upstream (the −35 region) and one at about 10 bases upstream (the −10 region or Pribnow box) of the start of

Figure 5.1: DNA sequence in the promoter region of the *E. coli lac* operon

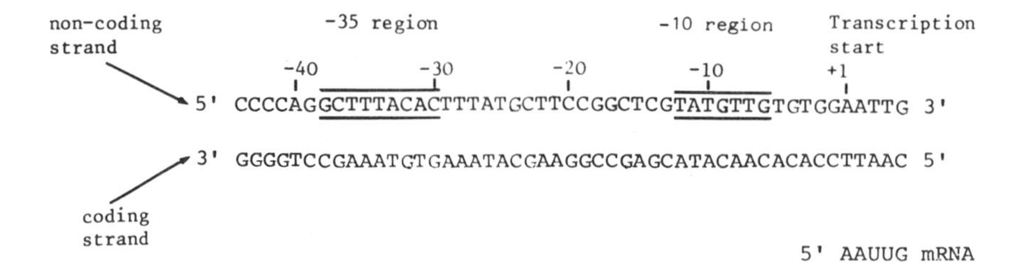

transcription (Rosenberg and Court, 1979). The strength of promoters seems to be determined partly by their degree of homology with the agreed or consensus sequences for the −10 (5′ TATAATG 3′) and −35 (5′ TTGACA 3′) regions. However, no natural promoter has yet been identified with perfect consensus sequences at both −35 and −10 positions. The spacing between the −35 and −10 regions, which ranges in natural promoters from 16 to 19 nucleotides, also seems to be critical in determining promoter strength. A spacing of 17 nucleotides would appear to be ideal. It should be noted that certain *E. coli* promoters, for example those for regulating genes involved in nitrogen metabolism, have features that differ from those described above (see section 8.2.2.c).

The vegetative promoters of *Bacillus subtilis* that have been sequenced so far correspond closely in the conserved −10 and −35 regions with those of *E. coli* (see, for example, Murray and Rabinowitz, 1982). However, *E. coli* genes are generally not expressed in *B. subtilis* whereas *Bacillus* genes are generally expressed in *E. coli* (see section 5.7). Moreover, sequences between the −10 and −35 regions and upstream of the −35 sequence are conserved in *B. subtilis* but not in *E. coli* promoters. This suggests that the RNA polymerase of *Bacillus* is more stringent in its requirements for promoter recognition than the corresponding enzyme of *E. coli.* The RNA polymerase holoenzymes of both species comprise several subunits designated α_2, β, β' and σ, although the sizes of the components and their functions vary. Recognition and selection of promoters in *E. coli* are dependent upon the direct interaction of the sigma (σ) factor with specific DNA sequences (Siebenlist *et al.*, 1980). The sigma subunit of *B. subtilis* RNA polymerase is smaller than that of *E. coli* and contains an additional subunit called the delta (δ) factor, which serves to enhance discrimination among DNA templates. Vegetative and sporulating cells of *B. subtilis* contain forms of polymerase that differ with respect to their promoter recognition specificity. This is due to interaction of the core enzyme with different sigma factors (Losick and Pero, 1981). Promoter recognition in vegetative cells is determined primarily by σ^{55}. A second polymerase containing σ^{28}, which has a different promoter specificity, is also present in vegetative cells, but its function is as yet unresolved. Two further polymerases, one containing σ^{37} and the other σ^{29}, are involved in the initiation and maintenance of sporulation respectively. These polymerases seem to recognise different but overlapping promoters of certain *Bacillus* genes (Tatti and Moran, 1985). The promoters recognised by the different polymerases of *B. subtilis* exhibit conserved regions at positions −10 and −35. However, the conserved sequences of some of these promoters differ from those found in *E. coli* (Table 5.1).

Promoters in *Streptomyces* have similar but somewhat more diverse consensus sequences compared with *E. coli* (Janssen *et al.*, 1985). Heterogeneity in promoter specificity of RNA polymerase has also been found in *Streptomyces* (Westpheling *et al.*, 1985). Two forms of RNA polymerase were identified. One contained the sigma factor σ^{35} and appeared to have specificity for the same consensus sequences as *E. coli* and *B. subtilis* σ^{55} RNA polymerase. The other contained sigma factor σ^{49} and recognised a consensus sequence related to that of *B. subtilis* σ^{37} RNA polymerase. It is possible that this second RNA polymerase is required for the expression of genes involved in differentiation and/or secondary metabolism. Promoter specificity modulated by RNA polymerase heterogeneity may pose problems for the expression of certain genes from native promoters in heterologous genetic backgrounds (see section 5.7).

Table 5.1: RNA polymerases and promoters of bacteria

Sigma factor of RNA polymerase	Conserved sequences in promoters* -35 region	-10 region	Comments
E. coli	TTGACA	TATAAT	
B. subtilis σ^{55}	TTGACA	TATAAT	Major form of polymerase in vegetative cells. Recognises *veg* promoter
σ^{28}	CTAAA	CCGATAT	Minor form in vegetative cells
σ^{37}	AGGNTT	GGNATTGNT	Present in vegetative cells. Involved in initiation of sporulation. Recognises *ctc* promoter
σ^{29}	ANTTNAAAA	CATATTNT	Replaces σ^{55} during sporulation. Can overlap with σ^{37} in specificity
S. coelicolor σ^{35}	TTGACA	TATAAT	May be normal vegetative form. Recognises *veg* promoter of *B. subtilis*
σ^{49}	TGATTGA	GGGCAGGGG	May be involved in differentiation and secondary metabolite production. Recognises *ctc* promoter of *B. subtilis*

N = any nucleotide.
*Note that these are consensus sequences and actual examples may vary.

The control of transcription in eukaryotes differs from that in prokaryotes. Eukaryotic nuclei contain three types of RNA polymerase, each transcribing a separate set of genes. RNA polymerase I acts on the ribosomal RNA (rRNA) transcription unit, RNA polymerase II transcribes the many protein-encoding genes, whereas RNA polymerase III is responsible for the synthesis of transfer RNA (tRNA) and 5S rRNA. Promoters recognised by polymerases II and III are identified as conserved sequences located a short distance from the transcription initiation site. Promoters for genes transcribed by polymerase I are less conserved, although there are detectable homologies between the transcription initiation regions of rRNA genes of closely related species. Genes transcribed by polymerase II may contain three types of control sequence located upstream of the transcription start site (see Figure 3.2) (Chambon *et al.*, 1984). About 30 bases upstream of the start site is a sequence, usually referred to as the **TATA box**, that is generally homologous to the sequence 5′ TATAA 3′. This sequence is recognised by RNA polymerase II and may be analogous to the −10 sequence of the prokaryotic promoter. Alternatively the role of the TATA box may be to fix the initiation site of transcription to a position just upstream of the ATG codon that is at the beginning of the coding sequence (Guarente, 1985). A second class of sequences forms the **upstream element** or **upstream activation site** (UAS) (also referred to as the CAAT box) and lies variable distances upstream of the TATA box). This element seems to determine the efficiency of transcription and may therefore be the analogue of the prokaryotic promoter. In yeast some UASs may be regulated by the levels of metabolites. For example, expression of the *CYC1* gene is affected over a 10-fold range by catabolite repression (see section 6.2.1.a) (Guarente, 1985). At variable distances further upstream of the upstream elements of mammalian cells and viruses are found **enhancer sequences**. These can stimulate transcription at a distance and may be tissue-specific modulators of expression for certain genes (Walker *et al.*, 1983). For a discussion of the possible mechanisms of upstream activation the reader is referred to Guarente (1985). Some genes, for example human β-globin (Charnay *et al.*, 1984) also have regulatory sequences that lie downstream of the transcription start point within the structural genes. Such a complexity of gene regulation makes the manipulation of expression in eukaryotes potentially more difficult than in prokaryotes. However, because the UAS can be spaced at variable intervals from the TATA box, it is possible to construct functional hybrid promoters, for example in yeast, which are controlled *in cis* by a particular UAS (Guarente, 1985).

5.1.2 Promoters used in expression vectors

Expression of genes that are normally controlled by weak promoters can be dramatically improved by placing such genes downstream of known strong promoters. Table 5.2 illustrates some of the promoters commonly utilised for this purpose. Ideally a strong promoter should be readily controllable. This allows expression of cloned genes at the optimum time during growth

Table 5.2: Some controllable promoters for use in expression vectors

	Promoter	Source	Operational control	
			Off	On
E. coli	λpL, λpR	Leftward and rightward early promoters of λ	30°C	>37°C (in cI_{857} host)
	lac	*E. coli lac* operon	—	IPTG in medium
	trp	*E. coli trp* operon	Tryptophan in medium	Indoleacetic acid in medium
	tac	*trp*-35 region *lac*-10 region hybrid	—	IPTG in medium
	phoA	*E. coli* alkaline phosphatase operon	Excess phosphate in medium	Phosphate-limited medium
	recA	*E. coli recA* gene	—	Mitomycin C in medium
	tet	Tn*10* tetracycline-resistance gene	—	Tetracyclines in medium
B. subtilis	*bla*	*Bacillus licheniformis* β-lactamase gene	—	β-lactams in medium
	cat	*Bacillus pumilis* chloramphenicol acetyl transferase	—	Chloramphenicol in medium
Streptomyces	*gyl*	*Streptomyces coelicolor* glycerol operon	Glucose in medium	Glycerol in medium
S. cerevisiae	*ADH*	Yeast repressible alcohol dehydrogenase (*ADR*) gene	High glucose in medium	Low glucose in medium
	GAL1	Yeast galactose utilisation operon	Glucose in medium	Galactose in medium
	GPD-PH05	Hybrid between yeast glyceraldehyde 3-phosphate dehydrogenase and alkaline phosphatase gene promoters	Excess phosphate in medium	Phosphate-limited medium

of a culture and prevents the uncontrolled synthesis of products that are potentially deleterious to the cell. Furthermore, a constitutive high level of transcription may interfere with plasmid replication and hence lead to instability of the vector (see section 5.4.3).

The most commonly used expression vectors employ the *lac*UV5 or *trp* promoter from *E. coli* or the pL promoter from bacteriophage lambda, all of which can be readily controlled. The *lac* promoter is regulated by *lacI* repressor and can be induced by addition of the gratuitous inducer isopropyl-β-D-thiogalactoside (IPTG) to the growth medium. The *trp* promoter is regulated by the *trp* repressor and can be induced by adding either 3-indolylacetic acid or indoleacrylic acid to the medium. Alternatively, the culture can be starved of tryptophan. The *tac* promoter, a hybrid formed by fusing the −35 region of the *trp* operon to the −10 region of the *lac* operon, has been employed for the efficient production of human growth hormone in *E. coli* (de Boer *et al.*, 1983). The *tac* promoter combines the consensus (TTGACA) −35 sequence from *trp* with the consensus (TATAAT) −10 sequence from *lac* producing an 'ideal' hybrid promoter. It is at least three and ten times stronger than the parental *trp* and *lac* promoters respectively, and is controlled by *lac* repressor. The *recA* promoter may also be used for high-efficiency expression of cloned genes (Shirakawa *et al.*, 1984). In this case expression is induced by addition of mitomycin C to the medium which elicits the SOS response and relieves *lexA* repression (see section 4.3.4).

The pL promoter of lambda is regulated by the lambda *c*I repressor. This promoter is conveniently regulated by having a gene encoding a temperature-sensitive repressor (for example, cI_{857}) resident either on a prophage in the chromosome of the host *E. coli* or on the vector molecule itself. The lambda pL and pR promoters are very useful regulators for the expression of cloned genes. They exhibit extremely tight control of transcription and there is very little lag between the uninduced and induced states. In the uninduced state the lambda repressor ensures that there is essentially no transcription from pL and yet once induced this promoter is extremely efficient. This is in contrast to promoters for 'metabolic' operons, where control is not tight and some transcription occurs even in the uninduced state. However, thermoinduction of *cIts* mutants in large-scale culture does present the difficulty and expense of raising fermenter temperature from noninducing (30°C) to inducing (>37°C).

A *cIts* λ lysogen may also be used to regulate gene expression *in vivo* through Int/Xis-mediated site-specific recombination at *att*λ sites (Backman *et al.*, 1984). A specialised plasmid, pKB730, designed for this purpose was constructed by cloning into pBR322 analogues of *att*λ*R* and *att*λ*L*, the gene II promoter from phage M13 and as a target sequence in this case a promoter-less T4 DNA ligase gene. The gene II promoter and ligase gene were functionally separated from each other, but were each located adjacent to an *att* site. In the presence of Int and Xis (resulting from induction of the λ lysogen) a site-specific rearrangement of the plasmid (involving the *att*λ sites) occurred. One product of the reaction possessed the ligase gene alongside the promoter in a functional configuration (Figure 5.2). Following the rearrangement, up to 20% of soluble cell protein was found to be T4 DNA ligase. A strategy of this type has applications where precise timing of activation of expression, for example of a lethal gene product, is required.

Figure 5.2: Regulation of gene expression by synchronous site-specific recombination. p, Promoter; →, direction of transcription; *oriV*, origin of vegetative replication; t_1, t_2, transcription terminators 1 and 2 respectively; Ap^r, resistance to ampicillin; *attL*, *attR*, hybrid attachment sites from left and right of λ integration site respectively; *attP*, *attB* attachment sites from λ and bacterial chromosome respectively; Int, λ integrase; Xis, λ excisionase

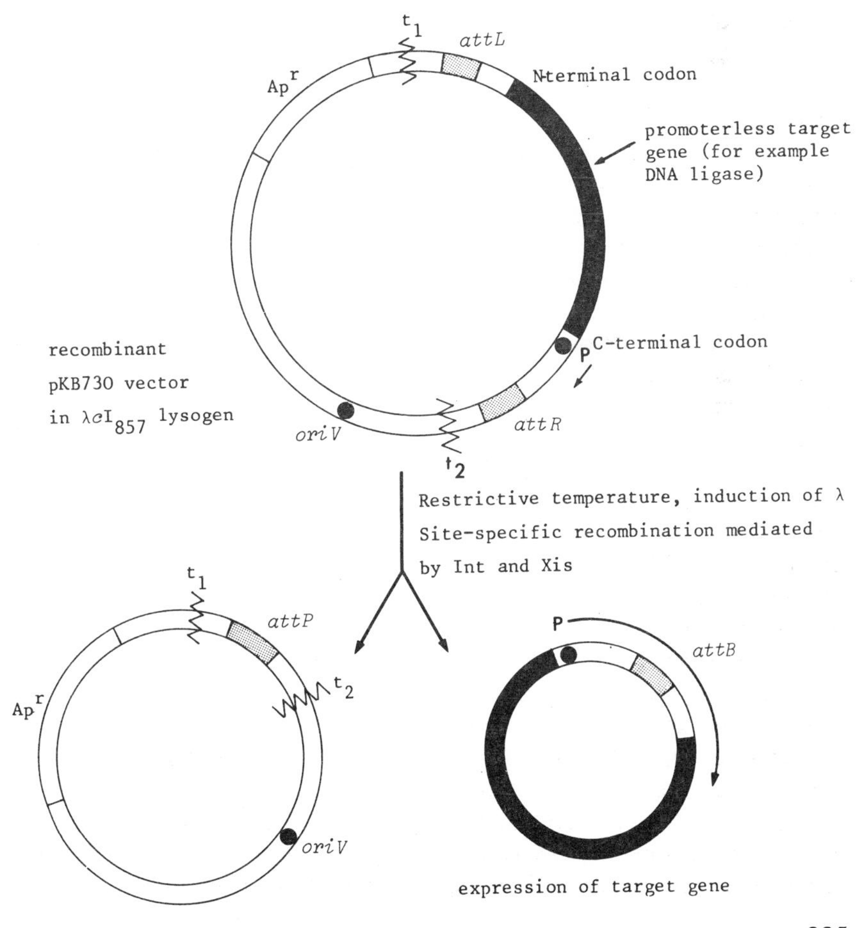

A more general solution to the problem of controlling potentially lethal gene expression may be to exploit the regulatory properties of anti-sense mRNA (Weintraub *et al.*, 1985). Anti-RNA species, which are complementary to mRNA (or other functional RNA), are produced by convergent, overlapping transcription of a gene sequence. Control of copy number in ColE1 plasmids (section 5.4.1) and of transcription in Tn*10* (section 2.1.1) provide examples of the role of small RNA species in regulating certain genes. Regulation of expression of a gene that is detrimental to the host organism could be achieved by placing the coding sequence downstream of a strong controllable promoter. Under non-inducing conditions any transcription that escapes control of that promoter and reads through the coding sequence could be negated by directing transcription in the opposite direction from the complementary DNA strand using an appropriately sited promoter (for example p*lac*) that has been induced. This type of control could occur simply by mechanical interference between RNA polymerase molecules as they transcribe in opposite directions. However, it seems equally likely that base pairing of the mRNA from the gene sequence with its complementary anti-mRNA species might inhibit subsequent translation.

Increases in transcription efficiency can be achieved by deleting regulatory sequences from control regions. For example, removal of the attenuator region (*trpa*) of the *trp* operon leads to a 7- to 8-fold increase in transcription efficiency from the *trp* promoter (Yanofsky, 1981; Tacon *et al.*, 1983). The amount of transcription of cloned genes may be increased further by placing the desired coding sequence downstream of tandem sequences comprising two or more identical (for example, p*trp*, Tacon *et al.*, 1983), or nonidentical (for example p*lac* and p*lpp*, (lipoprotein), Masui *et al.*, 1984) promoters.

5.1.3. Termination of transcription

The efficient and faithful termination of transcription is an important factor in optimising gene expression. Precise termination at the end of a cloned gene prevents long and wasteful transcripts being made. Interference with transcription of essential vector genes that lie adjacent to the passenger DNA is also prevented (see section 5.4.3). Transcriptional termination signals in bacterial DNA typically contain GC-rich sequences of variable length (from 3 to 11 consecutive GC bases) and an inverted repeat sequence immediately preceding the actual point of termination (Rosenberg and Court, 1979). There is substantial variation both in the length of the repeat and of the distance between repeats and hence in the stem length and loop size respectively, if the DNA or corresponding RNA molecule were to form intrastrand base-paired structures. Transcription terminators vary considerably in strength. Strong terminators that have been employed to regulate expression of cloned genes include the *trpa* terminator (Christie *et al.*, 1981) and the phage T7 early region terminator (Dunn and Studier, 1980). Where desired the DNA sequence for such a terminator may be inserted immediately downstream of the passenger DNA. Alternatively expression vectors are available in which the coding sequence of the passenger is automatically bracketed at its 5′ end by a strong promoter and its 3′ end by a strong transcription terminator.

5.2 TRANSLATION OF CLONED GENES

5.2.1 Control of translation

Production of large amounts of mRNA in a particular bacterial host is no guarantee that correspondingly large quantities of the appropriate protein will be synthesised. Eukaryotic mRNA contains translational initiation signals that are generally not recognised by prokaryotic ribosomes. It is therefore necessary to ensure that the coding sequence for a gene of interest is located downstream of an efficient ribosome-binding site and other signals appropriate to the host concerned. In *E. coli* the ribosome-binding site consists of an initiation codon (AUG or less commonly GUG) and a sequence typically of 3 to 9 nucleotides called the Shine-Dalgarno (S-D) sequence, which lies 3 to 11 nucleotides upstream of the translation start (Shine and Dalgarno, 1974) and is complementary to the sequence 5′ ACCUCC 3′ located in the 3′ end of *E. coli* 16S ribosomal RNA of the 30S ribosomal subunit. This complementarity permits ribosome binding by base-pairing between the S-D sequence on mRNA and the complementary sequence on 16S rRNA (Figure 5.3). The initiator tRNA, the 50S ribosomal subunit and the amino acyl tRNA molecules are added to this ribosome/mRNA complex for translation to proceed.

5.2.2 Translation in heterologous backgrounds

The S-D sequence is not identical for all mRNA species and the degree of complementarity with the 3′ end of 16S RNA varies from 2 to 8 nucleotides. Surprisingly, an S-D sequence of only 4 nucleotides complementarity was found to be more efficient that one of 8 nucleotides, which was in turn more efficient than one of 12 nucleotides in providing expression of human growth hormone in *E. coli* (de Boer *et al.*, 1983). The optimum spacing between the S-D sequence and the initiator codon, AUG, seems to be from 7 to 9 nucleotides. Precise spacing between the S-D sequence and the ATG

Figure 5.3: Base-pairing interactions of RNA species during the initiation of protein synthesis. fMet-tRNA^{fMet}, *N*-formylmethionyl-tRNA^{fMet} (initiator tRNA)

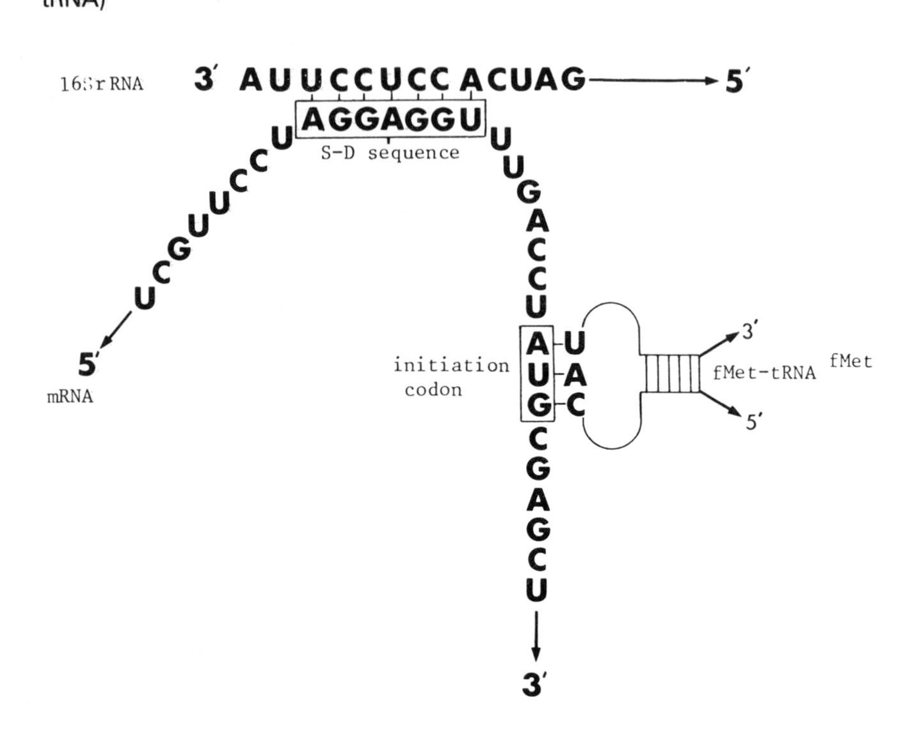

can be manipulated if one or more unique restriction sites lie between these sequences. Cleavage at such sites and removal of protruding nucleotide tails with S1 nuclease, or filling in of termini with DNA polymerase, followed by religation can be used to delete or insert known numbers of bases. It appears to be necessary to determine the ideal spacing each time a new gene is to be expressed in *E. coli.* This is probably because the secondary structure of individual mRNA molecules affects access of ribosomes to the S-D and AUG sequences. For example, abolition of secondary structure in the S-D-AUG region of a cloned murine heavy chain immunoglobulin gene can lead to dramatic increases in expression of the appropriate protein due to unmasking of the ribosome-binding site on mRNA (Wood *et al.*, 1984).

Many ribosome-binding sites contain one or more termination codons which may function to ensure that ribosomes translating the previous cistron dissociate and do not interfere with efficient initiation. Ribosome-binding sites for genes encoding capsid proteins from such phages as ϕX174, fd and MS2 and those for the ribosomal proteins L11, L12 and S12 contain all or part of a sequence called the RRUUURR sequence (where R = A or G). The RRUUURR sequence may occur in addition to, or instead of, the S-D sequence. The presence of RRUUURR in the control regions of genes that are translated with high efficiency suggests that the sequence may be of use in the construction of vectors for the efficient expression of eukaryotic genes in bacteria. Synthetic DNA sequences containing combinations of desirable features of *E. coli* ribosome-binding sites have been used to express efficiently the small tumour antigen of SV40 (Figure 5.4) (Jay *et al.*, 1982). Similar approaches are applicable to many other genes and hosts (Jay *et al.*, 1985).

Although the fundamental machinery for protein synthesis is probably similar in all prokaryotes, it has frequently proved difficult to express prokaryotic genes in heterologous species. Sequencing of ribosome-binding sites from *B. subtilis* suggests that the sequence GGAGG is essential and that more extensive complementarity between mRNA and the 3′ end of 16S rRNA is required than in *E. coli* (Murray and Rabinowitz, 1982).

5.2.3. Codon usage

With notable exceptions in mitochondria, ciliates and mycoplasmas, the same triplet codons encode the same amino acids in different organisms. The genetic code provides two or more options for all the amino acids except methionine and tryptophan, which each have a single codon. It is

Figure 5.4: A synthetic ribosome-binding site for directing expression of eukaryotic proteins in *E. coli*. 1, S-D sequence; 2, RRUUURR (GGTTTAA) sequence; 3, termination codon (TAA) in frame with the β-lactamase gene when cloned into the *Pst*I site on pBR322. The synthetic ribosome-binding site is inserted into recombinants such that the *Pst*I site is on the promoter proximal side. Target DNA bearing *Hin*dIII termini can be ligated to the *Hin*dIII terminus of the synthetic sequence. Alternatively the synthetic *Hin*dIII terminus can be end-filled with Klenow fragment and blunt-end ligated to appropriate target DNA. Translation of the target DNA depends on the presence of an initiator Met (ATG) codon being located about 3 to 11 bases downstream of the S-D sequence

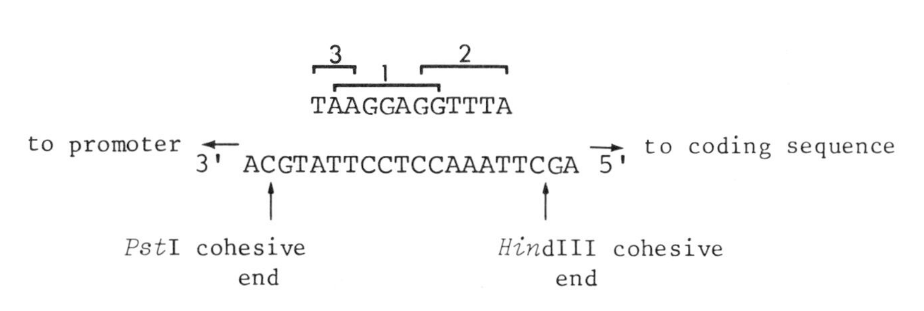

evident from sequence studies of many prokaryotic, eukaryotic and viral mRNA species that there are marked preferences in the use of alternative codons (Bibb *et al.*, 1984). Codon usage depends both upon the genome from which the message originates and the degree of expressivity of that message (Grantham *et al.*, 1981; Gouy and Gautier, 1982). For example, CCG (as opposed to CCA, CCC or CCU) is the favoured codon for proline in bacteria, and CUG (as opposed to CUA, CUC, CUU, UUA or UUG) is strongly favoured as the codon for leucine in both bacteria and animal cells. Furthermore, preference for both CCG and CUG is stronger in mRNA species from genes that are normally highly expressed. Codon preferences are believed to reflect molecular optimisation of codon-anticodon pairing energies.

Disparities between codon usage patterns of species may theoretically cause reduced efficiency of translation in heterologous backgrounds, for example when expressing a eukaryotic gene in *E. coli.* A requirement for 'rare' charged tRNA molecules would cause pauses in translation of foreign genes and increase the chances of incorporating incorrect amino acids into polypeptides (Old and Primrose, 1985). However, this has not proved to be a major problem in practice. Furthermore, if necessary, the natural coding sequence of a gene may be altered to optimise codon usage for a particular host, without affecting the amino acid sequence of the resultant translational product. Optimisation of codon usage is most conveniently achieved by *in vitro* synthesis of all or part of the appropriate coding sequence (see, for example, Goeddel *et al.*, 1979).

5.2.4 Stability of mRNA

Messenger RNA species differ considerably in their stability, which can lead to differential accumulation of specific proteins. Typical mRNA species, for example for the chloramphenicol acetyl transferase (*cat*) gene, have half-lives of 2 to 3 min in *E. coli.* However, mRNA from some genes, notably *ompA*, which encodes the highly abundant principal outer membrane protein I of *E. coli*, have much longer half-lives (up to 20 min). Stability of mRNA species, such as *cat* and *ompA* transcripts, is dependent on growth rate: half-life decreases with increasing culture doubling time (Nilsson *et al.*, 1984). In contrast other mRNA species, for example from the *bla* or lipoprotein (*lpp*) genes, are equally stable over a range of doubling times. mRNA from some heterologous genes may be particularly unstable during growth. Half-life is presumably determined, at least in part, by the secondary and tertiary RNA structure, which may be predisposed to degradation by cellular ribonucleases. Furthermore, mRNA structure may also influence the efficiency of translation, notably in eukaryotes (Vournakis, 1985). Restructuring of gene sequences by gene cloning and

DNA synthesis may overcome or reduce such potential problems of mRNA instability and expressivity.

5.2.5 Ensuring the correct reading frame

Where it is necessary to express a foreign gene as a fusion to a vector gene it is essential that the coding sequence of the former is inserted in the correct translational reading frame with respect to the latter (see sections 3.7.1.d and 3.9.5). The inserted DNA will only be in phase by chance in one out of three cases. It is, however, possible to shift the reading frame by one or two bases by trimming the ends of the coding fragment by limited exonucleolytic digestion (section 3.2.3) prior to ligation. A more efficient approach to obtaining correct fusions is to use a specialised series of three vectors into which the same insert DNA can be cloned. Each vector is essentially the same, except that the cloning site for insertion is located so that the junction between vector and insert is shifted by one base with respect to the next member of the series. Examples of such vectors have been described for cloning *Eco*RI fragments into a *lacZ* gene carried by phage λ or plasmid vectors (Charnay *et al.*, 1978) and *Hin*dIII fragments into a *trpE* gene on pBR322-based vectors (Tacon *et al.*, 1980). In each case, successful fusion should occur following insertion of any given foreign coding sequence into one of the three vectors of the series.

5.2.6 Termination of translation

Translation originating upstream of a ribosome-binding site used for the expression of cloned genes may lead to undesired gene fusion products or interfere with translation of the target sequence itself. Undesired C-terminal fusion polypeptides may also be produced if the target sequence contains no in-frame stop codon. A solution to these problems is to insert, on either side of the desired sequence, a synthetic oligonucleotide, for example 5′ TGATTGATTGA 3′ containing a translational stop (TGA) codon in all three reading frames (Pettersson *et al.*, 1983).

5.3. ISOLATION OF INDIGENOUS TRANSCRIPTION AND TRANSLATION SIGNALS

The well characterised promoters, terminators and ribosome-binding sites described above may function poorly, if at all, in certain microorganisms, including many of industrial importance. It may therefore be necessary to obtain strong transcriptional and/or translational control sequences that are indigenous to specific hosts. Suitable sequences for regulating transcription can be cloned by utilising specialised **promoter cloning vectors**, able to replicate in the host and containing a structural gene lacking a functional promoter. The structural gene should be preceded by one or more unique restriction sites for cloning. Gene expression resulting from the insertion of a promoter sequence in one of these sites should provide a positive trait for selection of desired clones (Figure 5.5). The insertion of a regulatory sequence at the cloning site will produce either a transcriptional (operon) or translational (gene) fusion, depending on whether the structural test gene cartridge possesses or lacks a functional ribosome binding site respectively.

Promoter probe vectors have been constructed for various species, including *E. coli* (Casadaban and Cohen, 1980), *B. subtilis* (Goldfarb *et al.*, 1981) and *Streptomyces* spp. (Bibb and Cohen, 1982). A frequent choice for the indicator gene in such vectors has been the chloramphenicol acetyl transferase (CAT) gene derived from Tn*9*. Expression of this gene usually confers chloramphenicol resistance on the host. The relative strength of cloned promoters can also be measured by a simple enzyme assay for chloramphenicol acetyl transferase.

5.4 OPTIMISING GENE DOSAGE

Increasing the number of copies of a gene is likely to increase the amount of the particular gene product that is manufactured by a host cell. Gene dosage may be increased automatically when a gene that is normally carried in single copy on the chromosome of a microorganism is cloned on a multicopy plasmid or phage vector. Further amplification of gene dosage may be possible by cloning multiple tandem copies of a gene into a vector. It should, however, be appreciated that raising gene dosage through the use of multicopy vectors does not lead to elevated expression of all genes. Autoregulatory mechanisms controlling the expression of some genes may mean that higher yields of gene product are obtained in low rather than high copy number vectors. If such processes regulate a gene of interest, they could be overcome by subcloning the appropriate coding sequence downstream of a suitable promoter not subject to autoregulatory control. Moreover, some genes, particularly those coding for regulatory proteins or proteins that are normally associated with the cell membranes, appear to be lethal to the host when cloned on high copy number, but not low copy number, plasmid vectors.

Figure 5.5: Isolation of promoters using promoter cloning vectors. p, Promoter; →, direction of transcription; *oriV*, origin of vegetative replication

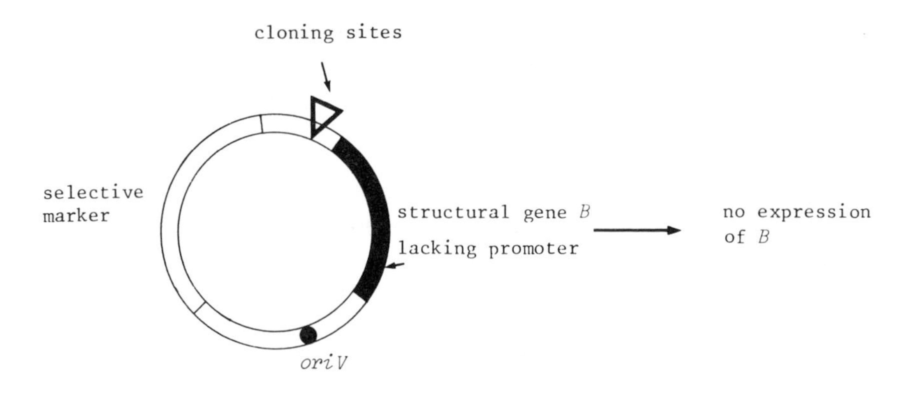

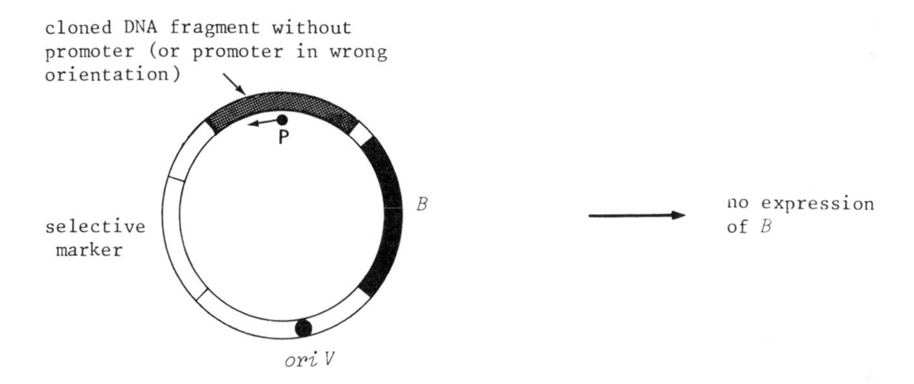

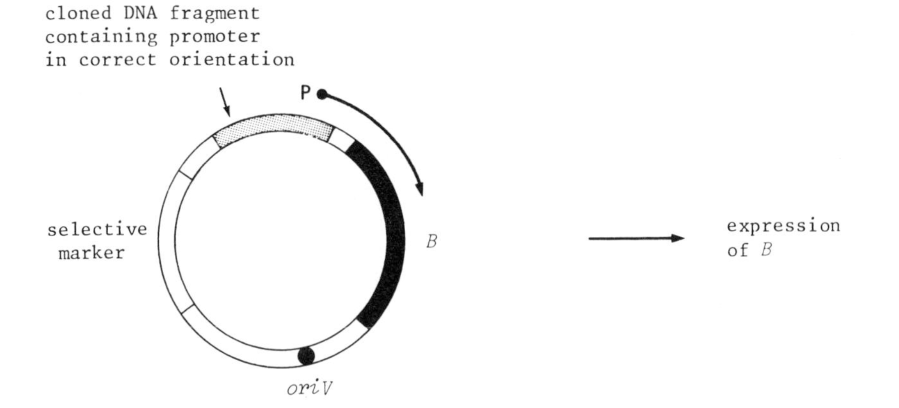

5.4.1 Amplification of vector copy number

Most viral genomes naturally undergo amplification as a consequence of the infectious cycle. By contrast, wild-type plasmids maintain an essentially constant and characteristic copy number. However, copy number of plasmid vectors can be increased by isolating variants with altered control of replication.

(a) Regulation of plasmid replication

Plasmid replication and copy number are normally strictly regulated. Two species of RNA molecules, RNA I and RNA II, encoded in the vicinity of the origin of vegetative replication (*oriV*), play a crucial role in controlling replication of ColE1 and other plasmids. The RNA II species of ColE1 is transcribed from a promoter upstream of *oriV* and terminates beyond it (Figure 5.6). The RNA II molecule is processed near the origin by RNase H to form a 555-base primer required for the initiation of DNA replication by DNA polymerase I (Tomizawa *et al.*, 1981). RNA I, a species 108 nucleotides in length, is transcribed from the same region as RNA II, but in the opposite direction (Figure 5.6). Several mutations (*cop*) that increase copy number map to the region that specifies RNA I. Furthermore, mutations in this region may alter the incompatibility properties of ColE1 (Lacatena and Cesareni, 1983). RNA I molecules can form cloverleaf stem-loop structures due to intrastrand base pairing (Figure 5.7). It has been proposed that base pairing between the loops on RNA I and the complementary structure on the nascent RNA II primer precursor and subsequent pairing along the entire length of RNA I prevents maturation of the primer transcript (Lacatena and Cesareni, 1981, 1983; Tomizawa, 1985). Negative regulation by RNA I probably occurs by inhibition of the formation of the RNA-DNA hybrid, which is a substrate for RNase H. RNA I only inhibits primer formation in homologous plasmids, suggesting that it is both a specific regulator of copy number and the major determinant of incompatibility. Binding of RNA I to the homologous RNA II is, however, inhibited by the RNA I species produced by a compatible plasmid. Inhibition is caused by the reversible interaction of RNA II with the heterologous RNA I, which competes with the normal reversible binding of homologous RNA I and II prior to their stable binding. However, the heterologous RNA molecules do not proceed to form stable complexes. As a consequence the copy numbers of two compatible plasmids are actually increased when present in the same cell (Tomizawa, 1985).

Another *trans*-acting regulatory element affecting copy number is encoded about 600 bp downstream of the replication origin, close to the mobility genes of ColE1 (Twigg and Sherratt, 1980). This element enhances the incompatibility exerted by RNA I and can be complemented by gene products of the compatible plasmid ColK. Regulation is effected

Figure 5.6: Genetic organisation of the replication and maintenance functions of ColE1. Genes and regions for noncoding RNA indicated on the inside of the circle are transcribed in an anticlockwise direction, whereas those on the outside of the circle are transcribed in a clockwise direction. Numbers above the map of replication region indicate nucleotide positions orientated with respect to the unique *Eco*RI site on ColE1 DNA according to Chan *et al.* (1985). pI and pII are the promoters for RNA I and RNA II respectively. (The 63 amino acid *rom* polypeptide is encoded by the ORF between nucleotides 1803 and 1614.) *cea*, Colicin E1 structural gene; *imm*, colicin immunity gene; *inc*, region that specifies the noncoding RNA I species and in which mutations affecting ColE1 incompatibility are located; *oriV*, origin of replication; *rom*, RNA I inhibition modulator; *mob*, genes for mobility polypeptides; *cer*, ColE1 oligomer resolution sequence; *exc*, region that confers resistance to conjugative transfer (by mobilisation) between mating pairs carrying homologous ColE1 plasmids

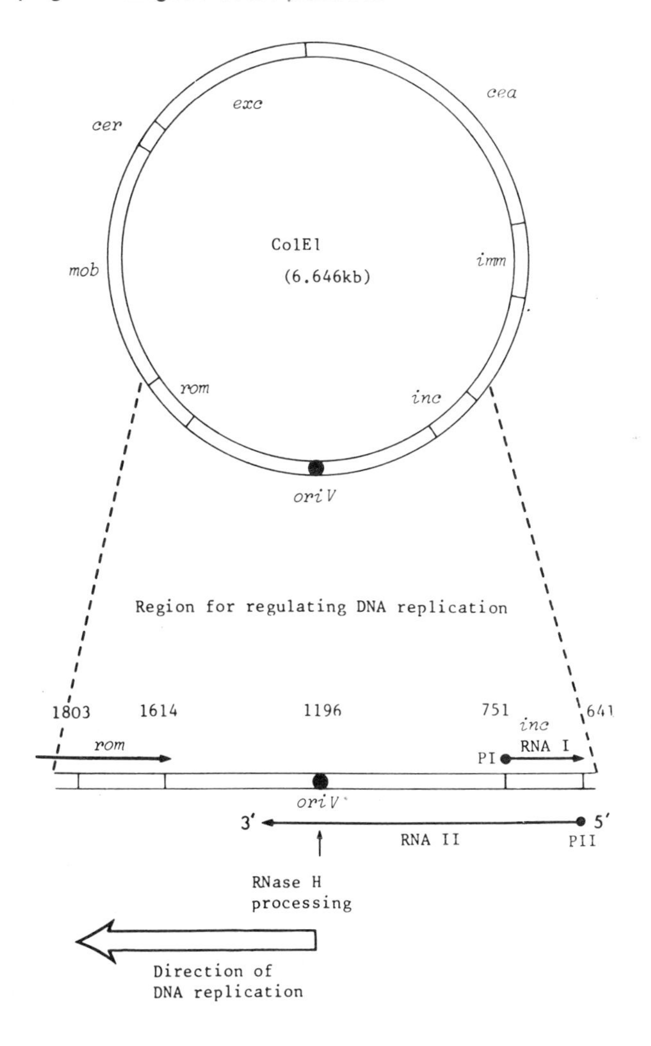

Figure 5.7: Model for the control of ColE1 replication. The arrangement of the possible folded structures of RNA I and RNA II is according to Tomizawa (1985). I′, II′ and III′ are the stem-loop structures that would form by base pairing between the palindromic sequences I, II and III respectively. Gaps in the base pairing in the stems indicate base mismatches. Primary interaction of the loops on RNA I and RNA II is indicated by solid lines. Subsequent interaction is believed to occur between the 5′ end of RNA I and complementary sequences in RNA II as indicated by broken lines. Interaction of RNA I and RNA II inhibits processing of the preprimer by RNaseH. For inhibition to be effective the RNA I molecule must bind to the nascent RNA II molecule before the RNA II transcript forms a hybrid with the template DNA. In the absence of RNA I the RNA II transcript forms a persistent hybrid with DNA after about 550 nucleotides have been synthesised

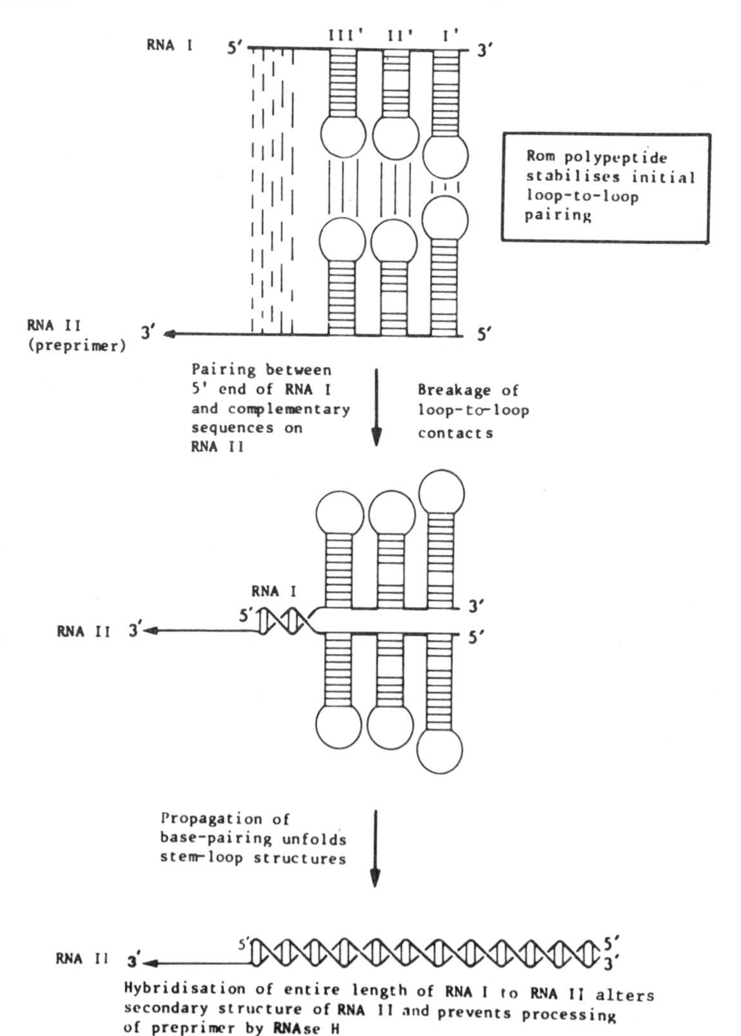

by a polypeptide of 63 amino acids encoded by a gene formerly called *rop* (*r*egulation *o*f *p*rimer) (Cesareni *et al.*, 1982). This region of the ColE1 genome is now known as the *rom* (*R*NA *o*ne inhibition *m*odulator) gene, which more clearly denotes its function (Chan *et al.*, 1985). RNA I is required both *in vivo* and *in vitro* for *rom* activity. The *rop*/*rom* gene product is thought to regulate plasmid replication by enhancing the initial reversible interaction between RNA I and the RNA II primer precursor (Tomizawa, 1985) (Figure 5.7). Amplification of ColE1 replicons, which occurs when host cells are treated with chloramphenicol, is explained in part by inhibition of synthesis of the Rom protein.

Analogous systems for regulating other plasmids have been found. For example, the low copy number plasmid R1 possesses a *copA* gene specifying an RNA species analogous to RNA I. The *copA* transcript hydridises with a complementary RNA species (encoded within the same segment of DNA) that specifies RepA protein required for plasmid replication. *copA*, an incompatibility determinant, may control copy number by inhibiting translation of the *repA* transcript (which may also prime DNA replication). *copB* encodes a protein which may repress initiation of transcription from the strong repA promoter. However, RepA is probably also expressed from the weaker constitutive copB promoter. A further replication control mechanism is found in plasmids such as F, R6K and pSC101. In these plasmids directly repeated DNA sequences of about 20 bp are clustered in the replication region and appear to be involved in the titration of Rep protein(s) (Terawaki and Itoh, 1985). The intensity of incompatibility towards a test plasmid seems to be proportional to the number of repeats present, suggesting that these sequences are Inc determinants. (For a thorough discussion of the regulation of plasmid replication see Scott, 1984; Sherratt, 1986.)

Plasmid mutants with elevated copy number may be isolated by selecting for derivatives that exhibit increased expression of their antibiotic resistance determinants. Such mutants of R1 have been isolated in which copy number is increased by two-fold or more. Furthermore, temperature-sensitive mutants can be isolated in which the normal copy number is maintained at 30°C but 'runaway' replication (where plasmid replication continues unabated until the host dies) occurs at 42°C (Uhlin *et al.*, 1979).

(b) Strategies for increasing vector copy number

Expression of certain cloned genes can be enhanced by using plasmid vectors with *cop* and/or *rom* mutations. ColE1 derivatives that are stably maintained at 500 or more copies per cell and that give enhanced expression have been constructed by using this approach (see, for example, Tacon *et al.*, 1983). However, the maintenance of such large numbers of plasmid molecules imposes a high genetical and physiological burden on the host and is likely to reduce the efficiency of large-scale fermentations. Moreover, continuous high-level expression of certain cloned genes may be deleterious to the host and further reduce growth efficiency. An ability to manipulate the replication machinery such that the vector can be maintained at low or high copy number at desired stages in a fermentation is therefore advantageous. One approach is to infect the culture with a virus vector at the time during growth that will give maximal expression of the cloned gene(s). However, it is difficult to ensure simultaneous infection of all cells in a culture under production conditions. An alternative is to use an inducible lysogen carrying, for example, a $\lambda c\mathrm{I}_{857}$-based vector (section 3.5.2). However, for ease of use and long-term stability of cloned genes, plasmid vectors are likely to prove superior to phage-based systems. The use of temperature-sensitive runaway-replication plasmid mutants (Uhlin *et al.*, 1979) provides a convenient method for modulating copy number by adjusting the operating temperature of the fermenter. Large amounts of T4 DNA ligase (20% of the total cellular protein) have been obtained within 3 h of temperature induction of a runaway replication plasmid vector that contained the ligase gene cloned downstream of λpL (Remaut *et al.*, 1983). An alternative approach is to use a plasmid vector, such as pMG411, that carries both a low and a high copy number origin of replication. In such a plasmid the native promoter for the high copy number replication primer (RNA II or its analogue) is replaced by a controllable promoter such as λpL (Yarranton *et al.*, 1984). In a $c\mathrm{I}_{857}$ lysogen the λpL promoter will not be operational at the permissive temperature (30°C). Hence plasmid replication from the high copy number origin will not occur. The vector will therefore be maintained using the low copy origin (Figure 5.8). When the temperature is raised sufficiently to inactivate the *c*I repressor, λpL will function to direct the synthesis of large quantities of RNA II primer at the multicopy origin. This in turn results in an increase in plasmid copy number and hence in the dosage of any genes cloned in the vector. However, in view of the operational difficulties that can be encountered when using temperature-sensitive repressors to control promoter activity under production conditions, metabolic promoters, whose activities rely on the addition of effector molecules to the fermenter (section 5.1.2), might ultimately prove more effective and cheaper for regulating plasmid copy number.

5.4.2 Tandem gene systems

Gene dosage may be increased by cloning several copies of a gene into the same vector. Tandem copies of DNA sequences are inherently unstable, due to recombinational looping out of homologous sequences. This problem can be obviated, in some circumstances, by the use of *recA*$^-$ or *recBC*$^-$ hosts (or their equivalents in organisms other than *E. coli*).

Figure 5.8: Use of a dual-origin plasmid vector to regulate plasmid copy number. Plasmid pMG411 is derived from a *rom*⁻ ColE1 plasmid in which the RNA II promoter has been replaced by λpL. The vector also carries a temperature-sensitive origin of replication from the low copy number plasmid pSC101. This origin can be used to maintain about four plasmid copies per chromosome equivalent at 30°C, but is nonfunctional above 37°C. pMG411 is maintained stably at low copy number in a λcI_{857} lysogen at 30°C using the pSC101*ts* origin. Raising the temperature to 42°C inactivates both the *cI* repressor and the pSC101*ts* origin. This permits efficient transcription of the ColE1 RNA II origin primer from pL and the plasmid is replicated to high copy number. →, Direction of transcription; ⇦, direction of DNA replication from ColE1 origin; Ap^r, resistance to ampicillin

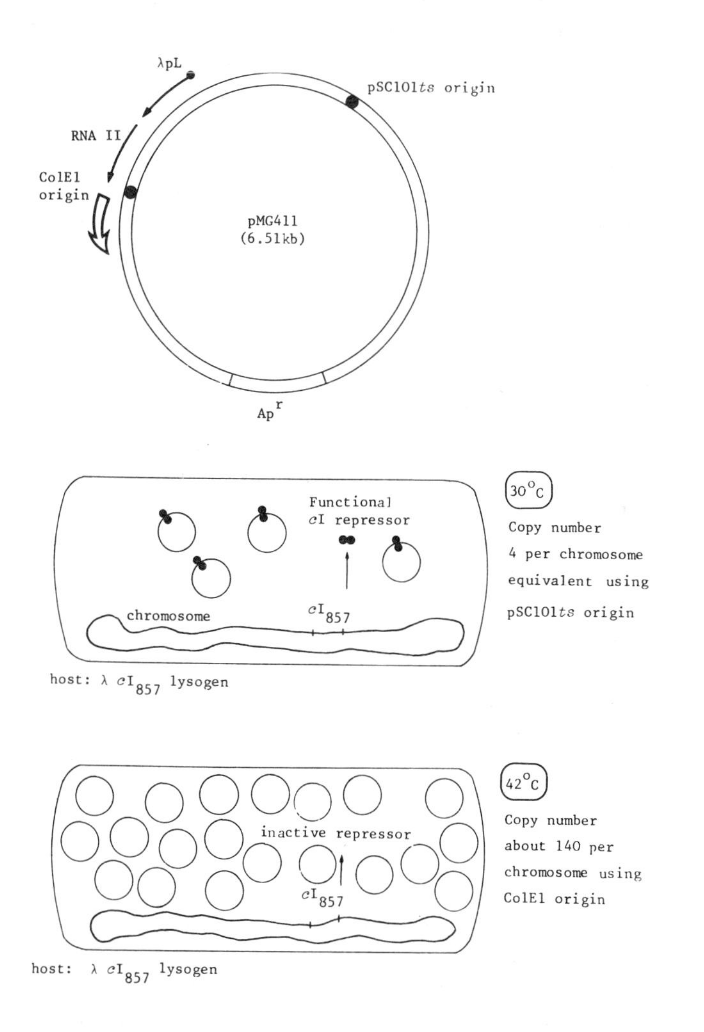

However, a drawback of recombination-deficient hosts is that they frequently exhibit reduced viability, which might offset gains in gene expression obtained from the stable maintenance of tandem gene copies. Furthermore, rearrangements in cloned DNA frequently occur by *recA*-independent processes.

Cloning of tandem copies of a gene, each complete with promoter and transcription terminator, is wasteful and increases vector size. Increased size of plasmid vectors not only reduces transformation efficiency (see section 3.5) but also increases instability (section 5.4.3). Economy in tandem gene systems can be achieved by constructing a homopolycistronic sequence (Figure 5.9) in which two or more gene coding sequences, each with its own ribosome-binding site, are bounded on one side by a promoter and on the other by a transcription terminator (Lee *et al.*, 1984). By dispensing with the intervening promoters and terminators from a tandem array of genes, it is possible to reduce plasmid size while still achieving high levels of gene expression. Multicopy plasmid vectors carrying the *E. coli* lipoprotein (*lpp*) promoter followed by up to four human leukocyte interferon genes and a transcription terminator have been found to be stable for >100 generations in a *recA*$^-$ *E. coli* host (Lee *et al.*, 1984). Furthermore the titres of interferon produced by the host were roughly proportional to the number of interferon genes per plasmid.

Figure 5.9: Optimisation of gene expression by construction of a homopolycistronic sequence. RBS, Ribosome binding site; p*A*, promoter for gene *A*; t*A*, transcription terminator for gene *A*; 8 , ribosome; ~~~, nascent polypeptide

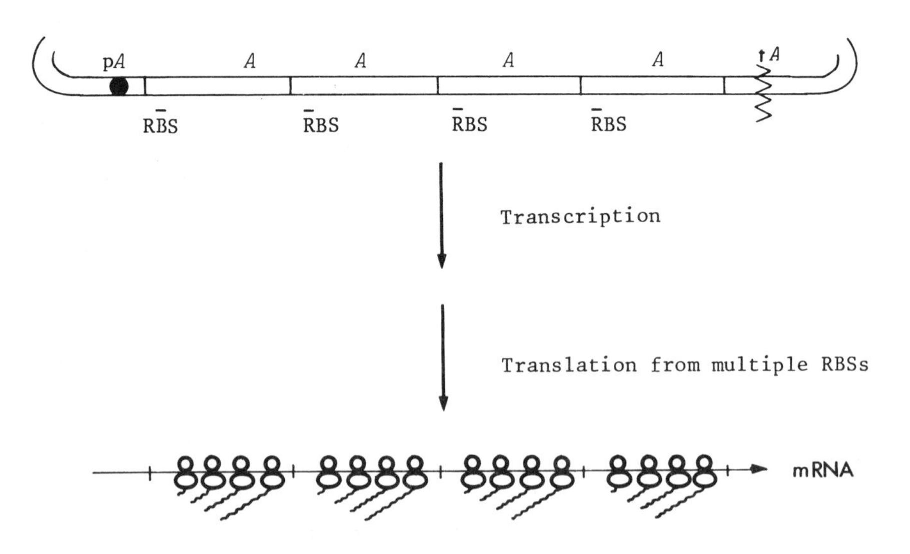

5.4.3 Stability of vectors

Carriage of plasmids imposes a burden upon the host. Plasmid-free cells that arise during growth may thus ultimately outgrow plasmid-containing cells. Most wild-type plasmids, including those with low copy number, are relatively stable in their natural hosts. However, the genetic manipulations necessary to produce cloning vectors have sometimes resulted in the creation of plasmids that are less stable than their parental replicons. Instability is accentuated during prolonged growth and may result in the displacement of plasmid$^+$ by plasmid$^-$ cells. For example, pBR322 and pBR325 can be lost from *E. coli* grown under conditions of glucose or phosphate limitation in a chemostat (Jones *et al.*, 1980; Noack *et al.*, 1981). This indicates that the high copy number of plasmids like pBR322 is of itself insufficient to ensure stability.

(a) The genetic basis of plasmid stability

Certain low copy number plasmids contain loci that are required for faithful segregation of plasmid copies to daughter cells. A DNA segment of 270 bp, designated *par*, is required *in cis* for the correct partitioning of the plasmid pSC101 (Meacock and Cohen, 1980). No *trans*-acting products are required for the functioning of *par*, nor does the *par* locus have any effect on plasmid incompatibility. In contrast, partitioning of the F plasmid seems to require a *cis*-acting locus *incD* (*sopC*) and two *trans*-acting gene products encoded by *incG* (*sopB*) and *sopA* (Ogura and Hiraga, 1983). Unlike *par*, the *incD* locus does contribute to F incompatibility. Partition loci, such as *par* and *incD*, are apparently analogous to the centromere of eukaryotic chromosomes and may represent sites for the attachment of plasmids to the host cell membrane, ensuring even distribution of DNA molecules at cell division (Meacock and Cohen, 1980; Austin and Abeles, 1983; Gustafsson *et al.*, 1983).

ColE1 and related multicopy plasmids, such as CloDF13, possess alternative partition mechanisms for ensuring stability. Plasmids such as pBR322, pAT153 and pACYC184 frequently form oligomers in *E. coli.* Oligomerisation occurs during replication or by interplasmidic recombination. Plasmid oligomers can be resolved to monomers by intraplasmidic recombination events (Fishel *et al.*, 1981; Laban and Cohen, 1981; Doherty *et al.*, 1983). There is strong correlation between oligomerisation and plasmid instability, presumably because replication control mechanisms limit the total number of plasmid origins of replication. Consequently the presence of oligomers will reduce the total number of independent plasmid molecules and increase the chances of unequal partitioning at division. However, ColE1, the parental replicon for pBR322 and pAT153, is maintained stably by host cells. ColE1 carries a *cis*-acting locus, *cer* (see Figure 5.6), which is responsible for reversing oligomerisation (Summers and Sherratt, 1984). The *Hae*IIB fragment of ColE1 confers a 100- to 1000-fold increased stability when inserted *in vitro* into pAT153. The segment of ColE1 DNA that carries *cer* was deleted in the construction of pBR322 and its derivatives. The *cer* locus provides the substrate for a site-specific recombination system that is independent of *recA*-, *recF*- and *recE*-mediated recombination events. The recombination system that is responsible for resolving plasmid dimers and higher order oligomers to monomers is encoded by an *E. coli* chromosomal gene(s) designated *xer* (Figure 5.10). The degree of expression of *xer* varies in different strains of *E. coli*. Therefore, the extent of oligomerisation and hence instability of ColE1 is strain dependent. This may be an important consideration when choosing a host strain for large-scale fermentations. A locus called *parB* on CloDF13 is analogous to *cer* and is required for the recombinational resolution of oligomers of plasmid copy number mutants but not of wild-type plasmids (Hakkaart *et al.*, 1984).

Instability of plasmid cloning vectors may be accentuated when transcription of cloned genes is directed from strong promoters. Such effects are probably due partly to readthrough transcription from a strong promoter interfering with the normal transcription processes at the origin of plasmid replication (Stuber and Bujard, 1982). The high-level expression of cloned genes may also place considerable demands on the protein-synthesising machinery of the host cell leading to a selective advantage for cells that have lost the vector plasmid (Caulcott *et al.*, 1985).

(b) Strategies for ensuring vector stability

Instability of plasmid cloning vectors is undesirable since the efficiency of fermentation will be reduced by plasmid loss. It is therefore essential to adopt strategies that ensure maintenance of the vector in all cells of a culture. A simple method of ensuring extrachromosomal maintenance is to incorporate a selective agent, normally an antibiotic appropriate to the resistance determinant(s) carried by the vector, into the culture medium. However, this approach is generally expensive. Furthermore, it may increase selective pressures for the development and dissemination of drug resistance in bacterial populations. An alternative is to incorporate a gene essential to the host into the vector. Cells that lose the vector will therefore be unable to grow. For example, growth of a streptomycin-dependent (*strM*) mutant of *E. coli* can be rendered independent of the antibiotic if the strain carries a multicopy plasmid bearing the appropriate wild-type, streptomycin-independent sequence. Loss of the plasmid renders the host dependent on the drug. Plasmid-free cells that arise will therefore stop growing in streptomycin-free medium. A similar strategy has been employed for improving yields of L-tryptophan by cloning the *trp* genes on a multicopy plasmid that carries the *serB*$^+$ locus. If the plasmid is lost from a *serB*$^-$ *E. coli*, the host becomes auxotrophic for serine and will therefore

Figure 5.10: Plasmid oligomer formation and resolution in ColE1 plasmids. (a) in a *xer*$^+$ host with *cer*$^+$ plasmids, oligomers are resolved, therefore plasmid-free daughter cells are generated with low probability. (b) In a *xer*$^+$ host with *cer*$^-$ plasmids (or a *xer*$^-$ host with *cer*$^+$ plasmids) oligomers are not resolved, therefore plasmid-free daughter cells arise with high probability. *cer*, Plasmid recombination (resolving) site; *xer*, chromosomal gene(s) for site-specific recombination; 1, 2, 3, 4, monomer, dimer, trimer, tetramer respectively; ●, origin of vegetative replication; ►, direction of plasmid sequences (all plasmid sequences are in direct repeat in these oligomeric molecules)

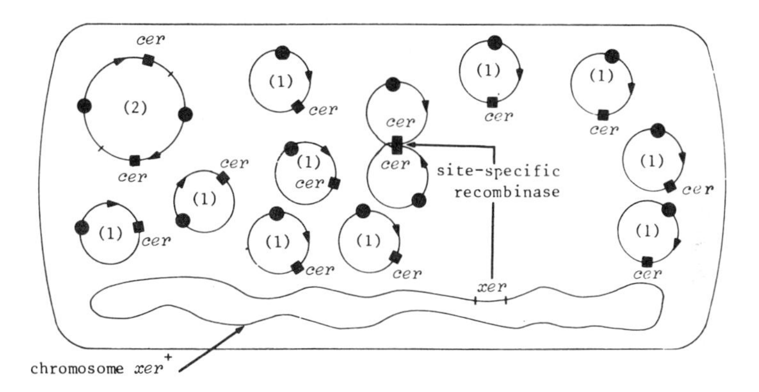

a. host (xer^+), ColE1 plasmids (cer^+) 14 origins ∿ 13 plasmid molecules

Cell division

plasmid-free cells generated with low probability

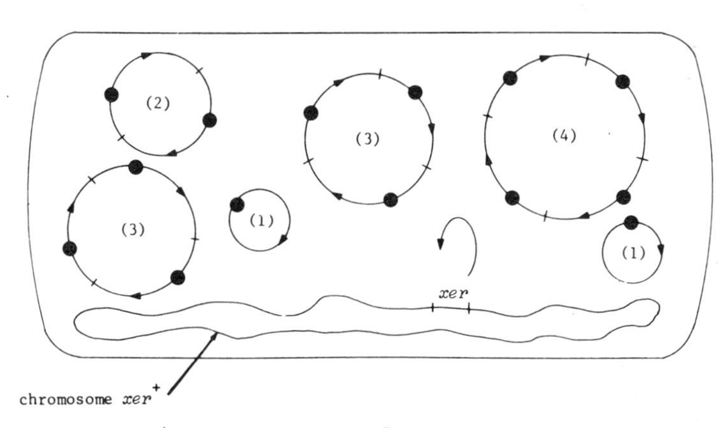

b. host (xer^+), ColE1 plasmids (cer^-) 14 origins ∿ 6 plasmid molecules

Cell division

plasmid-free cells generated with high probability.

be unable to grow in medium that lacks this amino acid (Anderson *et al.*, 1983).

A principal disadvantage of methods that select for retention of the vector is that they do not ensure maintenance of high copy number, since selection will generally be effective only when the last plasmid copy has been lost from a cell line. A genetic approach that builds stability functions into the vector is therefore likely to be more effective in the long term. Substantial increases in plasmid stability in batch and chemostat culture have been achieved by insertion *in vitro* of the *par* locus of pSC101 into multicopy vectors, such as pBR322 and pACYC184 (Jones *et al.*, 1980; Meacock and Cohen, 1980; Skogman *et al.*, 1983; Caulcott *et al.*, 1985). An alternative (or additional) measure is to clone the *cer* region of ColE1 into such vectors and propagate the plasmid in a xer^+ background. rom^- plasmid vectors are often used to maximise gene dosage. However, *E. coli* strains bearing rom^- plasmids grow more poorly than those containing rom^+ plasmids. Therefore, to combine stability with strong expression it might be desirable to use temperature-sensitive rom^- plasmid mutants, or provide Rom *in trans* using a compatible, low copy number plasmid. Stability of expression vectors that carry strong promoters can be enhanced by ensuring that the cloned DNA is bracketed on its promoter distal side by a strong transcriptional terminator. This prevents readthrough transcription from the 'expression' promoter from interfering with transcription of the replication primer (see section 5.1.3). Transcriptional repressors such as *lacI* and *tetR*, whose effects can be alleviated by appropriate inducers, can also be used to damp down transcription until required and hence stabilise a plasmid vector (Lambert and Reznikoff, 1985).

The problem of vector instability may be overcome by inserting the cloned sequence into the chromosome(s) of the host organism. The main disadvantage of such a strategy is a reduced flexibility to manipulate gene dosage. The limited number of gene copies that can be inserted makes the choice of efficient transcriptional and translational signals even more critical than for genes located on extrachromosomes. Tandem insertions of a gene into the same chromosome may also lead to instability of desired characteristics, due to recombination between homologous DNA sequences.

A variety of techniques can be used to ensure insertion of a cloned gene into the host chromosome. For example, insertion may be directed at *att*λ sites in *E. coli* using λ-based vectors (see section 2.4.2). Genes may be inserted at specific chromosomal loci in yeast using YIp vectors (section 3.5.5). In *B. subtilis*, insertion of genes carried on plasmid vectors may arise as a consequence of recombinational rescue following DNA uptake during transformation (section 2.2.1), provided that the incoming vector plasmid carries sequences homologous to the recipient chromosome (Figure 5.11). It is possible to rescue a second (identical or different)

Figure 5.11: Insertion of genes into chromosomes using scaffolding. (a) Integration of recombinant vector using homology provided by a gene on the vector and chromosome. (b) Integration of a second recombinant vector into scaffold provided by the first. ●, Plasmid origin of replication

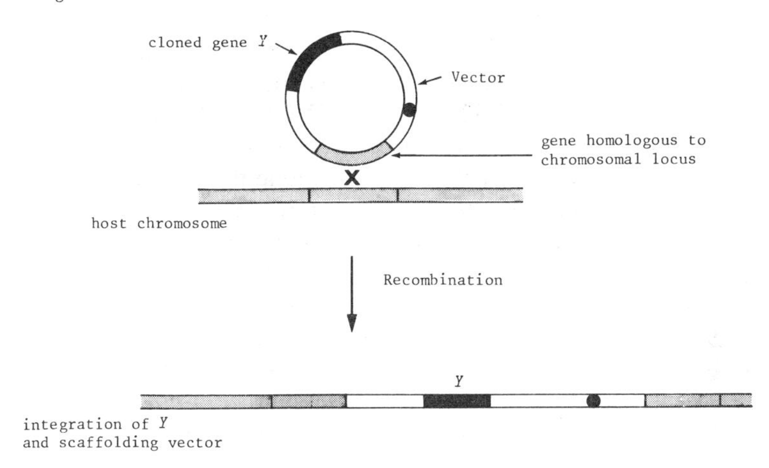

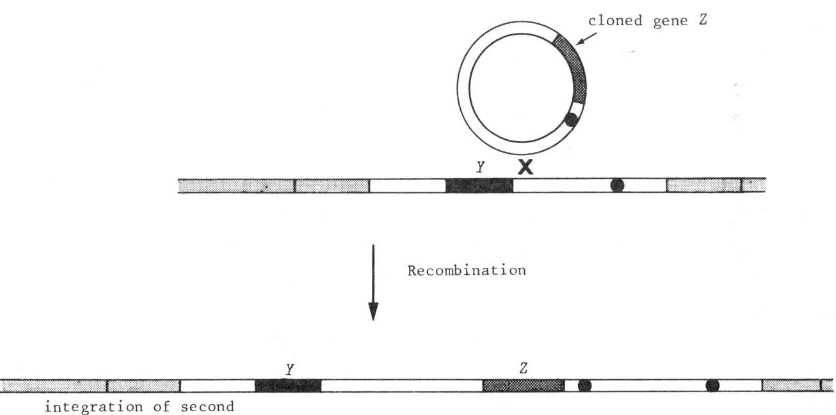

cloned gene using vector sequences to provide the homology for rescue. This technique, referred to as 'scaffolding' (Young, 1980), provides a means of inserting a series of identical or different genes at desired sites in the *B. subtilis* or other genome.

5.5. PROTEIN STABILITY

Most proteins that are synthesised naturally by *E. coli* (and probably other bacteria) are stable and turn over extremely slowly, except under adverse conditions, such as nutrient starvation (Mount, 1980). However, abnormal proteins produced, for example, by the incorporation of amino acid analogues and certain proteins expressed from heterologous DNA are rapidly degraded by proteolytic enzymes. Turnover of proteins from cloned genes varies greatly between different bacterial strains of the same species. *E. coli* and other bacteria contain a variety of intracellular proteases. Some bacteria, notably *Bacillus* spp., also produce large quantities of extracellular proteases. Such proteolytic enzymes can substantially reduce the yield of foreign proteins.

5.5.1 Removal of proteolytic activity by mutation

Mutants deficient in production of peptidases and/or proteases can be isolated for yeast and a number of bacteria. *E. coli* strains deficient in the Lon protease (*lon*$^-$ mutants) are particularly valuable hosts for expressing cloned genes. The *lon* gene product is an ATP-dependent serine protease. This enzyme acts *in vivo* as a regulator of certain operons by specific cleavage of regulatory proteins. One target of *lon* protease is the *sulA* gene product, which is an inhibitor of cell-division septation (Mizusawa and Gottesman, 1983). *sulA* is regulated by the SOS system (section 4.3.4) through LexA protein. Synthesis of SulA protein is therefore induced by DNA damage. Hence Lon$^-$ bacteria become filamentous following UV irradiation. Degradation of nonsense polypeptides and proteins encoded by some cloned genes is reduced in *lon*$^-$ mutants of *E. coli.* For example, expression of murine light chain immunoglobulin was substantially higher in *E. coli* B which is deficient in Lon protease than in *E. coli* HB101 which is *lon*$^+$ (Boss *et al.*, 1984).

5.5.2 Inhibition of proteases

Inhibition of the degradation of abnormal proteins occurs when *E. coli* is infected with bacteriophages, such as T4, T5 or T7. A potent inhibitor of protein degradation in *E. coli* is produced as an early function in T4 infection. The *pin* (*pro*teolysis *in*hibitor) gene encoding this inhibitor has been cloned and used to stabilise the expression of human fibroblast interferon without apparently affecting the turnover of normal proteins in *E. coli* (Simon *et al.*, 1983). Genes for protease inhibitors that function in other microorganisms could be incorporated into vectors used to direct high-level expression in these hosts.

5.5.3 Saturation of proteolytic activity

The problem of intracellular proteolysis can sometimes be overcome simply by overproducing the desired polypeptide, for example in a burst at the end of fermentation. The large amounts of foreign protein produced by enhancing transcription and translation may be sufficient to saturate the cellular capacity of the Lon and other proteases. Synthesis of large amounts of foreign protein (between 1 and 40% of total cellular protein with most expression systems) in *E. coli* is frequently accompanied by filamentation of the host and the production of refractile, crystalline inclusion bodies (Figure 5.12). Filamentation probably results from titration of the Lon protease by the excess of foreign protein. As a consequence, sufficient *sulA* product remains to inhibit septation and cell division. The inclusion bodies are masses of insoluble foreign proteins. These bodies may be formed even from normally soluble proteins when the physicochemical environment provided by the host interferes with normal folding and/or processing of the polypeptide. Aberrant disulphide bridges formed between cysteine residues in different polypeptide chains account in part for the formation of such insoluble inclusions (Shoemaker *et al.*, 1985). The formation of insoluble intracellular bodies can be advantageous, since proteolytic breakdown of the product is reduced. Recovery of product may also be facilitated, since the protein is concentrated in a form that can be purified relatively easily from broken bacterial cells. However, the protein in inclusion bodies not only may contain abnormal linkages but is also frequently denatured and may therefore lack properties of the native protein. Furthermore, the vigorous chemical extraction procedures that may be required to solubilise the inclusions for product recovery may result in loss of normal protein function. The yield of functional protein may, therefore, represent only a small fraction of the total protein in such bodies. For some proteins with particularly labile pharmacological or enzymatic activities, this may present a problem.

Figure 5.12: Filamentation and inclusion bodies in *E. coli* expressing high levels of the B chain of human relaxin under the control of the *trp* promoter. Approximate magnification ×3300 (Courtesy of H. Richards)

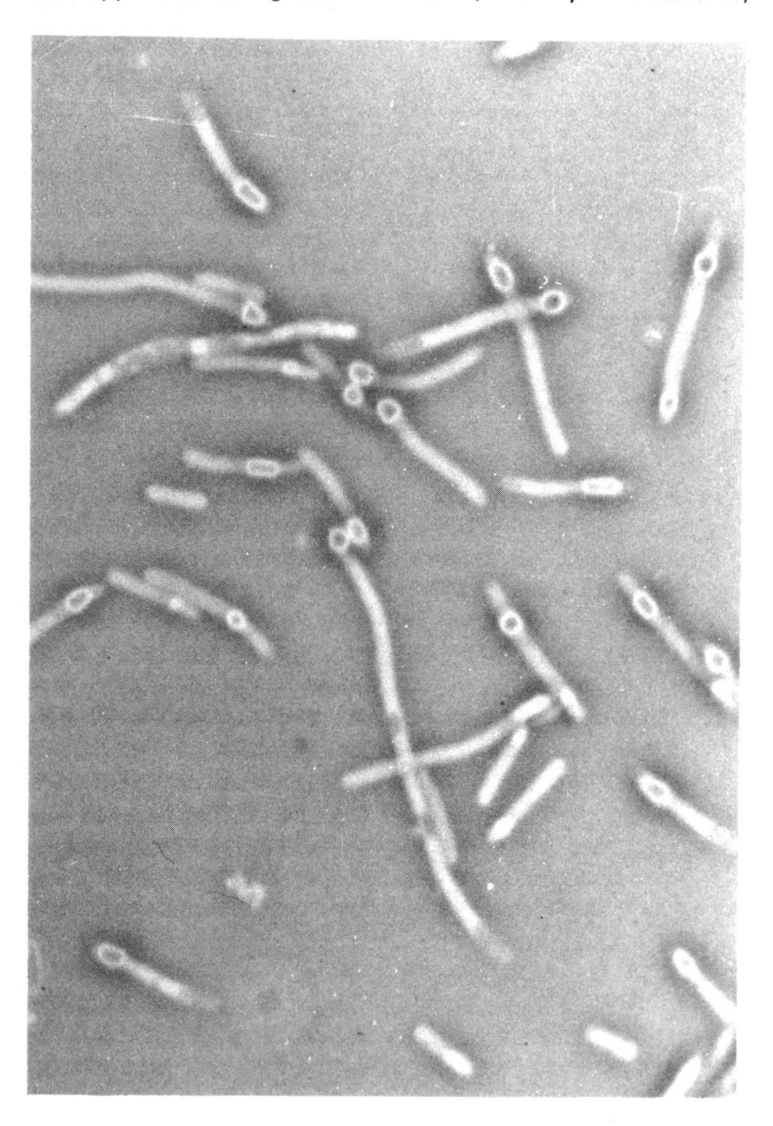

5.5.4 Polypeptide fusions

Some foreign proteins can be stabilised against proteolytic degradation by synthesis as **fused polypeptides**. By utilising an appropriate cloning strategy (see sections 3.9.5 and 5.2.5) the foreign protein can be synthesised as a fused product with a native stable protein (such as β-lactamase or β-galactosidase in *E. coli*). This approach is applicable where a fused protein is acceptable as an end product, or where the hybrid protein produced can be cleaved specifically to release active product. Release of the desired product may be achieved chemically by cyanogen bromide treatment which cleaves polypeptides on the carboxyl side of methionine residues. Alternatively proteolytic enzymes that cleave polypeptides at specific amino acid residues may be utilised to release product from fused proteins. For example, trypsin cleaves polypeptides on the carboxyl side of arginine and lysine residues. Such strategies rely on the fact that sensitive amino acid residues either do not lie within the desired protein sequence, or, if present, can be specifically protected by chemical modifications. For example, lysine residues can be protected against trypsin action by treatment with citracomic anhydride. Native protein can subsequently be recovered by removal of citracomic groups.

Stabilisation of the hormone somatostatin, which is degraded rapidly when synthesised in *E. coli*, has been achieved by inserting a synthetic somatostatin gene into the *lacZ* gene (Itakara *et al.*, 1977). The synthetic somatostatin coding sequence included an extra codon that specified Met at the N terminus of the polypeptide, and the gene itself contained no internal Met codons. It was therefore possible to purify native somatostatin from the N-terminal β-galactosidase peptide following cyanogen bromide treatment (Figure 5.13).

5.5.5 Exporting the protein

Intracellular proteolysis can be avoided by the rapid removal of foreign proteins through secretion from the cell. The accumulation of foreign polypeptides to potentially toxic concentrations in the host cell is also less likely by using this approach. Furthermore, certain polypeptides may be purified more economically from a large volume of growth medium that is relatively uncontaminated with undesired products than from concentrated cell extracts that would be heavily contaminated.

Compartmentalisation of proteins differs in Gram-negative and Gram-positive bacteria as a consequence of differences in the architecture of the cell envelope. In Gram-positive bacteria, such as *B. subtilis*, proteins are located in the cytoplasm, in or on the cytoplasmic membrane, or are exported to the growth medium. In Gram-negative bacteria, such as *E.*

Figure 5.13: Synthesis and recovery of somatostatin from a β-galactosidase fusion polypeptide. p*lac*, Promoter from the *lac* operon

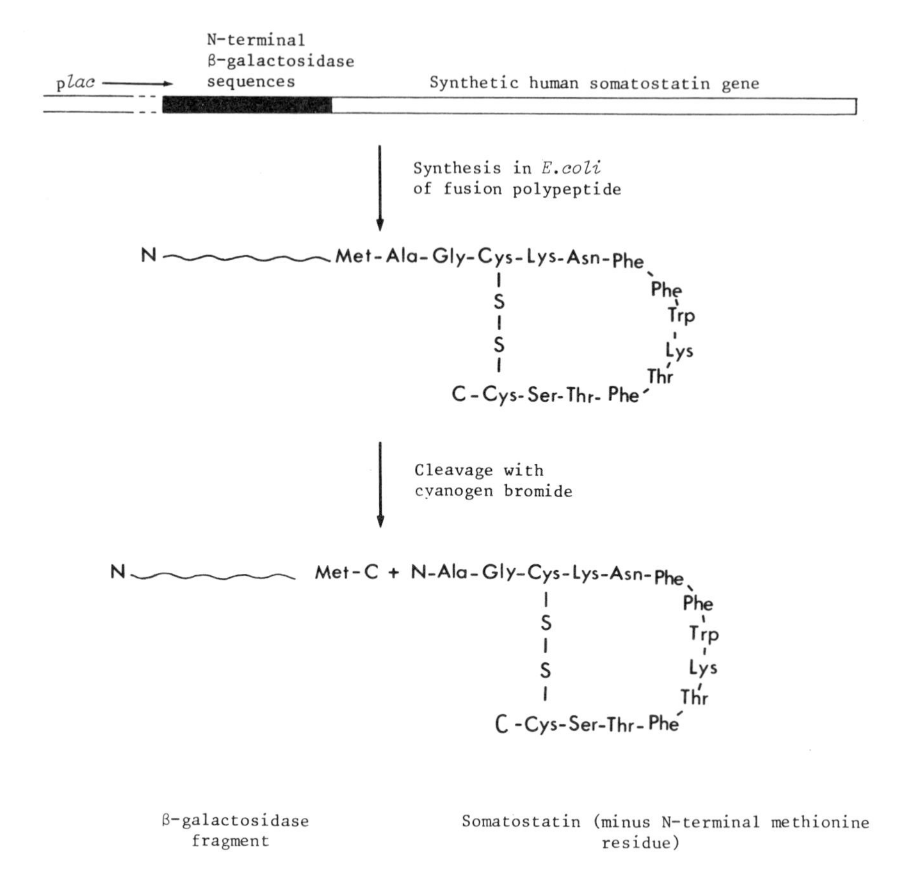

coli, some proteins are further compartmentalised to the inner or outer membrane or to the periplasmic space between the membranes (Figure 5.14).

Polypeptides that are exported across membranes generally possess a **signal or leader sequence** of 15 to 70 amino acid residues that is located at the N terminus and is removed by proteolytic cleavage during, or shortly after, translocation (Randall and Hardy, 1984). Exported proteins are therefore generally translated as precursors that are 2000 daltons or more larger than the mature polypeptide. The signal polypeptides of proteins found in the periplasm or the outer membrane of *E. coli* normally share a number of features, such as basic residues (arginine or lysine) at the N terminus followed by a sequence of at least 8 uninterrupted hydrophobic or neutral amino acids (with a tendency to form an α-helix) and a signal

Figure 5.14: Diagrammatic representation of the localisation of proteins in Gram-positive and Gram-negative bacteria

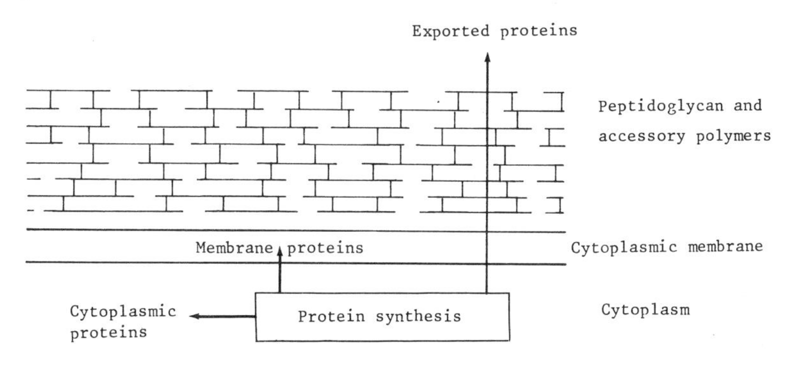

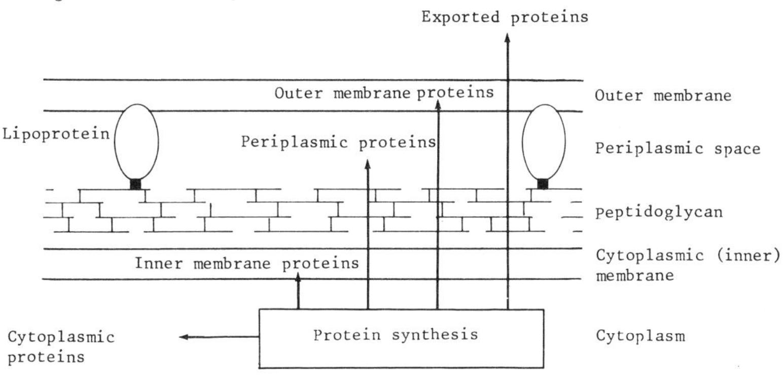

peptidase cleavage site (for a review see Pugsley and Schwartz, 1985). The two enzymes capable of processing signal peptides in *E. coli* are the leader peptidase and the lipoprotein signal peptidase. These enzymes are guided to the cleavage site by the conserved consensus sequence, –A–X–B↓, where A is Ala, Gly, Ser, Leu, Val or Ile; X is any residue and B is principally Ala, Gly or Ser (Perlman and Halvorson, 1983).

Several models have been proposed to explain protein export. The signal hypothesis (Blobel and Dobberstein, 1975) postulates (from experiments in eukaryotic systems) that as soon as the signal sequence is translated it interacts with a signal interaction particle, blocking further elongation until the resulting complex interacts with a membrane-associated docking protein. The nascent polypeptide chain would then be extruded through a proteinaceous pore in the membrane as elongation proceeds. The signal sequence would be removed by a leader peptidase on the far side of the membrane. An alternative model called the membrane-triggered folding hypothesis has been proposed (Wickner, 1980). This postulates that the leader peptide might fold, while nascent, to allow interaction with the membrane, hence triggering translocation of the protein through the membrane. In this case translocation and removal of the signal sequence would be driven by proton motive force. Features of both models have been incorporated into a working model for transport of bacterial proteins (Randall and Hardy, 1984).

The process of asymmetric integration of proteins into membranes (for example insertion of proteins into the outer membrane of *E. coli*) is related to the process of translocation across membranes (Silhavy *et al.*, 1983). In most cases integration involves transfer of only a specific domain(s) of the polypeptide, leaving the other domains untranslocated. In principle this process is believed to operate like complete translocation, but to involve 'stop-transfer' or 'membrane-anchor' sequences to halt the process of translocation initiated by a signal sequence. Other so-called 'sorting sequences' have been postulated to ensure correct protein traffic subsequent to translocation (Silhavy *et al.*, 1983). Thus it is the structural sequence of the protein and not the signal sequence that determines the ultimate location of an exported polypeptide (Tommassen *et al.*, 1983). Furthermore, C-terminal sequences of secreted proteins, such as those of the glycerol-phosphate phosphodiesterase of *Salmonella typhimurium*, may be essential for release into the periplasm (Hengge and Boos, 1985).

There are many similarities between protein transport in prokaryotes and eukaryotes. The major difference seems to be the requirement in bacteria for a trans-membrane electrochemical gradient. Certain eukaryotic proteins, for example insulin (Talmadge *et al.*, 1980), are exported by *E. coli* using their native signal sequences. In *E. coli*, prokaryotic, eukaryotic and hybrid signal sequences seem to function and to be correctly processed. In the case of some proteins such as ovalbumin, export has been observed without cleavage of a signal polypeptide (Baty *et al.*, 1981). In this case a sequence internal to the N terminus of the polypeptide and that does not have the characteristics of a signal sequence appears to be required for export. Proteins that are exported, for example from Gram-positive bacteria, may not be fully released when produced in *E. coli* but may be transferred to the periplasm or outer membrane.

The export of recombinant proteins is generally assisted if the vector used ensures that the desired polypeptide is secreted using the normal protein export system of the host. **Secretion vectors** appropriate to various hosts have been devised for this purpose. Such a vector has been constructed for *B. subtilis* by linking the α-amylase gene of *Bacillus amyloliquefaciens* to the *Bacillus* plasmid vector pUB110 (Ulmanen *et al.*,

1985). The coding sequence for the foreign protein is cloned downstream of the promoter, ribosome-binding site and signal sequence for α-amylase (Figure 5.15). Replacing the coding sequence for mature α-amylase with foreign DNA does not impair the transcription or translation of the resultant hybrid gene. However, a marked difference is observed between the yield of secreted hybrid and native protein. This can be attributed to differential susceptibility to exoproteases produced by *Bacillus* (Chang *et al.*, 1983; Ulmanen *et al.*, 1985). The problem can be alleviated by using exoprotease-defective *B. subtilis* strains or by using an alternative host/ secretion vector combination.

Even when using a secretion vector, particular fused polypeptides with signal sequences may fail to be exported because the internal amino acid sequence of the hybrid protein is inappropriate. For example, a fusion protein produced by joining a chemically synthesised α-neo-endorphin gene downstream of the region encoding the signal sequence and N terminus of the alkaline phosphatase gene of *E. coli* was processed, but not exported to the periplasm (Ohsuye *et al.*, 1983). This was found despite the fact that native alkaline phosphatase is normally periplasmic and that the fusion polypeptide produced was present at about 1.3×10^6 molecules per cell (equivalent to about 60% of total cellular protein). This suggests that amino acid residues normally on the C-terminal end of alkaline phosphatase and that were replaced by endorphin sequences are essential for export of the protein. A further general possibility is that hydrophobic or other domains may be present fortuitously in fusion or other foreign polypeptides causing them to jam in membranes. In certain cases these problems could be overcome by restructuring the coding sequence for the desired protein to remove such domains (where this could be carried out without adversely affecting other properties). High-level expression of secreted proteins *per se* can also cause problems if the export machinery becomes saturated or

Figure 5.15: Expression of foreign genes in *Bacillus subtilis* using a secretion vector. p, Promoter; RBS, ribosome binding site; SS, signal sequence of *Bacillus amyloliquefaciens* α-amylase. Vector plasmid is pUB110

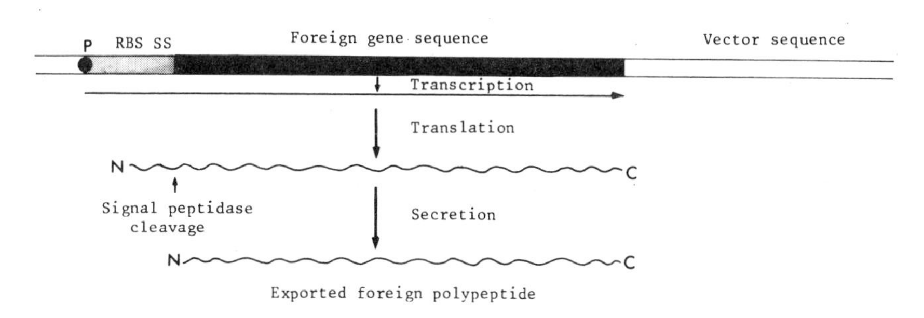

otherwise jammed by excessive export. This may, in turn, lead to accumulation of protein in the cell and/or toxic effects on the host.

5.6 ASSISTING PRODUCT RECOVERY

Specific proteins can be isolated and purified relatively easily from genetically manipulated microorganisms on a small scale. However, the methods used may be unsuitable and/or uneconomic on a large scale. Host-vector systems that facilitate recovery of proteins from large-scale cultures are therefore of considerable benefit. In the majority of cases proteins encoded by recombinant DNA molecules will have no special properties that will allow specific purification in a single step from the mass of cellular protein. The synthesis of polypeptide fusions between the desired protein and a readily purified protein may provide a means of effective product recovery. For example, plasmid vectors have been constructed that allow the fusion of any gene sequence to the gene for *S. aureus* protein A (Uhlen *et al.*, 1983). Fusion to protein A allows purification by means of affinity chromatography with IgG, since protein A binds specifically to this immunoglobulin. A further generally applicable strategy for the purification of specific polypeptides has been devised by Sassenfeld and Brewer (1984). They utilised recombinant DNA techniques to add a synthetic DNA sequence coding for five additional arginine residues on to the C-terminus of a synthetic human β-urogastrone gene. Since β-urogastrone has a C-terminal arginine residue, this resulted in a fused protein with a six-residue polyarginine tail (Figure 5.16). Because arginine is the most basic amino acid, the resulting polypeptide bound strongly to the cationic exchanger SP-Sephadex at acid pH. In contrast, most bacterial proteins are acidic and should bind poorly to this exchanger under the same conditions. Consequently, fused polyarginine-urogastrone could be freed of most contaminating proteins by a single purification on an SP-Sephadex column. β-Urogastrone was then released (minus its natural C-terminal arginine residue) from the fusion by cleavage with carboxypeptidase B, which specifically digests C-terminal arginine and lysine residues from proteins. The digested protein was then rechromatographed on the Sephadex column, where it eluted at a different salt concentration from that required for any very basic contaminating proteins that happened to copurify initially with the fusion protein during the first chromatography step. The purified urogastrone was found to have normal activity. This strategy is likely to be of general utility because the addition of a run of very basic amino acids to the C terminus of a protein is unlikely to cause interference with normal protein folding. (In contrast location of such a sequence at the N terminus could cause interference since the amino terminus is synthesised first, whereas the carboxy terminus of a protein is the last to be

Figure 5.16: Purification of C-terminal polyarginine fusion polypeptides

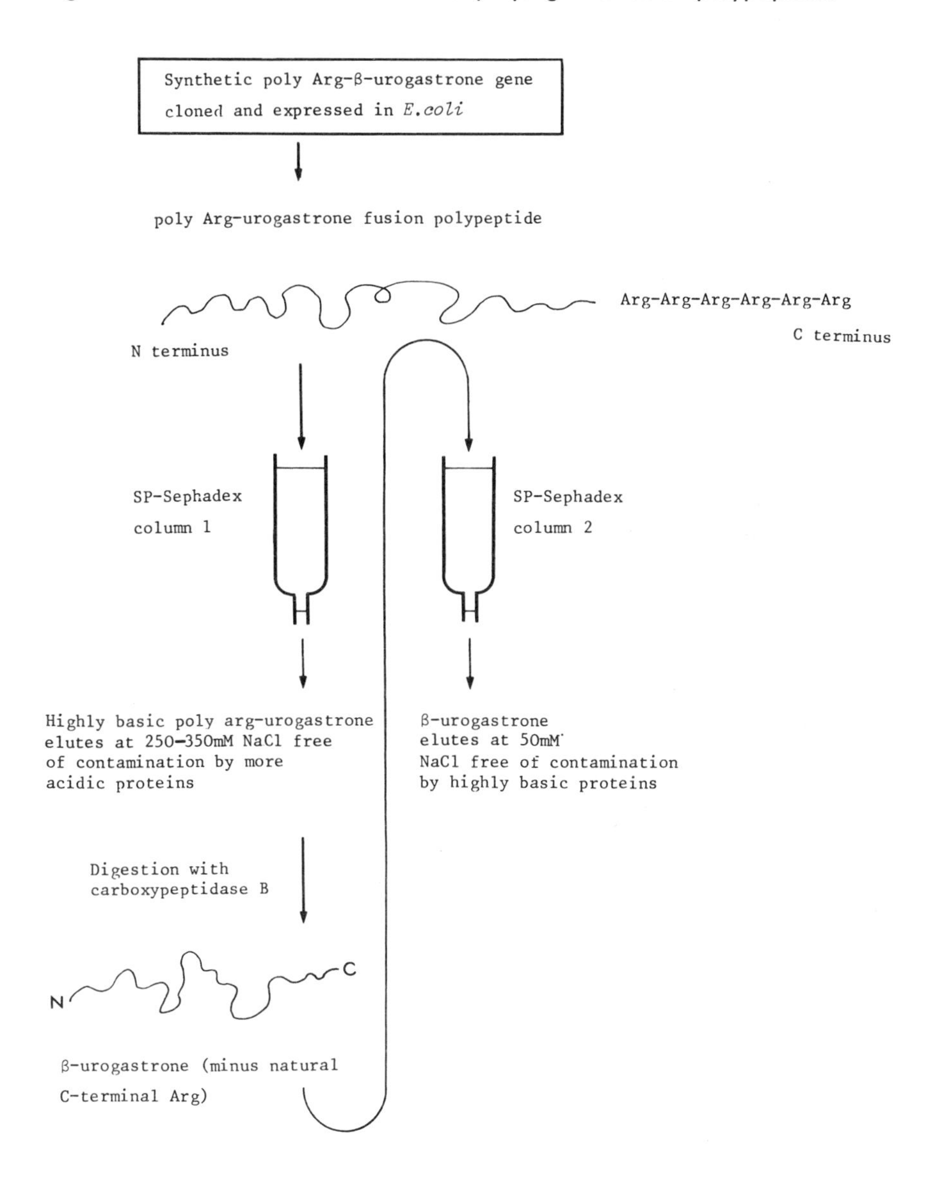

made.) Even if protein folding were to be altered, treatment with carboxypeptidase B should return the protein to its normal configuration. Carboxypeptidase B also has advantages in that it is relatively cheap and is active over a wide range of chemical and physical conditions likely to be encountered during the preliminary stages of protein extraction from host cells.

5.7 CHOICE OF HOST-VECTOR SYSTEM

The choice of host organism to be used for the production of any given polypeptide depends not only on the physicochemical properties of the protein, but also on the economics of the host-vector and recovery systems. Indeed, for small polypeptides of 20 residues or less it is normally cheaper to synthesise the required product chemically than to make it biologically. It is therefore not possible to devise a universal system that is appropriate for the large-scale production of all polypeptides. Factors that must be taken into account when deciding on the host-vector combination are outlined in Table 5.3. The aim in most cases is to minimise cost while maximising yield. Even when the same host organism is used, the cost of inducing efficient expression from controllable promoters may vary considerably. For example, induction of a gene controlled by the *tetA* promoter and *tetR* repressor of Tn*10* (through addition of tetracycline to the medium) in *E. coli* costs about 1000 times less than inducing an equivalent level of expression of that gene from the *lac* promoter (by adding IPTG) (De La Torre *et al.*, 1984).

In recent years *E. coli* has received much attention as a host for producing foreign proteins. This stems largely from the ease of genetic manipulation of this organism and the considerable knowledge available concerning

Table 5.3: Factors that determine choice of host and vector for production of foreign polypeptides

Safety	Potential virulence of host Containment of host Potential toxicity of product
Fermentation	Culture growth rate Cell density achieved Media and metabolic inducer costs Physical properties of cells (frothing, sedimentation, etc.) Ease of application of inducer
Stability	Stability of cloned coding DNA sequence(s) Stability of vector in host Toxicity of product to host
Expression	Optimised gene dosage (vector copy number) Control of expression Efficient transcription Efficient translation Level of fully induced expression Ratio of induced : uninduced gene expression (especially for lethal product) Stability of product (for example susceptibility to degradation)
Recovery	Acceptability of protein as an inclusion body, or as an exported product to periplasm or medium

its physiology. Historically *E. coli* has not been cultured on an industrial scale, and other organisms may have distinct advantages where large-scale fermentations are required. For example, *Methylophilus methylotrophus*, which is grown on a large scale on substrates such as methanol and ammonia (see section 6.3.2), is capable of producing mammalian peptides from cloned cDNA (Hennam *et al.*, 1982).

A major problem encountered with *E. coli* and most other Gram-negative bacteria is the release of endotoxin into the growth medium. Endotoxins are normal lipopolysaccharide components of Gram-negative cell walls. Even at concentrations of less than 1 pg per ml, endotoxins exert potent biological effects on man and most animals. The elimination of endotoxin from aqueous solutions is difficult and expensive. An alternative approach is to avoid the problem by using Gram-positive bacteria or other hosts that do not produce endotoxin.

A further problem encountered with *E. coli* is that, in common with most enterobacteria, it excretes very few proteins into the medium. Those that are fully exported are mainly toxins, such as haemolysins (HLYs). The remaining secreted proteins are normally periplasmic or integrated into the outer membrane (Silhavy *et al.*, 1983). Human β-endorphin has been secreted across both cytoplasmic and outer membranes as a fusion to part of the *E. coli ompF* outer membrane protein (Nagahari *et al.*, 1985). (The *ompF* gene product is a protein that forms porin channels in the outer membrane through which hydrophilic solutes can pass.) This was achieved by cloning the β-endorphin sequence in a secretion vector downstream of the *ompF* promoter, signal sequence and N-terminal coding sequence. The precise mechanism by which the resulting fusion polypeptide is exported to the culture medium does, however, remain unresolved and it is possible that partial lysis of the host population is involved. One possible alternative Gram-negative host to *E. coli* is *Serratia marcescens*, which is exceptional among Enterobacteriaceae in its ability to secrete large amounts of exoprotease into the medium (Schmitz and Braun, 1985). The signal and other sorting sequences of this protein might be exploited to ensure efficient export of fused hybrid proteins expressed from secretion vectors in *S. marcescens* or other bacteria.

Bacillus subtilis has several advantages as a host for expression of foreign genes, notably its ability to secrete many proteins. However, exploitation of *Bacillus* commercially has been relatively disappointing due to plasmid vector instability, to frequent deletion of cloned genes (Hahn and Dubnau, 1985) and to lower levels of expression than might be expected (Espinosa *et al.*, 1984). Many of these drawbacks may, however, be overcome by using bacilli other than *B. subtilis* 168, which has been used traditionally for genetic studies.

Streptomyces spp. possess a number of properties making them attractive hosts for the production of foreign proteins. These prokaryotes are already grown on a large scale for the production of secondary metabolites and enzymes. Furthermore, they are not pathogenic to man and do not secrete endotoxin. *Streptomyces lividans* and probably other streptomycetes can efficiently transcribe genes from a wide range of bacteria including *Bacillus* spp., *E. coli* and *Serratia marcescens* (Bibb and Cohen, 1982; Jaurin and Cohen, 1984). This is in contrast to *Bacillus* spp. which are very inefficient at transcribing genes from other than close relatives such as streptococci. In the scale of transcription preference *E. coli* seems to be intermediate, being able to transcribe *Bacillus* promoters of the σ^{55} class, but unable to transcribe from most *Streptomyces* promoters (Bibb and Cohen, 1982). A general rule that seems to be emerging is that a gene is more likely to be expressed heterologously if it is introduced into a host with a genome of higher GC content than that of the originating organism.

Saccharomyces cerevisiae has a number of potential advantages for the production of commercially important polypeptides. It is non-pathogenic, no endotoxin is released into the growth medium and it has been an essential and acceptable component of foods and beverages for many centuries. Unlike bacteria, yeasts remove the N-terminal methionine residue (fMet) from polypeptides. (The presence of a terminal methionine is undesirable in heterologously produced eukaryotic polypeptides used as pharmaceuticals. The additional residue may alter the protein sufficiently for it to be recognised as foreign by the immune system of the host to which it is administered. This may reduce efficiency of treatment.) Yeasts also glycosylate proteins in a manner that is similar to that in mammalian cells. Yeast recognises signal sequences from higher eukaryotes and can secrete polypeptides into the medium. *Saccharomyces cerevisiae* naturally secretes few polypeptides, but some proteins, such as the mating factors, are exported and their genes can be used as the basis of secretion vectors. Furthermore, certain foreign proteins such as the TEM β-lactamase encoded by pBR322, can also be secreted by yeast (Roggenkamp *et al.*, 1981). For some proteins, for example human γ interferon, the levels of expression that can be achieved are greater in *S. cerevisiae* than in *E. coli* (Derynck *et al.*, 1983). Furthermore, some proteins that are accumulated as inclusion bodies in *E. coli*, for example calf chymosin (rennin), are secreted in yeast and hence may be purified more easily (Mellor *et al.*, 1983). Yields of foreign proteins achieved in yeast are affected strongly by the promoter used. For example, the yeast phosphoglycerate kinase (*PGK*) promoter is about 500 times more efficient at directing the synthesis of heterologous genes than the tryptophan (*TRP1*) promoter (Mellor *et al.*, 1985). Yields of proteins are also much higher with multicopy 2 μm plasmid-based YEp vectors than with either low or high copy number ARS vectors (section 3.5.5). Even when both an efficient promoter and a 2 μm vector are used, the total foreign protein may only amount to 1-5% of total cellular protein, which is below the theoretical maximum expected (50% of cell protein for PGK). Mellor and co-workers (1985) have concluded that poorer than expected expression of human interferon-α-2 (IFNα2) in yeast was due both to a rapid turnover of heterologous protein and to a reduced level of IFNα2-specific mRNA. Poor transcription of heterologous coding sequences could result from the absence in such genes of specific yeast sequences that are required for efficient transcription.

For large polypeptides or for proteins whose activity depends on particular post-translational modifications, such as glycosylation, production in microorganisms may be inappropriate and/or uneconomic. Many proteins of therapeutic value are glycoproteins which cannot be produced in bacteria. In such cases the proteins may be produced in either plant or animal cells.

BIBLIOGRAPHY

Anderson, D.M., Herrmann, K.M. and Somerville, R.L. (1983) *Escherichia coli* bacteria carrying recombinant plasmids and their use in the fermentative production of L-tryptophan. US Patent No. 4, 371, 614

Austin, S. and Abeles, A. (1983) Partition of unit-copy miniplasmids to daughter cells. I. P1 and F miniplasmids contain discrete, interchangeable sequences sufficient to promote equipartition. *J. Mol. Biol.* **169**, 357-72

Backman, K., O'Connor, M.J., Maruya, A. and Erfle, M. (1984) Use of synchronous site-specific recombination *in vivo* to regulate gene expression. *Biotechnology* **2**, 1045-9

Baty, D., Mercereau-Puijalon, O., Perrin, D., Kourilsky, P. and Lazdunski, C. (1981) Secretion into the bacterial periplasmic space of chicken ovalbumin synthesized in *Escherichia coli. Gene* **16**, 79-87

Bibb, M. and Cohen, S.N. (1982) Gene expression in *Streptomyces*: construction and application of promoter-probe plasmid vectors in *Streptomyces lividans. Mol. Gen. Genet.* **187**, 265-77

Bibb, M.J., Findlay, P.R. and Johnson, M.W. (1984) The relationship between base composition and codon usage in bacterial genes and its use for the simple and reliable identification of protein coding sequences. *Gene* **30**, 157-66

Blobel, G. and Dobberstein, B. (1975) Transfer of proteins across membranes. I. Presence of proteolytically processed and nonprocessed nascent immunoglobulin light chains on membrane bound ribosomes of murine myeloma. *J. Cell. Biol.* **67**, 852-62

Boss, M.A., Kenton, J.H., Wood, C.R. and Emtage, J.S. (1984) Assembly of functional antibodies from immunoglobulin heavy and light chains synthesized in *E. coli. Nucl. Acids Res.* **12**, 3791-806

Casadaban, M.J. and Cohen, S.N. (1980) Analysis of gene control signals by DNA fusion and gene cloning in *E. coli. J. Mol. Biol.* **138**, 179-207

Caulcott, C.A., Lilley, G., Wright, E.M., Robinson, M.K. and Yarranton, G.T. (1985) Investigation of the instability of plasmids directing the expression of Metprochymosin in *Escherichia coli. J. Gen. Microbiol.* **131**, 3355-65

Cesareni, G., Muesing, M.A. and Polisky, B. (1982) Control of ColE1 DNA replication: the *rop* gene product negatively affects transcription from the replication primer promoter. *Proc. Natl Acad. Sci. USA* **79**, 6313-17

Chambon, P., Dierich, A., Gaub, M-P., Jawouolev, S., Jongstra, J., Krust, A., Le Pennec, J-P., Oudet, P. and Reudelhuber, T. (1984) Promoter elements of genes coding for proteins and modulation of transcription by estrogens and progesterone. In *Recent progress in hormone research, vol. 40* (R.O. Greep, ed.), pp. 1-42. Academic Press, New York

Chan, P.T., Ohmori, H., Tomizawa, J. and Lebowitz, J. (1985) Nucleotide sequence and gene organization of ColE1 DNA. *J. Biol. Chem.* **260**, 8925-35

Chang, S., Ho, O., Gray, S., Chang, Y. and McLaughlin, J. (1983) Functional expression of human interferon genes and construction of partition-proficient plasmid vector in *B. subtilis.* In *Genetics of industrial microorganisms* (Y. Ikeda and T. Beppu, eds), pp. 227-31. Kodansha, Tokyo

Charnay, P., Perricaudet, M., Galibert, F. and Tiollais, P. (1978) Bacteriophage lambda and plasmid vectors allowing fusion of cloned genes in each of the three translational phases. *Nucl. Acids Res.* **5**, 4479-94

Charnay, P., Triesman, R., Mellon, P., Chao, M., Axel, R. and Maniatis, T. (1984) Differences in human alpha- and beta-globin gene expression in mouse erythroleukemia cells: the role of intragenic sequences. *Cell* **38**, 251-63

Christie, G.E., Farnham, P.J. and Platt, T. (1981) Synthetic sites for transcription/termination and a functional comparison with tryptophan operon termination sites *in vitro. Proc. Natl Acad. Sci. USA* **78**, 4180-4

de Boer, H.A., Comstock, L.J. and Vasser, M. (1983) The *tac* promoter: a functional hybrid derived from the *trp* and *lac* promoters. *Proc. Natl Acad. Sci. USA* **80**, 21-5

De La Torre, J.C., Ortin, J., Domingo, E., Delamarter, J., Allet, B., Davies, J.E., Bertrand, K.P., Wray, L.V. and Reznikoff, W. (1984) Plasmid vectors based on Tn*10* DNA: gene expression regulated by tetracycline. *Plasmid* **12**, 103-10

Derynck, R., Singh, A. and Goeddel, D.V. (1983) Expression of the human interferon gamma cDNA in yeast. *Nucl. Acids Res.* **11**, 1819-37

Doherty, M.J., Morrison, P.T. and Kolodner, R. (1983) Genetic recombination of bacterial plasmid DNA. Physical and genetical analysis of the products of plasmid recombination in *Escherichia coli. J. Mol. Biol.* **167**, 539-60

Dunn, J.J. and Studier, F.W. (1980) The transcription termination site at the end of the early region of bacteriophage T7. *Nucl. Acids Res.* **8**, 2119-32

Espinosa, M., Lopez, P. and Lacks, S.A. (1984) Transfer and expression of recombinant plasmids carrying pneumococcal *mal* genes in *Bacillus subtilis. Gene* **28**, 301-10

Fishel, R.A., James, A.A. and Kolodner, R. (1981) *recA*-independent general genetic recombination of plasmids. *Nature (London)* **294**, 184-6

Goeddel, D.V., Kleid, D.G., Bolivar, F., Heynecker, H.L., Yansura, D.G., Crea, R., Hirose, T., Krasenzewski, A., Itakura, K. and Riggs, A.D. (1979) Expression in *Escherichia coli* of chemically synthesized genes for human insulin. *Proc. Natl Acad. Sci. USA*, **76**, 106-10

Goldfarb, D.S., Doi, R. and Rodriguez, R.L. (1981) Expression of Tn*9*-derived chloramphenicol resistance in *Bacillus subtilis. Nature (London)* **293**, 309-11

Gouy, M. and Gautier, C. (1982) Codon usage in bacteria: correlation with gene expressivity. *Nucl. Acids Res.* **10**, 7055-74

Grantham, R., Gautier, C., Gouy, M., Jacobzone, M. and Mercier, R. (1981) Codon catalog usage is a genome strategy for gene expressivity. *Nucl. Acids Res.* **9**, r43-r74

Guarente, L. (1985) Regulation of the *Saccharomyces cerevisiae CYC1* gene. In *Microbiology–1985* (L. Leive, ed.), pp. 379-83. American Society for Microbiology, Washington, DC

Gustafsson, P., Wolf-Watz, H., Lind, L., Johansson, K-E. and Nordstrom, K. (1983) Binding between the *par* region of plasmids R1 and pSC101 and the outer membrane fraction of the host bacteria. *EMBO J.* **2**, 27-32

Hahn, J. and Dubnau, D. (1985) Analysis of plasmid deletional instability in *Bacillus subtilis. J. Bacteriol.* **162**, 1014-23

Hakkaart, M.J., van den Elzen, P.J., Veltkamp, E. and Nijkamp, H.J. (1984) Maintenance of multicopy plasmid CloDF13 in *E. coli. Cell* **36**, 203-9

Hengge, R. and Boos, W. (1985) Defective secretion of maltose- and ribose-binding proteins caused by a truncated periplasmic protein in *Escherichia coli. J. Bacteriol.* **162**, 972-8

Hennam, J.F., Cunningham, A.E., Sharpe, G.S. and Atherton, K. (1982) Expression of eukaryotic coding sequences in *Methylophilus methylotrophus. Nature (London)* **297**, 80-2

Itakara, K., Hirose, T., Crea, R., Riggs, A.D., Heyneker, H.L., Bolivar, F. and Boyer, H.W. (1977) Expression in *Escherichia coli* of a chemically synthesised gene for the hormone somatostatin. *Science* **198**, 1056-63

Janssen, G.R., Bibb, M.J., Smith, C.P., Kieser, T. and Bibb, M. (1985) Isolation and analysis of *Streptomyces* promoters. In *Microbiology-1985* (L. Leive, ed.), pp. 392-6. American Society for Microbiology, Washington, DC

Jaurin, B. and Cohen, S.N. (1984) *Streptomyces lividans* RNA polymerase recognizes and uses *Escherichia coli* transcriptional signals. *Gene* **28**, 83-91

Jay, E., Rommens, J. and Jay, G. (1985) Synthesis of mammalian proteins in bacteria. In *Biotechnology, applications and research* (P.N. Cheremisinoff and R.P. Quellette, eds), pp. 388-400. Technomic Publishing, Lancaster

Jay, G., Seth, A.K., Rommens, J., Sood, A. and Jay, E. (1982) Gene expression: chemical synthesis of *E. coli* ribosome binding sites and their use in directing the expression of mammalian proteins in bacteria. *Nucl. Acids Res.* **10**, 6319-29

Jones, I.M., Primrose, S.B., Robinson, A. and Elwood, D.C. (1980) Maintenance of some ColE1-type plasmids in chemostat culture. *Mol. Gen. Genet.* **180**, 579-84

Laban, A. and Cohen, A. (1981) Interplasmidic and intraplasmidic recombination in *Escherichia coli* K-12. *Mol. Gen. Genet.* **184**, 200-7

Lacatena, R.M. and Cesareni, G. (1981) Base pairing of RNAI with its complementary sequence in the primer precursor inhibits ColE1 replication. *Nature (London)* **294**, 623-6

Lacatena, R.M. and Cesareni, G. (1983) Interaction between RNAI and the primer precursor in the regulation of ColE1 replication. *J. Mol. Biol.* **170**, 635-50

Lambert, P.F. and Reznikoff, W.S. (1985) Use of transcriptional repressors to stabilize plasmid copy number of transcriptional fusion vectors. *J. Bacteriol.* **162**, 441-4

Lee, N., Cozzitorto, J., Wainwright, N. and Testa, D. (1984) Cloning with tandem gene systems for high level gene expression. *Nucl. Acids Res.* **12**, 6797-812

Losick, R. and Pero, J. (1981) Cascades of sigma factors. *Cell* **25**, 582-4

Masui, Y., Mizuno, T. and Inouye, M. (1984) Novel high-level expression cloning vehicles: 10^4-fold amplification of *Escherichia coli* minor protein. *Biotechnology* **2**, 81-6

Meacock, P.A. and Cohen, S.N. (1980) Partitioning of bacterial plasmids during cell division: a *cis*-acting locus that accomplishes stable plasmid inheritance. *Cell* **20**, 529-42

Mellor, J., Dobson, M.J., Roberts, N.A., Tuite, M.F., Emtage, J.S., White, S., Lowe, P.A., Patel, J., Kingsman, A.J. and Kingsman, S.M. (1983) Efficient synthesis of enzymatically active chymosin in *Saccharomyces cerevisiae. Gene* **24**, 1-14

Mellor, J., Dobson, N.A., Roberts, N.A., Kingsman, A. and Kingsman, S.M. (1985)

Factors affecting heterologous gene expression in *Saccharomyces cerevisiae. Gene* **33**, 215-26

Mizusawa, S. and Gottesman, S. (1983) Protein degradation in *Escherichia coli*: the *lon* gene controls the stability of *sulA* protein. *Proc. Natl Acad. Sci. USA* **80**, 358-62

Mount, D.W. (1980) The genetics of protein degradation in bacteria. *Ann. Rev. Genet.* **14**, 279-320

Murray, C.L. and Rabinowitz, J.C. (1982) Nucleotide sequences of transcription and translation regions in *Bacillus* phage ϕ29 early genes. *J. Biol. Chem.* **257**, 1053-62

Nagahari, K., Kanaya, S., Munukata, K., Aoyagi, Y. and Mizushima, S. (1985) Secretion into the culture medium of a foreign gene product from *Escherichia coli*: use of the *ompF* gene for secretion of human β-endorphin. *EMBO J.* **4**, 3589-92

Nilsson, G., Belasco, J.G., Cohen, S.N. and van Gabain, A. (1984) Growth-rate dependent regulation of mRNA stability in *Escherichia coli. Nature (London)* **312**, 75-7

Noack, D., Roth, M., Geuther, R., Muller, G., Undisz, K., Hoffmeier, C. and Gaspar, S. (1981) Maintenance and genetic stability of vector plasmids pBR322 and pBR325 in *Escherichia coli* strains grown in a chemostat. *Mol. Gen. Genet.* **184**, 121-4

Ogura, T. and Hiraga, S. (1983) Partition mechanism of F plasmid: two plasmid gene-encoded products and a *cis*-acting region are involved in partition. *Cell* **32**, 351-60

Ohsuye, K., Nomura, M., Tanaka, S., Kabota, I., Nakazato, H., Shinagawa, H., Nakata, A. and Noguchi, T. (1983) Expression of chemically synthesized α-neo-endorphin gene fused to *E. coli* alkaline phosphatase. *Nucl. Acids Res.* **11**, 1283-94

Old, R.W. and Primrose, S.B. (1985) *Principles of gene manipulation*, 3rd Edn. Blackwell Scientific Publications, Oxford

Perlman, D. and Halvorson, H.O. (1983) Putative signal peptidase recognition site and sequence in eukaryotic and prokaryotic signal peptides. *J. Mol. Biol.* **167**, 391-409

Pettersson, R.F., Lundström, K., Chattopadhyaya, J.B., Josephson, S., Philipson, L., Käärianen, L. and Palva, I. (1983) Chemical synthesis and molecular cloning of a STOP oligonucleotide encoding UGA translation terminator in all three reading frames. *Gene* **24**, 15-27

Pugsley, A.P. and Schwartz, M. (1985) Export and secretion of proteins by bacteria. *FEMS Microbiol. Rev.* **32**, 3-38

Randall, L.L. and Hardy, S.J.S. (1984) Export of protein in bacteria. *Microbiol. Rev.* **48**, 290-8

Remaut, E., Tsao, H. and Fiers, W. (1983) Improved plasmid vectors with a thermoinducible expression and temperature-regulated runaway replication. *Gene* **22**, 103-13

Roggenkamp, R., Kusterman-Kuhn, B. and Hollenberg, C. (1981) Expression and processing of bacterial β-lactamase in the yeast *Saccharomyces cerevisiae. Proc. Natl Acad. Sci. USA* **78**, 4466-70

Rosenberg, M. and Court, D. (1979) Regulatory sequences involved in the promotion and termination of RNA transcription. *Ann. Rev. Genet.* **13**, 319-53

Sassenfeld, H.M. and Brewer, S.J. (1984) A polypeptide fusion designed for the purification of recombinant proteins. *Biotechnology* **3**, 76-81

Schmitz, G. and Braun, V. (1985) Cell bound and secreted proteases of *Serratia marcescens. J. Bacteriol.* **161**, 1002-9

Scott, J.R. (1984) Regulation of plasmid replication. *Microbiol. Rev.* **48**, 1-23

Sherratt, D. (1986) Control of plasmid maintenance. In *Regulation of gene expression — 25 years on* (39th Symp. Soc. Gen. Microbiol.) (I.R. Booth and C.F. Higgins eds), pp. 239-50. Cambridge University Press, Cambridge

Shine, J. and Dalgarno, L. (1974) The 3′-terminal sequence of *E. coli* 16S ribosomal RNA: complementarity to nonsense triplets and ribosome-binding sites. *Proc. Natl Acad. Sci. USA* **71**, 1342-6

Shirakawa, M., Tsurimoto, T. and Matsubara, K. (1984) Plasmid vectors designed for high efficiency expression controlled by the portable *recA* promoter-operator. *Gene* **28**, 127-32

Shoemaker, J.M., Brasnett, A.H. and Marston, F.A.O. (1985) Examination of calf prochymosin accumulation in *E. coli*: disulphide linkages are a structural component of prochymosin-containing inclusion bodies. *EMBO J.* **4**, 775-80

Siebenlist, U., Simpson, R.B. and Gilbert, W. (1980) *E. coli* RNA polymerase interacts homologously with two different promoters. *Cell* **20**, 269-81

Silhavy, T.J., Benson, S.A. and Emr, S.D. (1983) Mechanisms of protein localization. *Microbiol. Rev.* **47**, 313-44

Simon, L.D., Randolph, B., Irwin, N. and Binowski, G. (1983) Stabilization of proteins by a bacteriophage T4 gene cloned in *Escherichia coli*. *Proc. Natl Acad. Sci. USA* **80**, 2059-62

Skogman, G., Nilsson, J. and Gustafsson, P. (1983) The use of a partition locus to increase stability of tryptophan-bearing plasmids in *Escherichia coli*. *Gene* **23**, 105-15

Stuber, D. and Bujard, H. (1982) Transcription from efficient promoters can interfere with plasmid replication and diminish expression of plasmid specified genes. *EMBO J.* **1**, 1399-1404

Summers, D.K. and Sherratt, D.J. (1984) Multimerization of high copy number plasmids causes instability: ColE1 encodes a determinant essential for plasmid monomerization and stability. *Cell* **36**, 1097-1103

Tacon, W., Carey, N. and Emtage, S. (1980) The construction and characterization of plasmid vectors suitable for the expression of all DNA phases under the control of the *E. coli* tryptophan promoter. *Mol. Gen. Genet.* **177**, 427-38

Tacon, W.C.A., Bonass, W.A., Jenkins, B. and Emtage, J.S. (1983) Expression plasmid vectors containing *Escherichia coli* tryptophan promoter transcription units lacking the attenuator. *Gene* **23**, 255-65

Talmadge, K., Stahl, S. and Gilbert, W. (1980) Eukaryotic signal sequence transports insulin antigen in *Escherichia coli*. *Proc. Natl Acad. Sci. USA* **77**, 3369-73

Tatti, K.M. and Moran, C.P. (1985) Utilization of one promoter by two forms of RNA polymerase from *Bacillus subtilis*. *Nature (London)* **314**, 190-2

Terawaki, Y. and Itoh, Y. (1985) Copy mutant of mini-Rtsl: lowered binding affinity of mutated RepA protein to direct repeats. *J. Bacteriol.* **162**, 72-7

Tomizawa, J. (1985) Control of ColE1 plasmid replication: initial interaction of RNA I and the primer transcript is reversible. *Cell.* **40**, 527-35

Tomizawa, J-I., Itoh, T., Selzer, G. and Som, T. (1981) Inhibition of ColE1 RNA primer formation by a plasmid specified small RNA. *Proc. Natl Acad. Sci. USA* **78**, 1421-5

Tommassen, J., van Tol, H. and Lugtenberg, B. (1983) The ultimate localization of an outer membrane protein of *Escherichia coli* K-12 is not determined by the signal sequence. *EMBO J.* **2**, 1275-9

Twigg, A. and Sherratt, D.J. (1980) *Trans*-complementable copy number mutants of plasmid ColE1. *Nature (London)* **283**, 216-18

Uhlen, M., Nilsson, B., Guss, B., Lindbert, M., Gatenbeck, S. and Philipson, L. (1983) Gene fusion vectors based on the gene for staphylococcal protein A.

Gene **23**, 369-78
Uhlin, B.E., Molin, S., Gustafsson, P. and Nordstrom, K. (1979
temperature-dependent copy number for amplification of clonec
products. *Gene* **6**, 91-106
Ulmanen, I., Lundstrom, K., Lehtovaara, P., Sarvas, M., Ruohonen,
I. (1985) Transcription and translation of foreign genes in *Bacillus su*
aid of a secretion vector. *J. Bacteriol.* **162**, 176-82
Vournakis, J.N. (1985) Role of mRNA structure in the regulation of protein s
sis. In *Microbiology-1985* (L. Leive, ed.), pp. 473-9. American Society
Microbiology, Washington, DC
Walker, M.D., Edmund, T., Boulet, A.M. and Rutter, W.J. (1983) Cell-specific expression controlled by the 5′-flanking region of insulin and chymotrypsin genes. *Nature (London)* **306**, 557-61
Weintraub, H., Izant, J.G. and Harland, R.M. (1985) Anti-sense RNA as a molecular tool for genetic analysis. *Trends Genet.* **1**, 22-5
Westpheling, J., Ranes, M. and Losick, R. (1985) RNA polymerase heterogeneity in *Streptomyces coelicolor. Nature (London)* **313**, 22-7
Wickner, W. (1980) Assembly of proteins into membranes. *Science* **210**, 861-8
Wood, C.R., Boss, M.A., Patel, T.P. and Emtage, J.S. (1984) The influence of secondary structure on expression of an immunoglobulin heavy chain in *Escherichia coli. Nucl. Acids Res.* **12**, 3937-50
Yanofsky, C. (1981) Attenuation in the control of bacterial operons. *Nature (London)* **289**, 751-8
Yarranton, G.T., Wright, E., Robinson, M.K. and Humphreys, G.O. (1984) Dual-origin plasmid vectors whose origin of replication is controlled by the coliphage lambda promoter pL. *Gene* **28**, 293-300
Young, F.E. (1980) The impact of cloning in *Bacillus subtilis* on fundamental and industrial microbiology. *J. Gen. Microbiol.* **119**, 1-15

6

Microbial Strain Improvement and Novel Products

The tailoring of microbial strains for specific biotechnological purposes may require the application of genetic principles and techniques. Strains may be genetically modified in various ways in order to improve existing, desirable capabilities, eliminate undesirable qualities or add new properties. The major genetic routes to strain improvement are those of mutagenesis, gene transfer and genetic recombination. Basically an improvement programme involves:

(i) generation of required novel genotypes (in ways that do not reduce the overall fitness of the strain);
(ii) selection of desired genotypes;
(iii) evaluation of improved strains in large-scale production. Often it is necessary to modify the techniques applied to small-scale operations prior to use on a large scale, if performance of improved strains, selected in the laboratory, is to be maintained in commercial production.

The ultimate success of a programme relies not only upon effective genetic (and physiological) manipulations but also upon efficient process engineering.

This chapter considers the application of genetic techniques to strain improvement. Strain improvement programmes may be employed in the various sectors of industry where microorganisms are exploited (see Table 6.1 for examples). Each programme may involve mutation and selection, conventional breeding, protoplast fusion, *in vitro* recombinant DNA technology or a combination of these techniques. A number of programmes are described in order to illustrate the operational principles involved.

Table 6.1: Areas where microbial strain improvement is applicable

Area	Microbial activity	Example
Pharmaceuticals	Biosynthesis	Antibiotics, vitamins, enzymes, vaccines
	Bioconversion	Hormones
Commodity chemicals and energy production	Biosynthesis	Organic acids, bioplastics
	Catabolism	Alcohols
Foods and beverages	Biosynthesis	Amino acids, organic acids, vitamins, polysaccharides, enzymes, single cell protein, baker's yeast, brewer's yeast
Agriculture*	Biosynthesis	Microbial insecticides, herbicides, nitrogen-fixing inoculants, growth promoters
Effluent treatment†	Catabolism/bioconversion	Sewage treatment, clearing oil spills, pesticide detoxification
Microbial mining†	Acid production	Ore recovery
	Biosynthesis	Enhanced oil recovery
Environmental screening†	Various	Biosensors, mutagen testing, bioassay, diagnosis of disease

* See Chapter 8. † See Chapter 9.

6.1 STRATEGIES FOR STRAIN IMPROVEMENT

The choice of strategy for strain improvement depends not only upon the nature of the improvement required, which may itself be dictated by the type of process technology (whether a batch or continuous process), but also upon the extent of genetic and biochemical knowledge of the organism. Industrial programmes are additionally constrained by economic considerations. Where fundamental knowledge of the organism and/or process is limited, empirical approaches based on the application of random mutation and screening and/or breeding techniques have been used. However, where the genetics and/or biochemistry of the organism is well advanced, newer, more rational approaches involving the use of directed selection procedures and/or *in vitro* genetic manipulations may be applicable. These newer strategies afford greater precision and are thus potentially more effective for the generation and selection of desirable genotypes than empirical methods. Strategies that may be adopted to improve productivity or to develop new products are considered below.

6.1.1 Yield and quality of product

A major objective of many industrial programmes is to increase the yield of useful product. The procedures involved are generally more straightforward where the product is specified by a single gene or a few genes. Knowledge of the rate-limiting step(s) and of the thermodynamic limit of relevant pathways is important when devising strategies for improving yield of desired metabolites. Overall flux through a pathway might be increased in a number of ways, for example:

(i) By alleviating or bypassing the rate-limiting step: the restraint might be lifted by increasing the concentration of rate-limiting enzyme (by, for example, increasing gene dosage and/or the rate of gene expression of the appropriate gene(s)); alternatively the addition of new steps in a metabolic pathway could provide a bypass mechanism.
(ii) By increasing the concentration of precursors.
(iii) By diverting metabolism away from undesirable by-products.

Where the strategy involves increasing the expression of a specific gene(s), this may be accomplished by:

(i) Introducing strong transcriptional and/or translational signals by, for example: (1) self-cloning the target gene on a high-expression vector (see Chapter 5); or (2) directed insertion, upstream of the target gene, of an appropriate transposon that provides promoter activity (see section 2.1.3); alternatively existing expression signals may be modified to improve their efficiency.
(ii) Relieving genetic repression by introducing appropriate mutations (such as constitutive mutations). Transposon-mediated translocation of the target gene may divorce it from its normal control elements and result in increased gene expression.
(iii) Increasing gene dosage by, for example: (1) cloning and amplifying the gene on a multicopy or runaway replication plasmid or on a phage vector (see section 5.4.1); (2) induction of an appropriate lysogen generating a HFT lysate (see section 2.4.3); (3) construction of polyploid strains (see section 2.5.3).

Another approach to increasing product yield is to relieve **feedback (or end-product) inhibition**. This can be achieved by reducing the sensitivity of the relevant enzyme to the feedback inhibitor or by rapid removal of the effector from the cell. Where the biosynthetic pathway is branched, blocking of one branch channels intermediates solely to the desired end product, and can in turn eliminate concerted feedback inhibition which relies upon the coexistence of the products of both pathways.

Other strategies for yield improvement include:

(i) reducing susceptibility of the process organism to high concentrations of potentially toxic substrate, precursor or product;
(ii) inhibiting the activity of enzymes that may degrade the product;
(iii) facilitating product secretion.

Improvements in the quality of product may be effected through the addition of specific desirable characters and/or the elimination of undesirable properties. Undesirable by-products might be removed by blocking the relevant pathway(s) through mutation. Alternatively specific determinants could be eliminated by genetic segregation or their expression could be suppressed.

6.1.2 Growth efficiency

Improvements in production efficiency may be achieved by enhancing growth efficiency of the process organism. One approach is to increase the effectiveness of substrate utilisation by the organism. This may be achieved by identifying potentially energy-wasteful features of metabolism and altering them in ways that reduce energy demand. The existing function(s) could be inactivated and replaced with a heterologous one that is energetically more efficient. Alternatively, the substrate range of the organism may be extended by the acquisition of new enzymic capabilities. The effective use of cheaper and more abundant feedstock (such as wood products, grain or organic wastes) should help to reduce operational costs. At present such materials normally require considerable pretreatment to convert them into suitable substrates for the process organism. This increases the complexity and costs of the process. It would thus prove beneficial if abilities for effecting the necessary conversions could be acquired by the process organism.

6.1.3 Operational efficiency

Although engineering constraints can limit process efficiency, it can also be influenced markedly by operational conditions, such as temperature. Higher process rates may be achieved by performing reactions at higher temperature, through the use of thermostable biocatalysts. Fermentation at higher temperature should facilitate the recovery of volatile products and reduce the need to cool fermenters to remove heat accumulated during microbial metabolism.

Where product formation is the result of **bioconversion** involving a series of reactions carried out by different organisms, operational efficiency might be improved either by the use of mixed cultures, or by condensing the multistep process into one organism. Genetics may be used to stabilise associations between components of mixed cultures. Where a strain can be constructed that possesses all the necessary enzymes to carry out the conversions, it may be possible to perform the entire process in a single reactor.

Morphological features of the process organism can influence growth and operational efficiency by, for example, affecting the degree of frothing or flocculence in the fermentation vessel, or the effectiveness of nutrient consumption. The use of appropriate morphological variants should provide a means of reducing such operational problems.

6.1.4 Product recovery

Product recovery and purification contribute significantly to the economic success of biotechnological processes (for a discussion of problems associated with recovery of proteins see sections 5.5 and 5.6). Various strategies can be adopted in order to improve product recovery. Manipulations aimed at optimising product concentration should facilitate recovery from dilute solution. However, optimal conditions required for fermentation may deviate substantially from those required for product recovery. An approach in such cases may be to develop strains capable of carrying out the fermentation efficiently under conditions more favourable to product recovery. Microorganisms from extreme environments (including thermophiles, halophiles, acidophiles and alkalophiles) may prove useful in this respect. For example, product recovery by distillation should be facilitated if the specific fermentation is performed at higher temperatures. This might be achieved by engineering thermophiles to serve as process organisms. Furthermore, the morphology of the process organism can influence product recovery by affecting the ease of separation of broth and biocatalyst. Recovery might thus be improved by isolating appropriate morphological mutants. For example, mutants with an increased ability to flocculate should facilitate separation of cells from suspending medium following fermentation. Manipulations aimed at preventing the production of undesirable compounds (in the medium) that interfere with the efficiency of chemical extraction of the product of interest may assist its recovery.

6.1.5 Novel products

Novel products may be obtained by modifying existing natural products, by the formation of hybrid products or by artificial syntheses. It may be possible to activate 'silent' regions of a genome by mutation or recombination events, in turn expressing determinants that are not normally expressed and that may cooperate with existing active determinants to produce novel metabolites. The application of *in vitro* techniques should permit more rational design of product: the direct and deliberate manipulation of genes should enable the restructuring of proteins, such that they possess precisely defined mixtures of desirable chemical, physical and biological properties. Furthermore, addition of specific new steps to existing pathways or the creation of entirely new synthetic pathways should permit the generation of novel metabolites.

6.2 APPLICATION OF *IN VIVO* GENE TECHNOLOGIES

6.2.1 Mutation and selection

Mutation is widely used as a tool for inducing genetic variability. Many industrial strain improvement programmes still depend primarily upon the use of **mutation-selection** procedures. One of the main reasons for their continued use has been a lack of utilisable breeding and/or cloning systems for organisms of interest. Desired strain improvements may necessitate the induction of single or multiple genetic lesions. Where specific multiple mutations are required (for example for improving production of secondary metabolites), they may be obtained following repeated rounds of random mutagenesis and screening. Mutants isolated after one round of mutagenesis serve as the starting point for a new round of mutation. Since the effectiveness of a given mutagen is likely to decrease progressively with successive rounds of mutation, a variety of mutagens is often used (in addition to spontaneous mutagenesis) in such stepwise mutation programmes.

Isolation of improved strains following random mutagenesis is generally a slow process. One of the reasons for this is that there is an increasing likelihood, during the course of the mutation programme, that in selecting for one desirable property another desirable property might be impaired. The probability of introducing mutations that will reduce the overall fitness of the strain increases with each new round of mutagenesis. The use of directed mutagenesis techniques (*in vitro*) (section 4.9), which permit the generation of mutations at predetermined sites, may circumvent such problems. In addition, such techniques may prove more effective for inducing the subtle genetic changes that may be required in highly developed strains in order to introduce further improvements. However, the application of these newer techniques generally depends upon knowledge of the genetic alteration(s) required, and this information is not always available.

Despite the empirical nature of conventional mutagenesis procedures, they have been employed successfully for strain improvement particularly in the antibiotics and brewing industries. Notable examples are programmes for improving the production of the β-lactam antibiotics, penicillins and cephalosporins (see Elander, 1967, 1979; Elander *et al.*, 1976). Stepwise mutation programmes carried out in a number of laboratories to isolate superior penicillin-producing strains of *Penicillium chrysogenum* (see Figure 6.1) generated a series of strains giving up to a 55-fold increase in penicillin production relative to the starting strains. Such strain improvements coupled with improvements in fermentation methods enabled the first commercial production of penicillin. Continuing genetic programmes have led to further increases in productivity. Current production strains of *P. chrysogenum* have penicillin titres of 50000 units per ml or more. By comparison, the progenitor strain of present-day industrial strains, *P. chrysogenum* NRRL 1951, had a titre of about 100 units per ml. Once high-titre strains have been obtained it becomes increasingly more difficult to improve yield further. There is an even greater dependence upon effective screening procedures to detect variants exhibiting slight increases in product yield.

(a) Rational selection procedures

Programmes for strain improvement may involve either random screening of mutagenised cells for improved characteristics or rational screening. The application of **rational (directed) selection techniques** (Chang and Elander, 1979), which are based on known biochemical and/or genetic processes of the organism concerned, reduces the number of strains to be tested and so facilitates the isolation of desired strains. Rational screens may be direct (based on direct assay of the desired property) or indirect (based on assay of a character associated with the one of interest). Indirect rational screens generally serve as primary (pre) screens (see also section 4.13.1) and it is subsequently necessary to test selected strains for desired properties. A variety of rational selection procedures can be employed, examples of which are given below.

1. Resistance or tolerance to toxicity of end-product (for example an antibiotic or alcohol). Where the desired product is potentially toxic to the producer strain, productivity can sometimes be improved by the selection of strains with increased resistance or tolerance to the product.

(i) *Antibiotic resistance.* Antibiotics may inhibit the growth of those organisms that produce them. In such cases the efficiency of antibiotic production can sometimes be improved by using mutants that are resistant to high concentrations of the antibiotic produced. For example, chlortetracycline-resistant strains of *Streptomyces aureofaciens* have been isolated that produce increased yields of chlortetracycline compared with the sensitive parental strain (Katagiri, 1954). A

Figure 6.1: Stepwise mutation programme for improving penicillin production by *Penicillium chrysogenum*. Yield figures reflect genetic improvements together with improvements in fermentation technology. S, Spontaneous mutagenesis; X, X-ray radiation; UV, ultraviolet radiation; NM, nitrogen mustard. (Adapted from Elander, 1967)

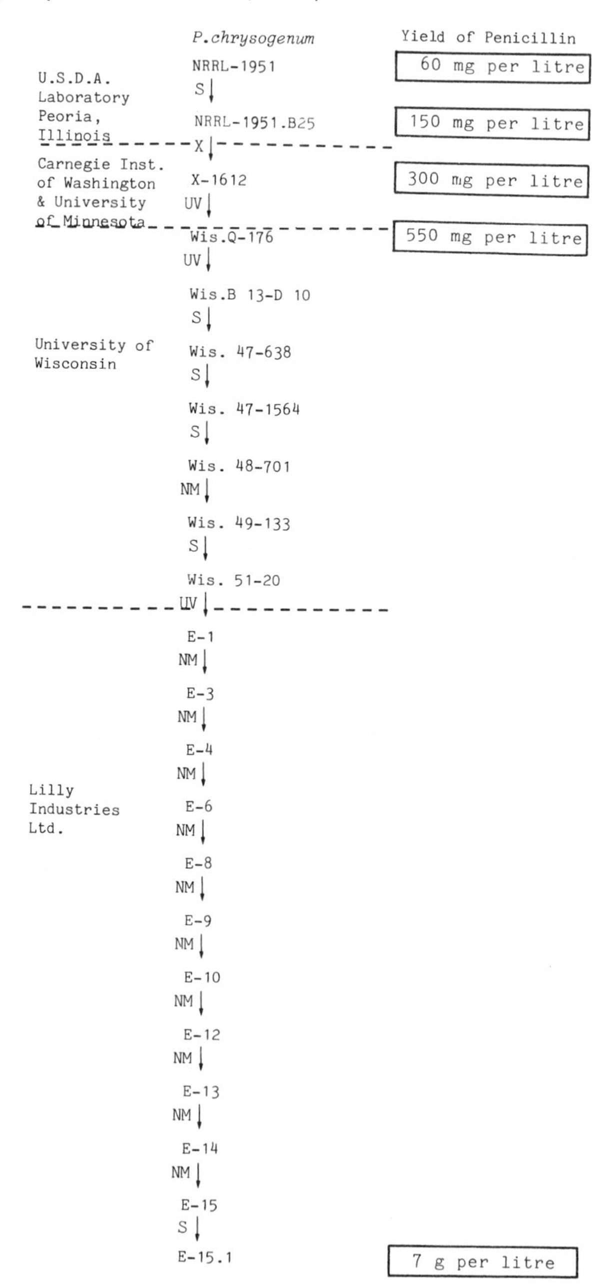

problem that may be encountered when using this approach is that antibiotic resistance may be associated with enzyme-mediated inactivation of the antibiotic. This problem can be overcome by selecting for mutants that are resistant to the antibiotic but are unable to produce the appropriate antibiotic-inactivating enzyme. Such selection techniques have been applied to the isolation of nocardicin A-resistant, β-lactamase-deficient strains of *Nocardia uniformis* that show improved production of the β-lactam compound nocardicin A (see Vournakis and Elander, 1983).

(ii) *Alcohol tolerance.* A major constraint on the fermentative production of ethanol (and other organic solvents) by microorganisms is that of the capacity of the producer strain to tolerate the alcohol produced. The viability of *Saccharomyces* used in the production of ethanol from crude sugar (or from starch that has been converted into sugar) is reduced as fermentation proceeds. At concentrations of ethanol of 12% (weight/volume), growth of wild-type yeasts in the fermenting mixture is inhibited. The selection of mutants with higher ethanol tolerance can result in strains that exhibit increased rates of alcohol production. A method of continuous selection, whereby the culture itself determines the intensity of selection (Brown and Oliver, 1982) has been applied to the isolation of ethanol-tolerant mutants of *Saccharomyces uvarum.* Such mutants produce alcohol at up to twice the rate of wild-type in the presence of 10% (weight/volume) ethanol. The procedure involves constant monitoring of the fermentation process. The rate of CO_2 produced during fermentation is used as an indicator of the rate of alcohol production. Alcohol concentration in the reaction vessel is adjusted in accordance with CO_2 output, such that as CO_2 production increases, more ethanol is introduced into the vessel. Conversely as CO_2 evolution decreases, flow of ethanol is reduced (see Figure 6.2). The frequency of ethanol addition thus increases as the tolerance of the organisms improves. In prolonged operation the vessel becomes colonised by mutants with elevated ethanol tolerance. The use of such ethanol-tolerant strains in the production of alcohol should improve productivity and hence facilitate product recovery by distillation, in turn reducing production costs. Ethanol-tolerant strains of *Zymomonas mobilis* have also been isolated. This organism, which can give higher yields of ethanol than yeast, shows promise for use in large-scale production of ethanol (Skotnicki *et al.*, 1982).

The method of continuous selection, using a feedback control system, should be applicable to the selection of strains with improved tolerance to any inhibitory condition of either their physical or chemical environment.

Figure 6.2: Continuous culture apparatus for selection of ethanol-tolerant mutants. The fermentative activity of the culture in the vessel was monitored continuously by measuring the CO_2 concentration in the exit gas, using an infra-red CO_2 analyser. The signal from the analyser was fed to a potentiometric controller. When the concentration of CO_2 exceeded that determined by the set point of this controller, it operated a pump that introduced ethanol into the culture vessel (from Brown and Oliver, 1982)

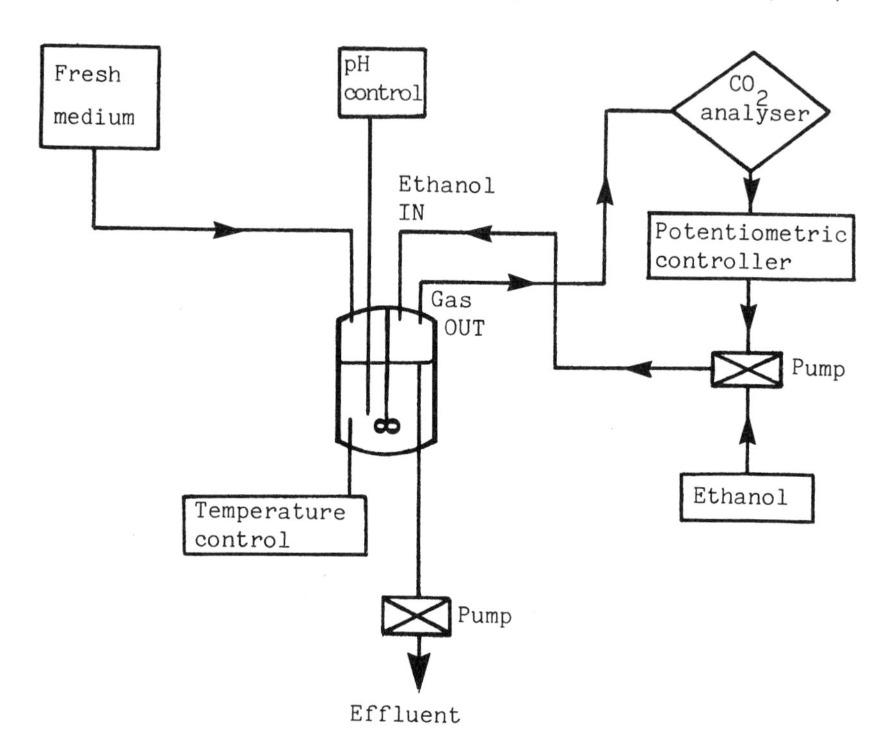

2. Phage resistance. Susceptibility of microorganisms to phage attack can have detrimental effects on fermentations. The problem may be avoided by employing mutants that are resistant to the phage(s) involved (see Klaenhammer, 1984). Such mutants can be obtained by using the particular phage as selective agent. However, this sometimes results in organisms with altered permeability properties (see also(6)) that may affect growth. Phage-resistant mutants have also been isolated by using the amino acid lysine (a component of bacterial cell walls) instead of phage as the selective agent. A number of mutants resistant to inhibitory concentrations of lysine proved to be phage-resistant. These mutants did not exhibit altered permeability to medium substrates, and growth was normal (Lein, 1983).

3. Resistance to toxic analogues. Productivity may be increased by deregulating metabolism using **toxic analogues** of regulatory metabolites (or using high concentrations of regulatory products themselves, where they are sufficiently toxic). Such analogues can act as false feedback effectors and in turn inhibit growth of sensitive organisms. Mutants resistant to these analogues often possess enzymes that no longer respond to feedback inhibition (or are much reduced in their ability to do so) and thus overproduce the antagonised metabolites. End-product analogues have been used to select mutants that overproduce amino acids (see Yamada, 1977). One example is the overproduction of lysine by *Corynebacterium glutamicum* mutants that are resistant to the lysine analogue S-(β-aminoethyl)-L-cysteine (AEC) (Sano and Shiio, 1970). Such mutants possess an altered form of aspartate kinase (a cardinal enzyme in lysine biosynthesis, see Figure 6.3) that functions in the production of lysine but is no longer sensitive to concerted feedback inhibition by lysine and threonine (even when lysine is present in excess). As a consequence lysine is overproduced. Growth of parental strains is inhibited by AEC, due to lysine starvation.

Mutants resistant to amino acid analogues have also been used for the overproduction of antibiotics having amino acids as precursors. For example, *Pseudomonas fluorescens* strains that overproduce D-tryptophan (a precursor of the antifungal agent, pyrrolnitrin), selected as mutants resistant to either fluorotryptophan or methyltryptophan, exhibit improved pyrrolnitrin production (see Elander, 1982). Other examples include analogue resistant strains of *Penicillium chrysogenum* and *Cephalosporium acremonium* (*Acremonium chrysogenum*) that are superior β-lactam producers. Biosynthesis of penicillins or cephalosporins involves the amino acids α-aminoadipic acid, cysteine and valine as precursors (Figure 6.4). Analogues of these amino acids have been used to select strains of *P. chrysogenum* and *C. acremonium* that overproduce penicillin and cephalosporin respectively.

The synthesis of several antibiotics, including penicillin, chloramphenicol and cycloheximide, may be subject to feedback inhibition by the appropriate antibiotic. The antibiotic may inhibit activity of the antibiotic synthetase (see Martin and Demain, 1980). The concentration of antibiotic that appears to be required for inhibition of its own synthesis correlates well with the production capacity of the producer strain. Removal of such feedback regulation may therefore improve antibiotic production.

4. Auxotrophic mutants. Nutritional mutants possessing blocks in specific biosynthetic pathways have been exploited in the overproduction of microbial products. Overproduction of various amino acids, including L-proline and L-valine, whose syntheses occur via branched pathways, has been achieved by the isolation of appropriate auxotrophic mutants (see Nakayama, 1973; Yamada, 1977). The underlying principle is to block one

Figure 6.3: Lysine biosynthesis in *Corynebacterium glutamicum*. In wild-type strains, activity of aspartate kinase is subject to inhibition by excess lysine and threonine. Overproduction of lysine can be achieved by a mutation that alters aspartate kinase such that it is insensitive to feedback inhibition; or by a mutation that inactivates homoserine dehydrogenase, in turn preventing the accumulation of threonine and so alleviating feedback inhibition of aspartate kinase

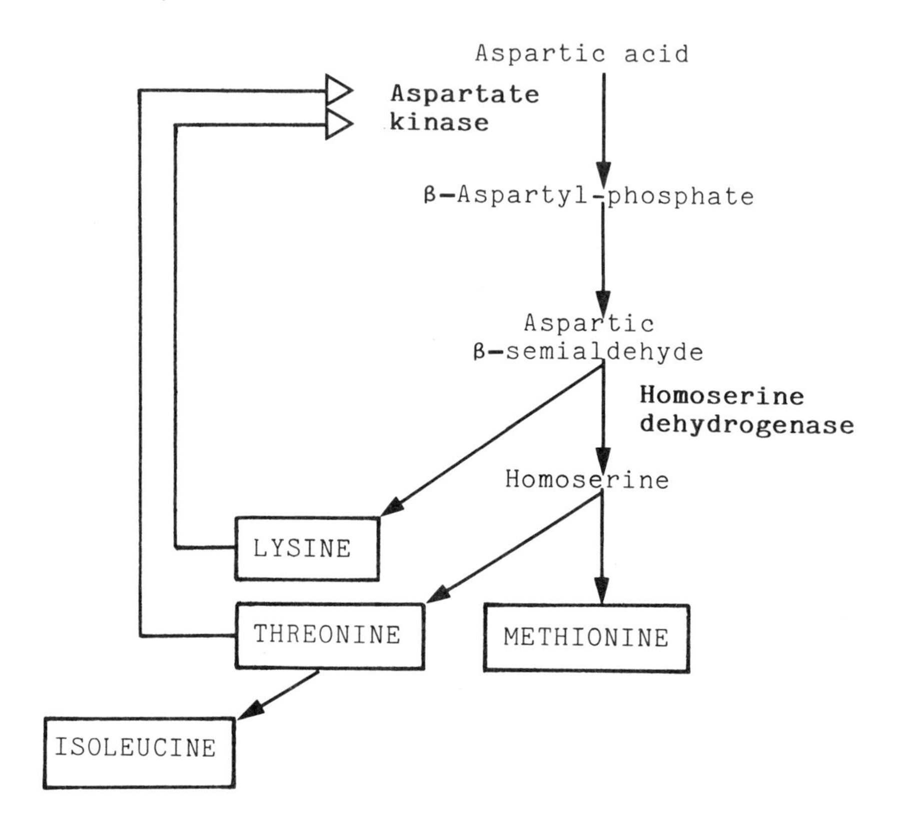

part of the branched pathway such that flow of metabolites is diverted away from undesirable products. For example, homoserine auxotrophs of *C. glutamicum* have been selected that overproduce lysine. Such mutants possess a defective homoserine dehydrogenase and are therefore unable to synthesise homoserine (see Figure 6.3). Addition of sufficient homoserine to support growth, but not to inhibit the activity of aspartate kinase, results in overproduction of lysine (Kinoshita *et al.*, 1958). Mutants of *Bacillus* spp. blocked in pentose metabolism have been used for the production of D-ribose. Transketolase-deficient mutants (isolated as strains requiring shikimic acid,see Figure 6.5) were shown to excrete large amounts of the pentose into the culture medium (Sasajima and Yoneda, 1984).

Figure 6.4: Penicillin and cephalosporin biosynthesis in *Penicillium chrysogenum* and *Cephalosporium acremonium (Acremonium chrysogenum)* respectively

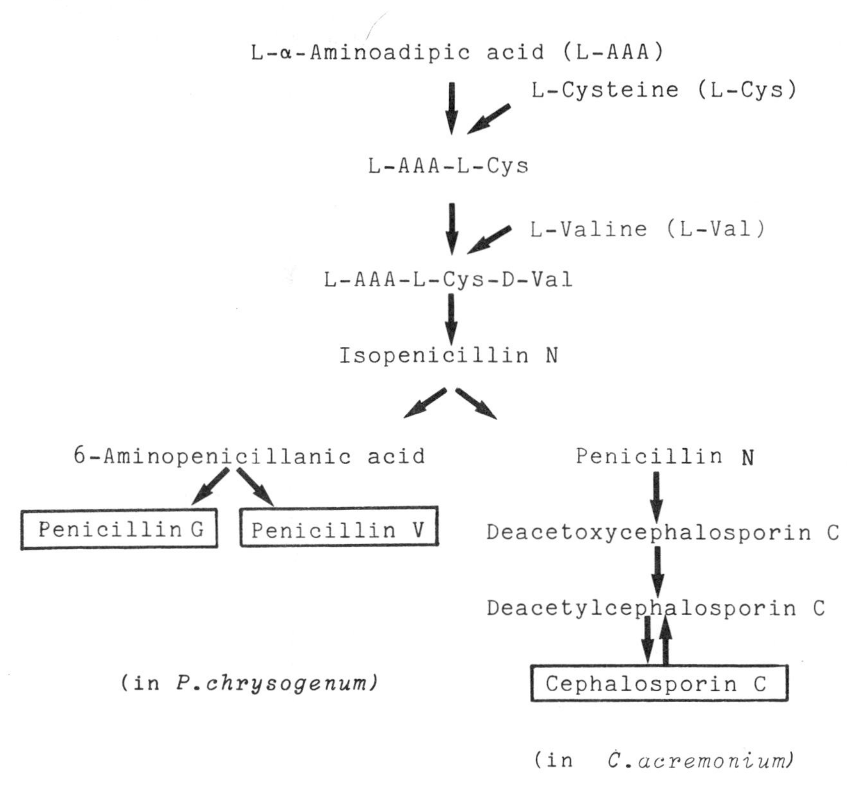

Often mutants that possess both nutritional defects and regulatory defects removing feedback inhibition are more effective as production strains than mutants with a single defect. A combination of auxotrophy and analogue resistance has been used to enhance production of amino acids by *C. glutamicum.* For example, a methionine auxotroph resistant to the threonine analogue, α-amino β-hydroxyvaleric acid (AHV) and to the lysine analogue, AEC (obtained by a stepwise programme, see Figure 6.6), showed an increased ability to produce threonine relative to the wild-type (Kase and Nakayama, 1972).

There are cases where auxotrophs have been used successfully for improving antibiotic production. For example, the selection of certain strains of *Streptomyces* with defects in pyruvate and aspartate metabolism resulted in increased production of the β-lactam, cephamycin (Godfrey, 1973).

Figure 6.5: Reactions involved in pentose phosphate and aromatic amino acid pathways

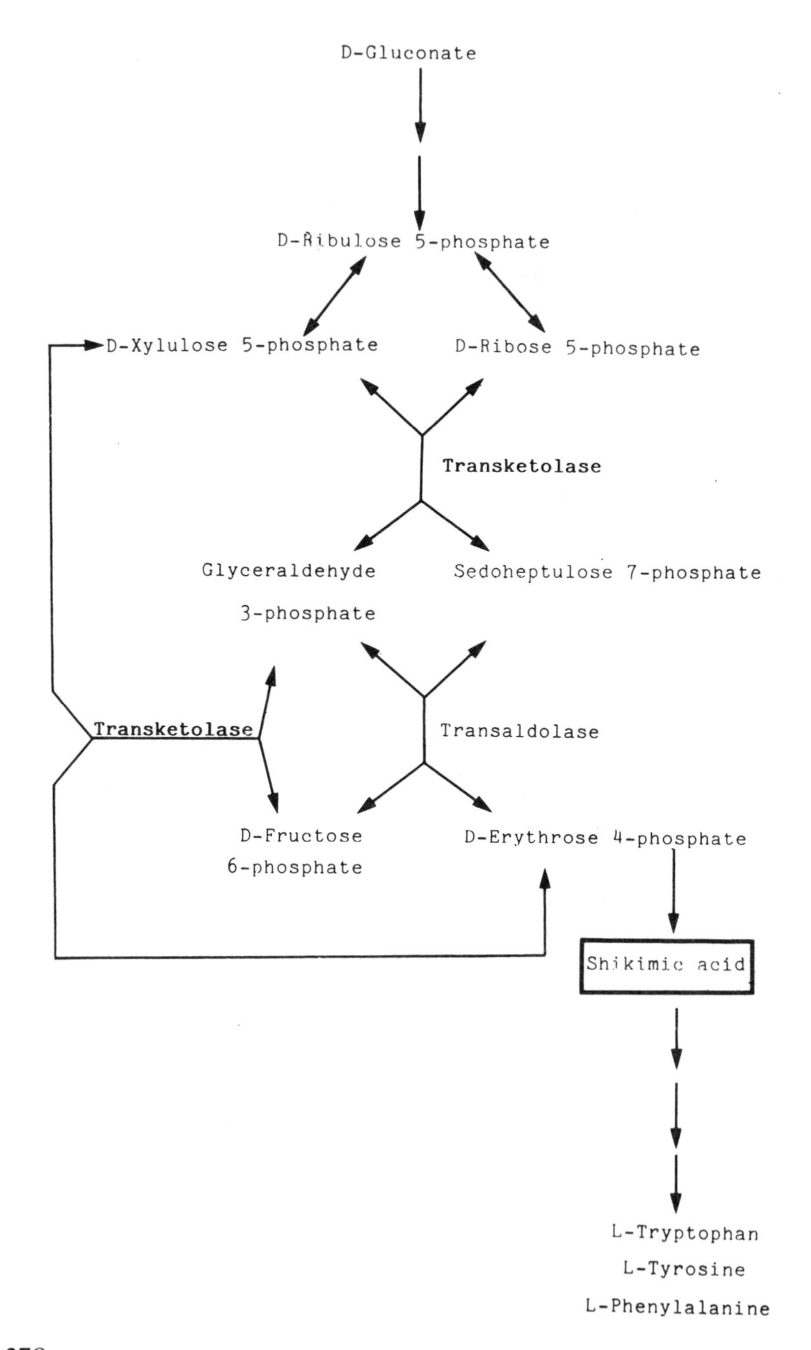

Figure 6.6: Stepwise mutation programme to obtain improved threonine producers of *Corynebacterium glutamicum*. Met , AHV^r, AEC^r mutants produced 9.5 mg/ml threonine in medium containing methionine (100μg/ml). N-MNNG, *N*-methyl-*N'*-nitro-*N*-nitrosoguanidine; AHV^r, resistant to threonine analogue, α-amino-β-hydroxyvaleric acid; AEC^r, resistant to lysine analogue, *S*-(β-aminoethyl)-L-cysteine; Met , methionine requiring (adapted from Kase and Nakayama, 1972)

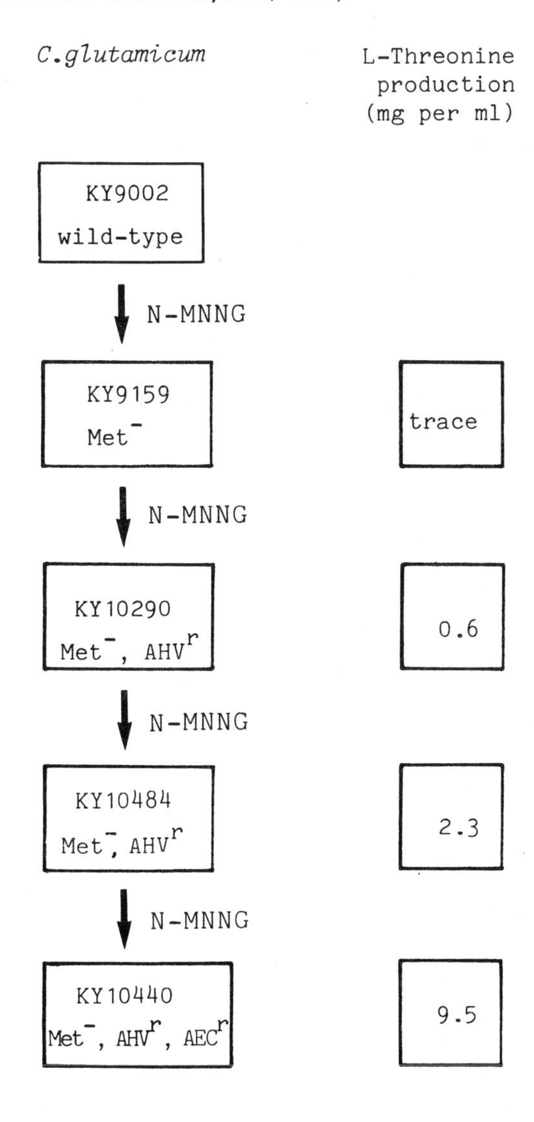

5. Revertants. In certain cases reversion or suppression of nutritional mutations to prototrophy can result in superior production strains. Suppression of a primary (auxotrophic) mutation by a second mutation can sometimes generate feedback-resistant mutants with increased production capacity. For example, suppression of a cysteine auxotroph of *Streptomyces lipmanii* and of a methionine auxotroph of *Streptomyces viridifaciens* has led to improved production of cephamycin and chlortetracycline respectively (Dulaney and Dulaney, 1967; Godfrey, 1973). Revertants of antibiotic non-producing mutants have been isolated that produce increased amounts of antibiotic relative to the original producer strain. A number of such revertants of *S. viridifaciens* has been obtained that overproduce chlortetracycline (Dulaney and Dulaney, 1967).

6. Permeability mutants. Mutants with altered cell wall/membrane permeability enabling increased efflux of products may exhibit increased product yield. Rapid removal of regulatory metabolites from the cell may, for example, relieve feedback inhibition thereby resulting in overproduction of particular products. Permeability mutants of *P. chrysogenum*, selected as strains resistant to polyene antibiotics (which act on the cell membrane), have been used for increasing penicillin titre (Luengo *et al.*, 1979). In addition, *Streptomyces fradiae* mutants that are resistant to actinophage infection, due to alterations in the cell envelope, show improved antibiotic titre (Perlman and Hall, 1976).

7. Resistance to catabolite repression. Where biosynthesis is sensitive to **carbon catabolite repression**[1] (Demain, 1982), yield of product may be enhanced by the selection of mutants that are insensitive to such repression. Such mutants can utilise non-repressing substrates in the presence of toxic analogues of repressing ones. One example is the isolation of cellulase-hyperproducing strains of the cellulolytic fungus, *Trichoderma reesei*, by selection for mutants insensitive to 2-deoxyglucose (Ghosh *et al.*, 1982). 2-Deoxyglucose mediates catabolite repression but is not metabolised. Strains that can utilise nonrepressing carbon sources in the presence of 2-deoxyglucose are therefore resistant to catabolite repression.

---

[1] Carbon catabolite repression is repression, by readily utilisable carbon sources, of genes encoding enzymes for certain catabolic pathways. In *E. coli* carbon catabolite sensitive operons can be positively regulated by catabolite activator protein (CAP) (or cyclic AMP receptor protein (CRP)). CAP is active in the presence of cAMP. Binding of cAMP-CAP to the promotor region assists initiation of transcription of the relevant operon. In the presence of glucose the intracellular concentration of cAMP is reduced. This renders CAP defective in binding in the promoter region, which in turn reduces the efficiency of transcription (for reviews see Pastan and Adhya, 1976; Ullmann and Danchin, 1980; Busby, 1986). Mutation in the promoter region may render the operon insensitive to catabolite repression.

8. Phosphate-deregulated mutants. Where synthesis of metabolites is sensitive to inorganic phosphate (Demain, 1982) fermentation must normally be carried out in low (growth-limiting) phosphate media. The isolation of phosphate-deregulated mutants (by selection for resistance to toxic analogues of phosphate, such as arsenate) should, however, permit the use of complex media with high phosphate content. Martin and co-workers (1979) have isolated *Streptomyces griseus* mutants that are deregulated in phosphate control of the synthesis of the polyene macrolide antibiotic, candicidin. Such mutants show improved yield of candicidin in both phosphate-supplemented and nonsupplemented media.

(b) Novel product formation by mutation

1. Direct mutation. The modification of microbial products can be effected directly by mutation. Such modified products typically possess the basic structural features of the parent compound, but may have altered functional groups that influence biological activity. Microbial enzymes with new and improved capabilities, such as novel substrate specificities, can be obtained by mutation. Appropriate modifications to specific enzymes of metabolic pathways can result in the production of various novel metabolites. Examples of antibiotic derivatives obtained include those of tetracyclines, β-lactams, and rifamycins (Sebek, 1976).

2. Mutational biosynthesis (mutasynthesis). Mutants blocked in antibiotic synthesis have been used to generate novel antibiotics by the technique of **mutational biosynthesis** (Nagaoka and Demain, 1975). Such mutants (**idiotrophs**) are unable to synthesise specific antibiotic precursors. Addition of the normal precursor to the culture can result in synthesis of the natural antibiotic. However, if modified precursors are added, new antibiotics may be elaborated (Figure 6.7). The success of this technique depends upon the ability of the precursor analogue (**mutasynthon**) (Rinehart and Stroshane, 1976) to enter the cell and permit completion of the synthetic pathway, in turn generating a product with antibiotic properties (**mutasynthetic**). Mutational biosynthesis enables a particular portion of an antibiotic to be modified independently of the rest. Various new antibiotics including derivatives of aminocyclitol antibiotics, novobiocins, actinomycins and β-lactams have been prepared using this methodology (Daum and Lemke, 1979).

6.2.2 Microbial breeding

Strain improvement can be effected by bringing desirable genes together in one strain through microbial breeding. Several desirable characters may be acquired simultaneously by an individual using this approach. The cumula-

Figure 6.7: Mutational biosynthesis. (a) A mutant blocked in antibiotic synthesis, such that one precursor of the antibiotic is not made, will synthesise an incomplete antibiotic. Addition of that precursor to the medium can lead to production of the original antibiotic. Addition of a modified precursor may result in synthesis of a new antibiotic (mutasynthetic). (b) For example, *Streptomyces fradiae* can produce neomycin from glucose and ribose (i). A blocked mutant that is unable to produce the neomycin precursor 2-deoxystreptamine from glucose will produce the modified antibiotic 6-deoxyneomycin B when precursor analogue is supplied (ii).

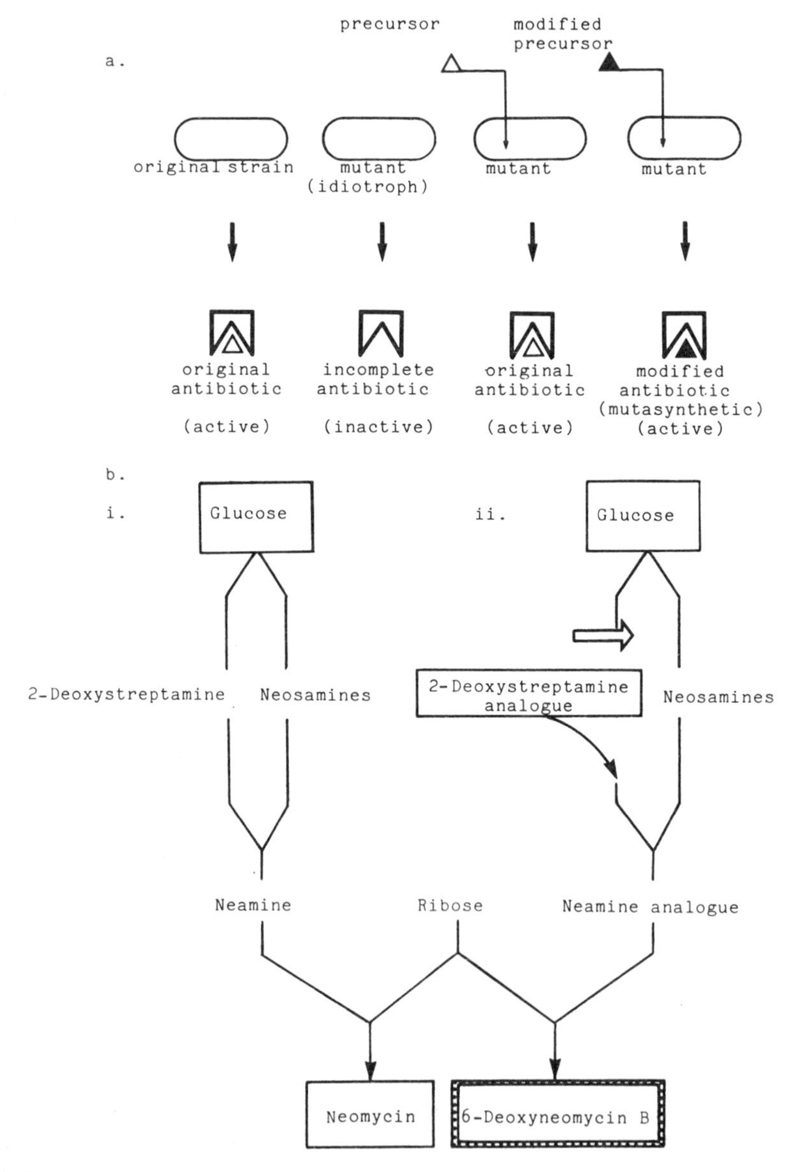

tive effect of this may be much greater than the effect of acquiring a single character. Furthermore, undesirable characters may be bred out. Accordingly breeding affords an attractive alternative/addition to mutagensis for strain improvement, particularly where the target for improvement is polygenic in nature. However, the transfer of desirable characters through breeding may be accompanied by the transfer of non-beneficial properties. Genetic recombination effects the rearrangement of genes (or parts of genes) that have been brought together by gene transfer. Recombination techniques can thus be applied to create new genotypes for strain improvement and for novel production formation. Moreover, the ability to combine sets of genes from distantly related organisms increases the scope for generating novel products from hybrids.

Gene transfer and genetic recombination can be exploited not only in strain development *per se*, but also to provide basic knowledge about genome organisation in species of interest and about the genetics of specific microbial processes. This, in turn, enables the application of more rational (and potentially more successful) breeding programmes. Examples of programmes involving parasexual and sexual crossing of fungi, bacterial gene transfer and protoplast fusion are described below.

(a) Use of the parasexual cycle in breeding programmes

Classical breeding programmes exploiting the parasexual cycle (see section 2.6) have been of limited success. This has probably been largely due to the use of parental strains that had already been subjected to repeated mutation-selection programmes. Structural rearrangements in homologous chromosomes, effected by the successive rounds of mutagenesis, probably accounted for the general lack of recombination found. The adoption of controlled breeding programmes (see below) using carefully selected strains with high recombining abilities has, however, circumvented some of the problems associated with the classical methods.

1. Yield improvement. Basically there are three types of cross that can be performed in order to improve productivity (see Figure 6.8):

(i) **sister cross** involving parents that apparently differ only in those markers required to force the heterokaryon and detect segregants;
(ii) **ancestral cross** involving parents marked as in (i), but in addition one parent is derived from the other, but differs from it by mutations that increase titre;
(iii) **divergent cross** involving parents marked as in (i), but in addition possessing different titre-increasing mutations.

The isolation of haploid segregants from a divergent cross, where the degree of divergence is not too large (i.e. using parental strains that differ

Figure 6.8: Genetic lineage and main types of cross in fungal breeding. The dashed lines represent serial mutation and selection of haploid strains for increased product formation. W, Wild-type strain; A, common ancestor for variants B and C. Sister and divergent crosses are normally made to combine the better features of two strain lines. The ancestral cross is normally made to restore vigour to a strain line.

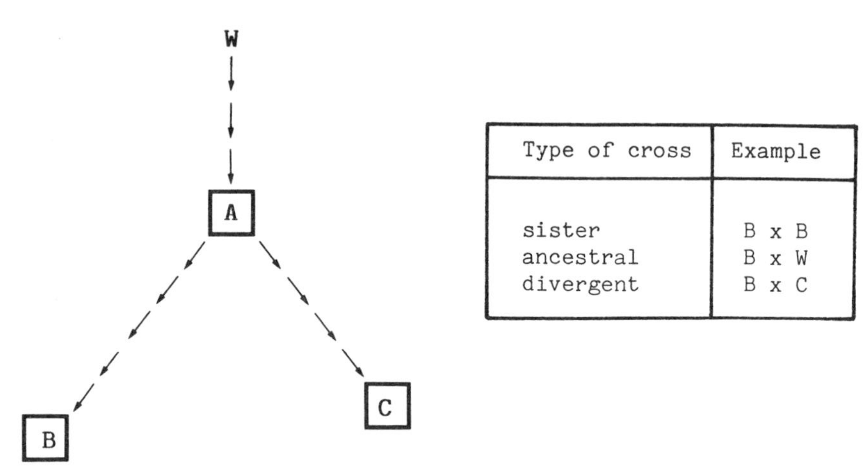

Type of cross	Example
sister	B x B
ancestral	B x W
divergent	B x C

by only a few mutational steps, such that recombination barriers are largely overcome) is normally the preferred approach to obtaining strains with enhanced yield (Ball, 1978). Furthermore such strains are likely to be amenable to additional mutation and recombination techniques, with a view to further improvements. Haploidisation agents (see section 2.6) are often used to promote the production of desired haploid segregants. Treatment with such agents might be preceded by treatment with agents that increase mitotic crossing-over in order to assist isolation of useful and stable segregants. A problem with the use of diploid segregants is that they tend to be unstable and further segregation can occur. This may lead to loss of desirable characters. However, diploids have sometimes proved useful for yield improvement. For example, stable diploids of *Penicillium chrysogenum* that exhibit enhanced penicillin production relative to the original haploid parent have been obtained from a sister cross (Elander *et al.*, 1973).

A number of conditions need to be satisfied for the performance of controlled breeding programmes:

(i) use of marked parental strains with markers that are not titre-modifying and that are distributed between different haploidisation groups;
(ii) use of parental strains with similar yields;
(iii) use of stable parental strains;
(iv) recombinants should be readily selected.

Ball (1973a,b) used the parasexual cycle to improve the penicillin titre of *Penicillium chrysogenum.* The objective was to recombine two yield-increasing alleles, *t2* and *t5.* The genetic programme is outlined in Figure 6.9. Many of the haploid segregants of genotype *brw t2*; *nic t5* had enhanced penicillin titre compared with the parental strains. The parasexual cycle has been used in other programmes, for example to improve citric acid production by industrial strains of *Aspergillus niger* (Azevedo and Bonatelli, 1982).

2. Improvement in properties other than yield. The parasexual cycle can be used to introduce advantageous characters into already high-yielding production strains. Such characters include those enabling improved efficiency of substrate utilisation (Elander *et al.*, 1973), enhanced sporulation ability (Calam *et al.*, 1976), or improved filtration efficiency (Ball *et al.*, 1978). The transfer of such qualitative characters from relatively low-yielding ancestral strains into production strains often results in an initial reduction of titre in the recombinants. This is due to the segregation of the titre genes. Much of the lost titre can, however, normally be regained either by backcrossing with the production strain or by mutagenesis. Production of amyloglucosidase by *Aspergillus niger* has been improved by transferring efficient broth filtration characteristics of a low-yielding strain to a related high-yielding strain through parasexual crossing (Ball *et al.*, 1978). The mycelial morphology exhibited by the low-yielding strain in submerged culture enabled more efficient filtration of the fermentation broth during product recovery (since the broth was less viscous) than did that of the high-yielding strain. Transfer of this characteristic to strains with high amyloglucosidase titre facilitated broth filtration and decreased filtration time.

(b) Uses of fungal mating systems

The sexual cycle has played a relatively minor role in strain improvement programmes. This has been due largely to the lack of effective sexual cycles in many industrial strains of fungi. The development of hybridisation programmes for improving yeasts has been hampered by the polyploid or aneuploid nature of various industrial strains. Such strains are prone to poor sporulation or to low spore viability and show poor mating ability. Conventional mating techniques can generally be used where haploid derivatives of production strains are available. In other cases '**rare mating**' has been used to enhance capabilities of industrial yeasts (Johnston and Oberman 1979; Stewart and Russell, 1981). The rare mating technique involves the mixing together of 'nonmating' strains at a high cell density and selection of the few true hybrids which form (Spencer and Spencer, 1983). This allows the hybridisation of commercial strains of undetermined ploidy with haploid laboratory strains. Rare mating techniques can be used

Figure 6.9: Genetic programme for isolation of haploid recombinants of *Penicillium chrysogenum* with increased penicillin titres. Yield-increasing allele t_2 was allocated to haploidisation group I, as marked by the spore colour marker *brw*; yield-increasing allele t_5 was allocated to haploidisation group III, as marked by *nic*. Penicillin titres (units/ml) are given. *brw*, Brown spore colour; *bgn*, bright green spore colour; *whi*, white spore colour; *nic*, requirement for nicotinamide; *thi*, requirement for thiamine; *rib*, inability to grow on ribose as sole carbon source

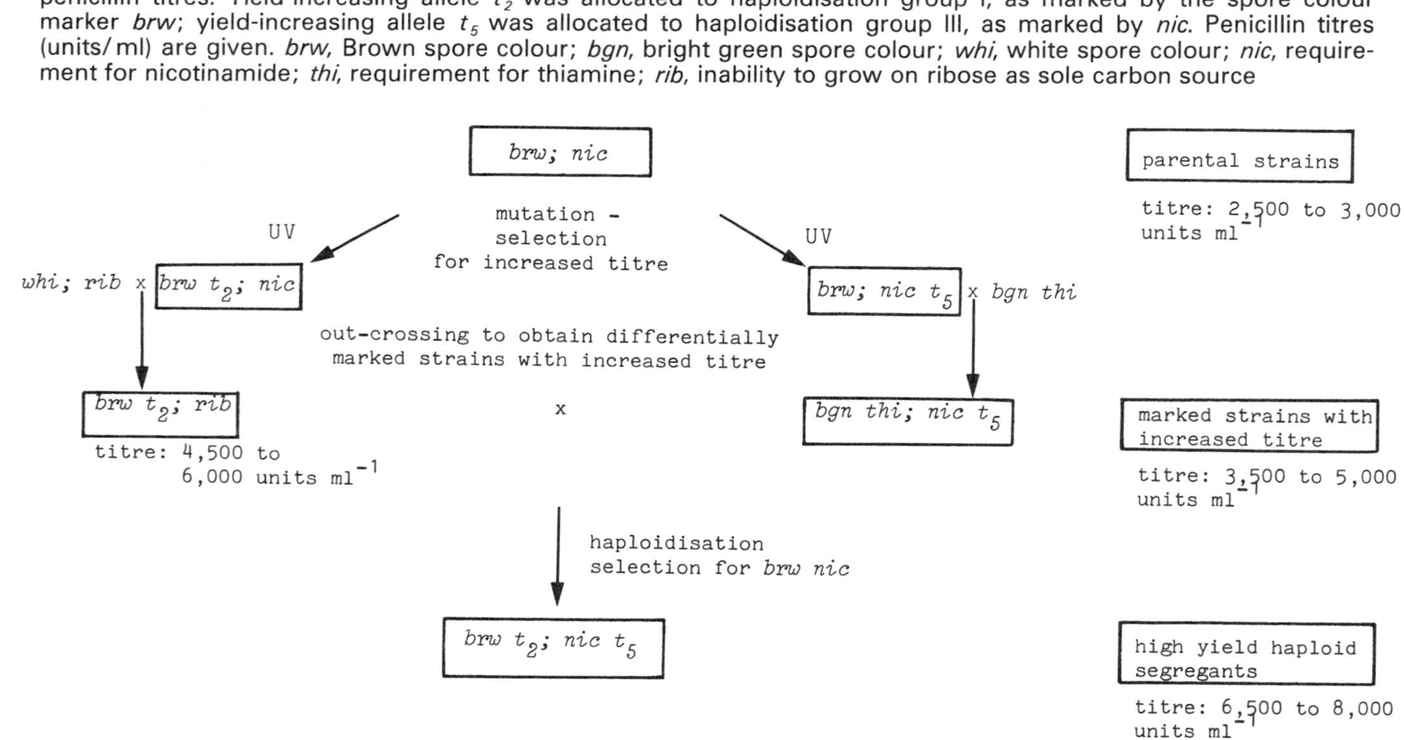

to transfer nuclear and/or cytoplasmic genetic elements. Transfer of cytoplasmic components, for example mitochondrial DNA, or plasmids (including those carrying cloned genes), in the absence of nuclear genes (a process termed **cytoduction**) (Zakharov and Yarovoy, 1977) depends upon the presence in parental strains of a mutation *kar* (karyogamy-defective), which impairs nuclear fusion.

1. Improvement of industrial yeasts

(i) Baker's yeast. Major objectives of programmes for improving baker's yeast are: (a) improvement of quality in terms of storage stability, dough fermentation and osmotic resistance to salts and sugars; (b) improvement of quantity in terms of yield of yeast per unit of substrate. Substantial improvements in both quality and quantity of baker's yeast have been achieved through mating (see, for example, Figure 6.10 and Burrows, 1979). Improved yields have generally been obtained with polyploid, as opposed to diploid, hybrids of baker's yeast. This is due to the larger cell size (of the polyploid) which facilitates separation and reduces cell loss.

(ii) Brewer's yeast. Programmes for brewing yeasts aim to improve characters such as (a) sugar utilisation; (b) alcohol tolerance and production; (c) flavour of brew; and (d) fermentation rate. The two main types of beer are lager and ale, fermented with strains of *Saccharomyces uvarum* (*carlsbergensis*) and *S. cerevisiae* respectively. In the brewing industry, hybrids formed between the lager yeast *S. carlsbergensis* and *S. diastaticus* have been used in the production of low-carbohydrate (light) beer. *S. diastaticus* is capable of fermenting dextrins (which comprise a substantial proportion of the carbohydrate in wort) and is hence superattenuating[2], but the beer produced by this yeast is unpalatable. However, by crossing *S. diastaticus* with *S. carlsbergensis* and subsequently with wild yeast, hybrids have been obtained capable of producing palatable beer. The hybridisation scheme outlined in Figure 6.11 (see Emeis, 1971) resulted in improved fermentation activity among hybrids: the degree of fermentation was 81% for the parent strain of *S. carlsbergensis*, 96% for hybrid I and approaching 100% for hybrid II.

Unpalatable beer produced by hybrids from matings between lager yeasts and dextrin fermenters is due to enzymatic decarboxylation of ferulic acid (a wort constituent) to 4-vinyl guaiacol, which imparts a phenolic flavour. A single dominant nuclear gene, designated *POF* (phenolic off-flavour), is responsible for production of 4-vinyl guaiacol by the non-brewing strains. Elimination of the Pof^+

[2] The degree of attenuation of brewer's wort depends upon its carbohydrate content and upon the ability of the yeast to ferment the different saccharides.

Figure 6.10: Hybridisation scheme for improving baker's yeast (Distillers Co. Ltd). Strains A and C were commercial baker's yeast and strain B a brewer's yeast. Selected hybrids H1 to H6 are shown

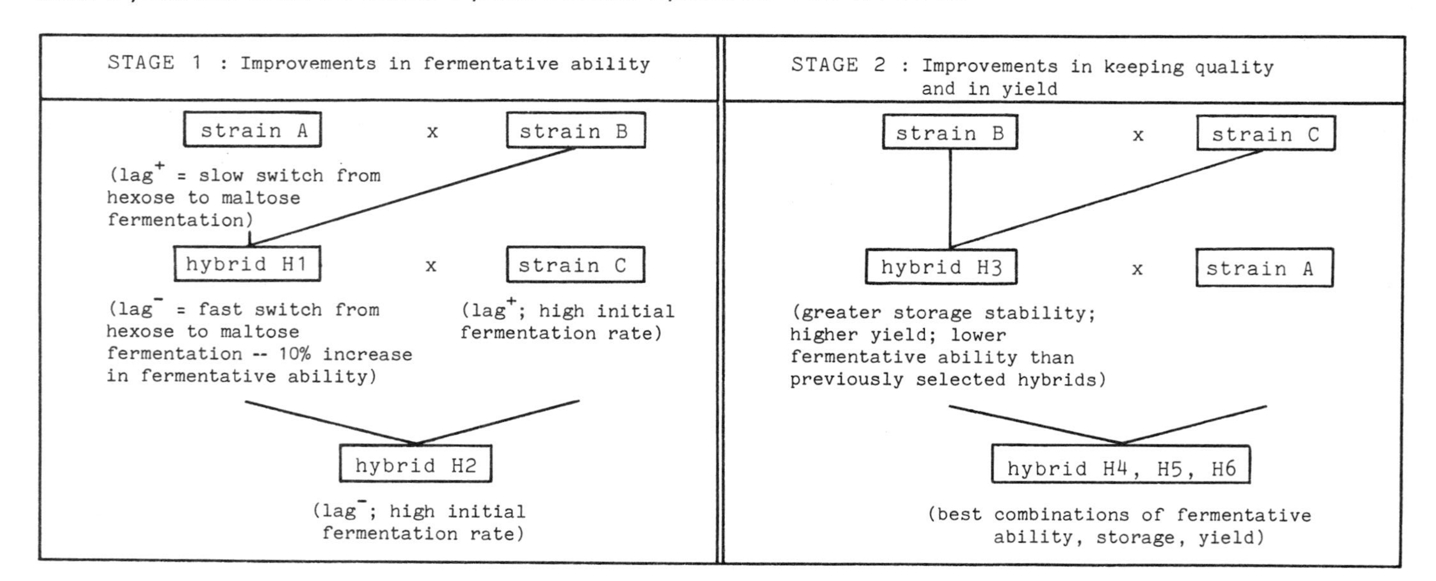

Figure 6.11: Hybridisation scheme to obtain *Saccharomyces* strains for the production of low carbohydrate beer. Strain 1 was a brewing yeast. The degree of fermentation was 96% for hybrid I and almost 100% for hybrid II (compared with 81% for the parent strain of *Saccharomyces carlsbergensis*). Intercrossing strains of hybrid II produced improved hybrid III

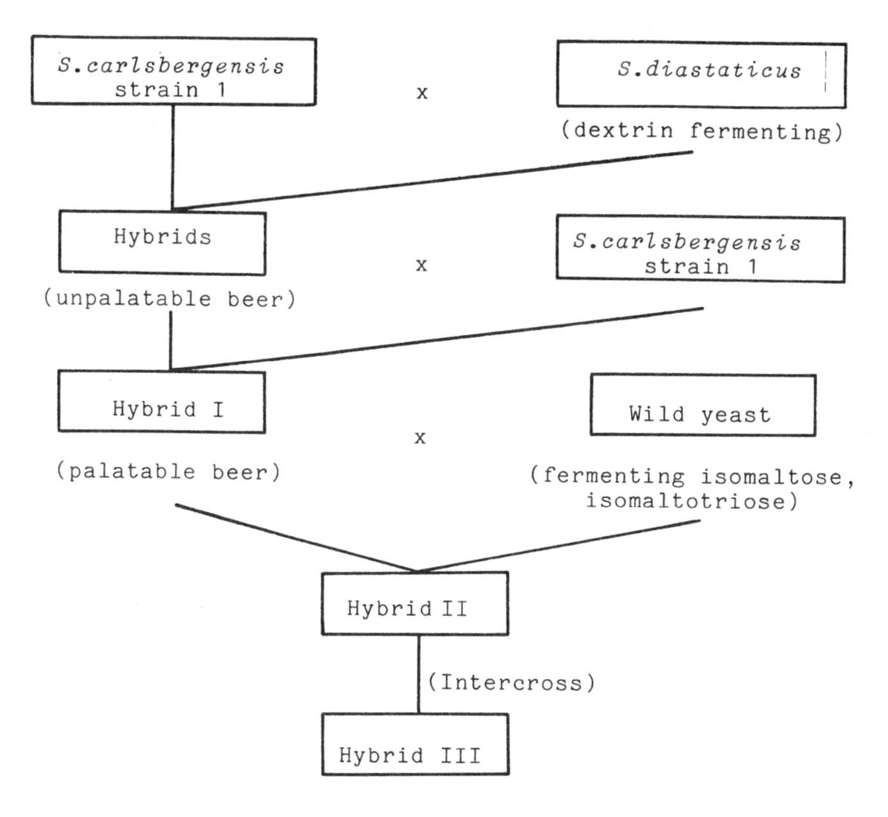

phenotype can be effected by genetic segregation. Segregants that have lost the ability to produce 4-vinyl guaiacol (Pof$^-$) can be isolated following sporulation of hybrids. Alternatively Pof$^-$ dextrin-fermenting parental strains may be used in the matings. Another approach is to suppress, rather than remove, the undesirable character. Replacement of the mitochondrial DNA of the dextrin-fermenting parent with that of an ale yeast, by cytoduction, has led to a reduction in the amount of 4-vinyl guaiacol produced by hybrids (Stewart and Russell, 1981).

A characteristic that has been transferred to brewing strains by rare mating is zymocidal (killer) activity (Hammond and Eckersley, 1984; Stewart *et al.*, 1985). Killer activity is associated with some strains of *Saccharomyces* (and other yeast genera). Killer strains secrete a proteinaceous toxin that is lethal to sensitive yeast strains. The killer character of *Saccharomyces* spp. is determined by two cytoplasmic double-stranded RNA (dsRNA) elements: M ds RNA (killer plasmid) encodes killer toxin (zymocide) production and immunity to the toxin. L dsRNA (also present in many non-killer yeast strains) encodes a capsid protein that encapsidates the two RNA elements. Zymocidal yeasts are a problem in fermentation systems. An infection of the cell population can totally eliminate all the brewing yeast from the fermenter. Rare mating in conjunction with *kar* strains has been used to introduce the cytoplasmic L and M dsRNA into brewing strains. Such strains acquired both resistance to killing by a zymocidal yeast and zymocidal activity themselves, enabling elimination of contaminating wild yeasts from the fermentation mixture.

Mating techniques have also been applied to strain development in the wine-making industry and for industrial chemical and single cell protein (SCP) production.

(c) Bacterial gene transfer

The development of industrially important bacterial strains possessing specific combinations of genetic traits may be accomplished by exploiting one of the natural mechanisms of genetic exchange (see Chapter 2).

1. Transduction. Strains of *Serratia marcescens* exhibiting improved production of amino acids (for example histidine, arginine, threonine or isoleucine) have been constructed by transduction (Kisumi, 1982). Such strains possess multiple mutations resulting in complete deregulation of the biosynthesis of the specific amino acid. Individual regulatory mutations (affecting amino acid biosynthesis) have been combined in a single strain by a series of transductional events.

2. Transformation. The transformation system of *Bacillus* has been used for improvement of enzyme production by *Bacillus* spp. Mutations in a number of genes resulting in the overproduction of the starch-degrading enzyme α-amylase have been selected and introduced into a single strain by repeatedly transforming with chromosomal DNA and selecting for an improved ability to degrade starch. By this route α-amylase-hyperproducing strains of *Bacillus* have been developed (Hitotsuyanagi *et al.*, 1979).

3. Conjugation. A common use of conjugation in strain improvement is in the transfer of bacterial plasmids that encode functions of biotechnological interest. A number of specialised microbial functions, including the production of antibiotics, the degradation of organic compounds and nitrogen fixation, may be specified by plasmids (see also Chapters 8 and 9). Transfer of plasmids between unrelated organisms permits the construction of strains with totally novel abilities. Such strains may be used for specific purposes, such as degradation of pollutants and bioconversions.

Conjugative plasmids can also be used to mobilise the chromosome. Recombinants resulting from the matings can be screened for useful traits. This approach has been used to obtain strains of *Streptomyces* with increased antibiotic yield or that produce new antibiotics. For example, in interspecific matings between two aminoglycoside producers, an auxotrophic strain of *Streptomyces rimosus* forma *paromomycinus* producing mainly paromomycin and a trace of neomycin, and an auxotroph of *Streptomyces kanamyceticus* producing kanamycin, prototrophic recombinants that yielded large amounts of neomycin were obtained (Mazières *et al.*, 1981). Furthermore, in matings between mutants of *Nocardia mediterranea*, one producing rifamycin B and the other rifamycin W, recombinants were obtained that produced novel rifamycins (Schupp *et al.*, 1981).

(d) Uses of protoplast fusion in strain improvement

Protoplast fusion (see section 2.7) affords an effective method for the intermixing of genes of two (or more) strains. In some cases barriers that restrict exchange by conventional mating (see section 2.8) may be circumvented by using protoplast fusion techniques. Protoplast fusion technology may thus extend the uses of the gene pool in breeding, in turn facilitating the construction of hybrids with desired combinations of genetic traits. Protoplast fusion is a particularly valuable technique since it can obviate the need to introduce selective markers into parental strains in those cases where the recombination frequency is high. Furthermore, inactivation of parental strains in the fusion enables recombinants to be selected directly (see section 2.7.1). Protoplast fusion enables improved strains from independent lineages (often developed by random mutation and screening) to be fused, with a view to achieving hyperproduction of useful metabolites. In addition, the technology is attractive for the development of novel microbial products, since it enables interspecies and intergeneric gene transfer. Fusions between distantly related antibiotic producers, for example, may permit the elaboration of new antibiotics.

1. Fungal protoplast fusion. Fungal protoplast fusion has proved useful in facilitating parasexual recombination (see section 2.6). Hamlyn and Ball (1979) obtained strains of *Cephalosporium acremonium* (an organism difficult to manipulate by conventional genetic techniques) with improved cephalosporin titre by using protoplast fusion techniques. Growth rate and sporulation were also improved compared with the parents. Protoplast fusion has been applied to the development of fast-growing, low *p*-hydroxypenicillin V producing strains of *Penicillium chrysogenum* (Elander, 1981). Such strains are useful in fermentations since *p*-hydroxypenicillin V interferes with the chemical ring expansion of penicillins.

Protoplast fusion techniques have also found application in the genetic manipulation of industrial strains of yeast. Since such strains are often polyploid and frequently asexual and asporogenous, protoplast fusion, which depends neither upon mating type characteristics nor upon the ploidy of the strains involved, provides an attractive method for generating hybrids. Furthermore, in most fusions recombinants show improved sporulation compared with the parents.

Protoplast fusion has been used in programmes for the improvement of brewer's yeast. An approach to the development of brewer's yeast with anticontaminant properties is to introduce killer plasmids from killer yeasts. Secretion of active toxin by brewing strains that have acquired such plasmids should eliminate wild yeast contaminants. Furthermore, brewing killer strains are immune to the toxin and hence are not displaced from fermentation by any contaminating wild killer yeasts. Protoplast fusion using inviable donor protoplasts provides an effective route for transferring the killer plasmid to the brewer's yeast, while avoiding hybrid formation (Ouchi *et al.*, 1982). Killer hybrids would be undesirable since they are likely to possess inferior brewing qualities relative to the parental brewer's yeast.

2. Bacterial protoplast fusion. Protoplast fusion has been used for the hybridisation of *Streptomyces* species. One example is the use of protoplast fusion in the development of improved carbapenem-producing strains of *Streptomyces griseus* (subsp. *cryophilus*) (Kitano *et al.*, 1982). Mutants defective in sulphate transport were unable to produce sulphated carbapenem antibiotics, but produced unsulphated carbapenems of increased potency and in higher amount compared with the parental strain. Reintroduction of the sulphate transport system into the sulphate transport-negative mutants by protoplast fusion resulted in the formation of recombinants that produced high yields of the sulphated carbapenem antibiotics. Protoplast fusion has also been applied to the development of amino acid overproducing strains of *Brevibacterium.* For example lysine-auxotroph fusants, obtained between a lysine- and threonine-producing strain of *Brevibacterium lactofermentum* and a lysine auxotroph, showed improved threonine production (Enei and Hirose, 1984).

6.3 APPLICATIONS OF *IN VITRO* RECOMBINANT DNA TECHNOLOGY

Conventional methods of gene transfer and recombination permit the alteration of specific genetic traits by combining sets of genes from two (or more) microbial strains. However, such alterations are not always achieved without adversely affecting other desirable genetic characters. *In vitro* recombinant DNA technology (see Chapters 3,4 and 5), which provides a more controlled approach to gene manipulation, can permit the alteration of specific traits without detrimental effects on other characters. The specific gene(s) can be cloned and reintroduced into the host without cotransfer of extraneous characters. The hybrid constructed may differ from the parent strain only in the character(s) specified by the cloned DNA. Homospecific, heterospecific or chemically synthesised genes may be used in order to alter or replace existing capabilities of the process organism. An ability to mutagenise DNA in a directed manner *in vitro* and replace in the organism of interest by self-cloning offers further scope for fashioning strains in highly specific ways.

The use of *in vitro* genetic techniques in conjuction with *in vivo* techniques should greatly facilitate strain improvement for large-scale low-cost production of various commodities. Furthermore such methodologies afford considerable potential for the development of novel products.

6.3.1 Improvements in yield of microbial product

Yield can sometimes be enhanced by cloning genes for the desired product in plasmid or phage expression vectors (see Chapter 5) *in vitro*. This is relatively straightforward where a single gene or a few genes is involved. Increased production of commercial enzymes, such as certain restriction enzymes and T4 DNA ligase, has been obtained by cloning and amplification of the requisite genes in bacteria.

The situation is, however, more complex for multigene products (primary and secondary metabolites). For primary metabolites, such as amino acids, it has proved possible to clone complete operons. Amino acid hyperproducing strains of *E. coli* have been constructed by using this approach. For example, strains with improved threonine productivity have been obtained by cloning the threonine (*thr*) operon into pBR322 (Debabov, 1982). The operon was isolated from threonine overproducers (selected as strains resistant to threonine analogues). Cloning and amplification of the *thr* genes resulted in increased threonine production. A similar rationale has been applied to the construction of other amino acid overproducers (see, for example, Bloom *et al.*, 1983; Enei and Hirose, 1984).

In the case of secondary metabolites, such as antibiotics, sets of interacting biosynthetic enzymes involving perhaps 10 or more genes are required to generate the final product. Yield may be improved by amplifying the gene for the rate-limiting enzyme or by cloning the complete set of biosynthetic genes. Genes involved in biosynthesis of several antibiotics, including the isochromanequinone antibiotic actinorhodin in *Streptomyces coelicolor* (Malpartida and Hopwood, 1984), the aminoglycoside streptomycin in *Streptomyces griseus* (Ohnuki *et al.*, 1985), the macrolide erythromycin in *Streptomyces erythreus* (Stanzak *et al.*, 1986), the herbicide bialaphos in *Streptomyces hygroscopicus* (Davies, 1985), and cephalosporins in *Cephalosporium acremonium* (Samson *et al.*, 1985), have been isolated. Fragments of genomic DNA of *S. coelicolor* (in the size range 15 to 30 kb) were cloned on the low copy number plasmid vector pIJ922. The whole set of genes (*act*) responsible for actinorhodin biosynthesis appeared to be clustered on a continuous segment of *S. coelicolor* DNA. Such cloned DNA was capable of complementing all available classes of actinorhodin nonproducing mutants of *S. coelicolor* and of directing the synthesis of actinorhodin when introduced into other species, such as *Streptomyces parvulus* and *Streptomyces peucetius.* Introduction of an additional gene set for actinorhodin synthesis into an existing producer strain resulted in increased (30 times the amount produced by wild type) antibiotic production.

In some cases strains with increased antibiotic resistance exhibit enhanced antibiotic production (see section 6.2.1). By cloning appropriate aminoglycoside resistance genes on multicopy vectors and introducing them into aminoglycoside-producing strains, antibiotic yield has been enhanced (Davies, 1985). Cloned antibiotic resistance genes may prove useful not only for enhancing antibiotic production, but also as probes for detecting antibiotic biosynthetic genes where these are linked to resistance genes.

6.3.2 Improvements in growth and operational efficiency of process organism

In vitro techniques have been applied to the development of thermostable catalysts, with a view to their use in improving the efficiency of product formation. One example is the production of a heat stable α-amylase by strains of *Bacillus subtilis.* The native α-amylase produced by *B. subtilis* is denatured by heat. However, by inserting the α-amylase gene from a thermophilic bacterium into *B. subtilis,* high yields of heat-stable α-amylase have been obtained. The thermostability of this genetically engineered α-amylase should permit the conversion of starch into glucose to proceed at a higher temperature than that required when the native enzyme is employed. This should, in turn, result in more rapid starch hydrolysis.

A major factor influencing overall efficiency of microbial processes is the effectiveness of substrate utilisation by the process organism. Reductions in process costs might be achieved by using cheaper substrates and/or by improving the efficiency of substrate assimilation. A notable example of a programme for increasing the efficiency of substrate conversion into cellular carbon by reducing energy expenditure was that devised by ICI for the production of single cell protein (SCP) from the obligate methylotroph,

Methylophilus methylotrophus AS1. The programme involved a combination of *in vivo* and *in vitro* genetic techniques in order to improve the efficiency of nitrogen assimilation, which, in turn, improved the conversion of methanol to cellular carbon (Windass *et al.*, 1980). In *M. methylotrophus* nitrogen assimilation involves a two-stage process requiring the enzymes glutamine synthetase (GS) and glutamate synthase (GOGAT) for conversion of ammonia to glutamate. This is an energy-consuming process in which ATP is required for the reaction catalysed by GS. A similar, but energy-conserving, process operates in *E. coli* in which glutamate dehydrogenase (GDH) catalyses the conversion of ammonia and 2-oxoglutarate into glutamate (Figure 6.12). In order to improve the efficiency of nitrogen assimilation in *M. methylotrophus* the energetically wasteful GS/GOGAT pathway was replaced by the GDH pathway, by introducing the *E. coli gdh* gene into an *M. methylotrophus* mutant that possessed a temperature-sensitive mutation in the gene for GOGAT. The genetic methodology employed is outlined in Figure 6.13. The *E. coli gdh* gene was initially cloned *in vivo* on RP4′*gdh* plasmids. The formation of such plasmids was promoted by Mu. RP4::Mu*cts* was introduced into a strain of *E. coli* lysogenic for Mu and carrying a functional *gdh* gene. This strain was then mated with a Mu lysogen deficient in both GDH and GOGAT and therefore requiring glutamate for growth. Progeny of this mating that carried RP4′*gdh* were identified by an ability to grow in the absence of glutamate (by utilising ammonia). The *gdh* gene was then cloned *in vitro* in *E. coli* on the IncQ plasmid, pTB70, and the resultant recombinant (pTB70*gdh*) was mobilised by RP4 to the Ts mutant of *M. methylotrophus*. At the nonpermissive temperature, transconjugants of *M. methy-*

Figure 6.12: NH_4^+ assimilation in (i) *Escherichia coli*, and (ii) *Methylophilus methylotrophus*. Conversion of ammonia to glutamate by glutamate dehydrogenase (GDH) in *E. coli* does not require ATP, whereas one of the enzymes, glutamine synthetase (GS), involved in the conversion to glutamate in *M. methylotrophus* does

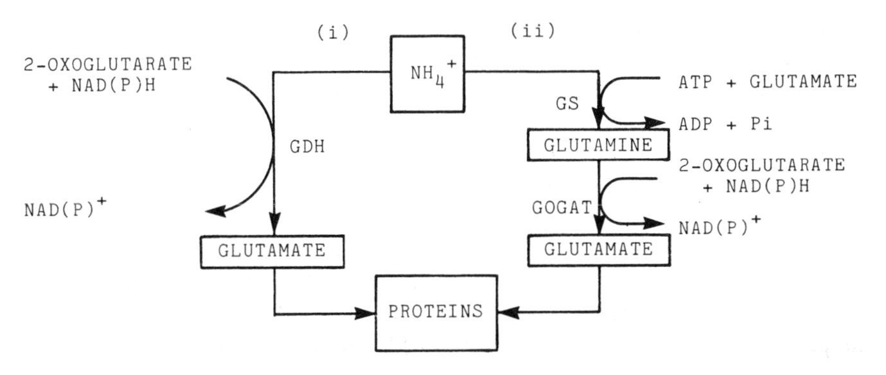

Figure 6.13: Scheme for cloning the *gdh* gene of *Escherichia coli* in *Methylophilus methylotrophus*. *gdh*, Gene for glutamate dehydrogenase; *gltB*, gene for glutamate synthase (GOGAT); *recA*, gene for general recombination; *rpoB*, gene for β subunit of RNA polymerase (*rpoB*⁻ confers resistance to rifampicin)

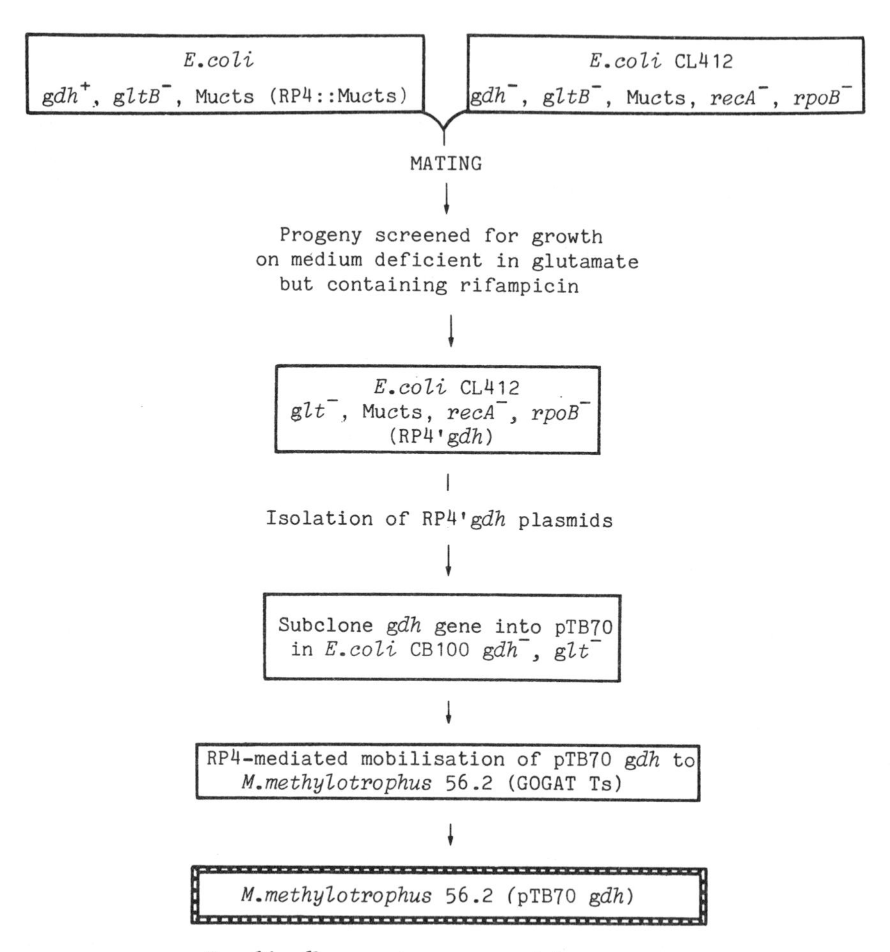

lotrophus used the *E. coli* GDH for glutamate synthesis. Utilisation of this energetically more efficient pathway for ammonia assimilation by the methylotroph resulted in a 4 to 7% higher carbon conversion efficiency than in the parental strain.

New brewing yeasts capable of fermenting substrates can be developed by using recombinant DNA techniques. Barley, which is used in beer making, has a high glucan content. Glucan formation can cause filtration problems during production and can effect precipitation of some polysaccharides during cold storage of stout beers. The Guinness brewery uses malted barley in its stout-brewing process. Malted barley contains the enzyme, β-glucanase, which degrades β-glucan. Less expensive unmalted barley has little β-glucanase activity. The enzyme can be added to the brew to remove glucans. However, this relatively expensive practice, employed by some brewers, could be avoided by the use of β-glucanase-producing yeasts. Furthermore, such yeasts should permit a range of cheaper unmalted barleys to be utilised in brewing. The gene for β-glucanase from *B. subtilis* has been cloned in *S. cerevisiae* (see Henahan and Klausner, 1983). This is an important step in the engineering of commercial brewing yeasts for the degradation of contaminating substances during beer production.

6.3.3 Protein engineering

The technology of protein engineering permits the tailoring of proteins in highly specific ways. By using techniques such as site-directed mutagenesis (section 4.9) and gene synthesis (section 3.1.3) (in conjunction with detailed knowledge of the three-dimensional structure of the protein from X-ray crystallography), specific amino acid sequences may be modified, such that a protein is better suited to a particular biotechnological application(s). For example, an enzyme may be altered in its tolerance to environmental conditions, its substrate specificity and its catalytic efficiency. The thermostability of phage T4 lysozyme (a disulphide-free enzyme) has been improved through the introduction of a disulphide bond at a specific site in the molecule (Perry and Wetzel, 1984). A Cys codon has been substituted for an Ile codon at position 3 in the cloned lysozyme gene by using oligonucleotide mutagenesis (section 4.9.3d) so that a disulphide bond could be formed between Cys3 and another cysteine residue (Cys97) in the enzyme. The crosslinked protein showed increased stability at elevated temperatures, but possessed the same activity as the wild-type enzyme. The pH dependence of enzyme catalysis has also been modified by changing the electrostatic environment of the active site. Substitution of the amino acid serine (Ser) for aspartate (Asp) at position 99 in the enzyme subtilisin from *Bacillus amyloliquefaciens*, by oligonucleotide mutagenesis, has been shown to alter the pH dependence of the catalytic reaction (Thomas *et al.*, 1985). The mutation (Asp→Ser 99) effects modification of a single surface charge in the vicinity of the active site of the enzyme. Specific modifications in surface charge of enzymes may permit pH dependence of catalysis to be changed in ways that improve enzymic activities in industrial processes.

Fusion of different enzymes or protein domains to produce multiple activities on a single protein is also possible using this technology. Site-directed mutagenesis has been used to probe structure and function relationships of the enzyme, tyrosyl-tRNA synthetase from *Bacillus stearothermophilus* (Winter and Fersht, 1984; Fersht *et al.*, 1985). The enzyme catalyses the aminoacylation of tyrosine-tRNA ($tRNA^{Tyr}$) in a two-step process involving the formation of an enzyme-bound tyrosyl-adenylate. By specifically deleting the C-terminal structural domain through mutation, the synthetase has been separated physically into the two functional domains: one (the N-terminal domain) catalyses activation of the amino acid, the other (the C-terminal domain) binds tRNA (Waye *et al.*, 1983). Such techniques that enable physical separation of functional domains provide the opportunity to engineer novel enzyme activities by combining unrelated protein domains.

In vitro **gene fusion** permits the production of hybrid proteins with multiple functions. A DNA chimera encoding a hybrid enzyme with two active sites can be constructed by in-frame fusion of the structural genes for two enzymes. Intervening stop signals at the 3′ end of one gene and the promoter region at the 5′ end of the other are removed prior to ligation. Bülow and co-workers (1985) have prepared such a hybrid enzyme by fusing the *lacZ* gene (for β-galactosidase) and the *galK* gene (for galactokinase). The hybrid (bifunctional) enzyme was capable of catalysing the hydrolysis of lactose and subsequent phosphorylation of galactose to galactose-1-phosphate. (Conversion of lactose to galactose-1-phosphate is normally carried out by the separate native enzymes operating sequentially.) The technique of gene fusion should find application in the provision of customised hybrid proteins, such as bi- or multifunctional enzymes (for catalysis of sequential reactions) or antigen-enzyme conjugates (for ELISA). Purification and immobilisation of the hybrids should prove simpler than for the individual protein components. In addition, gene fusion should prove a useful tool in the study of protein-protein interactions and substrate channelling. (For other examples of the use of protein engineering technology see Chapter 7.)

6.3.4 Development of novel products

In vitro techniques can be used to create new products by scrambling genes from different sources. Parts of different biosynthetic pathways may be combined to give new metabolites. Novel (hybrid) antibiotics have been constructed by using this approach in *Streptomyces.* Genes (*act*) for biosynthesis of the isochromanequinone antibiotic actinorhodin in *Streptomyces coelicolor* A3(2) have been cloned (Malpartida and Hopwood, 1984; see also section 6.3.1) and subsequently introduced into *Streptomyces* spp. that produce other antibiotics of the isochromanequinone class. Introduction of *act* genes into *Streptomyces* sp. AM-7161, which produces medermycin, resulted in the production of two novel antibiotics, mederrhodins A and B (structurally intermediate between actinorhodin and medermycin). By transferring *act* genes into *Streptomyces violaceoruber* Tü22, which produces granaticin, a further hybrid antibiotic, dihydrogranatirhodin, was obtained (Figure 6.14) (Hopwood *et al.*, 1985). It should be noted that the introduction of a complete set of genes for actinorhodin

Figure 6.14: Structures of isochromanequinone antibiotics

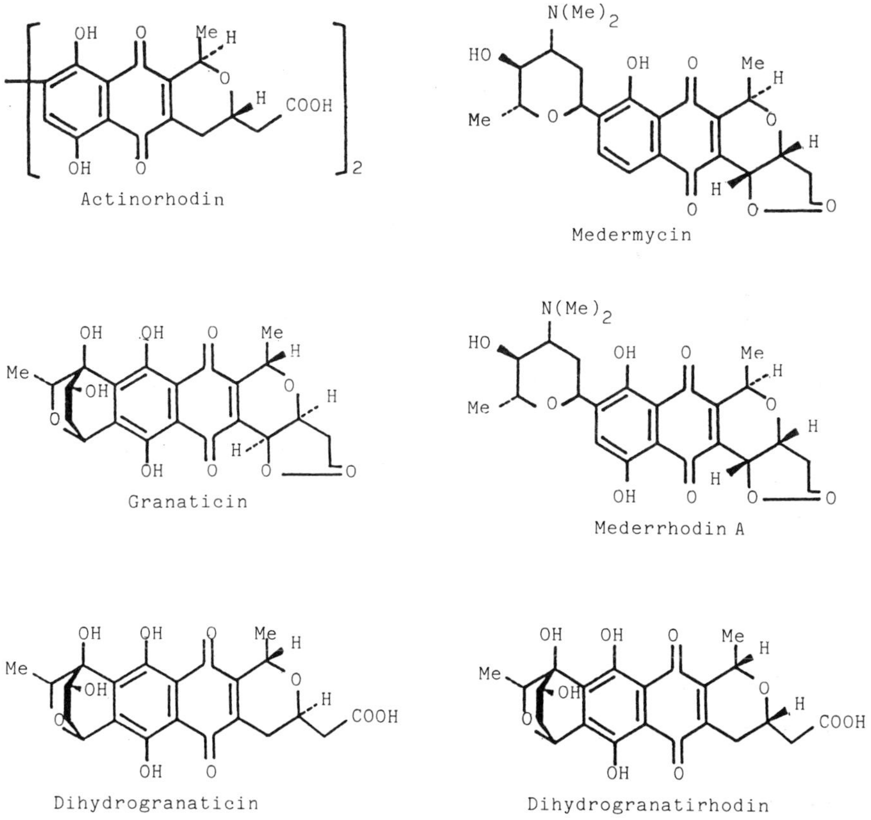

Actinorhodin

Medermycin

Granaticin

Mederrhodin A

Dihydrogranaticin

Dihydrogranatirhodin

biosynthesis into the medermycin producer resulted in the formation of the two parental antibiotics, but no novel compounds. Hybrid antibiotic synthesis required the introduction of a partial *act* clone. By contrast the granaticin-producer synthesised hybrid antibiotic when the complete *act* gene set was introduced. These differences may be attributed to the relative specificities of the biosynthetic enzymes for their substrates, regulation of gene expression or other factors. Although these hybrid antibiotics have no clinical or commercial importance, the methodology should prove useful in the construction of other novel metabolites (including antiviral or anti-cancer agents). One approach is to combine potentially useful metabolites and test the products. A more rational approach might be to combine compounds with specific chemical groups.

The activation of normally dormant endogenous genes may provide a means of obtaining novel antibiotics. Activation might be achieved simply through the cloning of random DNA fragments. Indeed transfer of cloned DNA fragments from the actinomycin-producing *Streptomyces antibioticus* to *Streptomyces lividans* has effected expression of a phenoxazinone synthase gene. This gene is presumed to be normally silent or expressed at a very low level in *S. lividans* (Jones, 1985). The cloned fragments may activate normally silent genes by a number of mechanisms. For example, if the specific gene(s) were subjected to repression, provision of operator sequences (on the cloned fragment) in sufficient amount to titrate all the repressor could lead to gene expression. Another mechanism may involve the provision of positive effectors of gene expression (such as σ factors) by the cloned DNA.

Horinouchi and colleagues (1983) have identified antibiotic regulatory genes, *afsA* and *afsB*, that may prove useful in activating a range of antibiotic genes for synthesis of novel bioactive metabolites in *Streptomyces*. *afsA* is required for synthesis of A-factor, a pleiotropic effector, which regulates streptomycin synthesis, streptomycin resistance and spore formation in *Streptomyces griseus* and is found in various actinomycetes. The *afsB* gene positively controls several independent genes involved in the synthesis of a number of metabolites, including A-factor. By screening actinomycetes in the presence of such pleiotropic regulatory genes, new and useful metabolites may be discovered.

One problem that may be encountered in the production of novel antibiotics is susceptibility of the producer strain to the desired product. This may be circumvented by combining and manipulating the necessary determinants in an antibiotic-insensitive host organism.

6.4 STRAIN DEGENERACY

A critical test of the success of an industrial strain improvement programme is whether performance of an improved strain under conditions of the screen is maintained under production conditions. Expression of genotype can be markedly influenced by the different conditions that prevail from screen to production. It is therefore important to ensure that the necessary adjustments to process conditions are made in order to elicit the best performance from the strain. Genetic instability constitutes a major threat to consistency of performance. Such instability, which may be caused by spontaneous mutation, recombination, or loss of chromosomes, plasmids or cloned genes (for a discussion of the instability of cloned genes, see section 5.4), can give rise to variants that are at a selective advantage under the prevailing conditions and thus supplant the original production strain. If the variant(s) does not possess the desirable characteristics of the production strain, degeneration of performance occurs. This degeneracy might be averted by adjusting environmental conditions in ways that either reduced the incidence of instability (for example by antimutagenesis, see section 4.12) or prevent expression of instability by reducing the selective advantage of variants. The problem of instability is exacerbated during prolonged fermentations. However, it should be possible to calculate, by kinetic modelling, the point at which a fermentation will no longer be economically profitable and harvest at that time. Alternatively instability may be reduced through the use of **immobilised** or non-growing cell systems (see, for example, Mosbach *et al.*, 1983).

BIBLIOGRAPHY

Azevedo, J.L. and Bonatelli, R. Jr (1982) Genetics of the over-production of organic acids. In *Overproduction of microbial products* (V. Krumphanzl, B. Sikyta and Z. Vaněk, eds) pp. 439-50. Academic Press, London and New York

Ball, C. (1973a) Improvements in penicillin productivity in *Penicillium chrysogenum* by recombination. In *1st international symposium on the genetics of industrial microorganisms, vol. II, Actinomycetes and fungi* (Z. Vaněk, Z. Hoštálek and J. Cudlin, eds), pp. 227-37. Elsevier, Amsterdam

Ball, C. (1973b) The genetics of *Penicillium chrysogenum.* In *Progress in industrial microbiology 12* (D.J.D. Hockenhull, ed.), pp. 47-72. Elsevier, Amsterdam

Ball, C. (1978) Genetics in the development of the penicillin process. In *Antibiotics and other secondary metabolites. Biosynthesis and production* (R. Hütter, T. Leisinger, J. Nüesch, and W. Wehrl, eds), pp. 165-76. Academic Press, London and New York

Ball, C., Lawrence, A.J., Butler, J.M. and Morrison, K.B. (1978) Improvement in amyloglucosidase production following genetic recombination of *Aspergillus niger* strains. *Eur. J. Appl. Microbiol. Biotechnol.* **5**, 95-102

Bloom, F., Smith, C.J., Jessee, J., Veilleux, B. and Deutch, A.H. (1983) The use of genetically engineered strains of *Escherichia coli* for the overproduction of free amino acids: proline as a model system. In *Advances in gene technology: molecular genetics of plants and animals.* Miami Winter Symposia, vol. 20 (K. Downey, R.W. Voellmy, F. Ahmad and J. Schultz eds) pp. 383-94. Academic Press,

London and New York

Brown, S.W. and Oliver, S.G. (1982) Isolation of ethanol-tolerant mutants of yeast by continuous selection. *Eur. J. Appl. Microbiol. Biotechnol.* **16**, 119-22

Bülow, L., Ljungcrantz, P. and Mosbach, K. (1985) Preparation of a soluble bifunctional enzyme by gene fusion. *Biotechnology* **3**, 821-3

Burrows, S. (1979) Baker's yeast. In *Economic microbiology 4. Microbial biomass* (A.H. Rose, ed.), pp. 31-64. Academic Press, London and New York

Busby, S.J.W. (1986) Positive regulation in gene expression. In *Regulation of gene expression — 25 years on* (39th Symp. Soc. Gen. Microbiol.) (I.R. Booth and C.F. Higgins, eds), pp. 51-77. Cambridge University Press, Cambridge

Calam, C. T., Daglish, L.B. and McCann, E.P. (1976) Penicillin: tactics in strain improvement. In *2nd international symposium on the genetics of industrial microorganisms* (K.D. MacDonald, ed.), pp. 273-87. Academic Press, London and New York

Chang, L.T. and Elander, R.P. (1979) Rational selection for improved cephalosporin C productivity in strains of *Acremonium chrysogenum* Gams. *Devel. Ind. Microbiol.* **20**, 367-79

Daum, S.J. and Lemke, J.R. (1979) Mutational biosynthesis of new antibiotics. *Ann. Rev. Microbiol.* **33**, 241-65

Davies, J. (1985) Recombinant DNA and the production of small molecules. In *Microbiology — 1985* (L. Leive, ed.), pp. 364-6. American Society for Microbiology, Washington, DC

Debabov, V. (1982) Construction of strains producing L-threonine. *Abstracts of the 4th international symposium on genetics of industrial microorganisms*, I-11-4, p. 9. Kyoto, Japan

Demain, A.L. (1982) Catabolite regulation in industrial microbiology. In *Overproduction of microbial products* (V. Krumphanzl, B. Sikyta and Z. Vaněk, eds), pp. 3-20. Academic Press, London and New York

Dulaney, E.L. and Dulaney, D.D. (1967) Mutant populations of *Streptomyces viridifaciens. Trans. NY Acad. Sci.* **29**, 782-99

Elander, R.P. (1967) Enhanced penicillin biosynthesis in mutant and recombinant strains of *Penicillium chrysogenum.* In *Induced mutations and their utilization* (H. Stubbe, ed.), pp. 403-23. Akademie-Verlag, Berlin

Elander, R.P. (1979) Mutations affecting antibiotic synthesis in fungi producing β-lactam antibiotics. In *Proceedings of the 3rd international symposium on genetics of industrial microorganisms* (O.K. Sebek and A.I. Laskin, eds), pp. 21-35. American Society for Microbiology, Washington, DC

Elander, R.P. (1981) Strain improvement programs in antibiotic-producing microorganisms: present and future strategies. In *Advances in biotechnology, vol. I, Scientific and engineering principles* (M. Moo-Young, C.W. Robinson and C. Vezina, eds), pp. 3-8. Pergamon Press, Oxford, Toronto and New York

Elander, R.P. (1982) Trends in the genetic improvement of antibiotic producing organisms. In *Trends in antibiotic research. Genetics, biosyntheses, actions and new substances* (H. Umezawa, A.L. Demain, T. Hata and C.R. Hutchinson, eds), pp. 16-31. Japan Antibiotics Research Association, Tokyo

Elander, R.P., Espenshade, M.A., Pathak, S.G. and Pan, C.H. (1973) The use of parasexual cycle in an industrial strain improvement programme with *Penicillium chrysogenum.* In *1st international symposium on genetics of industrial microorganisms, vol. II, Actinomycetes and fungi* (Z. Vaněk, Z.Hoštálek and J. Cudlin, eds), pp. 239-53. Elsevier, Amsterdam

Elander, R.P., Corum, C.J., DeValeria, H. and Wilgus, R.M. (1976) Ultraviolet mutagenesis and cephalosporin synthesis in strains of *Cephalosporium acremonium.* In *2nd international symposium on the genetics of industrial micro-*

organisms (K.D. Macdonald, ed.), pp. 253-71. Academic Press, London and New York

Emeis, C-C. (1971) A new hybrid yeast for the fermentation of wort dextrins. *Proc. Amer. Soc. Brew. Chem.* 58-62

Enei, H. and Hirose, Y. (1984) Recent research on the development of microbial strains for amino-acid production. In *Biotechnology and genetic engineering reviews*, vol. 2 (G.E. Russell, ed.), pp. 101-20. Intercept, Newcastle upon Tyne

Fersht, A.R., Shi, J.P., Knill-Jones, J., Lowe, D.M., Wilkinson, A.J., Blow, D.M., Brick, P., Carter, P., Waye, M.M.Y. and Winter, G. (1985) Hydrogen bonding and biological specificity analysed by protein engineering. *Nature (London)* **314**, 235-8

Ghosh, V.K., Ghose, T.K. and Gopalkrishnan, K.S. (1982) Improvement of *T. reesei* strain through mutation and selective screening techniques. *Biotechnol. Bioeng.* **24**, 241-3

Godfrey, O.W. (1973) Isolation of regulator mutants of the aspartic and pyruvic acid families and their effect on antibiotic production in *Streptomyces lipmanii. Antimicrob. Agents Chemother.* **4**, 73-9

Hamlyn, P.F. and Ball, C. (1979) Recombination studies with *Cephalosporium acremonium.* In *Proceedings of the 3rd international symposium on genetics of industrial microorganisms* (O.K. Sebek and A.I. Laskin, eds), pp. 185-91. American Society for Microbiology, Washington, DC

Hammond, J.R.M. and Eckersley, K.W. (1984) Fermentation properties of brewing yeast with killer character. *J. Inst. Brew. (London)* **90**, 167-77

Henahan, J. and Klausner, A. (1983) Guinness brewing new yeast. *Biotechnology* **1**, 462-4

Hitotsuyanagi, K., Yamane, K. and Maruo, B. (1979) Stepwise introduction of regulatory genes stimulating production of α-amylase into *Bacillus subtilis: construction of an* α-amylase extra hyper-producing strain. *Agric. Biol. Chem.* **43**, 2343-9

Hopwood, D.A., Malpartida, F., Kieser, H.M., Ikeda, H., Duncan, J., Fujii, I., Rudd, B.A.M., Floss, H.G. and Omura, S. (1985) Production of 'hybrid' antibiotics by genetic engineering. *Nature (London)* **314**, 642-4

Horinouchi, S., Hara, O. and Beppu, T. (1983) Cloning of a pleiotropic gene that positively controls biosynthesis of A-factor, actinorhodin and prodigiosin in *Streptomyces coelicolor* A3(2) and *Streptomyces lividans. J. Bacteriol.* **155**, 1238-48

Johnston, J.R. and Oberman, H. (1979) Yeast genetics in industry. In *Progress in industrial microbiology 15* (M.T. Bull, ed.), pp. 151-205. Elsevier, Amsterdam

Jones, G.H. (1985) Cloning, regulation and expression of the phenoxazinone synthase gene from *Streptomyces antibioticus.* In *Microbiology — 1985* (L. Leive, ed.), pp. 445-8. American Society for Microbiology, Washington, DC

Kase, H. and Nakayama, K. (1972) Production of L-threonine by analog-resistant mutants. *Agric. Biol. Chem.* **36**, 1611-21

Katagiri, K. (1954) Study on the chlortetracycline. Improvement of chlortetracycline-producing strain by several kinds of methods. *J. Antibiotics* **7**, 45-52

Kinoshita, S., Nakayama, K. and Kitada, S. (1958) L-lysine production using microbial auxotroph. *J. Gen. Appl. Microbiol.* **4**, 128-9

Kisumi, M. (1982) Transductional construction of amino acid-producing strains in *Serratia marcescens. Abstracts of the 4th international symposium on genetics of industrial microorganisms* I-11-2, p. 8. Kyoto, Japan

Kitano, K., Nozaki, Y. and Imada, A. (1982) Strain improvement in carbapenem antibiotic production by *Streptomyces griseus* subsp. *cryophilus. Abstracts of the 4th international symposium on the genetics of industrial microorganisms* 0-VI-

7, p. 66, Kyoto, Japan

Klaenhammer, T.R. (1984) Interaction of bacteriophages with lactic streptococci. *Adv. Appl. Microbiol.* **30**, 1-29

Lein, J. (1983) Strain development with non-recombinant DNA techniques. *Amer. Soc. Microbiol. News* **40**, 576-9

Luengo, J.M., Revilla, G., Lopez, M.J., Villanueva, J.R. and Martin, J.F. (1979) Penicillin production by mutants of *Penicillium chrysogenum* resistant to polyene macrolide antibiotics. *Biotechnol. Lett.* **1**, 233-8

Malpartida, F. and Hopwood, D.A. (1984) Molecular cloning of the whole biosynthetic pathway of a *Streptomyces* antibiotic and its expression in a heterologous host. *Nature (London)* **309**, 462-4

Martin, J.F. and Demain, A.L. (1980) Control and antibiotic synthesis. *Microbiol. Rev.* **44**, 230-51

Martin, J.F., Naharro, G., Liras, P. and Villanueva, J.R. (1979) Isolation of mutants deregulated in phosphate control of candicidin biosynthesis. *J. Antibiotics* **32**, 600-6

Mazières, N., Peyre, M. and Penasse, L. (1981) Interspecific recombination among aminoglycoside producing streptomycetes. *J. Antibiotics* **34**, 544-50

Mosbach, K., Birnbaum, S., Hardy, K., Davies, J. and Bulow, L. (1983) Formation of proinsulin by immobilized *Bacillus subtilis. Nature (London)* **302**, 543-5.

Nagaoka, K. and Demain, A.L. (1975) Mutational biosynthesis of a new antibiotic, streptomutin A, by an idiotroph of *Streptomyces griseus. J. Antibiotics* **28**, 627-35

Nakayama, K. (1973) Amino-acid production using microbial auxotrophic mutants. In *1st international symposium on the genetics of industrial microorganisms, vol. 1, Bacteria* (Z. Vaněk, Z. Hoštálek and J. Cudlin, eds), pp. 219-48. Elsevier, Amsterdam

Ohnuki, T., Imanaka, T. and Aiba, S. (1985) Self-cloning in *Streptomyces griseus* of an *str* gene cluster for streptomycin biosynthesis and streptomycin resistance. *J. Bacteriol.* **164**, 85-94

Ouchi, K., Nishiya, T. and Akiyama, H. (1982) UV-killed protoplast fusion as a method for breeding killer yeasts. In *Abstracts of the 4th international symposium on genetics of industrial microorganisms.* P-11-14, p. 105, Kyoto, Japan

Pastan, I. and Adhya, S. (1976) Cyclic adenosine-3′-5′-monophosphate in *E. coli. Bacteriol. Rev.* **40**, 527-51

Perlman, D. and Hall, T.C.C. (1976) Actinophage as a selective agent for increasing antibiotic production. *Abstracts of 172nd Meeting American Chemical Society,* 34

Perry, L.J. and Wetzel, R. (1984) Disulfide bond engineered into T4 lysozyme: stabilization of the protein toward thermal inactivation. *Science* **226**, 555-7

Rinehart, K.L. Jr and Stroshane, R.M. (1976) Biosynthesis of aminocyclitol antibiotics. *J. Antibiotics* **29**, 319-53

Samson, S.M., Belagaje, R., Blankenship, D.T., Chapman, J.L., Perry, D., Skatrud, P.L., VanFrank, R.M., Abraham, E.P., Baldwin, J.E., Queener, S.W. and Ingolia, T.D. (1985) Isolation, sequence determination and expression in *Escherichia coli* of the isopenicillin N synthetase gene from *Cephalosporium acremonium. Nature (London)* **318**, 191-4

Sano, K. and Shiio, I. (1970) Microbial production of L-lysine. III. Production by mutants resistant to S-(2-aminoethyl)-L- cysteine. *J. Gen. Appl. Microbiol.* **16**, 373-91

Sasajima, K-I. and Yoneda, M. (1984) Production of pentoses by microorganisms. In *Biotechnology and genetic engineering reviews,* vol.2 (G.E. Russell, ed.), pp.

175-213. Intercept, Newcastle upon Tyne

Schupp, T., Traxler, P. and Auden, J.A.L. (1981) New rifamycins produced by a recombinant strain of *Nocardia mediterranea. J. Antibiotics* **34**, 965-70

Sebek, O.K. (1976) Use of mutants for the synthesis of new antibiotics. In *Microbiology — 1976* (D. Schlessinger, ed.), pp. 522-5. American Society for Microbiology, Washington, DC

Skotnicki, M.L., Lee, K.J., Tribe, D.E. and Rogers, P.L. (1982) Genetic alteration of *Zymomonas mobilis* for ethanol production. In *Genetic engineering of microorganisms for chemicals.* Basic life sciences, vol. 19 (A. Hollaender, R.D. DeMoss, S. Kaplan, J. Konisky, D. Savage and R.S. Wolfe, eds), pp. 271-90. Plenum Press, New York

Spencer, J.F.T. and Spencer, D.M. (1983) Genetic improvement of industrial yeasts. *Ann. Rev. Microbiol.* **37**, 121-42

Stanzak, R., Matsushima, P., Baltz, R.H. and Rao, R.N. (1986) Cloning and expresion in *Streptomyces lividans* of clustered erythromycin biosynthesis genes from *Streptomyces erythreus. Biotechnology* **4**, 229-32

Stewart, G.G. and Russell, I. (1981) *Advances in biotechnology. Current developments in yeast research.* Pergamon Press, Oxford, Toronto and New York

Stewart, G.G., Bilinski, C.A., Panchal, C.J., Russell, I. and Sills, A.M. (1985) Genetic manipulation of brewer's yeast strains. In *Microbiology — 1985* (L. Leive, ed.), pp. 367-74. American Society for Microbiology, Washington, DC

Thomas, P.G., Russell, A.J. and Fersht, A.R. (1985) Tailoring the pH dependence of enzyme catalysis using protein engineering. *Nature (London)* **318**, 375-6

Ullmann, A. and Danchin, A. (1980) Role of cyclic AMP in regulatory mechanisms of bacteria. *Trends Biochem. Sci.* **5**, 95-6

Vournakis, J.N. and Elander, R.P. (1983) Genetic manipulation of antibiotic-producing microorganisms. *Science* **219**, 703-9

Waye, M.M.Y., Winter, G., Wilkinson, A.J. and Fersht, A.R. (1983) Deletion mutagenesis using an 'M13 splint': the N-terminal structural domain of tyrosyl-tRNA synthetase (*B. stearothermophilus*) catalyses the formation of tyrosyl adenylate. *EMBO J.* **20**, 1827-9

Windass, J.D., Worsey, M.J., Pioli, E.M., Pioli, D., Barth, P.T., Atherton, K.T., Dart, E.C., Byrom, D., Powell, K. and Senior, P.J. (1980) Improved conversion of methanol to single-cell protein by *Methylophilus methylotrophus. Nature (London)* **287**, 396-401

Winter, G. and Fersht, A.R. (1984) Engineering enzymes. *Trends Biotechnol.* **2**, 115-19

Yamada, K. (1977) Recent advances in industrial fermentation in Japan. *Biotechnol. Bioeng.* **19**, 1563-1621

Zakharov, I.A. and Yarovoy, B. Ph. (1977) Cytoduction as a new tool in studying the cytoplasmic heredity in yeast. *Biochemistry* **14**, 15-18

7

Medical and Veterinary Applications

The techniques of microbial genetics have many applications in human and animal medicine including those in the production of drugs, vaccines and other pharmaceuticals and in the diagnosis of disease. The use of gene manipulation to produce antibiotics and polypeptides has been considered in other chapters. This chapter is intended to highlight some specific examples of the exploitation of microbial genetic techniques in medicine.

7.1 PRODUCTION AND MANIPULATION OF PHARMACEUTICALS

It is possible to produce large quantities of many useful enzymes and other proteins heterologously in microorganisms. In some cases, however, the natural protein produced may not possess all the properties that are desirable to man. This is particularly so when such proteins are divorced from their natural cellular environment in order, for example, to be administered as pharmaceuticals. Furthermore, the native protein may not be patentable whereas a modified protein may be considered so. There can, therefore, be considerable advantages in re-engineering proteins to produce more desirable commodities. This may be achieved by conventional mutation procedures (see Chapter 4), although recombinant DNA technology, and in particular site-directed mutagenesis, has made a more rapid and rational approach possible.

Site-directed mutagenesis (section 4.9.3) has been used to study fundamental structure-function relationships in enzymes, notably the tyrosyl-tRNA synthetase from *Bacillus stearothermophilus* (see section 6.3.3) (Winter and Fersht, 1984; Fersht *et al.*, 1985). By systematic mutation of genes encoding the amino acid residues that constitute the active site of an enzyme it is possible, for example, to alter the specificity of that enzyme. The techniques developed for protein engineering have numerous applications in the manufacture of pharmaceutical products. In some cases modifications may be necessary simply to recover proteins with biological activity from genetically manipulated microorganisms. For example, human fibroblast interferon and human interleukin-2, produced by cloning in bacteria, were found to have relatively disappointing activities. Reduced activity was due to the formation of incorrect disulphide bonds between cysteine residues present in the native amino acid sequences of these proteins. The interferon polypeptide contains three cysteine residues at amino acid positions 17, 31 and 141. In the native protein, a single disulphide bond is formed between residues 31 and 141. In contrast, interferon produced in *E. coli* is a mixture of polypeptides containing disulphide bonds in each of the three possible combinations, between residues 17 and 31, 31 and 141 and 17 and 141. Formation of the correct disulphide bond could, however, be achieved by using *in vitro* mutagenesis techniques. The coding sequence for interferon was altered, so that the offending cysteine residue at amino acid position 17 was converted to a serine residue. Disulphide bonds were then only possible between the two cysteine residues that are linked when this polypeptide is produced naturally (Mark *et al.*, 1984). A similar strategy involving conversion of the cysteine residue at position 125 on interleukin-2 to a serine residue was equally successful (Wang *et al.*, 1984).

Another example of the use of *in vitro* techniques to modify therapeutic agents is provided by the synthesis of mutant α_1-antitrypsins (Courtney *et al.*, 1985; Rosenberg *et al.*, 1985). The anti-protease α_1-antitrypsin is the principal defence mechanism against damage caused to lung tissue during pulmonary emphysema and adult respiratory distress syndrome. Such damage is caused by leukocyte elastase, a protease capable of hydrolysing components of connective tissue. Oxidation (for example, by oxidants in tobacco smoke) of a methionine residue at position 358 in the antitrypsin polypeptide inactivates the protective function of this enzyme. Therefore an oxidation-resistant antitrypsin protein would have potential as a therapeutic agent in the treatment of emphysema and other acute respiratory conditions. Mutant genes specifying oxidation-resistant antitrypsin have been constructed by site-directed mutagenesis using mismatched primers annealed to a native α_1-antitrypsin gene cloned in single-stranded form in M13. The primers used were designed to produce a single T→C transition in the coding region of the active site of antitrypsin, such that the methionine residue at position 358 was replaced by a valine residue (Table 7.1). The resulting mutant genes were subsequently found to encode functional but oxidation-resistant α_1-antitrypsin in either *E. coli* (Courtney *et al.*, 1985) or yeast (Rosenberg *et al.*, 1985). Such altered proteins may have application in preventing damage to connective tissue caused by elastase. A second type of variant was produced by Courtney and co-workers (1985) in which the methionine at position 358 had been replaced by an arginine residue (Table 7.1). In this case the mutant protein no longer inhibited

Table 7.1: Products of site-directed mutagenesis of α_1–antitrypsin gene

Polypeptide	Amino acid position							Activity
	355	356	357	358	359	360	361	
(1) α_1–AT	Ala	Ile	Pro	Met*	Ser	Ile	Pro	
	GCC	ATA	CCC	ATG	TCT	ATC	CCC	Oxidation-sensitive anti-elastase
(2) α_1–AT (Met358 → Val)	Ala	Ile	Pro	Val	Ser	Ile	Pro	
	GCC	ATA	CCC	GTG	TCT	ATC	CCC	Oxidation-resistant anti-elastase
(3) α_1–AT (Met358 → Arg)	Ala	Ile	Pro	Arg	Ser	Ile	Pro	
	GCC	ATA	CCC	AGG	TCT	ATC	CCC	Anti-thrombin (no anti-elastase)

α_1–AT, α_1–antitrypsin (394 amino acid polypeptide); *, oxidation site.

elastase but inhibited thrombin, thus generating a potentially useful anti-coagulant agent.

7.2 VACCINES

Two basic approaches have been used traditionally in immunising against infectious diseases. These involve the administration of either live, attenuated strains or killed whole or parts of disease-causing microorganisms. Vaccination with living organisms has the advantage that if administered by the route employed in the natural infection, the immunity stimulated will include all the characteristics of the immune response induced by that infection itself. These include cell-mediated immunity, the production of circulating antibody and, perhaps more importantly, for many infections the production of secretory antibody at the site of infection. A major disadvantage of live vaccines is that it is possible that the vaccine strain will revert to, or acquire, virulence. The application of genetic techniques to producing novel vaccines of commercial value depends not only on the efficacy and safety of such vaccines, but also on their relative cost when compared with conventional killed or live vaccines.

7.2.1 Subunit vaccines

For many diseases, whole cell or virus vaccines are not administered because the pathogen is difficult or dangerous to cultivate, produces undesirable side-effects, or induces poor levels of protection. Conventional killed or live vaccines present a complex antigenic stimulus when used to immunise a host. Some dead vaccines are simply whole killed cells or extracts therefrom, and only a few of the resulting antibodies that are induced may actually be responsible for neutralising an invading pathogen. In many cases the important antigenic component(s) of the pathogen can be identified by immunological, genetic and other methods. If such components can be purified in quantity, they may be used as simple immunogens, capable of eliciting a protective response in the host. Vaccines of this type are referred to as **subunit vaccines**. In some cases the subunit concerned may be purified from large-scale cultures of the pathogen or from pooled clinical material. However, such purification procedures are potentially dangerous and sometimes inefficient. There is thus considerable impetus to isolate the gene or genes encoding the desired antigen by cloning, and subsequently to express such genes in a harmless microbial host from which the antigen can be recovered. The cloning of genes encoding specific antigens has a number of advantages in the production of vaccines. Potentially large amounts of the antigen can be produced, in organisms such as *E. coli*, *B. subtilis* or yeast, where problems due to contaminating harmful products may be reduced. Furthermore, *in vitro* techniques, particularly site-directed mutagenesis of toxin genes, may permit the production of artificial toxoid vaccines. (Toxoids are conventionally made by chemical inactivation of toxins.) Such products would not carry the risk of incomplete inactivation that may be encountered with conventional toxoids. One of the main difficulties in producing vaccines from genetically engineered microorganisms is a requirement, in some cases, to refold polypeptide antigens from an essentially denatured state. This is necessary to produce a protein that presents the antigenic site(s) in a similar way to the native antigen of the pathogen. It is of particular significance if a natural antigenic site is produced by juxtaposition of two regions of the same or different polypeptide(s).

(a) Vaccines against diseases caused by E.coli

Strains of *E. coli* can cause a number of human and animal diseases including urinary tract infections and diarrhoeas. An economically important *E. coli*-related disease is scours, a diarrhoeal disease of newborn piglets and calves. The disease is caused mainly by enterotoxigenic strains of *E. coli* (ETEC). Such strains colonise the small bowel and produce one or both of two classes of enterotoxins, heat-labile (LT) and heat-stable (ST) toxin. Both toxins are encoded by conjugative plasmids. The B subunit of LT is nontoxic, but is strongly immunogenic and can therefore be used as a vaccine component. ST is too small to be a good immunogen but can be made immunogenic by coupling it to a larger protein (for example LT-B). In order to produce disease, ETEC must attach themselves to the lining of the lumen of the bowel. These pathogens produce one of a number of hair-like proteinaceous adhesins called fimbriae or pili, which are responsible for attaching the bacteria to the intestinal lining (see Dougan and Morrissey, 1985). Such fimbrial colonisation factors are fairly specific for the host species, for example K88 and 987P fimbriae are found in pig-pathogenic strains, K99 fimbriae are found in pig, calf and lamb pathogens, and CFA/I and CFA/II in strains pathogenic for humans (Levine *et al.*, 1983). Some of these fimbrial adhesins, such as K88 and K99, are encoded on conjugative plasmids that may additionally encode ST or LT. Purified K88, K99 or 987P fimbriae have been used for parental (by injection) inoculation of pregnant sows and cows. In such cases newborn animals are protected by anti-fimbrial antibodies secreted in colostrum. The success of such antibodies in protecting against infections has opened up the general vaccination strategy of immunising against the primary adhesion event between a pathogen and its target cell. If adhesion can be blocked by specific antibodies, the subsequent events of a normal infection can be prevented. An attractive route for the production of vaccines is therefore to clone and express the gene(s) necessary to produce the adhesin (or at least that part of it that induces the production of protective antibodies) in *E. coli* K12.

Genes encoding one or more of the ETEC fimbrial antigens together with genes encoding the nontoxigenic B subunit of LT have been inserted into *E. coli* K12 in order to provide dead or live vaccine strains for veterinary use (Levine *et al.*, 1983; Winther and Dougan, 1984). *E. coli* K12 laboratory strains are essentially nonpathogenic. This facilitates both the production of nonreactogenic whole cell vaccines and the recovery of specific antigens free of contaminating pyrogens, such as endotoxin. In addition, K12 strains, unlike most wild strains, do not repress the synthesis of virulence-associated surface adhesins, such as K88 fimbriae (Kehoe *et al.*, 1981) or 987P fimbriae (Dougan and Morrissey, 1985). This allows increased fimbrial biosynthesis, which may be useful for improving the production of these antigens for use as subunit vaccines. Furthermore, the low virulence of K12 strains makes them possible carriers for live vaccines for bacterial and other diseases (section 7.2.3).

(b) Foot and Mouth Disease vaccines

The major immunogenic determinant of the capsid of Foot and Mouth Disease Virus (FMDV) is VP1, a polypeptide of 212 amino acids. cDNA cloning and subsequent expression of the FMDV VP1 gene in *E. coli* have been achieved by several groups (Boothroyd *et al.*, 1981; Kupper *et al.*, 1981; Kleid *et al.*, 1985). A fusion protein consisting of 190 amino acids from the *E. coli trpLE* gene sequence and amino acids 8 to 211 of VP1 was expressed in *E. coli* using a pBR322/*trp* promoter-based expression vector (Kleid *et al.*, 1985). When purified from the bacteria, this protein was found to be immunogenic for mice and cattle, when given as two separate injections at 28-day intervals. However, an even more effective vaccine could be produced as a fusion protein comprising the 190 amino acids from *trpLE* linked to amino acids 137 to 168 of VP1 in direct dimeric repeat (Figure 7.1). (Amino acids in the region of 130 to 170 of VP1 are known to be strongly immunogenic and highly variable between various isolates of FMDV.) This protein, containing only the immunogenic region of VP1, was capable of eliciting protective antibodies in cattle following a single injection and was about 200 times as potent as the fusion containing most (amino acids 8 to 211) of the viral polypeptide.

(c) Hepatitis vaccines

The production of vaccines against hepatitis B is not possible by conventional routes since the causative agent Hepatitis B Virus (HBV) cannot be propagated in tissue culture. However, vaccines against hepatitis B can be obtained from purified defective 22 nm particles of hepatitis B surface antigen (HBsAg) from the pooled plasma of infected humans. Unfortunately, the supply of plasma limits production of such vaccines and stringent precautions must be taken to ensure inactivation of all infectious agents, notably AIDS viruses, in plasma products (Tiollais *et al.*, 1985). Synthesis of HBsAg has been achieved by recombinant DNA techniques in a number of organisms including *B. subtilis* (Hardy *et al.*, 1981), *E. coli* (Tiollais *et al.*, 1985) and yeast (Mijanohara *et al.*, 1983; Valenzuela *et al.*, 1985a). HBsAg is not, however, assembled into particles with full immunogenicity in bacteria. Therefore, in order to produce large amounts of this protein for use as an immunising antigen, it must be synthesised in yeast or mammalian cells. Efficient expression of HBsAg in yeast, under the control of the alcohol dehydrogenase or other promoters, has the additional advantage that the resulting particulate protein can be used as a matrix for the presentation of other antigens. This has been achieved by inserting coding sequences for a second immunogen, for example the herpes simplex type 1 surface glycoprotein gD, into the gene for HBsAg (Figure 7.2). This procedure results in the synthesis and assembly in yeast of a hybrid virus particle containing both hepatitis B and herpes simplex surface antigens (Valenzuela *et al.*, 1985b).

7.2.2 Synthetic peptide vaccines

Only small parts of most native antigens are actually recognised by the immune system. This is because only certain regions of polypeptides, carbohydrates etc. constitute antigenic sites. Furthermore, by virtue of the relative hydrophobicity of the amino acids in a polypeptide, some domains are more likely to be exposed on the surface of a native protein and hence are more likely to constitute an antigenic site than others. In many cases antigenic sites are located close to, or at, hydrophilic regions of proteins.

Figure 7.1: Construction of TrpLE′-FMDV VP1 immunogenic fusion proteins. Blunt-ended ligation of the DNA fragment specifying codons 137 to 168 into pUC9 fortuitously produced a recombinant containing a direct dimeric repeat of this VP1 sequence. → indicates orientation of VP1 sequences; αα, amino acids

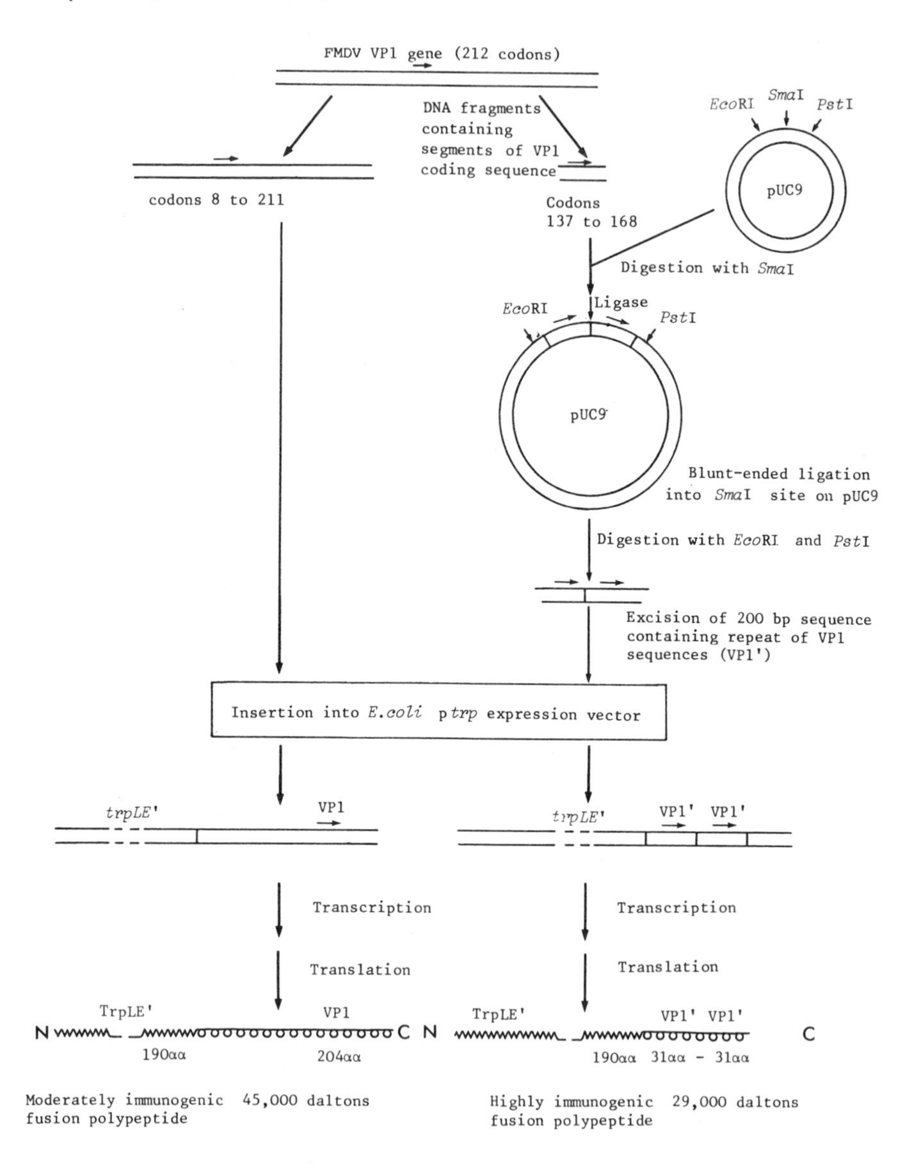

Figure 7.2: Synthesis of hybrid HBsAg-HSV1gD particles in yeast. The expression cassette carried by the YEp vector pC1/1 (a derivative of pJDB219) consists of 1047 base pairs from the 5′ flanking region of the yeast *GPD* (glyceraldehyde 3-phosphate-dehydrogenase) gene containing the strong *GPD* promoter (p), 780 base pairs of the HBsAg coding region and 900 base pairs of the 3′ flanking region of the yeast *GPD* gene containing the transcription terminator. The presurface (preS) region of HBsAg contains the receptor site of the virus and is therefore normally exposed, but is not essential for assembly of HBsAg particles. Foreign gene sequences can thus be inserted into the preS coding region without affecting assembly of particles. The final product is a complex multimeric lipoprotein resembling the hepatitis virus envelope and containing HBsAg and HSV1gD amino acid sequences that are highly antigenic when injected into humans or animals. ⟶, Direction of transcription

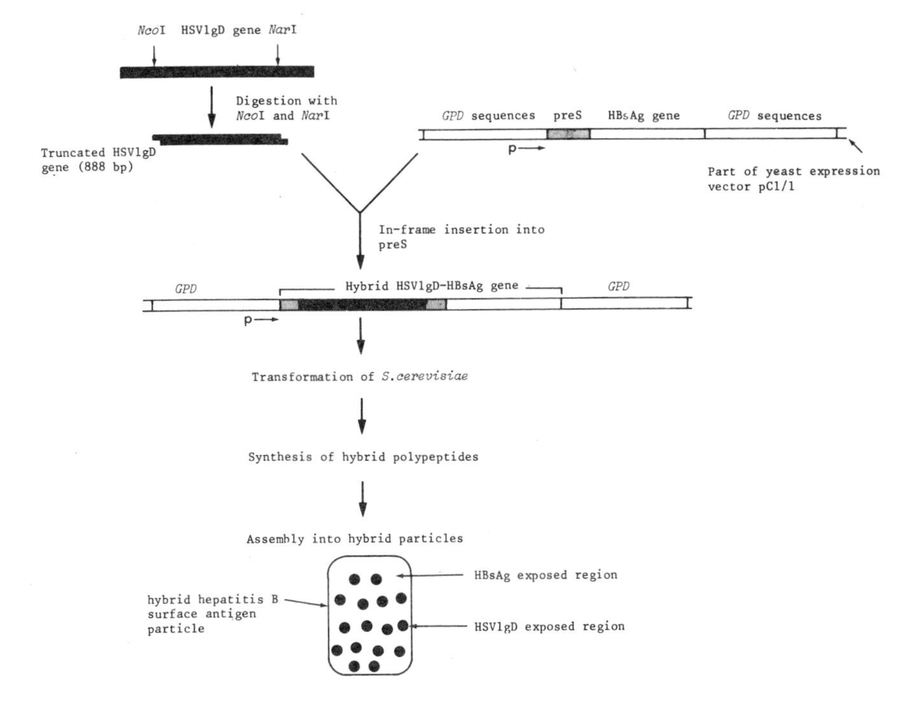

The location of such antigenic sites can be predicted by computer analysis of the amino acid sequence. However, certain sites are located in hydrophobic regions of polypeptides, for example in the haemagglutinin of influenza virus (Atassi and Webster, 1983). Once a relevant antigenic site has been identified, a synthetic peptide that includes the site can be synthesised, provided that the amino acid sequence is known. (This information may be obtained by peptide sequencing of the purified antigen or by DNA sequencing of the relevant gene.) This peptide can be used as a **synthetic peptide vaccine**. The advantage of such vaccines is that they are free of any competing antigens and of any contaminating and potentially harmful products present in biologically produced vaccines (Lerner, 1982; Brown 1985). Synthetic peptides corresponding to the species-specific circumsporozoite protein (CSP) of *Plasmodium* have been obtained and show promise as antimalarial vaccines (Ellis *et al.*, 1983). However, many synthetic peptides do not induce antibodies that react with the native parental protein. This may occur because the peptide(s) may not adopt the native configuration of the antigen. Furthermore, the natural immunogen may consist of two or more regions of a polypeptide(s) that are juxtaposed *in vivo.* These problems can, however, be overcome by using appropriate adjuvants or carriers to ensure that the peptide is presented to the vaccinated host in the most natural configuration.

Assuming that suitable antigens can be prepared by recombinant DNA technology or by synthesis of the appropriate peptide, problems may be encountered in using the vaccine due to the natural genetic diversity of pathogens of the same species. Such diversity may arise as a natural consequence of a parasite undergoing different stages in its life cycle, as is the case in malaria. In such instances, vaccines may have to be composed of antigens corresponding to more than one stage in the life cycle (Ravetch *et al.*, 1985). Many pathogens exhibit antigenic variation in their surface structures and possess genetic mechanisms for ensuring such diversity (Saunders, 1986). A good example of antigenic variation is exhibited by *Neisseria gonorrhoeae*, the causative agent of gonorrhoea. This bacterium produces proteinaceous hair-like structures (pili or fimbriae), which protrude from the surface of the bacterium and act as adhesins, enabling the bacteria to adhere to mucosal surfaces. Pilin, the structural subunit of the pili, varies in its biochemical and antigenic properties, both between and within different isolates of *N. gonorrhoeae.* The pilin molecule consists of three distinct domains, an N-terminal constant, a central semivariable and a C-terminal hypervariable region. A vaccine composed of whole pili or pilin would appear to be a viable proposition because antibodies raised against this surface structure would be expected to inhibit the initial binding of gonococci in the course of the disease. However, the variable parts of the pilin molecule appear to be immunodominant over the conserved region. Thus antibodies raised against whole pili or pilin will tend to be strain specific. A cocktail of different pilus antigens would therefore be required to provide protection against infection by different strains of *N. gonorrhoeae.* Alternatively, synthetic peptides that correspond to the constant and semivariable regions of the molecule (Rothbard *et al.*, 1985), or to short regions of conserved sequence that are found around the cysteine residues of the hypervariable region, may prove suitable as vaccine components.

7.2.3 Live attenuated vaccines

Attenuated microbial strains that are conventionally employed as vaccines have generally been selected as fully or partially avirulent derivatives, generated during long-term laboratory subculture. As such, they probably arise fortuitously as a result of the random accumulation of mutations, usually at multiple sites, during propagation.

(a) Disabling mutations for production of oral vaccines

The deliberate introduction of disabling mutations into live vaccine strains is a potentially valuable approach to vaccination since the vaccine is more likely to induce long-lasting immunity than a 'dead' vaccine. Very promising indications for protection against typhoid have been obtained by oral administration of live, attenuated *Salmonella typhi* strains that are completely defective in galactose epimerase (*galE*$^-$) (Wahdan *et al.*, 1982). *S. typhi* Ty21a, the vaccine strain used, is defective in the ability to make lipopolysaccharide (LPS), by virtue of a nitrosoguanidine-induced mutation in *galE* (Figure 7.3). Lack of LPS makes Ty21a colonies appear rough. In the presence of galactose *galE* mutants produce smooth LPS. However, galactose-1-phosphate and UDP-galactose accumulate (due to absence of the epimerase) and this may cause bacterial death. In the gut of humans or animals, such strains metabolise the available galactose and die after only a few cell divisions. However, the limited propagation of strain Ty21a is sufficient to induce immunity in patients to whom the vaccine has been administered.

A similar approach, exploiting transposition mutagenesis, has been adopted by Hosieth and Stocker (1981). They constructed non-reverting mutants of *Salmonella typhimurium* defective in the *aroA* (5-enolpyruvylshikimate 3-phosphate synthase) gene. These mutants were obtained by inserting Tn*10* into the chromosome of the weakly pathogenic strain LT2 of *S. typhimurium* and selecting for tetracycline-resistant *aroA*$^-$ mutants (Figure 7.4). Nonreverting mutants were obtained by selecting for tetracycline-sensitive derivatives in which imprecise excision of Tn*10* (section 4.7.1) had produced deletion of part of the *aroA* gene. The mutations were then transduced using phage P22 to virulent strains of *S. typhimurium* and *S. dublin*. *aroA*$^-$ mutants of *Salmonella* are rendered avirulent because they become auxotrophic for para-aminobenzoic acid and 2,3-dihydroxybenzoate, both of which are not available in vertebrate tissues. The *aroA*$^-$ mutation blocks one of the enzymatic steps in the conversion of shikimate to chorismic acid, which is the final product of the common aromatic amino acid biosynthetic pathway (Figure 7.5). Chorismate is subsequently utilised to manufacture enterochelin, *p*-aminobenzoic acid and the aromatic amino acids phenylalanine, tryptophan and tyrosine. Enterochelin (2,3-dihydroxy-N-benzoyl-L-serine) allows bacteria to acquire essential

Figure 7.3: Relationship of galactose epimerase to biosynthesis of lipopolysaccharide and utilisation of galactose. *galE*$^-$ mutants are unable to make galactose-containing components of lipopolysaccharide and cannot use galactose as a carbon source. The *galE* product (UDP galactose 4-epimerase) is required for both anabolic and catabolic epimerisations between galactose and glucose. Shaded rectangle, block caused by *galE*$^-$ mutation

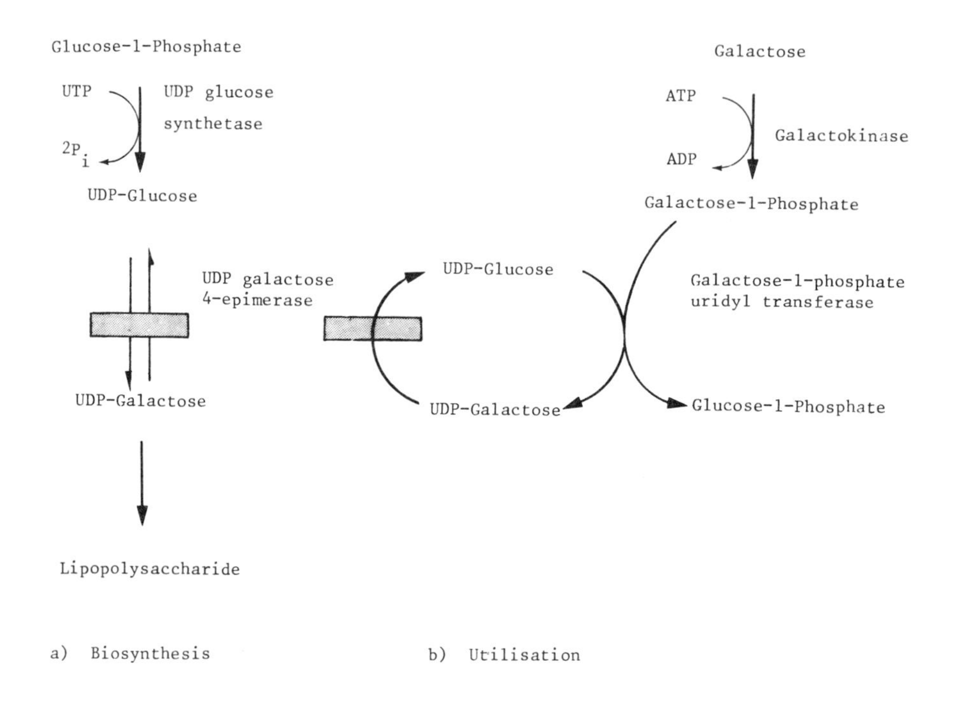

iron in the low free iron environment of animal tissue. *aroA*$^-$ derivatives do not make enterochelin and are thus unable to invade. When administered orally to calves, such *aroA*$^-$ strains produced more effective immunity than the conventional vaccine for salmonellosis, which is administered parentally (by injection) and consists of avirulent rough *Salmonella* strains (Robertsson *et al.*, 1983).

galE$^-$ and *aroA*$^-$ avirulent mutants of other pathogens are potentially useful as live vaccines for other diseases. Furthermore, avirulent enterobacteria may be useful carriers of plasmid-borne or cloned antigenic determinants. Temporary carriage in the host gut can induce protective antibodies against more than one pathogen. For example, a potentially useful vaccine strain for use against both typhoid and *Shigella* infections was constructed by transfer of a 180kb (120 Md) plasmid encoding the Form I (0) antigen of *Shigella sonnei* into *S. typhi* Ty21a (Formal *et al.*, 1981). A bivalent live vaccine for typhoid and cholera-*E. coli*-related

Figure 7.4: Construction of *Salmonella typhimurium* live vaccine strains by Tn*10*-induced mutagenesis of the *aroA* gene. *aroA*, 5-enolpyruvylshikimate-3-phosphate synthase gene; Tc^r, Tc^s, resistance, sensitivity to tetracycline, respectively; Phe^-, Trp^-, Tyr^-, growth requirement for phenylalanine, tryptophan and tyrosine, respectively

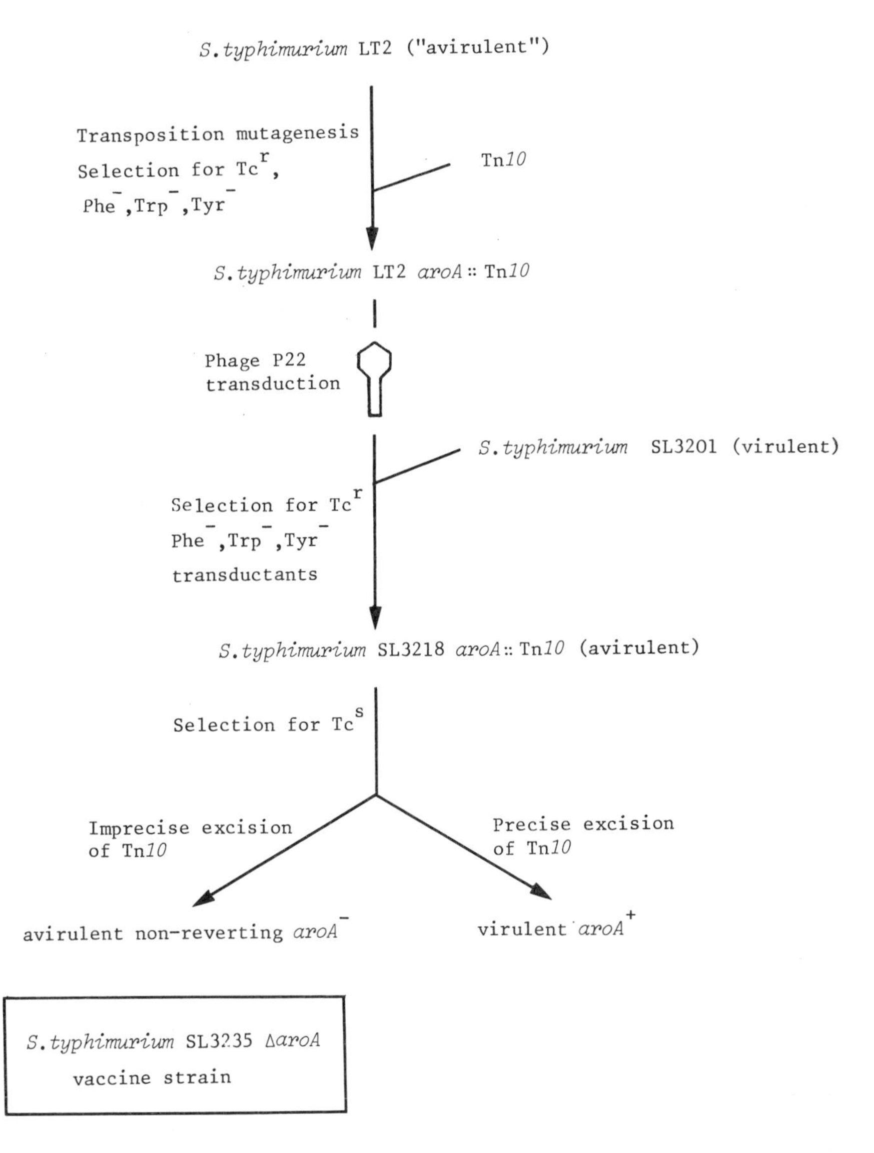

Figure 7.5: Relationship of the common aromatic amino acid biosynthetic pathway to synthesis of *p*-aminobenzoate and enterochelin. Shaded rectangle, blocked in *aroA*⁻ mutant

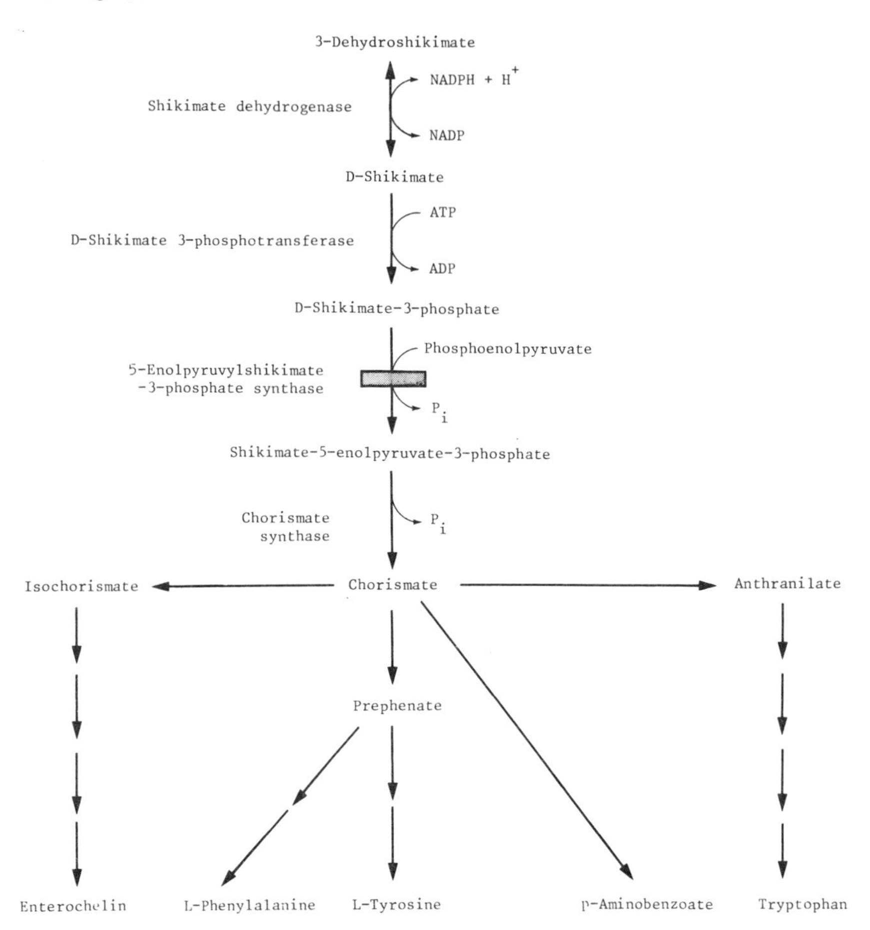

diarrhoeas has also been constructed by introduction of a recombinant plasmid carrying the gene for the non-toxic B subunit of *E. coli* heat-labile enterotoxin (LT) (see section 7.2.1) into *S. typhi* Ty21a (Clements and El-Morshidy, 1984). (The LT toxin produced by many enterotoxigenic *E. coli* is antigenically and structurally related to cholera toxin (CT). Immunisation with the LT-B subunit produces protective immunity to both LT and CT.)

(b) Application of reversed genetics to the production of anticholera vaccines

Cholera toxin is composed of two subunits. The A subunit (a 27000 daltons protein) activates adenyl cyclase and is responsible for causing the massive fluid loss characteristic of cholera. The B subunit (consisting of five identical 11600 daltons polypeptides) is responsible for binding the entire toxin to the mucosal epithelial surface and hence delivering the A subunit to the membrane of target cells. Toxoid vaccines (produced by inactivating the cholera toxin without destroying its antigenicity) or parentally administered killed whole cell vaccines are largely ineffective in conferring long-term immunity to cholera (Levine *et al.*, 1983; Finkelstein, 1985). This suggests that bacterial antigens other than the toxin, present during the natural infection, are important in inducing immunity in the host. In addition there is always a risk that a toxoid may retain toxic properties. This has been circumvented by administering purified B subunit of CT or LT to induce antitoxin antibodies. However antibacterial antibodies will still be required to give complete protection. These could be induced by simultaneous administration of killed *Vibrio cholerae* or by use of attenuated vaccine strains.

A nitrosoguanidine-induced mutant of *V. cholerae*, Texas Star SR, which is defective in the production of the CT-A subunit but capable of producing large amounts of the B subunit, gives some protection against cholera (Honda and Finkelstein, 1979; Levine *et al.*, 1983). Unfortunately the mutations involved in the construction of this strain were ill-defined, and patients administered with this live vaccine strain exhibited symptoms of mild diarrhoea. In an attempt to produce a safer, more effective vaccine strain, Mekalanos and co-workers (1983) cloned the *ctx* operon, which encodes the cholera toxin (Figure 7.6). In classical[1] strains of *V. cholerae* there are two identical distantly linked copies of this operon. About 70% of El Tor[1] strains contain only one copy of *ctx*, whereas the remainder contain two or more tandemly repeated copies. The *ctx* operon from the classical strain Ogawa 395 was cloned on a plasmid vector. Most of *ctxA* (subunit A coding) and half of *ctxB* (subunit B coding) was removed by BAL31 deletion from an *Xba*I site in the operon. Synthetic *Eco*RI linkers were added to the termini and the DNA religated. The deleted region was subsequently replaced by the insertion of a 1.4 kb *Eco*RI fragment carrying a kanamycin-resistance determinant (Figure 7.6). The resulting recombinant plasmid was mobilised to *V. cholerae* Ogawa 395 selecting for kanamycin resistance. Homogenotisation (section 4.10) was exploited to replace one copy of the wild-type resident *ctx* sequence by the mutant sequence. The donor plasmid was then displaced by the introduction of an incompatible plasmid into the mutated strain. A fortuitous *in vivo* recombination event was subsequently responsible for the spontaneous replacement of the remaining intact *ctx* sequence with a second *ctx*-

[1] Classical and El Tor strains are different biotypes of *V. cholerae.* A principal difference is the production of a soluble haemolysin (which lyses red blood cells) by El Tor vibrios.

Figure 7.6: Use of reversed genetics to create the non-toxigenic cholera live vaccine strain 0395-NT. RK2013 is a plasmid with IncP broad host range *tra* genes but a ColE1 origin of replication. It is therefore not cotransferred to *Vibrio cholerae* with the mobilised recombinant plasmid. P, promoter; →, direction of transcription; *ctxA, ctxB,* structural genes for the A and B subunits of cholera toxin; Gm^r, resistance to gentamicin; Km^r, resistance to kanamycin

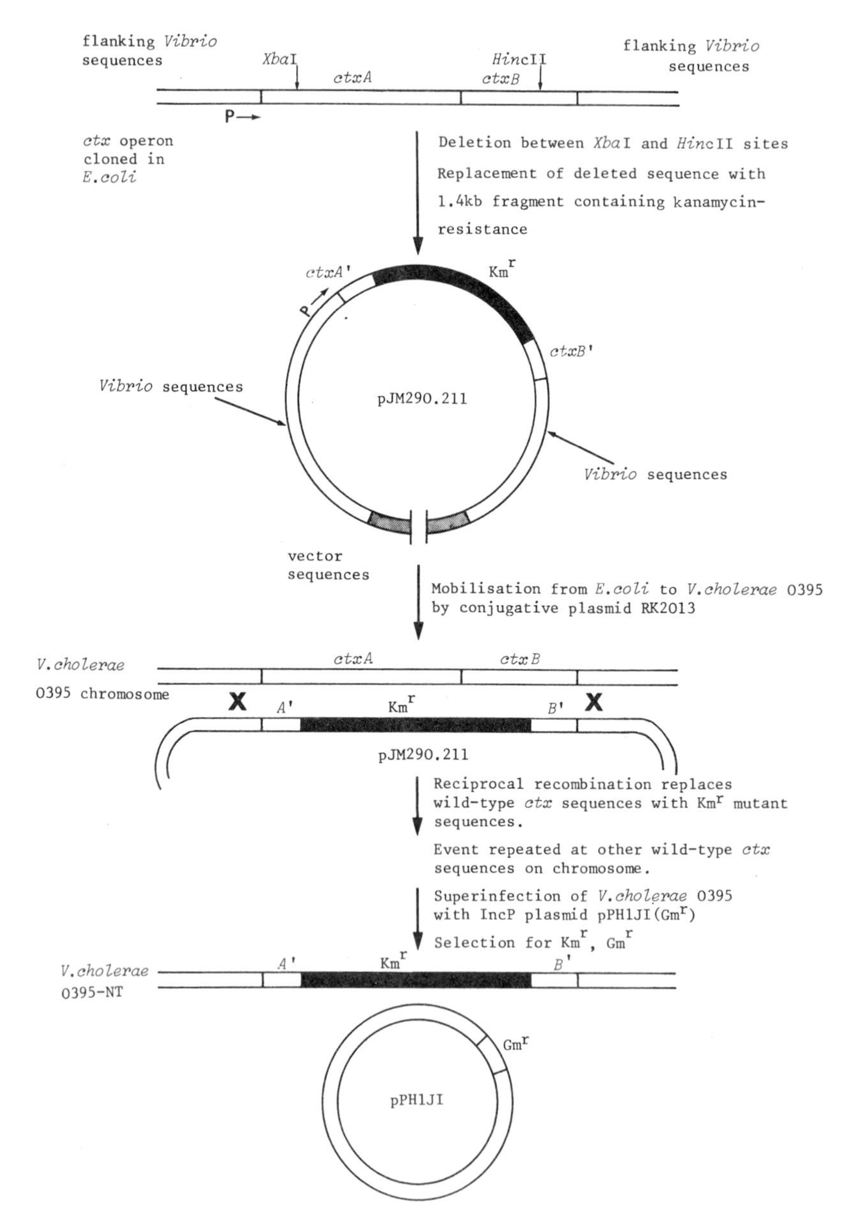

kanamycin-resistance mutant sequence (Figure 7.6). Essentially similar attenuated derivatives of an El Tor strain were constructed by Levine and co-workers (Levine *et al.*, 1983; Kaper *et al.*, 1984). They replaced part of the *ctx* operon with either a *bla* gene or (in order to avoid possible complications with the administration of an antibiotic-resistant vaccine strain) a mercury-resistance gene. The resulting strains of *V. cholerae* were found to be effective as vaccines in protection experiments. Potential vaccine strains have also been generated by cloning the *ctxB* gene in an *E. coli* expression vector under the control of a *trp* promoter (Levine *et al.*, 1983). Production of attenuated strains of *V. cholerae* provides good examples of the use of reversed genetics for production of mutant strains with considerable potential as live vaccines. Such strains would be very difficult to construct by more conventional procedures.

(c) Minicells as vaccines

Minicells derived either from bacteria (*E. coli* or *B. subtilis*) carrying cloned genes encoding immunogenic antigens or from minicell-producing (*min*) mutants of a pathogen *per se* might be used as vaccines. Since minicells do not replicate and will not survive long in the body, they may provide an antigenic stimulus to the host by inoculation via natural routes of infection (most probably the gut) without the attendant problems of administering infective, live organisms (Khachatourians, 1985). However, there are formidable problems in producing minicells that are totally free of parental nucleated cells. Generally, live nucleated cells will be present at 10^{-5} to 10^{-8} per minicell following purification. Thus the vaccine could be contaminated with potentially pathogenic bacteria. This would limit the use of minicell vaccines to diseases where the infective dose of parental minicell-producing cells necessary to produce disease was high.

(d) Vaccinia virus recombinants as live vaccines

Vaccinia virus recombinants are attractive as vehicles for vaccination because they are not only capable of expressing foreign proteins, but also retain infectivity. Vaccinia has been used successfully in vaccination to eradicate smallpox. Although there may be some side-effects, such as post-vaccinial encephalitis, inoculation with the virus is relatively safe. Furthermore, the vaccine strains are stable, easily administered and acceptable to recipients. The vaccinia genome is a double-stranded DNA molecule of 187 kb, which makes the virus too large to handle as a conventional cloning vector. However, vectors that contain only part of the genome, that retain the ability to infect animal cells and that express foreign genes can be constructed (Smith *et al.*, 1984). Vaccinia virus recombinants can be formed, without recourse to manipulation of the entire genome, by transforming vaccinia virus-infected cells with specially constructed plasmid recombination (insertion) vectors. Such insertion vectors contain the desired foreign coding sequence fused in the correct orientation to a vaccinia promoter and flanked by nonessential regions of vaccinia DNA (Figure 7.7). The plasmid vectors contain an origin of replication, enabling amplification in *E. coli* and a selectable marker, such as a *bla* gene. Foreign sequences are inserted into the vector at a cluster of unique cloning sites located just downstream of a vaccinia transcription start signal. Homologous recombination occurs in transformed cells resulting in the site-specific insertion of the foreign gene and vaccinia promoter into the nonessential region of the resident vaccinia genome. The site of insertion is

Figure 7.7: Use of vaccinia insertion (recombination) vectors for construction of recombinant virus vectors expressing foreign genes. *tk* thymidine kinase gene; p vaccinia promoter; → direction of transcription; Ap^r, ampicillin-resistance determinant; *oriV*, *E. coli* plasmid origin of replication

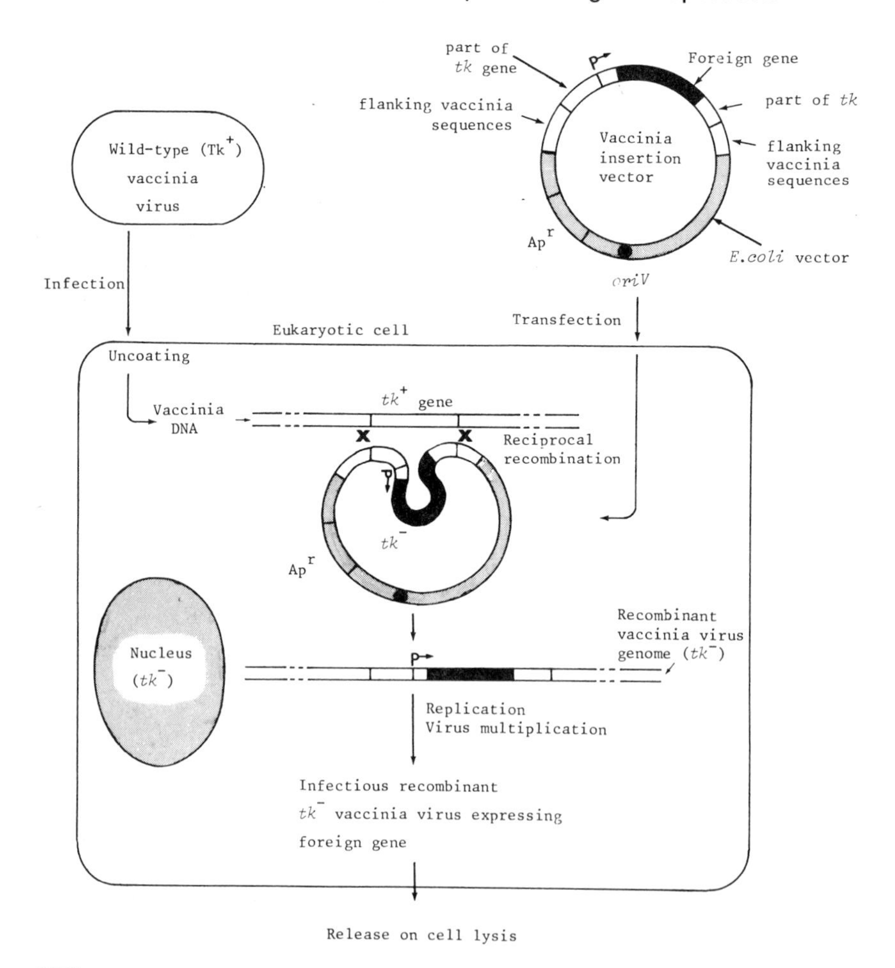

usually the vaccinia *tk* (thymidine kinase) gene. TK^- recombinant viruses can be selected readily because (unlike TK^+ viruses) they will replicate in TK^- host cells in the presence of 5-bromodeoxyuridine (Moss *et al.*, 1983). tk^- vaccinia recombinants have the additional advantage that they possess reduced virulence for animals compared with tk^+ viruses (Buller *et al.*, 1985). The efficiency of the recombinational rescue procedure is not high. At best only about 0.5% of the progeny vaccinia virions are recombinant. However, specific recombinants can be identified by using the *tk* gene as a target, by immunological screening, or by using DNA-DNA hybridisation (Smith *et al.*, 1984).

Several foreign genes, whose products are actual or potential immunogens, have been cloned in vaccinia, including influenza virus haemagglutinin (Panicalli *et al.*, 1983), hepatitis B virus surface antigen (Paoletti *et al.*, 1984), vesicular stomatitis virus glycoprotein, herpes simplex virus glycoprotein, rabies virus G glycoprotein (Buller *et al.*, 1985) and *Plasmodium knowlsei* surface antigen (Smith *et al.*, 1984). The capacity for insertion of foreign DNA without deleterious effects on vaccinia virus replication is high. More than 25 kb of DNA can be inserted into vaccinia virus DNA. This is higher than for other animal vectors (Smith and Moss, 1983). Therefore, two or more of the genes encoding appropriate immunising antigens can be cloned into vaccinia virus to form multivalent live virus vaccines. This is particularly advantageous where pathogens exhibit antigenic variation during infections. For example, genes encoding all the potential variable surface antigens produced by the different stages in the life cycle of malarial parasites might be cloned in vaccinia virus to provide a multivalent live vaccine against malaria (Ravetch *et al.*, 1985). Monovalent and multivalent vaccinia viruses have been shown to induce appropriate humoral antibodies and cell-mediated immunity in animals and to protect against challenge with the relevant pathogen(s). However, there are some problems associated with the use of vaccinia recombinant vaccines, not the least of which is that a large proportion of the human population has already been immunised with the virus during past smallpox vaccinations. Fortunately, this will be a rapidly diminishing problem, since the need for such vaccination no longer exists due to the eradication of the disease. Moreover, high-level expression of heterologous genes in vaccinia may still be successful even in the immune or partially vaccinia-immune host. A remaining problem will be to ensure the safety of the vaccine strains since the virus might mutate to increased virulence during or after administration. However, insertional inactivation of the vaccinia *tk* gene causes a dramatic reduction in the pathogenic potential of the virus, and reversion of the resulting TK^- viruses to TK^+ seems to occur extremely rarely (Buller *et al.*, 1985).

7.3 DIAGNOSIS OF DISEASE

The specificity of nucleic acid hybridisations makes possible the identification of particular DNA sequences in clinical material. Sequences of as little as 13 base pairs should, in theory, be sufficient to detect unique genes or subfragments therefrom diagnostic for particular conditions or specific infectious agents (Kelker, 1985). DNA is stable to many harsh treatments, including formalisation, and so can be recovered and characterised even from long-dead tissues (Pääbo, 1985).

7.3.1 Diagnosis and characterisation of infectious agents

DNA probes can be used to identify specific plasmids or resistance genes in clinical isolates of bacteria. Such probes may be used for rapid determination of the resistance properties of bacterial pathogens, in turn facilitating the choice of appropriate antibiotic therapy (Yang, 1985). Specific probes for some of the many different resistance genes are also important tools in studying the epidemiology of drug resistance. The probes may be hybridised to Southern blots of plasmid preparations or, more conveniently, used in direct colony- (section 3.7.2.b) or spot- (where small volumes of liquid culture are lysed *in situ* on a nitrocellulose membrane) hybridisation of clinical isolates. A probe consisting of a 1 kb *Eco*RI-*Hin*fI fragment bearing part of the TEM-1 *bla* gene from pBR322 was found to hybridise only to plasmids specifying TEM-1, TEM-2 or OXA-2 β-lactamases (Cooksey *et al.*,1985).[2] Another probe, specific for the gentamicin 2″-*O*-adenyltransferase [AAD(2″)] gene, has been constructed (Groot Obbink *et al.*, 1985). This probe hybridises with DNA from bacteria producing AAD (2″), but not with that from bacteria producing other aminoglycoside antibiotic inactivating enzymes.

At present the number of probes specific for resistance genes is limited. Furthermore, it is not certain whether probe technology will ever totally replace conventional antibiotic sensitivity testing in routine diagnostic laboratories. The basic problem is the number of different probes that would have to be used to encompass all the possible resistance phenotypes that are encountered in particular bacterial species of clinical interest. The use of probes is also relatively expensive and in many instances, as with most conventional methods, some growth of the bacterial isolate is required. Despite these limitations, there are a number of applications in clinical microbiology where DNA probes have considerable advantages over traditional methods. These applications include the identification of

[2] The TEM-1 enzyme is the most commonly distributed plasmid-mediated β-lactamase. TEM-2 is an evolutionary variant of the same enzyme.

viruses that may be difficult to cultivate in routine laboratories, and the rapid identification of fastidious or slow-growing bacteria (such as *Mycobacterium tuberculosis* and *M. leprae*) and of pathogens in tissues or body fluids (for example, *Neisseria meningitidis* in cerebrospinal fluid). Furthermore, the sensitivity of nucleic hybridisation techniques makes them particularly useful in the detection of pathogens, such as *Salmonella* species, in foodstuffs.

7.3.2 Diagnosis of genetic diseases

Single base-pair substitutions, inversions, duplications, insertions and deletions in human genes can cause genetic disease. Detection of these aberrations in genomic DNA may be useful in antenatal diagnosis of inherited diseases. In many instances genetic disease may be indicated by polymorphisms in either the gene(s) associated with the condition or in sequences that are in close proximity (see, for example, Shows *et al.*, 1982). Heritable disorders can sometimes be diagnosed by monitoring restriction fragment length polymorphisms (RFLPs). RFLPs can arise as a consequence of changes in genomic DNA sequence that generate or abolish a recognition site for a restriction endonuclease. Chromosome rearrangements, such as insertions and deletions, may not alter the recognition site *per se* but alter its position with respect to other sites. These polymorphisms are detected by measuring differences in size of restriction fragments that hybridise to a specific DNA probe (which may be homologous to DNA sequences at or close to the allele concerned) in a Southern blot (section 3.8.2) (Figure 7.8). An RFLP is not normally a direct marker for the presence of a gene defect, but may be used as a linkage marker if the restriction site(s) concerned is located near to the defective allele. RFLPs can be used predictively to screen for genetic defects in families, but may be informative only in certain cases. An advantage of examining genetic defects by using RFLPs is that a particular pathological condition may be diagnosed even where the precise biochemical defect responsible is not known. For example, the defective gene responsible for Huntington's chorea is linked to polymorphisms in *Hin*dIII restriction sites on human chromosome 4 (Gusella *et al.*, 1983). In the case of cystic fibrosis the defective locus has been localised to human chromosome 7 by establishing close linkage between *Taq*I and *Msp*I polymorphisms at the *met* oncogene locus on this chromosome (Wainwright *et al.*, 1985; White *et al.*, 1985).

RFLPs are found all over the human (and other) genome. Human DNA contains hypervariable **minisatellite** regions which are dispersed over the genome (Jeffreys *et al.*, 1985a). Each region consists of tandem repeats of a short DNA segment containing a common core sequence, similar to the *chi* sequence of *E. coli* (section 3.5.2). Many minisatellites are highly poly-

Figure 7.8: Detection of restriction fragment length polymorphisms in human DNA. DNA fragment length variants should segregate in a codominant Mendelian fashion. An individual can be homozygous or heterozygous depending on the variant carried by each chromosome homologue. DNA from the heterozygote therefore produces two different hybridising bands, one from each chromosome homologue. ↓, Restriction site; ⌐*¬, labelled probe

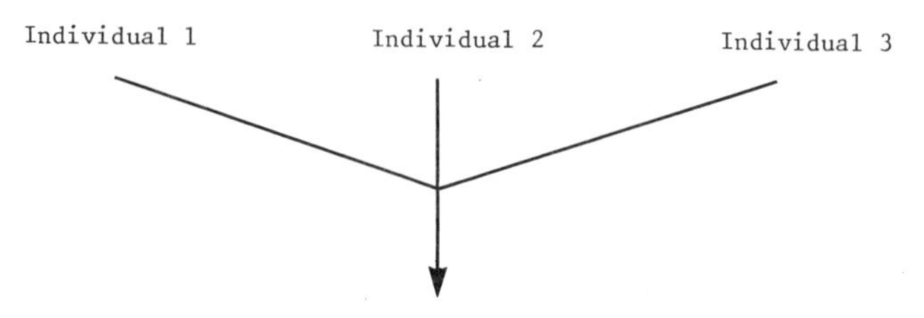

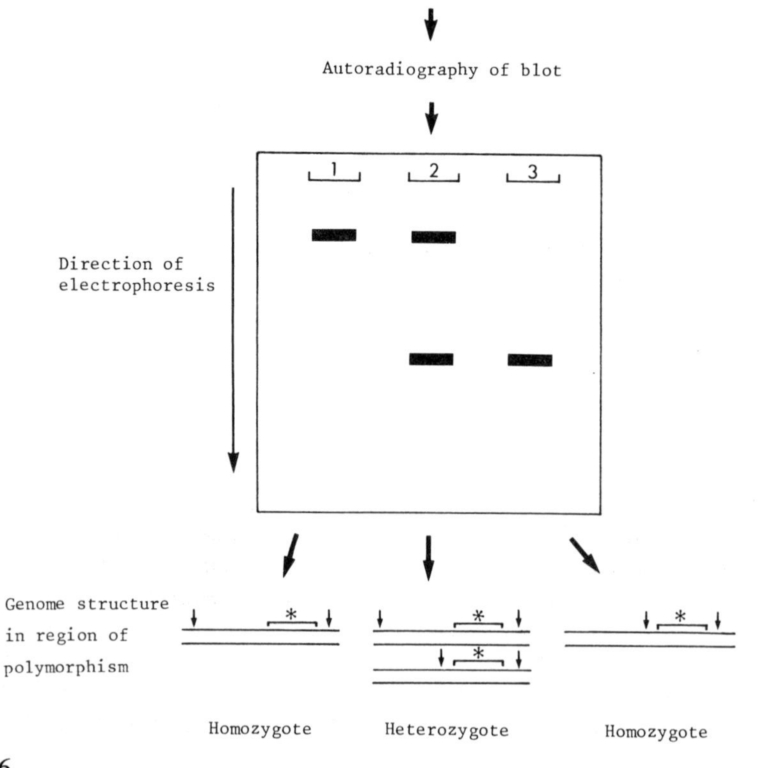

Figure 7.9: Detection of single base-pair substitution mutations in genomic DNA. Autoradiograph of duplexes formed between labelled probe DNA including the wild-type allele, and genomic DNA. Lane 1, homoduplex DNA from an individual homozygous for the wild-type allele; lane 2, heteroduplex DNA from an individual homozygous for a mutant allele; lane 3, homoduplex and heteroduplex DNA from an individual heterozygous for that mutant allele. ↓, Restriction site; *, position of point mutation

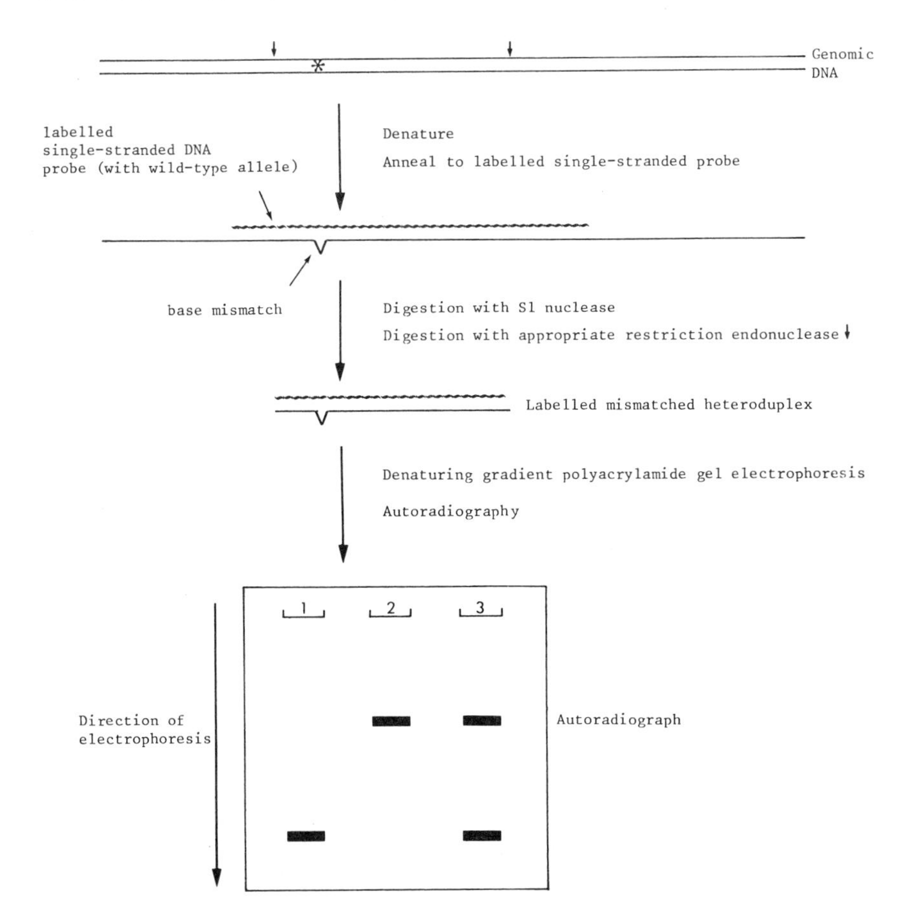

morphic due to allelic variation in the number of tandem repeats. Such variation probably arises by mitotic or meiotic unequal exchanges in which the core sequence may serve as a recombination signal. Minisatellite length polymorphism can be detected using a restriction enzyme that does not cleave the repeat unit. Hybridisation probes consisting of tandem repeats of the core sequence can detect many highly polymorphic minisatellites simultaneously and provide a DNA 'fingerprint' that is completely individual-specific and is stable both somatically and through the germ line (Jeffreys *et al.*, 1985b). These fingerprints are derived from a large number of autosomal loci and are of use in human linkage analysis and the identification of individuals in, for example, forensic medicine (Gill *et al.*, 1985) or maternity and paternity testing (Jeffreys *et al.*, 1985c). The general principle of using RFLPs to prove identity of individuals is also applicable to the identification of plant varieties and of strains of microorganisms (which maybe useful, for example, in disputes over patents).

Many base substitutions do not lead to alterations in restriction fragment length. In certain cases, where the DNA sequence surrounding the base change is known, it is possible to detect specific mutations (for example, in the α_1-antitrypsin gene (Kidd *et al.*, 1983)) using specific synthetic oligonucleotide hybridisation probes. However, some genetic diseases, such as β-thalassaemia, can result from many different single base-pair changes and would require a large number of such probes to be used for diagnosis. An alternative procedure that enables the detection of some, but not all, single base changes has been developed. This exploits the differential melting behaviour of wild-type (homoduplex) and mismatched mutant/wild-type (heteroduplex) DNA molecules ((Myers *et al.*, 1985). Differences between these types of molecule can be detected by electrophoresis through a polyacrylamide gel containing a denaturing gradient. Heteroduplex molecules containing a single base mismatch will exhibit different mobility to wild-type homoduplexes on such gels. Single base substitutions in DNA can be detected by annealing a specific radiolabelled DNA probe to total genomic DNA containing the suspected mutation (Figure 7.9). The duplexes formed are digested with S1 nuclease to eliminate DNA that has not reassociated, and subsequently with a restriction endonuclease to cut out a small duplex region containing the mismatch. (Fortunately S1 nuclease does not seem to cleave all single base mismatches. Furthermore, if the mutation of interest is a short deletion or addition, S1 nuclease will probably cleave the heteroduplex to produce two specific fragments, which may be diagnostic.) Not all single base mismatched heteroduplexes formed from β-thalassaemic globin genes can be detected by this method. This probably reflects the specific location of mismatches within domains of DNA with differential melting properties. However, it has been estimated that between 25 and 40% of all possible single base changes in genes should be detectable using this approach.

BIBLIOGRAPHY

Atassi, M.Z. and Webster, R.G. (1983) Localization, synthesis and activity of an antigenic site on influenza virus hemagglutinin. *Proc. Natl Acad. Sci. USA* **80**, 840-4

Boothroyd, J.C., Highfield, P.E., Cross, G.A.M., Rowlands, D.J., Lowe, P.A., Brown, F. and Harris, T.J.R. (1981) Molecular cloning of foot-and-mouth disease virus genome and the nucleotide sequences of the structural protein genes. *Nature (London)* **290**, 800-2

Brown, F. (1985) Peptides as the next generation of foot-and-mouth disease vaccines. *Biotechnology* **3**, 445-8

Buller, R.M.L., Smith, G.L., Cremer, K., Notkins, A.L. and Moss, B. (1985) Decreased virulence of recombinant vaccinia virus expression vectors is associated with a thymidine kinase-negative phenotype. *Nature (London)* **317**, 813-15

Clements, J.D. and El-Morshidy, S. (1984) Construction of a potential live oral vaccine for typhoid fever and cholera-*Escherichia coli* related diarrhoeas. *Infect. Immun.* **46**, 564-9

Cooksey, R.C., Clark, N.C. and Thornsberry, C. (1985) A gene probe for TEM type β-lactamases. *Antimicrob. Agents Chemother.* **28**, 154-6

Courtney, M., Jallat, S., Tessier, L-H., Benavente, A., Crystal, R.G. and Lecocq, J-P. (1985) Synthesis in *E. coli* of α_1-antitrypsin variants of therapeutic value for emphysema and thrombosis. *Nature (London)* **313**, 149-51

Dougan, G. and Morrissey, P.M. (1985) Molecular analysis of the virulence determinants of enterotoxigenic *Escherichia coli* isolated from domestic animals: applications for vaccine development. *Vet. Microbiol.* **10**, 241-57

Ellis, J., Ozaka, L.S., Gwadz, R.W., Cochrane, A.H., Nussenzweig, V., Nussenzweig, R.S. and Godson, N. (1983) Cloning and expression in *E. coli* of the malarial sporozoite surface antigen from *Plasmodium knowlesi.* *Nature (London)* **302**, 536-8

Fersht. A.R., Shi, J.P., Knill-Jones, J., Lowe, D.M., Wilkinson, A.J., Blow, D.M. and Winter, G. (1985) Hydrogen bonding and biological specificity analysed by protein engineering. *Nature (London)* **314**, 235-8

Finkelstein, R.A. (1985) Vaccines (?) against the cholera-related enterotoxin family. In *Microbiology — 1985* (L. Leive, ed.), pp. 114-18. American Society for Microbiology, Washington, DC

Formal, S.B., Baron, L.S., Kopecko, D.J., Washington, O., Powell, C. and Life, C.A. (1981) Construction of a potential bivalent vaccine strain: introduction of *Shigella sonnei* form I antigen into the *galE Salmonella typhi* Ty21a vaccine strain. *Infect. Immun.* **34**, 746-50

Gill, P., Jeffreys, A.J. and Werrett, D.J. (1985) Forensic application of DNA fingerprints. *Nature (London)* **318**, 577-9

Groot Obbink, D.J., Ritchie, L.J., Cameron, F.H., Mattick, J.S. and Ackerman, V.P. (1985) Construction of a gentamicin resistance gene probe for epidemiological studies. *Antimicrob. Agents Chemother.* **28**, 96-102

Gusella, J.F., Wexler, N.S., Conneally, P.M., Naylor, S. L., Anderson, M.A., Tanzi, R.E., Watkins, P.C., Ottina, K., Wallace, M.R., Sakaguchi, A.Y., Young, A.B., Shoulson, I., Bonilla, E. and Martin, J.B. (1983) A polymorphic DNA marker genetically linked to Huntington's disease. *Nature (London)* **306**, 234-8

Hardy, K., Stahl, S. and Kupper, H. (1981) Production in *B. subtilis* of hepatitis B core antigen and of major antigen of foot and mouth disease virus. *Nature (London)* **293**, 481-3

Honda, T. and Finkelstein, R.A. (1979) Selection and characteristics of a *Vibrio cholerae* mutant lacking the A (ADP-ribosylating) portion of the cholera enterotoxin. *Proc. Natl Acad. Sci. USA* **76**, 2052-6

Hosieth, S.K. and Stocker, B.A.D. (1981) Aromatic-dependent *Salmonella typhimurium* are non-virulent and effective as live vaccines. *Nature (London)* **291**, 238-9

Jeffreys, A.J., Wilson, V. and Thein, S.L. (1985a) Hypervariable 'minisatellite' regions in human DNA. *Nature (London)* **314**, 67-73

Jeffreys, A.J., Wilson, V. and Thein, S.L. (1985b) Individual-specific 'fingerprints' of human DNA. *Nature (London)* **316**, 76-9

Jeffreys, A.J., Brookfield, J.F.Y. and Semeonoff, R. (1985c) Positive identification of an immigration test-case using human DNA fingerprints. *Nature (London)* **317**, 818-19

Kaper, J.B., Lockman, H., Baldini, M.H. and Levine, M.M. (1984) Recombinant nontoxigenic *Vibrio cholerae* strains as attenuated cholera vaccine candidates. *Nature (London)* **308**, 655-8

Kehoe, M., Sellwood, R., Shipley, P. and Dougan, G. (1981) Genetic analysis of K88 mediated adhesion of enterotoxigenic *Escherichia coli. Nature (London)* **291**, 122-6

Kelker, N. (1985) Nucleic acid hybridization applied to the identification of infectious agents. In *Microbiology — 1985* (L. Leive, ed.), pp. 158-60. American Society for Microbiology, Washington, DC

Khachatourians, G.G. (1985) The use of anucleated minicells in biotechnology: an overview. In *Biotechnology, applications and research* (P.N. Cheremisinoff and R.P. Quellette, eds), pp, 309-19. Technomic Publishing Co., Lancaster

Kidd, V.J., Wallace, R.B., Itakura, K. and Woo, S.L.C. (1983) α_1-Antitrypsin deficiency detection by direct analysis of the mutation in the gene. *Nature (London)* **304**, 230-4

Kleid, D.G., Dowbenko, D.J., Bock, L.A., Hoatlin, M.E., Jackson, M.L., Patzer, E.J., Shire, S.J., Weddell, G.N., Yansura, D.G., Morgan, D.O., McKercher, P.D. and Moore, D.M. (1985) Production of recombinant vaccines from microorganisms: vaccine for Foot-and-Mouth Disease Virus. In *Microbiology — 1985* (L. Leive, ed.), pp. 405-8. American Society for Microbiology, Washington, DC

Kupper, H., Keller, W., Kurz, C.H., Forss, S., Schaller, H., Franze, R., Strohmaier, K., Marquardt, O., Zaslavsky, V.G. and Hofschneider, H. (1981) Cloning of cDNA of the major antigen of foot-and-mouth disease virus and expression in *E. coli. Nature (London)* **289**, 555-9

Lerner, R.A. (1982) Tapping the immunological repertoire to produce antibodies of predetermined specificity. *Nature (London)* **299**, 592-6

Levine, M.M., Kaper, J.B., Black, R.E. and Clements, M.L. (1983) New knowledge on pathogenesis of bacterial enteric infections as applied to vaccine development. *Microbiol. Rev.* **47**, 510-50

Mark, D.F., Lu, S.D., Creasey, A.A., Yamamoto, R. and Lin, L.S. (1984) Site specific mutagenesis of the human fibroblast interferon gene. *Proc. Natl Acad. Sci. USA* **81**, 5662-6

Mekalanos, J.J., Swarz, D.J., Pearson, G.D.N., Harford, N., Groyne, F. and De Wilde, M. (1983) Cholera toxin genes: nucleotide sequence, deletion analysis and vaccine development. *Nature (London)* **306**, 551-6

Mijanohara, A., Toh-E, A., Nozaki, C., Hamada, F., Ohtomo, N. and Matsubara, K. (1983) Expression of hepatitis B surface antigen in yeast. *Proc. Natl Acad. Sci. USA* **80**, 1-5

Moss, B., Smith, G.L. and Mackett, M. (1983) Use of vaccinia as an infectious molecular cloning vector. In *Gene amplification and analysis*, vol. 3 (T.K. Papas, M. Rosenberg and G. Chirikjian, eds), pp. 201-13. Elsevier, New York

Myers, R.M., Lumelsky, N., Lerman, L.S. and Maniatis, T. (1985) Detection of single base substitutions in total genomic DNA. *Nature (London)* **313**, 495-8

Pääbo, S. (1985) Molecular cloning of ancient Egyptian mummy DNA. *Nature (London)* **314**, 644-5

Panicalli, D., Davis, S.W., Weinberg, R.L. and Paoletti, E. (1983) Construction of live vaccines by using genetically engineered poxviruses: biological activity of recombinant vaccinia virus expressing influenza virus haemagglutinin. *Proc. Natl*

Acad. Sci. USA **80**, 5364-8

Paoletti, E., Lipinskas, B.R., Samsonoff, C., Mercer, S. and Panicalli, D. (1984) Construction of live vaccines by using genetically engineered poxviruses: biological activity of vaccinia virus recombinants expressing the hepatitis B virus surface antigen and the herpes simplex virus glycoprotein gD. *Proc. Natl Acad. Sci. USA* **81**, 193-7

Ravetch, J.V., Young, J. and Poste, G. (1985) Molecular genetic strategies for the development of anti-malaria vaccines. *Biotechnology* **3**, 729-40

Robertsson, J.A., Lindberg, A.A., Hosieth, S.K. and Stocker, B.A.D. (1983) *Salmonella typhimurium* infection in calves: protection and survival of virulent challenge bacteria after immunization with live or inactivated vaccines. *Infect. Immun.* **41**, 742-50

Rosenberg, S., Barr, P.J., Najarian, R.C. and Hallewell, R.A. (1985) Synthesis in yeast of a functional oxidation-resistant mutant of human α_1-antitrypsin. *Nature (London)* **312**, 77-80

Rothbard, J.B., Fernandez, R., Wang, L., Teng, N.N.H. and Schoolnik, G.K. (1985) Antibodies to peptides corresponding to a conserved sequence of gonococcal pilins block bacterial adhesion. *Proc. Natl Acad. Sci. USA*, **82**, 915-19

Saunders, J.R. (1986) The genetic basis of phase and antigenic variation in bacteria. In *Antigenic variation in infectious diseases* (T.H. Birkbeck and C.W. Penn, eds), pp. 57-76. Society for General Microbiology/IRL Press, Oxford

Shows, T.B., Sakaguchi, A.Y. and Naylor, S.L. (1982) Mapping the human genome, cloned genes, DNA polymorphisms, and inherited disease. In *Advances in human genetics*, vol. 12 (H. Harris and K. Hirschhorn, eds), pp. 341-452. Plenum Press, New York and London

Smith, G.L., Mackett, M. and Moss, B. (1984) Recombinant vaccinia viruses as new live vaccines. *Biotechnol. Genet. Engng Rev.* **2**, 383-407

Smith, G.L. and Moss, B. (1983) Infectious poxvirus vectors have capacity for at least 25,000 base pairs of foreign DNA. *Gene* **25**, 21-8

Tiollais, P., Pourcel, C. and Dejean, A. (1985) The hepatitis B virus. *Nature (London)* **317**, 489-95

Valenzuela, P., Coit, D. and Kuo, C.H. (1985a) Synthesis and assembly in yeast of hepatitis B surface antigen particles containing the polyalbumin receptor. *Biotechnology* **3**, 317-20

Valenzuela, P., Coit, D., Medina-Selby, A., Kuo, C.H., van Nest, G., Burke, R.L., Urdea, M. and Graves, P.V. (1985b) Antigen engineering in yeast: synthesis and assembly of hybrid hepatitis B surface antigen-herpes simplex 1 gD particles. *Biotechnology* **3**, 323-6

Wahdan, M.H., Serie, C., Cerisier, Y., Sallam, S. and Germanier, R. (1982) A controlled field trial of live *Salmonella typhi* Ty21a oral vaccine against typhoid: three-year results. *J. Inf. Dis.* **145**, 292-5

Wainwright, B.J., Scambler, P.J., Schmidtke, J., Watson, E.A., Law, H.L., Farrall, M., Cooke, H.J., Eiberg, H. and Williamson, R. (1985) Localization of cystic fibrosis locus to human chromosome 7cen-q22. *Nature (London)* **318**, 384-5

Wang, A., Lu, S.D. and Mark, D.F. (1984) Site-specific mutagenesis of the human interleukin-2 gene: structure-function analysis of the cysteine residues. *Science* **224**, 1431-3

White, R., Woodward, S., Leppert, M., O'Connell, P., Hoff, M., Herbst, J., Lalouel, J-M., Dean, M. and Woude, G.V. (1985) A closely-linked genetic marker for cystic fibrosis. *Nature (London)* **318**, 382-4

Winter, G. and Fersht, A.R. (1984) Engineering enzymes. *Trends Biotechnol.* **2**,115-19

Winther, M.D. and Dougan, G. (1984) The impact of new technologies on vaccine

development, *Biotechnol. Genet. Engng Rev.* **2**, 1-39
Yang, H-L. (1985) R-plasmid identification using biotinylated DNA probes. In *Microbiology — 1985* (L. Leive, ed.), pp. 161-4. American Society for Microbiology, Washington, DC

8

Plant Technology

There is considerable scope for applying the principles of microbial genetics to agriculture for the manipulation of plants and of agriculturally important microorganisms. A basic objective of plant breeders is the design of plant varieties that are more valuable as crop plants, as biomass or as sources of pharmaceuticals, enzymes or other useful commodities. There are several routes for developing genetically superior plants. Variability can be induced by conventional mutagenesis. However, plants regenerated from *in vitro* cell culture[1] can exhibit phenotypic and genetic variation. Such tissue culture variability (somaclonal variation), which is dependent upon such factors as the plant species, the genotypes used, type of explant for culture and the period of culture, may be useful for plant improvement (Lorz, 1984). Conventional plant breeding techniques permit the formation of hybrids that have inherited desirable traits from their parents. However, undesirable characters may also be acquired and expressed in the hybrids. Elimination of such unwanted traits can be achieved by repeated backcrossing of first-generation hybrids with one parent (a process termed introgressive breeding), until the desired plant variety is obtained. The use of conventional plant breeding techniques typically depends upon sexual compatibility of the species involved. Sexual incompatibility can sometimes be overcome by using protoplast fusion techniques. Generally, however, somatic hybrids obtained by fusions between species that cannot be crossed sexually are sterile, thereby precluding their integration into a breeding programme. In those rare cases where fertile somatic hybrids have been

[1] *In vitro* cell and tissue culture techniques involve the culturing of roots, stems or leaves from a plant on a growth medium containing appropriate nutrients and hormones. This technology permits the selection and propagation of individual cells with desirable traits. Regeneration of whole plants may be induced by exposure of the clonally propagated cells to medium with specific hormones. In this way large numbers of essentially identical plants can be grown.

generated, provision of a useful plant variety again depends upon generations of introgressive breeding. Somatic hybridisation can, however, provide new genetic variation, due to such phenomena as organelle segregation and mitotic recombination that can occur in somatic hybrids prior to plant regeneration. Such novelty affords scope for the development of new plant varieties that are not possible through sexual hybridisation (Evans, 1983). The high degree of inbreeding associated with conventional breeding programmes reduces 'hybrid vigour' (heterosis). This in turn is likely to increase the susceptibility of plants to major outbreaks of disease, and lead to loss of important genes from the gene pool. The availability of *in vitro* recombinant DNA techniques for introducing a desired gene or set of genes directly into a useful plant variety could circumvent the necessity for lengthy backcross programmes and may overcome mating barriers. Vector systems, including those based on Ti (tumour-inducing) plasmids of *Agrobacterium* and on plant DNA viruses, have been designed for this purpose. Plant genes might be cloned and modified in microorganisms and subsequently transferred back to the parent plant to effect beneficial changes. However, realisation of the full potential of this technology for plant engineering will depend not only upon the development of effective host-vector systems, but also upon a detailed knowledge of plant molecular genetics, which is wanting at present.

Improvements in efficiency of agricultural processes may additionally be achieved through the genetic manipulation of microorganisms, such as nitrogen-fixing bacteria and insecticide-producing organisms, that are important to plants. Furthermore genetically engineered microorganisms can be utilised for the manufacture of useful plant products (such as enzymes and food additives).

This chapter describes the adaptation of microbial genetics techniques to the management of crop plants.

8.1 PLANT GENETIC ENGINEERING

The development of effective recombinant DNA techniques for manipulating plant genomes depends upon the availability of suitable vectors for cloning desired genes, efficient procedures for introducing recombinant vectors into plant cells, and a means of expressing and stably maintaining the transferred genes. Ideally such genes should be transmitted both somatically and sexually to the offspring. In addition the vector system should permit gene cloning in organisms such as *E. coli*, so that recombinant molecules can be readily constructed and manipulated in these hosts prior to introduction into plant cells. Genetic vehicles may be derived from naturally occurring plant vectors, such as Ti (**tumour-inducing**) and Ri (**root-inducing**) plasmids of *Agrobacterium*, plant viruses or viroids; alternatively artificial vectors may be developed by utilising components of the plant genome.

There are several potential routes for introducing recombinant vectors into plants, for example:

(i) transformation/transfection of plant protoplasts using naked DNA or DNA encased in liposomes (as in microbial cells DNA uptake is promoted by PEG and cations such as Ca^{2+});
(ii) protoplast fusion between plant protoplasts and microbial protoplasts that carry the recombinant vector;
(iii) microinjection techniques using, for example, plant cells, gametes or seeds;
(iv) infection of whole plants, plant cells or protoplasts via the routes of entry of plant pathogens (such as *Agrobacterium* and plant DNA viruses).

It should, however, be noted that the application of these delivery techniques can be limited by the difficulty in regenerating cells or protoplasts of particular species as whole plants; or by the host range of specific plant pathogens.

8.1.1 Ti plasmids of *Agrobacterium* as vectors for plant engineering

(a) Ti plasmids and crown gall disease

Many dicotyledonous plants are susceptible to crown gall disease, which is caused by pathogenic strains of the soil bacterium *Agrobacterium tumefaciens* (De Cleene and Deley, 1976). Infection of wounded plant tissue by the pathogen results in the formation of crown gall tumours. Crown gall tissues synthesise one or more amino acid derivatives called **opines**, which can be used by *Agrobacterium* as sole carbon and/or nitrogen source. Tumour induction, opine synthesis in transformed plant cells and catabolism of opines by *Agrobacterium* depend upon the presence of Ti plasmids (Van Montagu and Schell, 1982). Ti plasmids are classified according to the type of opine (for example octopine, nopaline or agropine) that is synthesised in the gall. Ti plasmids are conjugative and can transfer to avirulent strains of *Agrobacterium.* Such transfer is induced by the opines.

A specific region of the Ti plasmid, the T-DNA (transferred/transforming DNA), which is between 15 and 30 kb, is transferred to plant cells during infection and becomes integrated into the plant nuclear DNA, where it is expressed (Chilton *et al.*, 1980; Willmitzer *et al.*, 1980). (Regions of the Ti plasmid far from the T-DNA may also be transferred to the plant cell (Virts and Gelvin, 1985).) T-DNA is responsible for induction and maintenance of the tumorous state and for opine synthesis in the plant cell. Part of the T-DNA region (the common or core sequence) is conserved in both octopine and nopaline-type Ti plasmids. Genes for opine catabolism are situated outside the T region of the Ti plasmid (Figure 8.1). Once inserted into the plant DNA, the T-DNA behaves like a normal Mendelian marker.

The precise mechanics of T-DNA transfer and integration are unclear. However, mobilisation of the DNA from bacterium to plant cell requires the *vir* (virulence) genes of the Ti plasmid. The *vir* region is also located outside the T-DNA on the plasmid (Figure 8.1). Separation of T-DNA of the Ti plasmid from the *vir* genes, by incorporating the two regions into independent replicons, still permits transfer of T-DNA into the plant cell. Strains of *A. tumefaciens* harbouring the two replicons (one carrying T-DNA, the other the *vir* genes) are fully tumorigenic (De Framond *et al.*, 1983). *vir* regions of nopaline-type and octopine-type Ti plasmids are homologous and encode equivalent functions (Engler *et al.*, 1981). The *vir* region of an octopine plasmid can effect transfer of T-DNA from a nopaline plasmid. At least six separate complementation groups: *virA, B, C, D, E* and *G*, and *pinF* are organised as a single regulon in the *vir* region of the octopine-type plasmid pTiA6. When *Agrobacterium* is co-cultivated with plants, expression of *virB, C, D, E, G* and *pinF* is induced to high levels. This activation probably initiates the process of T-DNA transfer and integration (Stachel *et al.*, 1986). Activation of *vir* genes (and production of T-DNA circles (see below)) appears to be mediated by acetosyringone (AS) and α-hydroxyacetosyringone (OH-AS), which occur in exudates of wounded and metabolically active plant cells (Stachel *et al.*, 1985). AS and OH-AS are probably responsible in nature for the recognition of susceptible plant cells by *Agrobacterium.* Such *vir* signal molecules (AS and OH-AS) might thus be useful in extending the range of plants that can be transformed by *Agrobacterium.* Sequences (called 'integration sequences') bordering T-DNA on Ti plasmids also appear to be necessary for the transfer process. Twenty-five base-pair imperfect repeat sequences, flanking the T-DNA of nopaline-type plasmids, apparently serve as border signals defining the functional ends of this DNA. Similar sequences have been found at the border of octopine T-DNA (Figure 8.2). However, deletion of one of the T-DNA border sequences of nopaline-type plasmids attenuates but does not abolish tumour induction, suggesting that alternative borders can substitute when a normal border is removed. There seems to be a more stringent requirement for the right border than for the left border sequence for transfer and integration of T-DNA into the plant cell. The right border region can transfer DNA from various positions around the Ti plasmid, even where there is no left border sequence with which to interact (Caplan *et al.*, 1985).

Figure 8.1: Maps of an octopine-type (pTiAch5) and a nopaline-type (pTiC58) Ti plasmid. (a) Genetic maps. *tra*, Transfer functions; *agc*, agrocinopine catabolism; *agr*, agropine catabolism; *inc*, incompatibility; *noc*, nopaline catabolism; *occ*, octopine catabolism; *ori*, origin of replication, *vir*, virulence. (b) T-regions with restriction maps (fragments are numbered). The shaded blocks indicate regions of homologous DNA (adapted from Engler *et al.*, 1981)

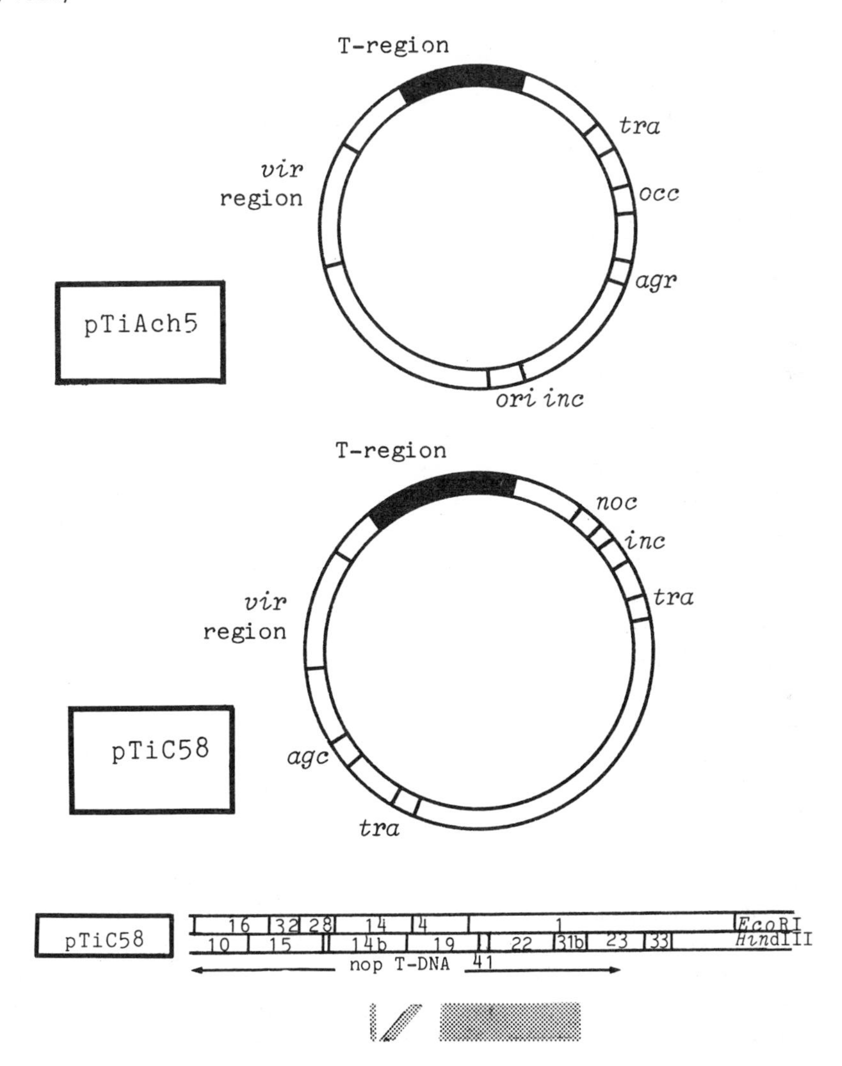

Figure 8.2: Border sequences of T regions. Twenty-five base-pair imperfect repeats border T-DNA in a nopaline-type Ti plasmid (pTiT37); 12 of the matching bases are contiguous. A similar sequence is found at the left border of an octopine-type Ti plasmid (Simpson *et al.*, 1982). Mismatched bases are in bold (adapted from Yadav *et al.*, 1982)

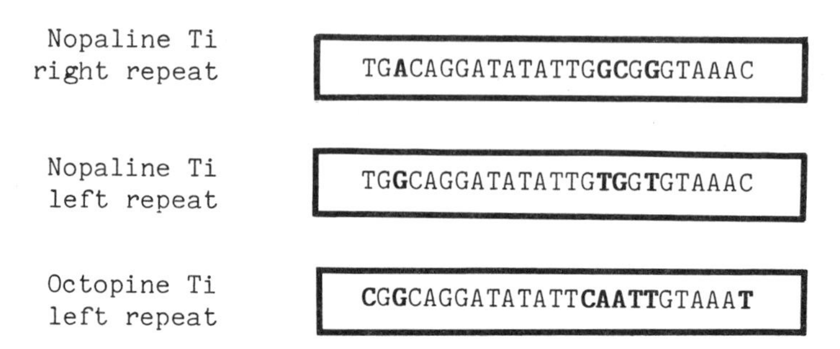

An early event in the transfer/integration process may be replication of the T-DNA. The replicative process may begin at the right border and proceed to the end of the left T-DNA region (Caplan *et al.*, 1985). T-DNA circles may be intermediates or by-products in the transfer process. The circular junction site occurs precisely within the 25 base-pair border sequence. Circles might, therefore, arise by a site-specific recombination involving the right and left border sequences (Koukolíková-Nicola *et al.*, 1985). With the exception of the border sequences, none of the T-DNA functions appears essential for the transfer mechanism (Bevan and Chilton, 1982).

Organisation of the T-DNA in the plant genome can differ for the different classes of Ti plasmid. Two distinct types of T-DNA (a left T-DNA region, TL, containing the core sequence, and a right T-DNA region, TR), which are not continuous in the Ti plasmid, may insert independently into the nuclear DNA of certain octopine tumour cells: all octopine tumour lines contain TL-DNA, whereas TR-DNA may additionally be present. The number of copies of both TL- and TR-DNA may vary. A more simple arrangement of T-DNA is found in nopaline tumours: a unique continuous segment of T-DNA is found (colinear with Ti plasmid DNA), although fusion fragments deriving either from tandem copies or circular forms of T-DNA may additionally be present (Lemmers *et al.*, 1980). T-DNA from octopine type or nopaline-type Ti plasmids can integrate at multiple sites in the plant genome and can cause insertion mutations.

Transformation of plant cells with T-DNA effects hormonal imbalances resulting in the development of cells with abnormal growth and differentiation properties. Such cells can proliferate on phytohormone-free medium (which does not support the growth of normal cells) and remain undifferentiated calli. Regeneration of normal plants from crown gall tumour cells is not normally possible. However, spontaneous deletion of that part of the T-DNA controlling tumour functions allows the regeneration of healthy plants, that stably maintain and express remaining T-DNA sequences (De Greve *et al.*, 1982). Importantly plants containing such modified T-DNA are fully susceptible to further infection with agrobacterial strains carrying Ti plasmids.

(b) Development of vectors using Ti plasmids

Agrobacterium and its Ti plasmids provide an effective means of transferring DNA to plants. Certain features of Ti plasmids make them particularly attractive as gene vectors for dicotyledonous plants. Such features include:

(i) The T-DNA (which is transferred to the plant cell) integrates into the plant genome and is stably transmitted through the divisions of mitosis and meiosis.
(ii) Genes (for example, the nopaline synthase (*nos*) gene) encoded by the T-DNA possess promoters that function in plant cells.
(iii) Foreign DNA inserted into the T-region is cotransferred and integrated into the plant genome as part of the T-DNA.
(iv) T-DNA can be transferred to many dicotyledonous plants, due to the broad host range of *Agrobacterium*.

The large size (commonly around 200 kb) and tumour-inducing ability of wild-type Ti plasmids make them difficult to manipulate experimentally and hence limit their practical use in genetic engineering. Provision of effective experimental vectors thus depends upon suitable modification of such properties. Various strategies have been adopted towards this goal:

1. Intermediate vector strategy. The large size and large number of restriction sites on Ti plasmids make it unlikely that an adequate number of unique restriction sites will be available within the T-DNA for cloning purposes. Direct insertion of foreign genes into the Ti plasmid is not, therefore, normally possible. Foreign genes can, however, be inserted into Ti plasmids through the agency of **intermediate vectors** (Matzke and Chilton, 1981). Such vectors contain a cloned subfragment of the T region. Foreign DNA inserted at one of the unique restriction sites in this T-DNA subfragment can be cloned in *E. coli.* Recombinant vectors, comprising a broad host range IncP-1 plasmid such as pRK290 (a derivative of RK2) and the engineered T-DNA (carrying a selectable marker), may be introduced into *Agrobacterium,* where pRK290 is stably maintained. Homologous recombination between the pRK290 derivative (intermediate vector) and the resident Ti plasmid effects insertion of engineered T-DNA into the Ti plasmid. Depending upon whether single or double cross-overs occur, recombination will result either in integration of the modified pRK290 (producing a cointegrate) or exchange of the wild-type region of the resident plasmid for the counterpart region on the intermediate vector respectively. Selection for the recombinant Ti plasmid involves introducing a plasmid that is incompatible with the intermediate vector (Figure 8.3). An alternative and simpler procedure involves the use of narrow host range ColE1-type plasmids, which can be mobilised into but are not maintained in *A. tumefaciens*, as intermediate vectors (Comai *et al.*, 1983a). Foreign DNA, contained within T-DNA that has been cloned in such a plasmid, can be inserted into the Ti plasmid by recombinational rescue. Homologous recombination involving a double cross-over event between modified T-DNA sequences of the recombinant ColE1-type vector and wild-type sequences of the Ti plasmid rescues the foreign DNA (Figure 8.3) Where a single cross-over occurs during recombination, a cointegrate of the two replicons is generated. Single and double cross-over events can be distinguished by scoring for the presence or absence of a marker on the vector portion of the incoming plasmid (in addition to a marker carried within the homologous region). By using such intermediate vector strategies foreign genes can be inserted at any specific site in the T-DNA of the Ti plasmid.

2. Mini-Ti strategy (binary vector strategy). Intermediate vector strategies for inserting foreign genes into the T-DNA region rely upon recombination between the vector and resident Ti plasmid. In order to simplify procedures, mini-Ti plasmids that enable genes to be transferred to plant cells without recourse to such recombination have been constructed. Such mini-Ti plasmids contain only T-DNA (in particular the 25 bp border sequences) of wild-type Ti plasmids. Shuttle vectors, capable of replicating in *E. coli* and *Agrobacterium*, have been constructed by integrating a mini-Ti replicon with a broad host range plasmid, such as pRK290 (Figure 8.4). In the presence of a 'helper' plasmid (for example pAL4404), which contains the *vir* genes, T-DNA can be transferred from the shuttle vector to plant cells (De Framond *et al.*, 1983). Replication of mini-Ti shuttle vectors in *E. coli* permits initial manipulations to be performed in this host. Resultant recombinant vectors can then be introduced into strains of *Agrobacterium* carrying *vir* genes, and engineered T-DNA can thence be inserted into the plant genome. Thus by using such 'helper' strategy, mini-Ti shuttle vectors can be employed directly to introduce foreign genes into plant cells. Although such binary vectors are smaller than wild-type Ti plasmids, they are still too large (around 70 kb) for easy handling in cloning experiments (commonly used plasmid vectors are between 3 and 15 kb (see section 3.5)). Smaller mini Ti-shuttle vectors can be obtained by deleting appropriate sequences from the T-DNA.

3. Deletion of tumour-controlling functions. An important requirement

Figure 8.3: Scheme for intermediate vector strategy to insert DNA sequences into the Ti plasmid of *Agrobacterium tumefaciens*. Method (i): Foreign DNA inserted at a unique site in a T-DNA fragment is cloned in *E. coli*. The T-DNA carrying the foreign DNA is then cloned in a broad host range plasmid, e.g. pRK290, that is able to replicate in *A. tumefaciens*. The recombinant vector is transferred to *Agrobacterium*. Recombinants between this plasmid and the resident Ti plasmid are selected in the presence of another plasmid that is incompatible with pRK290 and that may subsequently be maintained. Method (ii): Foreign DNA contained within T-DNA is cloned in *E. coli* on a ColE1-type vector. The recombinant vector is transferred to *A. tumefaciens* where the vector is not maintained. Foreign DNA can be rescued by homologous recombination between T-DNA sequences on the vector and wild-type sequences of the resident Ti plasmid. Method (ii) provides a simpler approach in which a single mating transfers the original vector plasmid to *A. tumefaciens* (adapted from Comai *et al.*, 1983a)

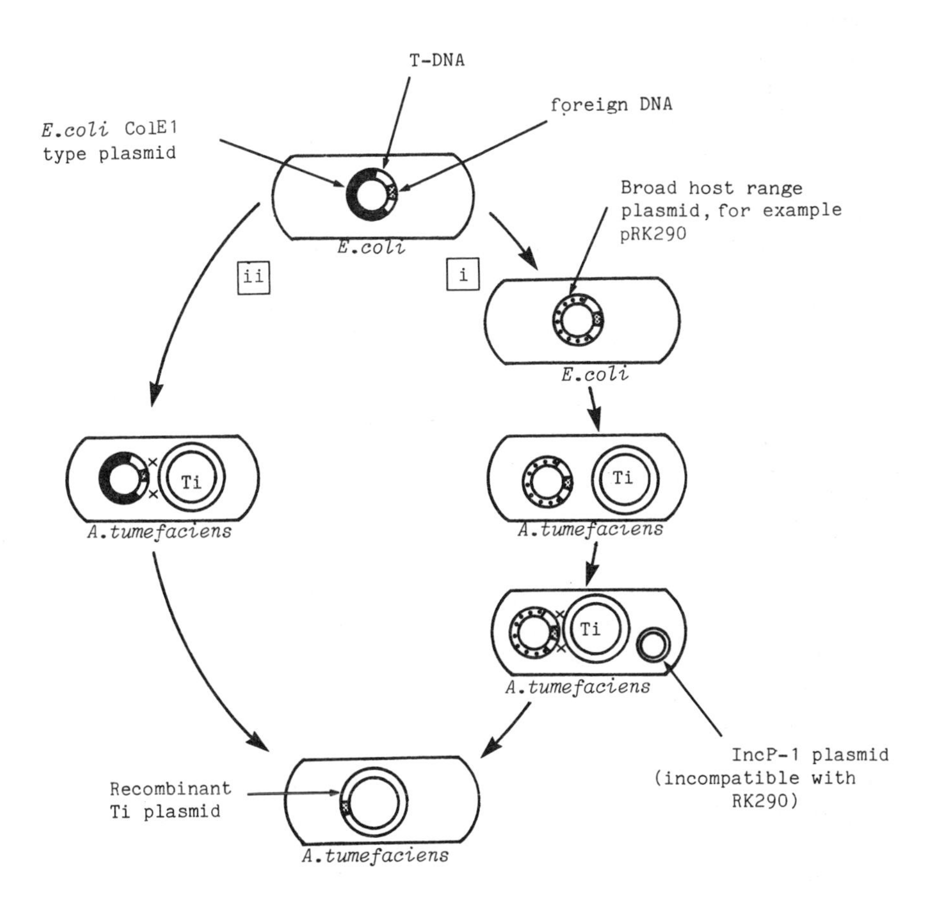

Figure 8.4: Formation of mini Ti/pRK. (a) Map of T-DNA region of nopaline-type Ti plasmid, pTiT37 showing *Eco*RI and *Kpn*I restriction endonuclease cleavage sites. Numbers refer to fragments produced following digestion with the appropriate enzyme. (b) Formation of mini Ti/pRK by cointegration. Mini Ti::*Kpn*11 is a ColE1 replicon containing the entire T-DNA of pTiT37 (including the 25 bp repeats) plus the next continguous *Kpn*I fragment (*Kpn*I fragment 11) on the right side of T-DNA. Mini Ti::*Kpn*11 contains all of the *Eco*RI fragment 1. pRK290::*Eco*RI-1 consists of pTiT37 *Eco*RI fragment 1 cloned into pRK290. The bacterial host was *E. coli* DS989, which carries a thermosensitive DNA polymerase I. At the restrictive temperature, ColE1 plasmids, such as mini Ti::*Kpn*11, are lost (since replication of such plasmids requires PolI activity), whereas IncP-1 group plasmids such as pRK290::*Eco*RI-1 are stable (since replication of such plasmids is not dependent upon PolI). Selection for Km^r and Ap^r markers on mini Ti::*Kpn*11 allows isolation of cointegrate plasmid mini Ti/pRK. Cointegrate formation occurs through homologous recombination involving *Eco*RI-1 fragments of mini Ti::*Kpn*11 and pRK290::*Eco*RI-1. Mini Ti/pRK can transfer its T-DNA to plant cells when helped by pAL4404, a replicon containing the *vir* region. Ap^r, resistance to ampicillin: Km^r, resistance to kanamycin; Tc^r, resistance to tetracycline; ●, ColE1 replication origin; black shaded area, *Eco*RI-1 fragment (from De Framond *et al.*, 1983)

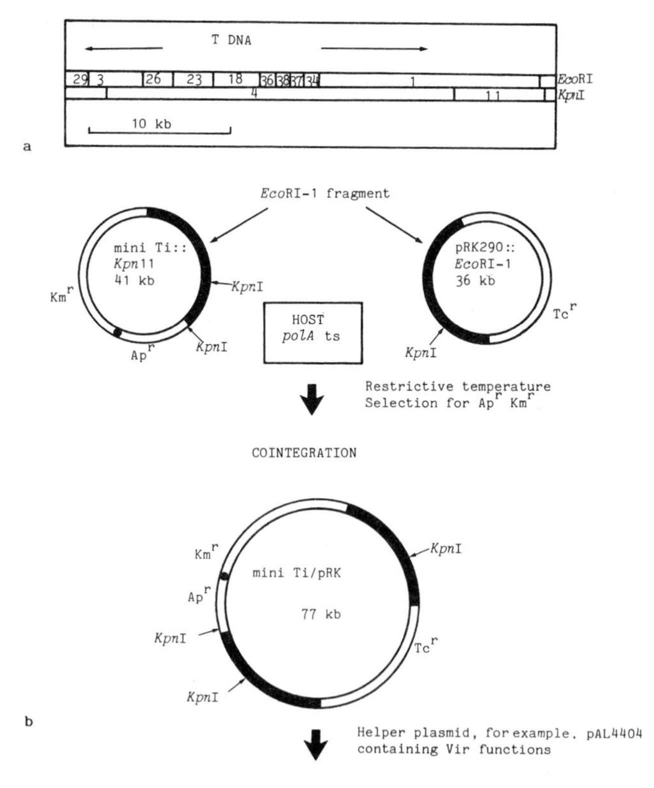

for the use of Ti plasmids as gene vectors is the regeneration of normal plants that stably maintain and express the transferred DNA. It is therefore essential that plant cells containing T-DNA redifferentiate into whole plants. This has been accomplished by using T-DNA possessing insertion mutations at specific sites. For example, by inserting foreign genes at the 'rooty' locus of T-DNA, root proliferation occurs and redifferentiation of transformed plant cells is not blocked. Furthermore, the regenerated plant contains copies of the foreign genes in all its cells (Barton *et al.*, 1983). An alternative approach, and one which conveniently reduces the size of the vector, is to delete the T-DNA sequences responsible for tumour formation (Leemans *et al.*, 1981). Such disarmed T-DNA can be transferred to plant cells and normal fertile plants can be regenerated.

4. Some cointegrate vectors. Chimeric genes that function as dominant selectable markers (see below, 6) have been used in the design of cointegrate vectors for transforming plant cells. The intermediate vector plasmid, pMON128 (a derivative of pMON120 containing a chimeric kanamycin-resistance gene), has been used to insert foreign DNA into the T-DNA of the octopine-type Ti plasmid pTiB6S3 by cointegration (Horsch *et al.*, 1984). pMON128 contains: a segment of pBR322 DNA (for replication in *E. coli*); an intact nopaline synthase (*nos*) gene; a functional nopaline T-DNA right border; a Tn7 segment with the spectinomycin-streptomycin (*spc/str*) resistance determinant (for selection of cointegrates in *A. tumefaciens*); a T-DNA region. LIH (*L*eft *I*nside *H*omology), homologous with the resident Ti plasmid (enabling recombination between pMON128 and TiB6S3); the chimeric kanamycin resistance gene (nopaline synthase-neomycin phosphotransferase) (to confer kanamycin resistance on transformed plant cells); and unique restriction sites for inserting foreign genes. Recombination of pMON128 with the Ti plasmid (via a single cross-over in the homologous LIH regions) generates the cointegrate pTiB6S3::pMON128, in which the nopaline T-DNA right border is positioned between the chimeric antibiotic resistance gene and tumour genes (Figure 8.5). This vector system is designated an SEV (*S*plit *E*nd *V*ector) system, because T-DNA borders are present on separate plasmids prior to recombination. T-DNA may be transferred into plant cells by using either the nopaline T-DNA right border or the octopine TL-DNA right border on the cointegrate. Utilisation of the nopaline T-DNA right border during infection effects transfer of a truncated T-DNA segment containing the kanamycin resistance gene but not the phytohormone biosynthetic tumour genes to plant cells. Such transformed cells can be regenerated into intact plants. However, where the octopine TL-DNA right border is utilised, transformants contain both the phytohormone biosynthetic and the kanamycin-resistance genes and cannot be regenerated into normal plants. Since the octopine right border is used preferentially, only about 10% of

Figure 8.5: Formation of SEV systems. (a) The pTiB6S3::pMON128 cointegrate. Recombination via a single cross-over in the homologous LIH regions of the Ti and pMON plasmids results in production o the cointegrate pTiB6S3::pMON128 plasmid. The nopaline T-DNA right border is located between the chimeric kanamycin resistance (nopaline synthase-neomycin phosphotransferase) gene and tumour genes. Utilisation of the nopaline T-DNA right border results in transfer of a truncated avirulent T-DNA segment to plants. Utilisation of the octopine TL-DNA right border results in transfer of the entire hybrid T-DNA region, including tumour genes. (b) The pTiB6S3-SE::pMON200 cointegrate. Recombination via a single cross-over in the homologous LIH regions of the Ti and pMON plasmids results in production of the cointegrate pTiB6S3-SE::pMON200 plasmid. A short avirulent SEV T-DNA region containing a chimeric kanamycin resistance (nopaline synthase-neomycin phosphotransferase) gene and nopaline synthase gene is transferred to plants by utilising the nopaline T-DNA right border. *ocs*, Octopine synthase; *nos*, nopaline synthase; pBR, segment of pBR322; *spc/str*^{r}, determinant for spectinomycin-streptomycin resistance; *tms*, *tmr*, tumour genes; Km^r, kanamycin resistance; ►, right border nopaline T-DNA; ◁,▷ left and right borders of octopine TL-DNA (adapted from Horsch *et al.*, 1984; Fraley *et al.*, 1985)

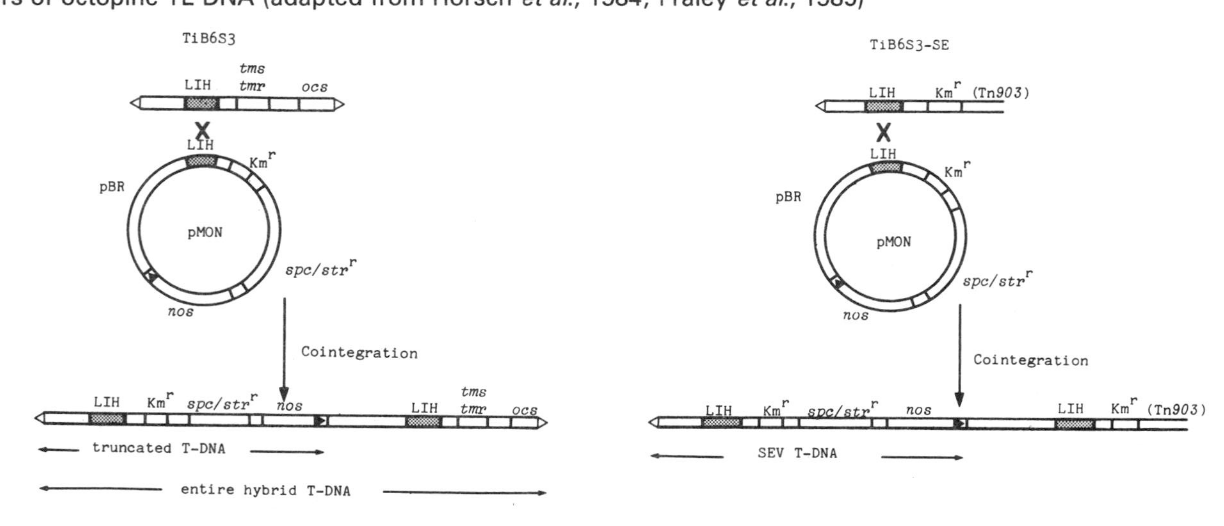

transformed calli carry truncated avirulent T-DNA. For this reason the pTiB6S3::pMON128 cointegrate system is relatively inefficient for transforming plant cells.

In order to improve efficiency an avirulent SEV system has been designed (Fraley *et al.*, 1985). Using this disarmed system all transformed colonies can potentially be regenerated into intact plants. This SEV system comprises a derivative of the octopine-type plasmid pTiB6S3 and an intermediate vector, such as pMON120 or pMON200. TL-DNA oncogenic functions, the TL right border sequence and all of TR-DNA were deleted from the Ti plasmid and replaced with the kanamycin-resistance marker from Tn*903* (*601*). The resultant avirulent plasmid, pTiB6S3-SE, contained the TL left border and a sequence of homologous DNA (LIH), permitting recombination with the intermediate vector. The intermediate vector pMON200 is a derivative of pMON120. pMON200 contains a modified chimeric kanamycin-resistance gene, a polylinker region with a number of unique restriction sites (facilitating insertion of foreign genes into the vector), an intact *nos* gene and a nopaline T-DNA right border sequence. Cointegration between pMON200 and pTiB6S3-SE at their homologous LIH regions results in the formation of a selectable avirulent T-DNA (Figure 8.5). The nopaline T-DNA right border sequence is utilised for the integration of SEV T-DNA into plant DNA.

Another cointegrate vector system involves a derivative of the nopaline-type plasmid, pTiT37. Oncogenic functions of this plasmid have been deleted and replaced with pBR322 sequences (Van Haute *et al.*, 1983). DNA fragments cloned in a pBR-type vector can be introduced into this pTiT37 derivative, pGV3850, by a single cross-over event. Utilisation of the resultant cointegrate to introduce DNA into the plant cells leads to the transfer of a direct repeat of pBR322 sequences. However, it is not clear whether such repeated sequences will enhance instability of the foreign DNA in plant DNA.

5. Some binary vectors. Since none of the functions of T-DNA (with the exception of the 25 bp border signals) appears essential for transfer, they may be eliminated and replaced with a desirable selectable marker, for example an antibiotic-resistance gene (see 6). By exploiting such a modified (avirulent) T region in conjunction with the mini-Ti strategy, small versatile binary vectors, carrying markers that can be selected for in transformed plants, have been constructed for plant engineering. One such vector, pEND4K (Klee *et al.*, 1985), contains a chimeric gene (nopaline synthase-neomycin phosphotransferase) conferring kanamycin resistance on transformed plant cells (Figure 8.6). The vector also contains a high copy ColE1 replication origin (facilitating plasmid isolation from *E. coli*), in addition to the replication and transfer origins of the broad host range plasmid, RK2. The α complementation sequence of pUC series plasmids is

Figure 8.6: Map of the binary vector pEND4K. The vector contains sequences from pUC19Cm, including the ColE1 replication origin and the alpha peptide of β-galactosidase with unique *Kpn*I, *Bam*HI, *Xba*I and *Sal*I sites, a chimeric nopaline synthase-NPTII (*nos-npt*II) gene, a *Bgl*II fragment of the cosmid vector pHC79 with the phage λ *cos* site, sequences from the broad host range plasmid pTJS75Km S (including an origin of replication and an origin of transfer), and the left (LB) and right (RB) border sequences of pTiA6 TL-DNA (adapted from Klee *et al.*, 1985)

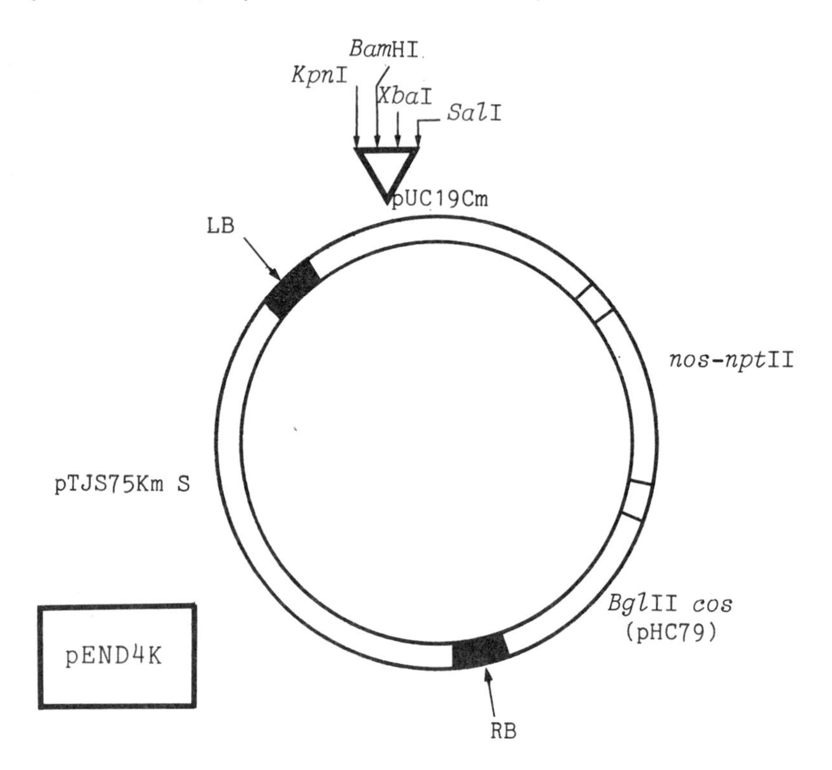

included. This permits cloning into a number of unique restriction sites and direct screening for insertions on agar plates containing Xgal (see section 3.5.2.c). Phage λ *cos* sites have also been incorporated to permit *in vitro* packaging of large (in excess of 30 kb) DNA inserts (see section 3.5.2.g). The right and left border sequences of the octopine-type plasmid pTiA6 TL-DNA, together with a source of *vir* genes, are utilised to transfer and integrate the foreign DNA into plant DNA. Other binary vectors with properties similar to pEND4K have been constructed (see Klee *et al.*, 1985). All these vectors transform plant cells efficiently and should prove useful in the shotgun cloning of plant genes. In conjunction with recombination-deficient *A. tumefaciens* hosts, the cosmid vectors should find application in the construction of plant gene banks. (Recombination-deficient mutants should prove beneficial, since repetitive plant DNA may be unstable in wild-type *A. tumefaciens* hosts.)

6. Expression of cloned genes. It is important that the foreign genes transferred to the plant cell are efficiently expressed. This can be achieved by using chimeric genes comprising promoter sequences that are known to function in plant cells and coding sequences from the foreign genes. Promoters derived from genes (for example the nopaline synthase (*nos*) gene) of wild-type Ti plasmids may be suitable for this purpose. Such promoters are functional in plant cells and the genes are expressed constitutively in all tissues of plants regenerated from T-DNA transformed cells. A hybrid gene containing the 5′ untranslated region of *nos*, bacterial coding sequences for neomycin phosphotransferase II (from Tn5) and the 3′ end of the *nos* gene with termination and polyadenylation signals has been transferred to and expressed in tobacco cells (Bevan *et al.*, 1983; Herrera-Estrella *et al.*, 1983). The consequent resistance of such transformed plant cells to kanamycin and G418 indicates the suitability of this gene as a dominant selectable marker for plant vectors. More efficient expression of foreign genes might be obtained by using stronger promoters such as those of the cauliflower mosaic virus (CaMV) genome (see section 8.1.3.a) in the construction of hybrid genes.

For many applications expression of foreign genes must be regulated and restricted to specific types of plant cell. This might be effected through the use of regulatory signals derived from organ-specific plant genes. For example, a chimeric gene comprising regulatory (5′ flanking) sequences of the nuclear gene for the small subunit ribulose 1,5-bisphosphate carboxylase of pea, the coding region of the bacterial gene for chloramphenicol acetyltransferase (*cat*) and the 3′ end of the *nos* gene has been constructed. Introduction of this chimeric gene into tobacco cells using a Ti plasmid vector resulted in light-inducible expression of *cat* in chloroplast-containing tissue (Herrera-Estrella *et al.*, 1984).

Foreign proteins might be encoded in the nucleus and transferred to a subcellular organelle, such as the chloroplast, by linking them to specific **transit peptides**. For example, the transit sequence of the gene for the small subunit of ribulose bisphosphate carboxylase has been joined to the gene (*npt*II) for neomycin phosphotransferase II and introduced into tobacco cells using a Ti vector. The fusion protein (from this chimeric gene) was translocated into chloroplasts and processed in a similar way to the small subunit polypeptide precursor. The transit peptide was removed and functional NPT II protein was found in the chloroplast (Van den Broeck *et al.*, 1985).

It should be possible to achieve expression of genes specifically in organelles, such as the chloroplast, by using vectors capable of integrating DNA into the chloroplast genome, in conjunction with chloroplast-specific regulatory sequences. Ti-plasmid vectors may prove useful for this purpose. Indeed De Block and co-workers (1985) have shown that a Ti plasmid vector containing a chimeric marker gene (*nos-cat*) provides a system for introducing genes into the chloroplast. When tobacco plants were transformed using this vector, the *nos-cat* gene could be detected in DNA from purified chloroplasts, and chloramphenicol acetyltransferase activity was associated with the chloroplast fraction. This suggests that the gene is being expressed when present in the chloroplast. By using chloroplast-specific promoters in place of the *nos* promoter (which appears to function in plant nuclei and prokaryotic systems), more effective gene expression and more stable chloroplast transformants may be obtained.

7. Potential use of Ti-based vectors in monocots. The observations of Hooykass-Van Slogteren and colleagues (1984) and of Hernalsteens and co-workers (1984) that Ti plasmids can be transferred to and expressed in certain monocots increases prospects for the use of Ti-based vectors in the engineering of monocotyledonous plants. Two monocot species, *Chlorophytum capense* (Liliaceae) and *Narcissus* cv. 'Paperwhite' (Amaryllidaceae), were infected with virulent strains of *A. tumefaciens* (Hooykass-Van Slogteren *et al.*, 1984). Opines were synthesised in the infected plant tissue, but in the absence of any sustained neoplastic growth. The failure of the Ti plasmid to induce tumorigenesis in these monocots is advantageous in respect of its potential use as a vector, since it would already be disarmed. Furthermore, the demonstration that nopaline and octopine synthase genes are transcribed in monocotyledonous plants suggests that the promoters from these genes function in a monocot genetic background. Such functional promoters could be of use in the development of vectors with selectable markers for monocots. Extension of the host range of T-DNA to other monocots to include the most important crop plants, the cereals, could ultimately lead to effective Ti-based vectors for a wide range of plant species of commercial interest.

8.1.2 Ri plasmids of *Agrobacterium* as vectors for plant engineering

While much attention has been focused on *Agrobacterium tumefaciens* and Ti plasmids, the Ri (root-inducing) plasmids of *Agrobacterium rhizogenes* may also provide effective vectors for plant engineering. *Agrobacterium rhizogenes* is the causative agent of hairy root disease, which results in the extensive formation of adventitious roots at or near the site of infection. The host range of *A. rhizogenes*, which includes several species of dicotyledonous plants, is apparently more restricted than that of the closely related organism, *A. tumefaciens*. Infection can be established in the laboratory by inoculation of wounded plant tissues with *A. rhizogenes* or by smearing the organism on sections of tissues. Susceptible plants generally respond with proliferation of roots and sometimes a tumour-like outgrowth (Spanò *et al.*, 1985).

The determinants for induction of hairy roots are encoded on large (commonly > 150 kb) Ri plasmids. These plasmids are classified according to the type of opine, for example agropine or mannopine, synthesised. Ri plasmids are conjugative and can confer the ability to produce hairy root symptoms on nonpathogenic recipient strains. Induction of hairy roots involves transfer of a specific segment (T-DNA) of the Ri plasmid to plant root cells, where it is incorporated stably into the genome (Bevan and Chilton, 1982; Chilton *et al.*, 1982). For agropine-type plasmids two T-DNA regions (TL and TR) can be independently transferred to plants. Both TL and TR promote root formation in transformed tissue. Genes in TL are responsible for the extreme hairy phenotype (White *et al.*, 1985). For mannopine-type plasmids a single T-DNA segment appears to be transferred. It is technically easier to regenerate whole plants from hairy roots than from *A. tumefaciens* transformed tissues (where it is necessary to alter/delete tumour genes from the T-DNA). Vectors based on Ri plasmids may thus provide attractive alternatives to Ti-based vectors for *in vitro* gene manipulation.

8.1.3 Plant DNA viruses as gene vectors

There are two main groups of plant DNA virus, the caulimoviruses and the geminiviruses (see Sheperd, 1979; Goodman, 1981). Caulimoviruses possess a double-stranded DNA genome of about 8000 base pairs, whereas geminiviruses contain a single-stranded, probably bipartite, genome of about 6000 bases. These viruses induce mosaic-mottle types of diseases in plants. The host range of the caulimovirus group includes members of the Amaranthaceae, Caryophyllaceae, Compositae, Cruciferae, Rosaceae and Solanaceae. However, individual viruses infect only a few closely related plants. On the other hand, members of the geminivirus group can often infect plants from a range of families, including monocots.

(a) Caulimoviruses

1. Properties of cauliflower mosaic virus (CaMV). CaMV, the type virus of the caulimovirus group, is an icosahedral double-stranded DNA virus that infects mainly members of the Cruciferae, although some strains can additionally infect the Solanaceae. The virus is transmitted in nature by aphids, but can also be transmitted mechanically by rubbing an inoculum on to plant leaves with an abrasive. Upon infection CaMV spreads systemically through the tissues of the plant, but is not seed-transmissible. Inclusion bodies, which are found in the cytoplasm of CaMV-infected plant cells, are repositories for virus particles.

The double-stranded circular genome of CaMV contains site-specific single-strand discontinuities (gaps) (see Figure 8.7). Generally three discontinuities are found, two in the (+) strand (these are not authentic gaps since no nucleotides are missing, instead they are regions of sequence overlap) and one in the (−) strand (equivalent to the absence of one or two nucleotides with respect to the complementary strand) (Franck *et al.*, 1980). The function of such discontinuities is not clear. It is proposed that the replication process occurs in both the nucleus and cytoplasm and prob-

Figure 8.7: Map of CaMV DNA. Double-stranded viral DNA (8024 bp) is represented by two outer rings containing the three S1 nuclease sensitive sites ('Gap' 1, 2 and 3). The inner circles indicate the positions of ORFs I to VIII. The coding regions are in all three reading frames. The large intergenic region (IGR) is shown. ORF II, Aphid transmissibility factor; ORF IV, viral coat protein; ORFVI, inclusion body protein. Unique *Bst*EII restriction site is indicated. Arrows indicate direction of ORFs

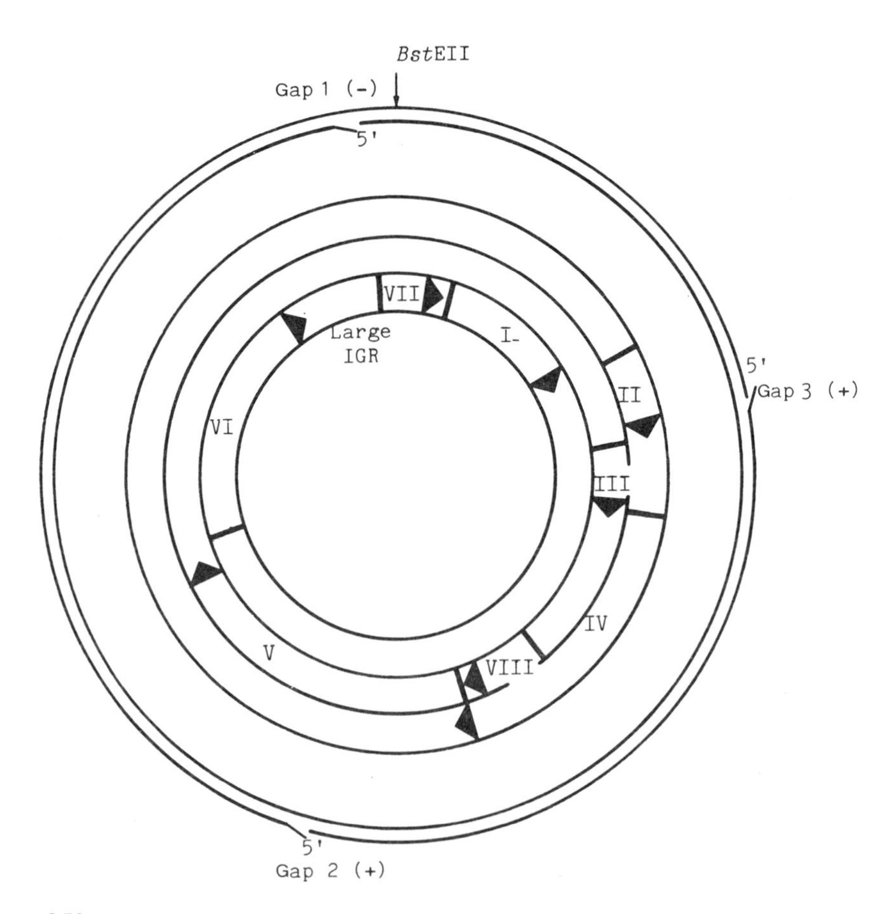

ably involves RNA as template for formation of the 'gapped' DNA by reverse transcriptase (see Pfeiffer and Hohn, 1983; Takatsuji *et al.*, 1986). Nucleotide sequence analysis indicates that there are eight possible open reading frames (ORFs) and two intergenic regions in the CaMV genome (Figure 8.7). With the exception of ORF II, which may encode an aphid transmissibility factor, the other major ORFs (I, III, IV, V and VI) appear essential for virus multiplication and spread in the plant host (Daubert *et al.*, 1983). ORF IV probably encodes viral coat polypeptide and ORF VI the major structural protein of the viral inclusion body. While transcription (which appears to be asymmetric with only the (−) strand encoding stable RNA transcripts) apparently occurs in the nucleus, there is no evidence for integration of CaMV DNA into the plant genome.

2. Towards the development of a plant gene vector using CaMV. Restriction maps of various isolates indicate a number of unique restriction sites that can be used for cloning purposes in the CaMV genome (Figure 8.8). When CaMV DNA is cloned on pBR322 in *E. coli* the single-stranded discontinuities, which are found in the native viral genome, are eliminated. However, viral DNA, excised from the recombinant plasmid by restriction at the cloning site, is fully infectious for plants. Furthermore, the progeny viral DNA molecules from such infections regain the single-strand discontinuities. This indicates that the single-strand breaks are not required for infectivity. On the other hand hybrid CaMV plasmids are not infectious. A possible explanation for this lack of infectivity is that the increased size of the DNA imposes a packaging constraint. Viral DNA, whether derived from the hybrid plasmids by restriction or directly from the virion, does not need to be circular to infect the plant. Linearisation of CaMV DNA with restriction enzymes does not significantly reduce infectivity as long as cohesive ends are present. Infectivity is related to the length of the cohesive single-stranded ends: enzymes that generate long (4 or 5 nucleotides) ends, whether 5′- or 3′-overlapping, produce linear DNA that is more infectious than DNA linearised by enzymes creating short (2 nucleotides) ends (Howell *et al.*, 1980; Lebeurier *et al.*, 1980). Recircularisation of the DNA can occur via ligation *in vivo.* The CaMV genome can thus be manipulated *in vitro* (as a plasmid) without affecting infectivity.

An important requirement for the use of CaMV as a gene vector is the elimination of pathogenic functions and the creation of space for insertion of foreign genes. In order to determine regions of CaMV DNA that are dispensable for virus function, linkers containing appropriate restriction sites have been inserted at a number of sites in the genome (Howell *et al.*, 1982; Daubert *et al.*, 1983; Dixon *et al.*, 1983). Whereas most of the insertion mutations destroyed viral infectivity, certain insertions in the large intergenic region, in ORF II, in ORF VII, in the amino-distal portion of ORF VI and close to the 3′ end of ORF IV, retained infectivity. The

Figure 8.8: Restriction endonuclease map of CaMV (Cabb B-S) DNA. Rare restriction sites and Gaps 1 (△) 2 and 3 (▲) are indicated

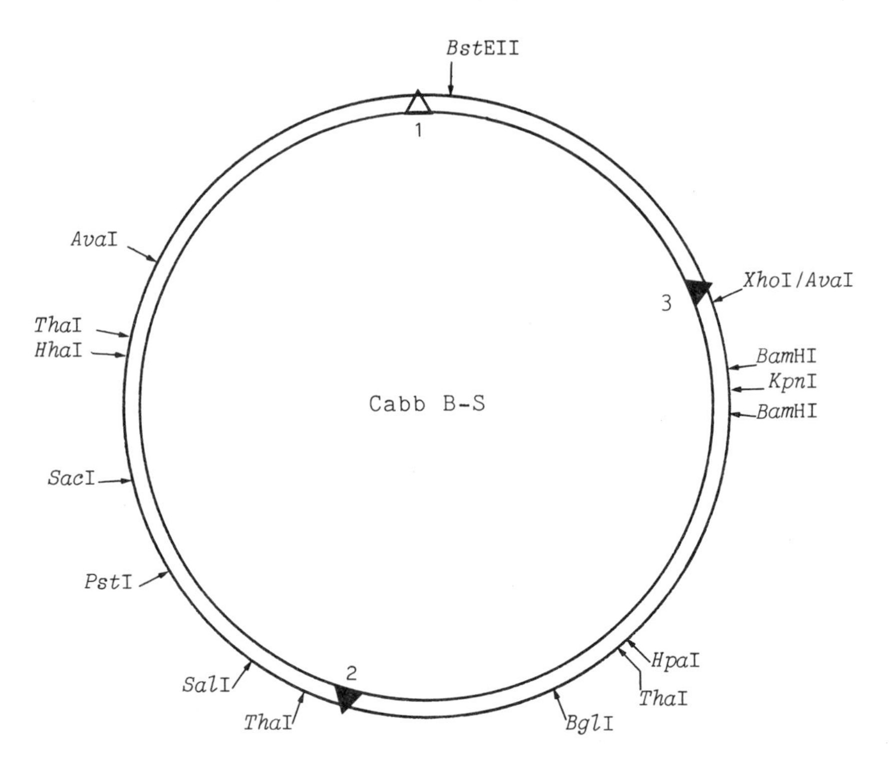

Note: the various CaMV isolates exhibit certain differences in the number and location of cleavage sites for some restriction enzymes; there exist variant and constant sites in the genome (see Hohn *et al.*, 1980)

amino-distal insertions in region VI reduced severity of viral symptoms, and insertion near the end of ORF IV retarded development of symptoms in plants. Insertion in ORF II destroyed aphid transmissibility.

The nonessential nature of ORF II has been exploited to clone the bacterial gene (*dhfr*) for dihydrofolate reductase (which confers resistance to methotrexate) in turnip plants (Brisson *et al.*, 1984). ORF II was deleted and replaced by the *dhfr* gene. Infection of turnip plants with the recombinant CaMV DNA resulted in expression of the *dhfr* gene. Furthermore the gene could be isolated from virus particles after two or three infective cycles, indicating that *dhfr* sequences were retained in the viral genome. Expression and maintenance of the *dhfr* gene in plants provides a convenient means of assessing the usefulness of CaMV as a vector, since drug resistance can be employed as a marker. Deletion of ORF II could provide space of up to 470 bp for foreign DNA. Extra space might be available by combining the ORF II deletion with deletions in ORF VII. Alternatively a 'helper' virus system might be employed to create more space for foreign DNA. In this system plants are coinfected with a non-infective viral genome (produced by deletion of a substantial part of the CaMV genome and replacement with foreign DNA) and helper viral DNA. It is presumed that the loss of functions in one genome could be complemented by the helper DNA. However, although it has been demonstrated that two noninfective viral DNAs, each bearing lesions in different regions of the genome, can co-operate to produce infection in plant cells, the rescue of dysfunctions occurs by recombination and not by complementation. Thereby deletions/insertions are eliminated and wild-type viral genomes are recovered (Howell *et al.*, 1982). Adoption of a helper virus approach for the introduction of foreign DNA into plant cells would thus demand suppression of such recombination processes.

Since the CaMV genome does not apparently integrate into the plant genome (and is not transmitted vertically), use of CaMV DNA as a vehicle may depend upon its encapsidation in the virus particle in order to permit movement through the vasculature of the plant. Such encapsidation could impose severe limits on the amount of DNA that can be inserted into the genome. On the other hand propagation of recombinant vectors in virion form might prove advantageous in cases where amplification and/or high rates of expression of inserted DNA are desired. Alternatively it may be possible to effect integration of CaMV DNA into the host genome through the agency of homologous recombination. Incorporation of an appropriate natural component of the plant genome into CaMV DNA (or the border sequences of T-DNA of *Agrobacterium* plasmids) may permit insertion in the plant DNA. Such inserts may be stably inherited through the seed.

CaMV thus possesses a number of features that make it potentially useful as a gene vector (Hohn *et al.*, 1982):

(i) CaMV can be introduced directly into plants.
(ii) The virus establishes systemic infection.
(iii) The CaMV genome can be manipulated *in vitro* and remain effective in transforming plant cells.
(iv) CaMV DNA possesses promoters that function efficiently in the plant cell.

Although the host range of CaMV is largely restricted to the Cruciferae, CaMV promoters can function in non-host plants (Odell *et al.*, 1985; Shewmaker *et al.*, 1985). However, the fidelity of CaMV DNA replication is apparently relatively low: the frequency of errors is about 1 per 10^3 to 10^4 nucleotides synthesised, based on RNA polymerase/reverse transcriptase activities. Thus foreign genes inserted into the CaMV genome would be likely to accumulate errors rapidly. This, together with the difficulty in eliminating unwanted characters encoded by CaMV DNA, poses problems for its utility as a gene vector. It may, therefore, prove more advantageous to exploit specific assets of CaMV, in particular its efficient gene expression signals, in the construction of hybrid vectors from different sources. CaMV promoters have been shown to function when integrated into the plant genome. CaMV has been introduced into the chromosomes of a number of different plant species (including plants that are not susceptible to the virus) by using the Ti plasmid. Integrated viral promoters directed the synthesis of two CaMV-derived transcripts (19S and 35S). The amounts of the two transcripts varied widely in the different species tested. Generally CaMV hosts showed higher transcript levels (Shewmaker *et al.*, 1985). Variation in the levels of expression of the same gene (directed by a specific promoter) in different plant species may limit the usefulness of a particular promoter in chimeric gene constructions.

(b) Geminiviruses

1. Properties. Geminiviruses are characterised by a twinned virion morphology. The isometric particles are frequently found in pairs, and each paired particle apparently contains a single-stranded covalently closed circular DNA molecule. The host range of members of this group includes major crop plants such as maize, beans, wheat and tomato, as well as tropical plants. Geminiviruses are transmitted in nature by leafhoppers and whiteflies. Only certain members are readily transmitted mechanically. Virus particles accumulate in the nuclei of infected plants and in some cases are limited to the nuclei of phloem parenchyma cells. However, although bean golden mosaic virus (BGMV) is associated with the phloem in intact plants, it can infect leaf mesophyll protoplasts in culture, indicating that the virus can enter other cell types.

The geminivirus genome is probably bipartite. Geminiviruses appear to consist of two populations of paired particles that differ only in the nucleotide sequence of their DNA molecules. The two DNAs (DNA 1 and DNA 2) from populations of cassava latent virus (CLV) have been cloned, restricted at the cloning site and transmitted mechanically to a plant host *Nicotiana benthamiana.* Infection of the plant host depends upon the presence of both DNA 1 and DNA 2. Infectious geminate virus particles could be isolated from such CLV-infected plants indicating that the CLV genome is bipartite (Stanley, 1983).

Little is known about the molecular events in the viral infective cycle. However, DNA replication and virus assembly probably occur in the nucleus. A double-stranded form of DNA may be a replicative intermediate.

2. Geminiviruses as potential plant gene vectors. One of the attractions for the use of geminiviruses as vectors is a broad host range that includes major crop plants. Furthermore, both double-stranded (presumed replicative form) and single-stranded forms of geminivirus DNA can infect whole plants and plant protoplasts. However, the low yield of geminivirus and the fact that the double-stranded DNA (required for many gene manipulation techniques) must be obtained from infective tissue make these viruses technically difficult to handle. The small capsid size may present problems for packaging modified DNA molecules. In addition, the lack of mechanical transmission and the restricted distribution of some geminiviruses to specific plant tissues are likely to hamper use of these viruses as gene vectors.

8.1.4 Possibilities for using plant RNA viruses and viroids as gene vectors

(a) RNA viruses

1. Genomic RNA. The genome of plant RNA viruses may be divided (multipartite) or undivided (monopartite). The RNA components of a divided genome are usually of smaller size than the single RNA molecule of 'monopartite' viruses and are thus probably more amenable to manipulation *in vitro.* The components of a divided genome can replicate autonomously in plant cells and may be packaged in a single virion or in distinct virions, depending upon the virus (for a review see Lane, 1979). The indispensability of the genomic RNA components for virus multiplication poses problems for the use of such RNAs as vectors. However, in some systems, for example bipartite viruses such as tobacco rattle virus (TRV), one of the RNA components is dispensable to certain aspects of multiplication. The RNA-2 of the TRV genome is dispensable for replication (but not for coat protein production) of the RNA-1 component. Thus RNA-1 (which contains all necessary determinants for its own replication) can replicate and spread, in unencapsidated form, in the absence of RNA-2 (which contains the information for packaging both RNA-1 and RNA-2). Such dispensability of RNA-2 has consequences for the potential use of these RNAs as plant vehicles. First, genes such as coat protein gene might be modified without affecting infectivity. Secondly, replication and spread of the RNA in an unencapsidated form might remove the limits on size of insert (presumably imposed during encapsidation of the viral genome) that could be incorporated into such a viral vector.

2. Satellite RNA. Satellite RNAs that are associated with plant viruses may also be candidates for plant gene vectors. Satellites are small RNA molecules (from about 250 b to 1.5 kb) and are dispensable to their companion viruses. Replication of satellite RNA depends upon functions encoded by the virus. Generally the satellite RNA is encapsidated in the 'helper' virus particle and may alter the pathogenicity of the virus. However, tobacco necrosis virus satellite (S-TNV) encodes a coat protein to encapsidate S-TNV separately from the helper virus particle. Furthermore S-TNV can be introduced into the host plant (as free or encapsidated RNA) in the absence of its helper viral RNA. The satellite can be rescued by helper viral RNA after introduction (Francki, 1985). Like their helper viruses, satellites are usually transmitted horizontally but not vertically. Such properties of S-TNV may make it attractive as a possible plant vehicle.

Use of satellites for gene cloning might be facilitated by generating double-stranded cDNAs of the RNA. The feasibility of this approach will, however, depend upon whether engineered cDNA is infectious for plants.

(b) Viroids

Viroids are small, single-stranded, unencapsidated RNA molecules with pathogenic properties (Diener, 1982; Riesner and Gross, 1985). It is not clear how viroids produce disease symptoms in infected plants. However, some viroids can cause very mild symptoms in certain hosts, indicating that viroid replication is not inextricably coupled to pathogenicity. The precise mechanics of viroid replication remain unclear. Replication presumably relies upon functions provided by the host plant or by endogenous viruses, and probably involves RNA intermediates.

Viroids possess a number of properties that may be worth exploiting in the development of plant vehicles:

(i) Viroids establish systemic infection in plants.
(ii) Viroids are mechanically transmissible.
(iii) Some viroids can be propagated vertically and are seed or pollen transmissible.
(iv) The host range includes a number of exotic plants, particularly those found in the Tropics.

However, use of viroids as plant vectors would necessitate control of the symptoms of viroid infection. Furthermore, manipulation of viroids might be easier if double-stranded cDNA of the viroid RNA could be used. The finding that some viroid cDNA clones are infectious (see Riesner and Gross, 1985) may make their use as vectors a feasible proposition.

8.1.5 Potential gene vectors derived from components of the plant genome

In addition to the use of natural vectors as molecular vehicles for plant engineering, various artificial vectors that incorporate components of the plant genome are being developed.

(a) Nuclear genomic components

Specialised regions of the plant genome might be utilised in vector construction. By employing strategies analogous to those used in the design of yeast vectors (see section 3.5.5), vehicles capable of autonomous replication in plants (cf. YRp vectors) might be obtained through the incorporation of an origin of replication from a plant chromosome into the vehicle. Stabilisation of such autonomously replicating elements might be effected through the use of mitotic stabilising elements (centromeric fragments). In addition integrating vectors (cf. YIp vectors) might be generated by inserting DNA sequences homologous to a resident region of plant genomic DNA into the vector.

(b) Mitochondrial components

Plant mitochondrial DNA and mitochondrial extrachromosomal elements have possibilities for use as plant gene vectors. Several extrachromosomal DNA elements have been found in mitochondria of *Zea mays* (Kemble *et al.*, 1980). Such elements apparently replicate autonomously in the mitochondria and may integrate reversibly into the mitochondrial genome.

(c) Transposable elements

A number of plant transposable elements including the Activator (*Ac*)-Dissociation (*Ds*) system, Suppressor-Mutator (*Spm*) element and Robertson's mutator (*Mu 1*) (Peacock *et al.*, 1983; Fedoroff, 1984) have been described on the basis of an ability to effect mutagenesis. The ability of such elements to insert at various sites in the plant genome may be exploited to introduce foreign DNA into the genome.

One potential vector system involves the *Ac-Ds* system of maize (Peacock *et al.*, 1984; Starlinger *et al.*, 1984). *Ac* is an autonomous transposable element, whereas *Ds* requires a copy of *Ac* elsewhere in the maize genome for transposition. Thus use of *Ds* as a gene vector depends upon the presence of *Ac*. One approach is to employ both *Ac* and *Ds* in constructing a vector. Alternatively recipient cells may harbour *Ac*, with *Ds* providing the vehicle for gene insertion. The *Ds* element contains short inverted repeat sequences that are required for its transposition, whereas much of the interior of the element seems unnecessary. Thus it is possible that any DNA sequence bracketed by the termini of *Ds* may act as a transposable element. The alcohol dehydrogenase gene (*adh*) system shows promise for use as a selectable marker for identifying transformed plant cells, provided that *adh*$^-$ recipients are available. Alcohol dehydrogenase activity is required for survival of maize cells under anaerobic conditions. Anaerobiosis thus provides an effective selection against cells lacking a functional *adh* gene. Foreign DNA might, therefore, be introduced into the plant genome *via* insertion into *DS* linked to the *adh* gene. Such vector systems based on transposable genetic elements may be of general use in plants. Furthermore, transposition mutagenesis might be used to create new genetic variation in plants and to identify and clone genes of agronomic importance.

8.1.6 Some applications of *in vitro* gene manipulations

Various types of gene vector are being developed for use in plant engineering, of which integrating vectors based on Ti plasmids are the most advanced. The transfer of Ti plasmids to certain monocots, as well as to a wide range of dicotyledonous plants, indicates the potential versatility of Ti-based vectors. Vectors derived from geminiviruses, which have a wide host range that includes monocots, or artificial vectors containing, for example, monocot transposable elements, should also provide useful plant vehicles. Bacterial plasmids, such as pBR322, that are capable of integrating randomly into plant chromosomes may provide a further means of transferring genes to plants.

The choice between an integrating and nonintegrating vector will depend upon its potential application. Alterations in the germplasm of the plant may demand the use of an integrating vector, whereas synthesis of large amounts of foreign gene product by plant cells might be best achieved by using a nonintegrating vector that is present in high copy number.

In vitro gene manipulation should assist the development of new plant varieties exhibiting herbicide, frost and disease resistance. Resistance to the herbicide glyphosate has been achieved by introducing the glyphosate-tolerant gene *aroA* of *Salmonella typhimurium* into plant cells, where proper expression occurred. Glyphosate is a broad-spectrum herbicide that interferes with synthesis of aromatic amino acids through an ability to inhibit the activity of 5-enolpyruvylshikimate 3-phosphate synthase (EPSP synthase) (section 7.2.3) in both bacteria and plants. A mutant *aroA* gene that encodes an altered EPSP synthase with increased resistance to glyphosate was isolated from *Salmonella typhimurium* (Comai *et al.*, 1983b). The gene has subsequently been transferred to the genome of turnip and tobacco by using a Ti plasmid of *A. tumefaciens* (Hiatt *et al.*, 1985) and an Ri plasmid of *A. rhizogenes* (Comai *et al.*, 1985), respectively. (The strain of *A. rhizogenes* employed had two T-DNA regions (TL and TR) in its Ri plasmid. The *aroA* construct, comprising coding sequences for the glyphosate-resistant EPSP synthase, a eukaryotic promoter and the polyadenylation signal, was recombined into the TL region.) The transformed plants exhibited significantly increased resistance to glyphosate when sprayed with the herbicide. Since the glyphosate-sensitive plant enzyme functions in the chloroplast whereas the transferred resistant EPSP synthase apparently worked in the cytoplasm, herbicide resistance might be increased by targeting the resistant enzyme to the chloroplast. This might be achieved by linking chloroplast-specific transit sequences (section 8.1.1) to the coding sequence for the resistant EPSP synthase. An alternative approach may be to insert the relevant genes into the chloroplast genome. By incorporating chloroplast-specific promoters plus regions of homology with chloroplast DNA, the gene may be directed into specific sites on the chloroplast genome and expressed. These tactics would be of general applicability in the engineering of herbicide-resistant crops where the herbicide blocks chloroplast-located amino acid biosynthetic functions.

Other herbicides, such as atrazine and diuron, interfere with the binding of plastoquinone to Q_B protein of photosystem II. Q_B protein is encoded by chloroplast DNA. A gene (mutated *psbA*) from the cyanobacterium *Anacystis nidulans* R2, encoding a diuron-resistant Q_B protein, has been cloned (Haselkorn and Golden, 1985). The coding sequence for this mutant Q_B protein might be used to construct a chimeric gene that can be transferred to plants and function in the chloroplast.

Transfer of genes specifying insecticidal toxins to plants should confer insect resistance. In these ways pests might be controlled without the need for expensive and potentially hazardous spraying programmes. The *Bacillus thuringiensis* gene for δ-endotoxin (see section 8.2.1.a) has apparently been transferred into tobacco plants, where toxin has been produced in the leaves (see Yanchinski, 1985). It has yet to be determined whether this toxin can be produced in sufficient quantity to control insects effectively.

Stress conditions, imposed by a variety of environmental factors, can result in crop damage and dramatically reduce crop yields, particularly where suboptimal conditions prevail. Provision of genetically engineered, stress-resistant plants that are capable of tolerating, for example, drought conditions, radical temperature changes and soils of high acidity and high salinity would thus be desirable.

Recombinant DNA technology may also be used to improve the efficiency of symbiotic nitrogen fixation and to extend the capability to plants (such as cereals) that currently do not benefit from the process. Increased nitrogen fixation is likely to be effected through genetic modifications to both the nitrogen-fixing bacterium (see section 8.2.2) and host plant cells. Alteration to or transfer of specific plant genes involved in the symbiosis may result in more effective plant-bacterial associations and may contribute to the creation of new associations. This in turn may enhance the efficiency of nitrogen fixation and improve crop yield, while reducing the demand for chemical fertilisers. A potentially more valuable objective might be the transfer to and expression in plants of the genes for nitrogen fixation (*nif*). However, the proper functioning of prokaryotic *nif* genes and their products in plant cells is likely to present many difficulties and is thus not an immediate prospect. Furthermore, the benefits from nitrogen-fixing plants might be negated by the possible adverse effects on crop yield effected by the high energy requirement of the process of nitrogen fixation. Unless such energy costs can be offset by increased photosynthesis, it may prove more beneficial to improve fertiliser assimilation by plants.

There are several possible approaches towards improving the photosynthetic abilities of plants through genetic engineering. For example, accelerated photosynthetic electron flow might be achieved by exchange of photosystem components between different plants. In addition, appropriate alteration to and/or exchange of genes for enzymes of the Calvin cycle between plant varieties might result in more efficient CO_2 fixation in engineered plants.

Genetic engineering technology can also be applied to improving the nutritional quality of plants. Storage proteins of cereal grains and other seeds that provide a dietary source for humans and animals are deficient in certain essential amino acids. The genes for a number of these proteins (including phaseolin, zein and hordein) have been cloned and well characterised (see Mantell *et al.*, 1985). By engineering such genes using site-directed mutagenesis (section 4.9) it may be possible to improve the amino acid composition of the storage proteins. This might be achieved either through the insertion of additional amino acids or by the replacement of existing ones with those that are nutritionally more beneficial. Alternatively determinants for new and nutritionally superior proteins might be introduced to improve protein content. However, these manipulations may have deleterious effects on the structural stability of the mRNA and protein concerned.

In vitro gene manipulation can also assist plant breeding through the provision of diagnostic aids. The detection of viruses and other pathogens is being made possible by using nucleic acid hybridisation techniques with suitable DNA probes (Owens and Diener, 1985). For example, probes comprising appropriate cloned viral genes can be used for rapid screening of sap extruded from plant tissue for various plant viruses. (Such tests are particularly important where crops are disseminated by cuttings, tubers or bulbs, since viruses are usually transmitted in such tissues rather than in seeds.)

The potential benefits to accrue from plant genetic engineering are likely to extend to areas other than agriculture. For example in the pharmaceutical industry there is scope for the development of novel pharmaceuticals from plants and for increased yields of existing plant products (such as alkaloids, steroids and oils).

Clearly foundations for the manipulation of crop plants using recombinant DNA techniques have been established. The combination of such techniques and existing plant breeding methods will provide an integrated approach to crop improvement. However, effective application of these technologies to yield new cultivars demands a thorough knowledge of plant biochemistry and molecular genetics. More information is still required about the life cycles of the various plant pathogens that are being exploited in vector design and about the molecular biology of agronomically important genes.

8.2 GENETIC MANIPULATION OF MICROORGANISMS OF AGRICULTURAL IMPORTANCE

8.2.1 Use of microorganisms to control plant pests and diseases

Certain plant pests and diseases can be controlled through the agency of microorganisms. This may involve a microorganism that is pathogenic to the pest and/or microbial products with pesticidal activity. The use of such microbial pesticides, rather than chemicals, in the field can enhance crop yield while reducing environmental pollution.

(a) Insect control

Various entomopathogenic microorganisms (bacteria, fungi and viruses) are used commercially to control insect pests that damage crops (Kurstak, 1982). A general feature of these microorganisms is their relatively narrow host range. The use of such agents therefore enables particular pest populations to be reduced, while natural predators and beneficial insects are preserved. However, the specificity afforded by such agents may be too great to permit control of mixed infestations of closely related plant pests. This factor, coupled with the often high costs of large-scale production, has limited the commercial attractiveness of microbial control agents. However, by appropriate genetic manipulations it should be possible to engineer microorganisms such that they show increased virulence for a broader range of insect pests, and to facilitate the commercial production of microbial insecticides. In these ways microbial control agents may have a more effective role in pest management programmes and in turn decrease dependence upon chemical pesticides.

1. Bacteria. Certain *Bacillus* spp. are used or show promise for use in insect control (Table 8.1). Of these *B. thuringiensis* is commercially the most important (see Kurstak, 1982). This species is pathogenic for the larvae of certain insect species, belonging principally to the orders Lepidoptera (butterflies and moths) and Diptera (two-winged flies and gnats), which include some serious crop pests. The major insecticidal

Table 8.1: Insecticidal spectrum of some *Bacillus* species

Bacillus species	Target insect	Mechanism of action
Bacillus thuringiensis var. *kurstaki*, *berliner*, *alesti*, *tolworthi*	Lepidopterous insects	Parasporal crystal toxin, disruption of midgut of larvae, paralysis
israelensis	Mosquito, blackfly	Parasporal crystal toxin, disruption of midgut of larvae, paralysis
Bacillus popilliae *Bacillus lentimorbus*	*Coleoptera* (beetles and weevils) especially Japanese beetle	Inhibits larval development ('milky disease')
Bacillus sphaericus	Mosquito	Toxin mediated, disrupts midgut of larvae
Bacillus moritae	Housefly, seedcorn maggot	Inhibits larval development

metabolite of *B. thuringiensis* is the delta (δ)-endotoxin, which is formed during the sporulation process and deposited in a crystalline form (parasporal body) within the sporangium (Figure 8.9). The crystals and spores are released following lysis of the wall of the sporangium. Upon ingestion by feeding larvae the proteinaceous crystals (protoxins) dissolve in the midgut, where conditions are alkaline. The protoxin is activated by the action of gut-juice proteases. Active δ-endotoxin induces gut paralysis with consequent cessation of insect feeding. Vegetative cells may invade the host, which has been weakened by the toxin, and cause a lethal septicaemia.

The insecticide preparations of *B. thuringiensis* used comprise spores and the protein crystals and are applied as dusts or in liquid form. The crystals alone may afford control, but the bacterium has a secondary role by invading the weakened insect host. Some sero-types of *B. thuringiensis* additionally produce a beta (β)-exotoxin, which has insecticidal activity. However, this compound is active against a wide range of organisms including not only insects but also vertebrates and some microorganisms, and is therefore generally considered unsuitable for commercial use. Thus strains of *B. thuringiensis* that do not produce this exotoxin are selected for insecticide manufacture. In addition certain exoenzymes, such as lecithinases and chitinases of *B. thuringiensis*, have toxic properties and support the pathogenesis. Inclusion of such enzymes in *B. thuringiensis* preparations may thus provide improved insect control.

A disadvantage of the use of the *B. thuringiensis* insecticide for field application is the short-term control afforded. The insecticide is inactivated by prolonged exposure to UV and does not persist or spread extensively. Thus to be effective several frequent applications are required. However, it

Figure 8.9: Parasporal crystal within the insect pathogen *Bacillus thuringiensis*

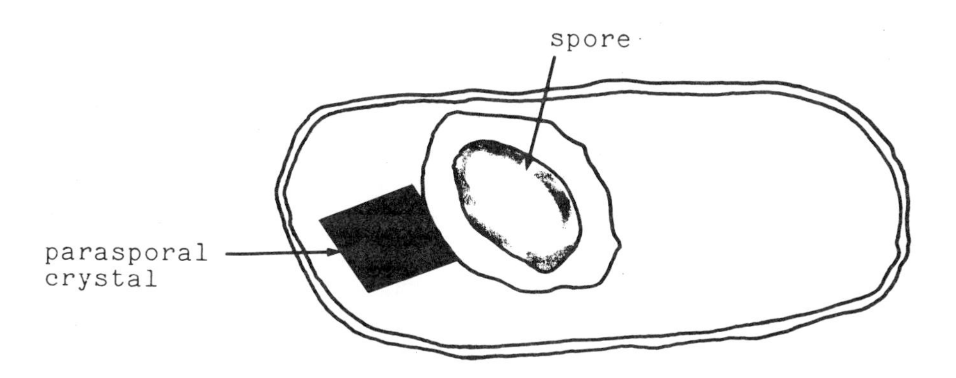

may be possible to enhance persistency (by, for example, increasing UV tolerance) and to improve adherence of the preparations, through specific gene manipulations. The development of strains with increased yield of crystal protein, higher potency or with altered host spectrum would also be beneficial. Production of the δ-endotoxin crystal generally appears to be plasmid-encoded. A number of different conjugative crystal-coding plasmids have been identified (see, for example, González *et al.*, 1981 1982; Dean, 1984; González and Carlton, 1984; Carlton and González, 1985). Strains of *B. cereus* that are capable of crystal synthesis have been constructed by plasmid transfer. Such manipulations might be extended to convert other strains into toxin producers and to improve the pathogenicity of existing producers. Crystal protein genes from *B. thuringiensis* have been cloned in a number of organisms, including *E. coli*, *B. megaterium*, *B. subtilis* and *Pseudomonas fluorescens*, where production of protoxin protein has been achieved (Held *et al.*, 1982; Beardsley, 1984; Ward *et al.*, 1984; Sekar and Carlton, 1985). One proposal is to use such engineered strains of *P. fluorescens* (which is found naturally in association with corn roots) to coat corn seeds, with a view to protecting corn plants from attack by black cutworm. Amplification of the cloned gene could lead to increased yield of crystal protein. Furthermore, novel toxins might be developed by manipulating the protoxin gene by *in vitro* mutagenesis (section 4.9), or by hybrid gene formation (section 4.11) using combinations of protoxin from different varieties.

Another strategy for insect control involves the use of antibiotics with insecticidal activity. *Streptomyces* spp. produce a number of these metabolites, such as the milbemycins (macrolide antibiotics) (Takiguchi *et al.*, 1980). Genetics may be used to increase the yield and insecticidal activity of such antibiotics, while reducing their toxicity to warm-blooded animals.

2. Fungi. Certain entomopathogenic fungi are used commercially for pest control, including *Beauveria bassiana* (for control of Colorado potato beetle) and *Verticillium lecanii* (for control of aphids) (Kurstak, 1982; Lisansky and Hall, 1983). The normal routes of infection of the pathogen are through the gut, the mouthparts and the cuticle. Death of the host may involve toxins or may be due to invasion. Since the fungus does not need to be ingested to cause infection, it can be used to control insects in non-feeding stages, often before much economic damage has been done. However, the spores, which are used as inocula in control programmes, are susceptible to desiccation. Furthermore, spore germination is sensitive to a variety of environmental conditions, including temperature and relative humidity. Such factors limit the effectiveness of these entomopathogenic fungi in the field. Strain improvement programmes aimed at increasing tolerance to climatic conditions should improve the ability of these fungi to survive and initiate infection and thus may enhance their efficacy as control agents. Mutants resistant to chemical pesticides would also be useful for the successful integration of fungal insecticides into pest management programmes.

3. Viruses. Viruses typically used or considered for use to control insect pests are members of the Baculoviridae (see Table 8.2) (Entwistle, 1983; Carter, 1984). In the nuclear polyhedrosis virus (NPV) and granulosis virus (GV) subgroups of this family the virus particles become embedded (occluded) in inclusion bodies, which protect the virions from the environment (both during transmission from host to host and in passage through the foregut of the insect). Such stability in the environment makes occluded baculoviruses particularly attractive as microbial control agents. Generally viral infection occurs via the gut when the insect ingests inclusion bodies. Virions are released following dissolution of inclusion bodies in the midgut. Early larval stages of the host are most susceptible to infection. The host range of the baculovirus group includes insect species of the orders Lepidoptera, Hymenoptera (sawflies, ants, bees and wasps) and Diptera. When used as an insecticide the virus is formulated with agents that protect against UV irradiation and that aid dispersal.

There are various genetic approaches to improving the efficacy and longevity of viral control agents. Field persistence can be increased by selecting strains that are more resistant to UV inactivation (Brassel and Benz, 1979). In addition, strains with increased virulence can be obtained by *in vivo* mutagenesis (Wood *et al.*, 1981). *In vitro* techniques could be used, for example, to incorporate a gene for an insect-specific toxin (such as a paralytic neurotoxin) into the baculovirus genome (Miller *et al.*, 1983). This could enhance virulence and extend the host range of the virus.

Table 8.2: Examples of insect pests controllable with baculoviruses

Target insect	Baculovirus*	Main crops
Colias philodice eurytheme	NPV	Alfalfa
Heliothis spp.	NPV	Cotton, corn, tobacco
Laspeyresia pomonella	GV	Apple, pear
Phthorimaea opercullela	GV	Potatoes
Pieris rapae	GV	Brassicas
Spodoptera spp.	NPV	Cotton, bananas

* GV, granulosis virus; NPV, nuclear polyhedrosis virus

(b) Weed control

Weeds may be controlled by introducing appropriate microorganisms that are plant pathogens (Hasan, 1980). Immunity of the crop plant to the pathogen must be assured. Rusts have been used in appropriate cases, for example *Uromyces rumicis* to control curly dock (*Rumex crispus*), *Phragmidium violaceum* to control wild blackberries (*Rubus constrictus* and *Rubus ulmifolius*), and *Puccinia chondrillina* to control skeleton weed (*Chondrilla juncea*). However, natural pathogens are often not sufficiently virulent to control the weeds effectively. A strategy for enhancing virulence is to identify and amplify the genes for pathogenicity. Furthermore the pathogens might be prevented from spreading into the general environment by, for example, incorporating a mutation(s) to increase cold sensitivity, such that the pathogen could not survive a winter. An advantage of the use of such engineered strains would be their specificity for particular weed species. Moreover toxic products would not accumulate in the soil, as often occurs following application of chemical herbicides.

Another approach to weed control is to utilise specific secondary metabolites with herbicidal activity, such as the herbicidins (nucleoside antibiotics) and herbimycins (benzoquinonoid ansamycins) produced by *Streptomyces* spp. (Arai *et al.*, 1976; Iwai *et al.*, 1980; Yoshikawa *et al.*, 1983). The genetic manipulation of producer strains with a view to increasing the yield and potency of the herbicides and to developing novel herbicides are current objectives.

(c) Disease control

One strategy for reducing crop losses caused by plant pathogens involves the use of microbial control agents (see Schroth and Hancock, 1981). Control may be effected by introducing a microorganism that is antagonistic to, competes with, or parasitises the disease-causing organism. A further control measure involves infection of the host with attenuated pathogenic strains in order to protect against subsequent infection by virulent strains (a phenomenon referred to as induced resistance). Examples of these various approaches to disease control are given below.

1. Antagonism-competition

(i) Crown gall disease. *Agrobacterium radiobacter* (strain K84) is used to control crown gall disease. This bacterium produces a plasmid-encoded bacteriocin (agrocin) that is effective against *A. tumefaciens*, the causative agent of the disease (Kerr, 1980). The efficacy of *A. radiobacter* as a control agent might be improved by increasing the yield of agrocin and enhancing the competitive ability of the producer strain.

(ii) Plant growth-promoting rhizobacteria. Specific rhizosphere-colonising strains of *Pseudomonas putida* are used as seed inoculants to promote growth and increase yield of crop plants (Schroth and Hancock, 1982). The organisms rapidly colonise the roots of a variety of crops and secrete siderophores that efficiently chelate iron in the soil. In this way iron near the roots of the plant is unavailable to potentially harmful endemic fungi and bacteria that require iron for growth (Kloepper *et al.*, 1980). Competition for essential nutrients is reduced and plant productivity thereby improved. *Pseudomonas* sp. strain B10 is an effective control agent for *Fusarium* wilt and take-all diseases. Pseudobactin, the siderophore produced by this pseudomonad, sequesters iron and inhibits growth of the pathogens involved (Neilands, 1982). Genes for pseudobactin biosynthesis have been cloned (Moores *et al.*, 1984) with a view to their transfer to and expression in other root-colonising microorganisms to confer plant growth-promoting ability upon them.

(iii) Frost protection. A considerable proportion of crops are lost annually due to frost damage. The development of effective agricultural frost-protection programmes would thus be beneficial. Frost damage is often promoted by bacteria (such as *Pseudomonas syringae* and *Erwinia herbicola*) that grow naturally on the surface of plants. Such bacteria serve as nucleation sites for the formation of ice crystals, which destroy developing plant tissues (Lindow, 1983). A proteinaceous substance that promotes ice formation at temperatures just below freezing point (about −2°C) is produced by the bacteria. Phosphatidylinositol also appears to be a component of the ice nucleation site. Genes (*ice*) responsible for the ice-nucleating ability of *P. syringae* have been cloned and expressed in *E. coli.* Increases in nucleating activity could be obtained by cloning the ice nucleation region (3.5 to 4.0 kb) on multicopy vectors (Orser *et al.*, 1985). Sequencing of an ice nucleation active gene (*inaZ*) of *P. syringae* has predicted a translational product containing a repeating octapeptide.

The repeated peptides appear to contribute individually to the ice nucleation process (Green and Warren, 1985). Removal of such ice nucleation active bacteria from plants reduces the temperature at which frost damage occurs. One approach to frost protection is therefore to kill the ice-nucleating bacteria by, for example, spraying with specific bacteriophages (see Miller, 1983). However, commercialisation of such a procedure may be hampered both by the specificity of the virus for particular bacterial hosts and by the development of phage-resistant bacteria. An alternative strategy involves the use of genetically engineered strains ('ice-minus' bacteria) that are defective in the synthesis of the component(s) promoting ice formation. The *ice* region has been specifically deleted in such strains of *P. syringae.* These *ice* deletion mutants do not incite frost injury to plants (Orser *et al.*, 1984). The introduction of large populations of 'ice-minus' bacteria into the field is expected to afford frost protection for crops by displacing wild ice-nucleating organisms.

A further strategy has been suggested whereby engineered 'ice-minus' bacteria would be applied in conjunction with a bacteriophage, whose receptor is the ice-nucleation site (Lenski, 1984). In this way conditions could be provided that would selectively favour 'ice-minus', phage-resistant bacteria. Furthermore, any phage-resistant strains that appeared among the endemic bacterial population would be likely to be 'ice-minus'. Thus by using such an integrated strategy it might be possible to release into the environment disarmed 'ice-minus' strains that would only compete favourably within the range of simultaneous bacteriophage application. Outside that range they would be competitively inferior. This would prevent the spread of engineered strains in the environment. However, the use of such a strategy requires the isolation of a phage for whom host resistance correlates with the 'ice-minus' phenotype.

2. Parasitism. Certain fungi, bacteria and bacteriophages are parasites of microorganisms that cause plant diseases. For example, *Trichoderma* spp. are parasitic on various fungi and have proven effective in the control of a number of soil-borne plant pathogens. The possibility of using bacteria, such as *Bdellovibrio,* bacteriocins and bacteriophages to control bacterial plant pathogens is often considered. However, factors such as the emergence of resistant hosts and the specific conditions necessary for attachment to the host limit the potential usefulness of these agents.

3. Cross-protection and induced resistance. Inoculation of a plant with an attenuated virus strain ('inducer') can cross-protect the host against more virulent strains of the same virus ('challenger') (Sequeira, 1984). The degree of protection afforded depends upon the concentration of both inducer and challenger. Such cross-protection has been exploited commercially to control specific plant diseases. For example, mild strains of tobacco mosaic virus (TMV) have been used to cross-protect tomato plants against virulent strains of the virus, with minimal effects on yield.

Identification of the underlying mechanisms of cross-protection should permit the development of inducer strains that provide increased protection. Alternatively the plant might be engineered to synthesise the specific viral gene products that elicit protection, thereby eliminating the need to inoculate the host with the inducing strains.

Induced resistance, which is a less specific phenomenon than cross-protection both in terms of the inducing agent and the range of pathogens that the plant is protected against, may also be applied to disease control. Plants can become less susceptible to certain diseases as a result of prior exposure to necrosis-inducing pathogens or to heat-killed, attenuated forms. Induced resistance may be localised or systemic. Use of this method of control has been limited due to the difficulty in achieving reliable initial inoculation of the plant. Furthermore, environmental conditions affect the development of inducing strains. This problem may be addressed by improving the tolerance of the inducer to changing climatic conditions. Alternatively the plant genes responsible for such resistance might be manipulated directly to ensure more efficient disease protection.

8.2.2 Nitrogen-fixing prokaryotes

Efficient growth of agricultural crops depends upon an adequate supply of usable nitrogen. This may be provided in nitrogenous fertilisers or through microbial reduction of atmospheric nitrogen. In order to reduce reliance on chemical fertilisers, attention has focused on increasing the supply of biologically fixed N_2 (see Alexander, 1984). Biological nitrogen fixation involves the ATP-dependent reduction of N_2 to ammonia by nitrogenase (Figure 8.10).

(a) Nitrogen-fixing associations

The ability to fix N_2 is found in a diversity of prokaryotic microorganisms (see Table 8.3) including free-living species that reduce N_2 for their own growth, and symbiotic species (endosymbionts) that supply ammonia to their plant hosts in exchange for photosynthate and an environment protected from competitors. Increases in crop yield and soil fertility might be achieved by genetic improvements in free-living or symbiotic nitrogen-fixing bacteria. However, large amounts of energy are required for nitrogen fixation and the amount of nitrogen fixed is related to the accessibility of available energy sources. Accordingly, free-living N_2 fixers (with the exception of the photosynthetic prokaryotes) do not normally fix large amounts

Figure 8.10: Nitrogenase and hydrogenase activity. Nitrogenase catalyses the conversion of molecular N_2 to ammonia. At least 2ATP are consumed per electron (e^-) transferred. H_2 is evolved as a side reaction. Hydrogenase catalyses the oxidation of the H_2 produced during N_2 fixation. The protons (H^+) and electrons (e^-) generated can be reused by the nitrogenase

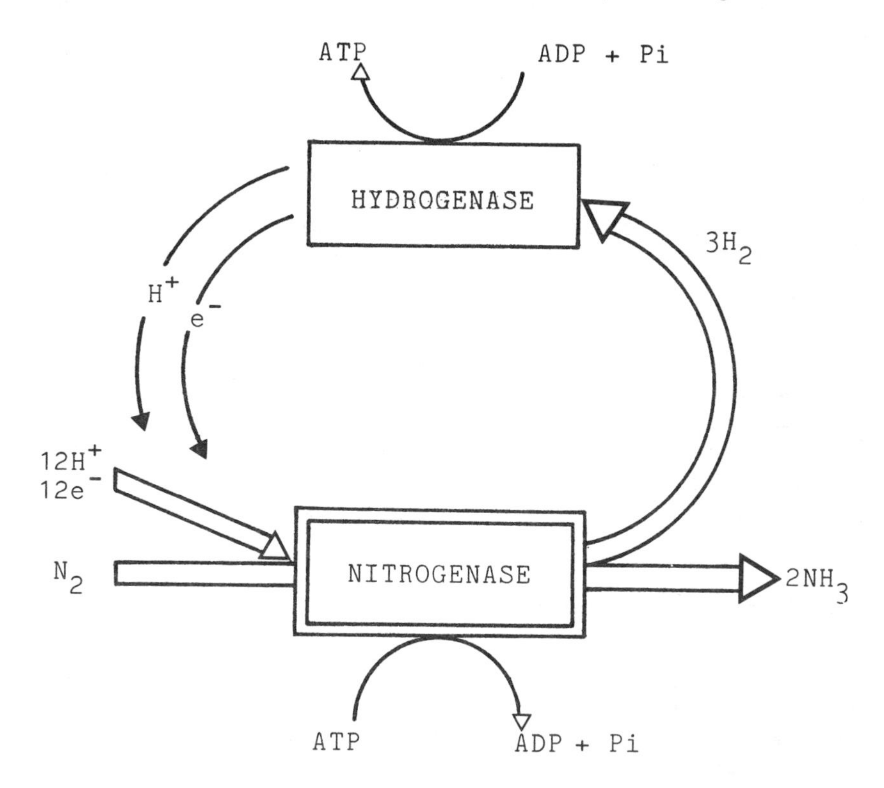

of N_2 (Beringer and Hirsch, 1984). Cyanobacteria can obtain the necessary energy for N_2 fixation from photosynthesis and can thus fix larger amounts of N_2. The development of superior N_2-fixing strains of cyanobacteria that export an excess of fixed N_2 may thus encourage their use in crop management programmes.

Certain N_2-fixing microorganisms grow in the rhizosphere of various crop plants. Here energy sources (present in root exudates) are relatively abundant and the N_2-fixing species have the potential to fix large amounts of N_2. However, in practice, the amount of N_2 fixed is generally low due to competition for energy sources from other indigenous microorganisms. Furthermore, only a portion of the fixed N_2 released by the N_2-fixing species is available directly to the plant because of microbial competition. A more beneficial association between plant and N_2-fixing bacteria might be established by enhancing both the access of N_2-fixing bacteria to energy

Table 8.3: Some nitrogen-fixing prokaryotes

Genus	Comments
Azotobacter	Free-living aerobic N_2 fixer
Azospirillum	Associations with plant roots
Rhizobium	Nodulates species of leguminous plants
Klebsiella	Free-living N_2 fixer
Frankia (actinomycete)	Nodulates shrubby dicotyledonous plants
Rhodopseudomonas, *Rhodospirillum* (photosynthetic bacteria)	Free-living N_2 fixers
Nostoc, Anabaena (cyanobacteria)	Free-living N_2 fixers, associations with certain plants and fungi

sources and the flow of fixed N_2 to the plant. These aspects have been addressed in a programme to develop associations between N_2-fixing species, such as *Azotobacter vinelandii* and cereal plants. Since *A. vinelandii* normally produces only sufficient ammonia for its own growth, feedback-resistant mutants (section 6.1.1) that excrete ammonia have been utilised. To ensure that the excreted ammonia is made available solely to the plant, the mutants have been engineered to associate with the roots of cereals. For a significant N_2-fixing association to occur the host plant should supply reduced carbon compounds to the bacterium in order to support the energy requirement for N_2 fixation. By selective breeding of varieties of corn from throughout the world, plants have been obtained with an increased ability to support bacterial N_2 fixation (Ela *et al.*, 1982).

The most important N_2-fixing associations in nature are symbiotic and involve in particular *Rhizobium* with leguminous plants and the cyanobacterium, *Anabaena azollae* with the aquatic fern, *Azolla* (which is used as a green manure in paddy fields). Rhizobia normally gain entry into the legume roots by penetrating the root hairs. A cellulose tube (infection thread) is produced which surrounds the bacteria, penetrates root cells and branches into newly divided cortical cells (Figure 8.11). This leads to the formation of the root nodule in which symbiotic N_2 fixation occurs. The bacteria, which are released from the infection thread, differentiate into bacteroids, and the genes for N_2 fixation (*nif*) are derepressed. The fixed N_2 does not appear to be assimilated by the bacteroids, but is instead released to be used directly by the plant (O'Gara and Shanmugan, 1976). The rhizobia induce synthesis by the host of large amounts of leghaemoglobin, an oxygen-binding protein which protects the oxygen-sensitive nitrogenase within the bacteroids. The plant can regulate N_2 fixation by limiting the amount of infected tissue and by restricting the flow of carbon compounds to the symbiont. Accordingly, genetic manipulation of the host

Figure 8.11: Infection of root hair of leguminous plant by *Rhizobium*. *Rhizobium* attaches to the root hair and enters the root hair through an infection thread. The bacteria infect an inner cortex cell, which swells and divides. A mass of infected cells (a root nodule) forms

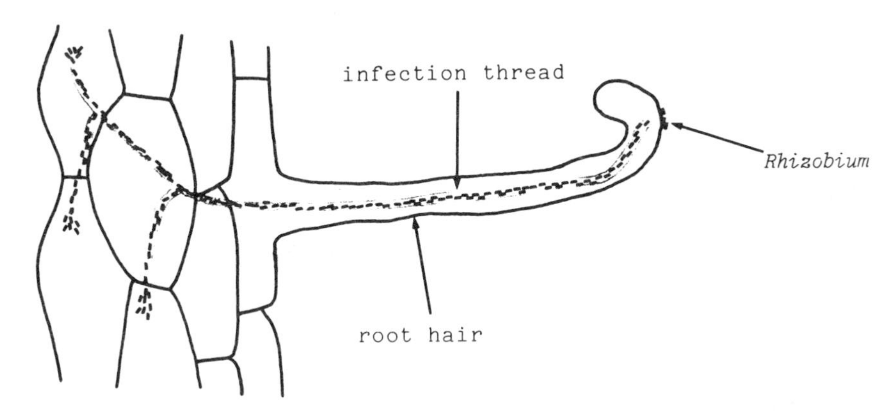

plant that results in accelerated rates of photosynthesis may enhance symbiotic N_2 fixation.

Induction of root nodule formation by rhizobia is specific in that certain species nodulate particular plant hosts; for example *R. meliloti* nodulates *Melilotus* (melilot) and *Medicago* (alfalfa); *R. phaseoli* nodulates *Phaseolus* (bean); *R. trifolii* nodulates *Trifolium* (clover), *R. japonicum* nodulates *Glycine* (soybean) and *R. leguminosarum* nodulates several legumes including *Pisum* (pea), *Lens* (lentil) and *Vicia* (vetch). The host-range specificity of nodulation appears to be plasmid-determined in various *Rhizobium* spp. and can be manipulated by plasmid transfer. For example, transfer of plasmid pRL1JI, which encodes both nodulation of pea plants and nitrogen fixation (Johnston *et al.*, 1978), from *R. leguminosarum* to *R. trifolii* and to *R. phaseoli* has enabled the recipients to nodulate pea. The *R. japonicum* plasmid pSym191 can broaden the host range of certain *Rhizobium* strains, such as *R. meliloti* and *R. leguminosarum*, by enabling them to nodulate soybeans (Appelbaum *et al.*, 1985). Acquisition of the nodulation plasmid of *R. trifolii* by *A. tumefaciens* and of nodulation genes (*nod*) by *Lignobacter* and *Pseudomonas aeruginosa* has rendered recipients capable of forming nodules or nodule-like structures on clover (Hooykaas *et al.*, 1981; Plazinski and Rolfe, 1985). Furthermore, *A. tumefaciens* and *E. coli* containing *nod* genes from *R. meliloti* formed pseudonodules on alfalfa (Hirsch *et al.*, 1984). Symbiotic associations might thus be extended by introducing determinants for root nodule formation into various strains. There is also potential for nodulating plant hosts other than legumes. Expression of early nodulation genes (root hair curling (Hac) genes) of *Rhizobium trifolii* has been achieved on maize and rice plants (Plazinski *et al.*, 1985). This opens up the possibility of developing a functional symbiotic N_2-fixing association with such monocotyledonous plants.

(b) Hydrogenase

In addition to catalysing the reduction of nitrogen to ammonia, the nitrogenase converts protons to H_2 with concomitant hydrolysis of ATP (Figure 8.10). This release of H_2 represents an inefficient use of energy. However, many free-living N_2-fixing species and certain symbiotic N_2-fixers contain a hydrogenase, which recycles the H_2. This H_2-uptake (Hup) system effects regeneration of ATP which can be utilised in N_2 fixation and other processes. Furthermore, oxygen consumption resulting from H_2 oxidation may protect the O_2-sensitive nitrogenase system (Eisbrenner and Evans, 1983). *Rhizobium* strains that contain the hydrogenase (Hup^+) fix N_2 more efficiently than strains (Hup^-) lacking this enzyme. Furthermore, yields of leguminous plants have been increased where Hup^+ strains are used as inocula (see, for example, Dejong *et al.*, 1982). In certain *Rhizobium* species, determinants for the Hup phenotype appear to be plasmid-borne and transmissible (Dejong *et al.*, 1982). Hup-specific sequences of *R. japonicum* have been isolated and shown to restore hydrogenase activity upon transfer to Hup^- mutants of the organism (Cantrell *et al.*, 1983; Haugland *et al.*, 1984; Harker *et al.*, 1985). Transfer of such determinants from highly active Hup^+ strains to other N_2-fixing strains (both Hup^- strains and strains exhibiting low H_2 uptake activity) may improve the efficiency of N_2 fixation, and in turn increase crop yields of appropriate host plants.

(c) Genetics of N_2 fixation and of root nodule formation

The molecular genetics of N_2 fixation has been extensively studied in *Klebsiella* (see Roberts and Brill, 1981; Ausubel *et al.*, 1982; Drummond, 1984). A cluster of 17 contiguous *nif* genes arranged in 7 or 8 transcription units has been identified in *K. pneumoniae* (Figure 8.12). All these *nif* operons, with the exception of *nifLA*, are positively controlled by the *nifA* gene product. Negative regulation is mediated by *nifL* product, which probably inactivates *nifA* product. It is proposed that the *nifL* gene product is active in the presence of NH_4^+ or O_2 and inactive in the absence of these molecules. Thus, when fixed N_2 is available, *nifA* expression is repressed and the remaining genes for N_2 fixation are not expressed. It is noteworthy that the *nifLA* promoter is atypical of most prokaryotic promoters. The −10 region of the *nifLA* promoter shows a possible Pribnow box, but the sequence 35 base pairs upstream of the transcription start shows no correlation with the −35 consensus sequence (Drummond *et al.*, 1983; Sundaresan *et al.*, 1983). However, this promoter contains the sequences GG at about position −24 and CC at about position −12 which are conserved in a number of bacterial promoters subject to nitrogen regu-

Figure 8.12: Arrangement of *nif* genes in *Klebsiella pneumoniae*. The *nifK* and *nifD* products are the α and β subunits of the nitrogenase molybdenum iron (MoFe) protein, and the *nifH* product is the subunit of the nitrogenase iron (Fe) protein. The *nifF* and *nifJ* products are involved in transporting electrons (e^-) to nitrogenase. The horizontal arrows ⇨ indicate the transcription units and direction of each transcript. The *nifA* and *nifL* products regulate expression of the other *nif* operons. The *nifA* product is a positive activator of *nif* transcription. The *nifL* product negatively controls the *nif* operons (except *nifLA*) (modified from Ausubel *et al.*, 1982; Drummond, 1984)

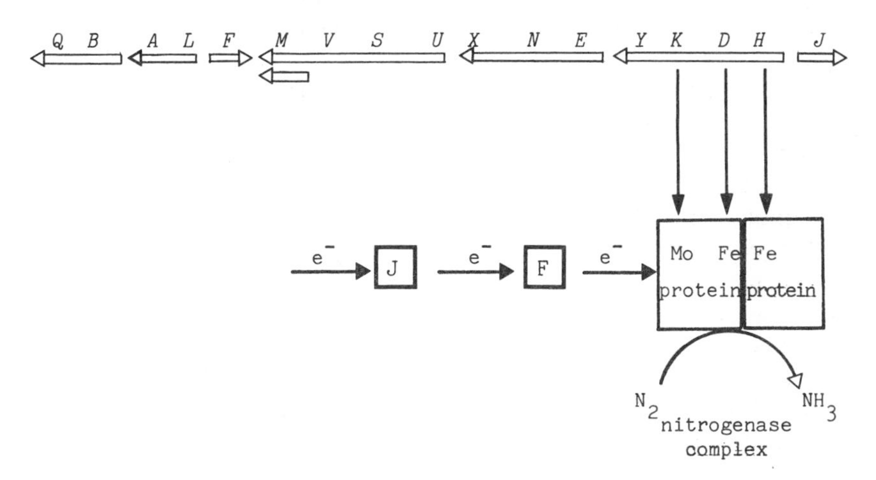

lation. *Klebsiella nif* genes have been transferred to and expressed in closely related genera, such as *Escherichia, Salmonella, Serratia, Erwinia* and *Pseudomonas,* but do not appear to be expressed in *Proteus, Agrobacterium* or *Saccharomyces.*

The conservation of DNA sequences for nitrogenase structural genes among widely divergent species has facilitated the identification of *nif* genes from other N_2-fixing species. Cloned *K. pneumoniae nif* genes have been used as hybridisation probes to detect and clone nitrogenase genes from organisms such as *R. meliloti, R. japonicum* and *Anabaena* 7120. *Rhizobium nif* genes appear to be plasmid-encoded and in some cases the plasmids are greater than 500 kb (megaplasmids). Furthermore, there is evidence for close linkage between *nif* and nodulation (*nod*) genes in certain *Rhizobium* spp. (Hombrecher *et al.*, 1981). Analysis of the genes for N_2 fixation and for nodulation in *Rhizobium* has been facilitated by the technique of transposition mutagenesis (see section 4.7.1). The transposon Tn*5* transposes to many sites in *Rhizobium* DNA, and the Tn*5* mutations have been mapped genetically by conjugation and transduction (Beringer *et al.*, 1984). Specific Tn*5*-mutagenised fragments have been cloned in *E. coli* and transferred back to wild-type *Rhizobium,* where they replace homologous DNA sequences effecting reversed mutagenesis (Ruvkun and Ausubel, 1981) (Figure 8.13). By using this technique a cluster of *nif/fix*[2] genes has been identified in a 14 kb region that surrounds the *R. meliloti nifHDK* genes on the symbiotic megaplasmid (Figure 8.14) (Pühler *et al.*, 1985). In addition, another gene (*fixF*) involved in symbiotic N_2 fixation has been located about 25 kb distant from this cluster (Aguilar *et al.*, 1985). It has been shown that the *nifHDK* promoter of *R. meliloti* is activated in *K. pneumoniae* by the *nifA* gene product (Sundaresan *et al.*, 1983). However, in the bacteroids *nif* genes are not repressed in the presence of 7.5 mM NH_4^+, a concentration that represses *K. pneumoniae nif* genes (presumably due to activation of the *nifL* repressor). This suggested that a *nifA*-like product could be involved in positively regulating the *Rhizobium nif* genes, although a *nifL*-like product does not appear to control the *nifA*-like product. Pühler and co-workers (1985) have demonstrated activation of *nifH* and *fixF* promoters of *R. meliloti* by the *fixD* gene product. The *fixD* DNA of *R. meliloti* hybridised to *Klebsiella nifA* DNA. Transcription of this *nifA*-like gene and consequent activation of *nif* genes appears to be initiated by a signal from the host plant.

Genes from both partners in the symbiotic N_2-fixing association appear to be involved in the formation of nodules on plant roots. Various nodulation mutants (Nod^-) of *Rhizobium* spp. have been isolated by Tn*5*-induced mutagenesis. Such mutants are proving useful in clarifying the molecular events governing symbiosis (see, for example, Pühler *et al.*, 1985).

Two sets of bacterial genes, for recognition of the appropriate legume and for nodule induction, have been identified. One cluster of nodulation genes, termed 'common' *nod* genes (*nodABC* and *D*), specifies general functions for nodulation and appears to be conserved in most *Rhizobium* spp. The other cluster determines host specificity of nodulation (*hsn* genes). Some interaction may occur between *nod* and *hsn* gene products in determining host specificity. *nod* and *hsn* genes may be coordinately regulated. In *R. meliloti* binding of *nodD* to the *nod* box (a conserved promoter sequence in front of the *nodABC*, *hsnABC* and *hsnD* transcription units) may result in initiation of transcription of nodulation genes. A plant-specific component appears to stimulate expression of these genes. The plant factor may modify the *nodD* gene product or may bind with it to the *nod* box. Conversely the nodulation genes appear to play a part in inducing cortical root cell division and possibly also in activating plant genes, such as leghaemoglobin and other early nodulin genes (Kondorosi and Kondorosi, 1986).

The availability of further information concerning the function and regulation of *nif*, *fix* and *nod* genes should permit more specific manipu-

[2] *fix* designates genes required for the process of nitrogen fixation.

Figure 8.13: Transposition mutagenesis of nitrogen fixation (*nif*) genes of *Rhizobium*. Tn*5* insertion mutations are introduced into *Rhizobium* DNA fragments cloned on a high copy number plasmid in *E. coli*. The Tn*5*-mutagenised fragments are then recloned into a low copy number, broad host range vector that can be mobilised into *Rhizobium*. Homologous recombination between wild-type *Rhizobium* genome sequences and the mutated cloned sequences results in transfer of the Tn*5* insertion from the vector plasmid to the *Rhizobium* genome. Colonies containing such Tn*5* insertions are selected in the presence of another IncP-plasmid. Tc^r, resistance to tetracycline; Tp^r, resistance to trimethoprim; Nm^r, resistance to neomycin (see Ruvkun and Ausubel, 1981).

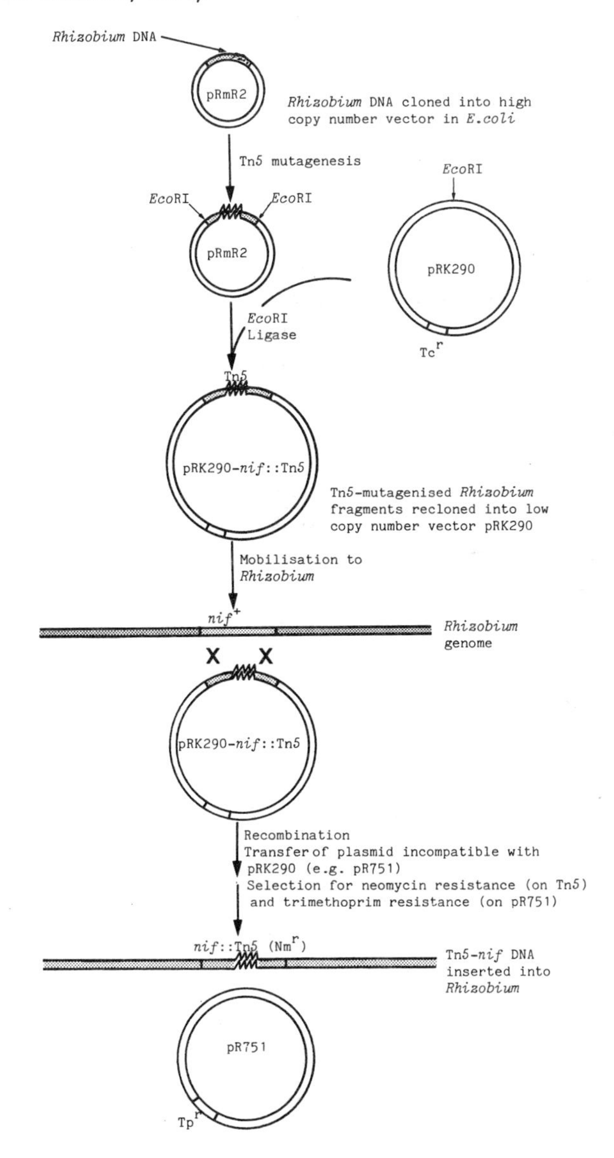

Figure 8.14: Organisation of some plasmid-encoded genes associated with nitrogen fixation and nodulation in *Rhizobium meliloti.* Coding regions of *nif* and *fix* genes are shown. *fixF* is about 25 kb distant from the main *nif-fix* cluster. The common *nod* genes (*nodABC* and *D*) are located close to *fixF* *hsn* genes (*hsnABC* and *D*) are located between the common *nod* genes and the *nif* genes. p, Promoter (adapted from Pühler *et al.*, 1985). Note: *nif* genes are located on the bacterial chromosome in *K. pneumoniae,* but on large plasmids in most *Rhizobium* species.

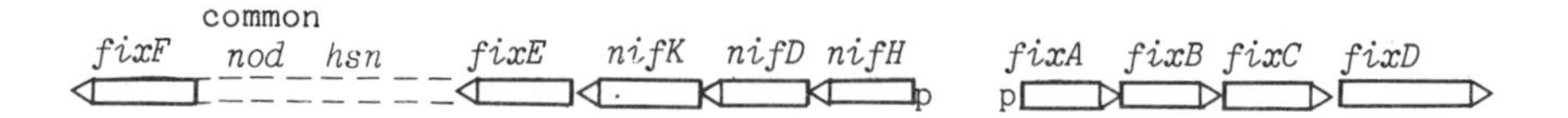

lations of the *Rhizobium* genome, with a view to improving the efficiency of N_2 fixation and to extending the range of N_2-fixing microorganisms. Inoculation of seeds with superior N_2-fixing strains of *Rhizobium* may result in improved crop yields and seed quality, provided that the host can effectively support bacterial N_2 fixation and that engineered organisms can outcompete indigenous rhizobia in the soil. It may be possible to develop host cultivars that exclude indigenous strains while admitting improved strains. However, strain development programmes for new N_2-fixing organisms are likely to encounter problems due to the stringent genetic and metabolic requirements of the N_2-fixation system. Features, such as the complexity of the *nif* regulatory system, inactivation of the nitrogenase by O_2, the requirement for energy as MgATP and for a suitable electron transport system for nitrogenase activity and the need to remove fixed N_2 from the vicinity of the nitrogenase, may all pose obstacles for efficient N_2 fixation, even if *nif* genes can be expressed in a new host.

BIBLIOGRAPHY

Aguilar, O.M., Kapp, D. and Pühler, A. (1985) Characterization of a *Rhizobium meliloti* fixation gene (*fixF*) located near the common nodulation region. *J. Bacteriol.* **164**, 245-54

Alexander, M. (1984) *Biological nitrogen fixation. Ecology, technology and physiology.* Plenum Press, New York

Appelbaum, E.R., McLoughlin, T.J., O'Connell, M. and Chartrain, N. (1985) Expression of symbiotic genes of *Rhizobium japonicum* USDA 191 in other rhizobia. *J. Bacteriol.* **163**, 385-8

Arai, M., Haneishi, T., Kitahara, N., Enokita, R., Kawakubo, K. and Kondo, Y. (1976) Herbicidins A and B, two new antibiotics with herbicidal activity. I. Producing organism and biological activities. *J. Antibiotics* **29**, 863-9

Ausubel, F.M., Brown, S.E., de Bruijn, F.J., Ow, D.W., Riedel, G.E., Ruvkun, G.B. and Sundaresan, V. (1982) Molecular cloning of nitrogen fixation genes from *Klebsiella pneumoniae* and *Rhizobium meliloti.* In *Genetic engineering 4* (J.K. Setlow and A. Hollaender, eds), pp. 169-98. Plenum Press, New York

Barton, K.A., Binns, A.N., Matzko, A.J.M. and Chilton, M-D. (1983) Regeneration of intact tobacco plants containing full length copies of genetically engineered T-DNA and transmission of T-DNA to R1 progeny. *Cell* **32**, 1033-43

Beardsley, T. (1984) Monsanto goes ahead with trials. *Nature (London)* **312**, 686

Beringer, J.E. and Hirsch, P.R. (1984) Genetic engineering and nitrogen fixation. In *Biotechnology and genetic engineering reviews*, vol. I (G.E. Russell, ed.), pp. 65-88. Intercept, Newcastle upon Tyne

Beringer, J.E., Ruiz Sainz, J.E. and Johnston, A.W.B. (1984) Methods for the genetic manipulation of *Rhizobium*. In *Microbiological methods for environmental biotechnology* (J.M. Grainger and J.M. Lynch, eds), pp. 79-94. Academic Press, London and New York

Bevan, M.W. and Chilton, M-D. (1982) T-DNA of the *Agrobacterium* Ti and Ri plasmids. *Ann. Rev. Genet.* **16**, 357-84

Bevan, M.W., Flavell, R.B. and Chilton, M-D. (1983) A chimaeric antibiotic resistance gene as a selectable marker for plant cell transformation. *Nature (London)* **304**, 184-7

Brassel, J. and Benz, G. (1979) Selection of a strain of the granulosis virus of the codling moth with improved resistance against artificial ultraviolet radiation and sunlight. *J. Invertebrate Pathol.* **33**, 358-63

Brisson, N., Paszkowski, J., Penswick, J.R., Gronenborn, B., Potrykus, I. and Hohn, T. (1984) Expression of a bacterial gene in plants by using a viral vector. *Nature (London)* **310**, 511-14

Cantrell, M.A., Haugland, R.A. and Evans, H.J. (1983) Construction of a *Rhizobium japonicum* gene bank and use in isolation of a hydrogen uptake gene. *Proc. Natl Acad. Sci. USA* **80**, 181-5

Caplan, A.B., Van Montagu, M. and Schell, J. (1985) Genetic analysis of integration mediated by single T-DNA borders. *J. Bacteriol.* **161**, 655-64

Carlton, B.C. and González, J.M. Jr (1985) The genetics and molecular biology of *Bacillus thuringiensis*. In *The molecular biology of the bacilli*, vol. II (D.A. Dubnau, ed.), pp. 211-49. Academic Press, London and New York

Carter, J.B. (1984) Viruses as pest-control agents. In *Biotechnology and genetic engineering reviews*, vol. 1 (G.E. Russell, ed.), pp. 375-419. Intercept, Newcastle upon Tyne

Chilton, M-D., Saiki, R.K., Yadav, N., Gordon, M.P. and Quétier, F. (1980) T-DNA from *Agrobacterium* Ti plasmid is in the nuclear DNA fraction of crown gall tumor cells. *Proc. Natl Acad. Sci. USA* **77**, 4060-4

Chilton, M-D., Tepfer, D.A., Petit, A., David, C., Casse-Delbart, F. and Tempé, J. (1982) *Agrobacterium rhizogenes* inserts T-DNA into the genomes of the host plant root cells. *Nature (London)* **295**, 432-4

Comai, L., Schilling-Cordaro, C., Mergia, A. and Houck, C.M. (1983a) A new technique for genetic engineering of *Agrobacterium* Ti plasmid. *Plasmid* **10**, 21-30

Comai, L., Sen, L.C. and Stalker, D.M. (1983b) An altered *aroA* gene product confers resistance to the herbicide glyphosate. *Science* **221**, 370-1

Comai, L., Facciotti, D., Hiatt, W.R., Thompson, G., Rose, R.E. and Stalker, D.M. (1985) Expression in plants of a mutant *aroA* gene from *Salmonella typhimurium* confers tolerance to glyphosate. *Nature (London)* **317**, 741-4

Daubert, S., Shepherd, R.J. and Gardner, R.C. (1983) Insertional mutagenesis of the cauliflower mosaic virus genome. *Gene* **25**, 201-8

Dean, D.H. (1984) Biochemical genetics of the bacterial insect control agent *Bacillus thuringiensis*: basic principles and prospects for genetic engineering. In *Biotechnology and genetic engineering reviews*, vol. 2 (G.E. Russell, ed.), pp. 341-63. Intercept, Newcastle upon Tyne

De Block, M., Schell, J. and Van Montagu, M. (1985) Chloroplast transformation by *Agrobacterium tumefaciens. EMBO J.* **4**, 1367-72

De Cleene, M. and Deley, J. (1976) The host range of crown gall. *Bot. Rev.* **42**, 389-466

De Framond, A.J., Barton, K.A. and Chilton M-D. (1983) Mini-Ti: a new vector strategy for plant genetic engineering. *Biotechnology* **1**, 262-9

De Greve, H., Leemans, J., Hernalsteens, J.P., Thia-Toong, L., De Beuckeleer, M., Willmitzer, L., Otten, L., Van Montagu, M. and Schell, J. (1982) Regeneration of normal and fertile plants that express octopine synthase, from tobacco crown galls after deletion of tumour-controlling functions. *Nature (London)* **300**, 752-5

Dejong, T.M., Brewin, N.J., Johnston, A.W.B. and Phillips, D.A. (1982) Improvement of symbiotic properties in *Rhizobium leguminosarum* by plasmid transfer. *J. Gen. Microbiol.* **128**, 1829-38

Diener, T.O. (1982) Viroids and their interactions with host cells. *Ann. Rev. Microbiol.* **36**, 239-58

Dixon, L.K., Koenig, I. and Hohn, T. (1983) Mutagenesis of cauliflower mosaic virus. *Gene* **25**, 189-99

Drummond, M.H. (1984) The nitrogen fixation genes of *Klebsiella pneumoniae*: a model system. *Microbiological Sciences* **1**, 29-33

Drummond, M., Clements, J., Merrick, M. and Dixon, R. (1983) Positive control and autogenous regulation of the *nifLA* promoter in *K. pneumoniae. Nature (London)* **301**, 302-7

Eisbrenner, G. and Evans, H.J. (1983) Aspects of hydrogen metabolism in nitrogen fixing legumes and other plant-microbe associations. *Ann. Rev. Plant Physiol.* **34**, 105-36

Ela, S.W., Anderson, M.A. and Brill, W.J. (1982) Screening and selection of maize to enhance associative bacterial nitrogen fixation. *Plant Physiol.* **70**, 1564-7

Engler, G., Depicker, A., Maenhaut, R., Villarroel-Mandiola, R., Van Montagu, M. and Schell, J. (1981) Physical mapping of DNA base sequence homologies between an octopine and a nopaline Ti plasmid of *Agrobacterium tumefaciens. J. Mol. Biol.* **152**, 183-208

Entwistle, P.F. (1983) Viruses for insect pest control. *Span* **26**, 59-62

Evans, D.A. (1983) Agricultural applications of plant protoplast fusion. *Biotechnology* **1**, 253-61

Fedoroff, N.V. (1984) Transposable genetic elements in maize. *Scientific American* **250**, 65-74

Fraley, R.T., Rogers, S.G., Horsch, R.B., Eichholtz, D.A., Flick, J.S., Fink, C.L., Hoffmann, N.L. and Saunders, P.R. (1985) The SEV system: a new disarmed Ti plasmid vector system for plant transformation. *Biotechnology* **3**, 629-35

Franck, A., Guilley, H., Jonard, G., Richards, K. and Hirth, L. (1980) Nucleotide sequence of cauliflower mosaic virus DNA. *Cell* **21**, 285-94

Francki, R.I.B. (1985) Plant virus satellites. *Ann. Rev. Microbiol.* **39**, 151-74

González, J.M. and Carlton, B.C. (1984) A large transmissible plasmid is required for crystal toxin production in *Bacillus thuringiensis* variety *israelensis. Plasmid* **11**, 28-38

González, J.M. Jr, Dulmage, H.T. and Carlton, B.C. (1981) Correlation between specific plasmids and δ-endotoxin production in *Bacillus thuringiensis. Plasmid* **5**, 351-65

González, J.M. Jr, Brown, B.J. and Carlton, B.C. (1982) Transfer of *Bacillus thuringiensis* plasmids coding for δ-endotoxin among strains of *B. thuringiensis* and *B. cereus. Proc. Natl Acad. Sci. USA* **79**, 6951-5

Goodman, R. (1981) Gemini viruses. *J. Gen. Virol.* **54**, 9-21

Green, R.L. and Warren, G.J. (1985) Physical and functional repetition in a bacterial ice nucleation gene. *Nature (London)* **317**, 645-8

Harker, A.R., Lambert, G.R., Hanus, F.J. and Evans, H.J. (1985) Further evidence that two unique subunits are essential for expression of hydrogenase activity in *Rhizobium japonicum. J. Bacteriol.* **164**, 187-91

Hasan, S. (1980) Plant pathogens and biological control of weeds. *Rev. Plant Pathology* **59**, 349-56

Haselkorn, R. and Golden, S.S. (1985) Mutation to herbicide resistance maps within the *psbA* gene of *Anacystis nidulans* R2. *Science* **229**, 1104-7

Haugland, R.A., Cantrell, M.A., Beaty, J.S., Hanus, F.J., Russell, S.A. and Evans, H.J. (1984) Characterization of *Rhizobium japonicum* hydrogen uptake genes. *J. Bacteriol.* **159**, 1006-12

Held, G.A., Bulla, L.A. Jr, Ferrari, E., Hoch, J., Aronson, A.I. and Minnich, S.A. (1982) Cloning and localization of the lepidopteran protoxin gene of *Bacillus thuringiensis* subsp. *kurstaki. Proc. Natl Acad. Sci. USA* **79**, 6065-9

Hernalsteens, J-P., Thia-Toong, L., Schell, J. and Van Montagu, M. (1984) An *Agrobacterium*-transformed cell culture from the monocot *Asparagus officinalis. EMBO J.* **3**, 3039-41

Herrera-Estrella, L., Depicker, A., Van Montagu, M. and Schell, J. (1983) Expression of chimaeric genes transferred into plant cells using a Ti-plasmid-derived vector. *Nature (London)* **303**, 209-13

Herrera-Estrella, L., Van den Broeck, G., Maenhaut, R., Van Montagu, M., Schell, J., Timko, M. and Cashmore, A. (1984) Light-inducible and chloroplast-associated expression of a chimaeric gene introduced into *Nicotiana tabacum* using a Ti plasmid vector. *Nature (London)* **310**, 115-20

Hiatt, W.R., Comai, L., Huang, L-J, Rose, R., Thompson, G. and Stalker, D. (1985) Introduction and expression in plants of a glyphosate resistant *aroA* gene isolated from *Salmonella typhimurium.* In *Molecular form and function of the plant genome* (L. van Vloten-Doting, G.S.P. Groot and T.C. Hall, eds), pp. 479-87. Plenum Press, New York

Hirsch, A.M., Wilson, K.J., Jones, J.D.G., Bang, M., Walker, V.V. and Ausubel, F.M. (1984) *Rhizobium meliloti* nodulation genes allow *Agrobacterium tumefaciens* and *Escherichia coli* to form pseudonodules on alfalfa. *J. Bacteriol.* **158**, 1133-43

Hohn, T., Hohn, B., Lesot, A. and Lebeurier, G. (1980) Restriction map of native and cloned cauliflower mosaic virus DNA. *Gene* **11**, 21-31

Hohn, T., Richards, K. and Lebeurier, G. (1982) Cauliflower Mosaic Virus on its way to becoming a useful plant vector. *Curr. Top. Microbiol. Immunol.* **96**, 193-220

Hombrecher, G., Brewin, N.J. and Johnston, A.W.B. (1981) Linkage of genes for nitrogenase and nodulation ability on plasmids in *Rhizobium leguminosarum* and *R. phaseoli. Mol. Gen. Genet.* **182**, 133-6

Hooykaas, P.J.J., Van Brussel, A.A.N., Den Dulk-Ras, H., Van Slogteren, G.M.S. and Schilperoort, R.A. (1981) Sym plasmid of *Rhizobium trifolii* expressed in different rhizobial species and *Agrobacterium tumefaciens. Nature (London)* **291**, 351-3

Hooykaas-Van Slogteren, G.M.S., Hooykaas, P.J.J. and Schilperoort, R.A. (1984) Expression of Ti plasmid genes in monocotyledonous plants infected with *Agrobacterium tumefaciens. Nature (London)* **311**, 763-4

Horsch, R.B., Fraley, R.T., Rogers, S.G., Sanders, P.R., Lloyd, A. and Hoffmann, N. (1984) Inheritance of functional foreign genes in plants. *Science* **223**, 496-8

Howell, S.H., Walker, L.L. and Dudley, R.K. (1980) Cloned cauliflower mosaic virus DNA infects turnips (*Brassica rapa*). *Science* **208**, 1265-7

Howell, S.H., Walker, L.L. and Walden, R.M. (1982) The rescue of *in vitro* generated mutants of the cloned cauliflower mosaic virus genome in infected plants. *Nature (London)* **293**, 483-6

Iwai, Y., Nakagawa, A., Sadakane, N. and Omura, S. (1980) Herbimycin B, a new benzoquinonoid ansamycin with anti-TMV and herbicidal activities. *J. Antibiotics* **33**, 1114-19

Johnston, A.W.B., Beynon, J.L., Buchanan-Wollaston, A.V., Setchell, S.M., Hirsch, P.R. and Beringer, J.E. (1978) High frequency transfer of nodulating ability between strains and species of *Rhizobium. Nature (London)* **276**, 634-6

Kemble, R.J., Gunn, R.E. and Flavell, R.B. (1980) Classification of normal and male-sterile cytoplasms in maize II. Electrophoretic analysis of DNA species in mitochondria. *Genetics* **95**, 451-8

Kerr, A. (1980) Biological control of crown gall through production of agrocin 84. *Plant Dis.* **64**, 25-30

Klee, H.J., Yanofsky, M.F. and Nester, E.W. (1985) Vectors for transformation of higher plants. *Biotechnology* **3**, 637-42

Kloepper, J.W., Leong, J., Teintze, M. and Schroth, M.N. (1980) Enhanced plant growth by siderophores produced by plant growth-promoting rhizobacteria. *Nature (London)* **286**, 885-6

Kondorosi, E. and Kondorosi, A. (1986) Nodule induction on plant roots by *Rhizobium. Trends Biochim. Sci.* **11**, 296-9

Koukolíkova-Nicolá, Z., Shillito, R.D., Hohn, B., Wang, K., Van Montagu, M. and Zembryski, P. (1985) Involvement of circular intermediates in the transfer of T-DNA from *Agrobacterium tumefaciens* to plant cells. *Nature (London)* **313**, 191-6

Kurstak, E. (1982) *Microbial and viral pesticides.* Marcel Dekker, New York

Lane, L.C. (1979) The nucleic acids of multipartite, defective, and satellite plant viruses. In *Nucleic acids in plants*, vol. II (T.G. Hall and J.W. Davies, eds), pp. 65-110. CRC Press, Fla

Lebeurier, G., Hirth, L., Hohn, Th. and Hohn, B. (1980) Infectivities of native and cloned DNA of cauliflower mosaic virus. *Gene* **12**, 139-46

Leemans, J., Shaw, C., Deblaere, R., De Greve, H., Hernalsteens, J.P., Maes, M., Van Montagu, M. and Schell, J. (1981) Site-specific mutagenesis of *Agrobacterium* Ti plasmids and transfer of genes to plant cells. *J. Mol. Appl. Genet.* **1**, 149-64

Lemmers, M., DeBeuckeleer, M., Holsters, M., Zambryski, P., Hernalsteens, J.P., Van Montagu, M. and Schell, J. (1980) Internal organization, boundaries and integration of Ti plasmid DNA in nopaline crown gall tumors. *J. Mol. Biol.* **144**, 353-76

Lenski, R.E. (1984) Releasing ice-minus bacteria. *Nature (London)* **307**, 8

Lindow, S.E. (1983) The role of bacterial ice nucleation in frost injury to plants. *Ann. Rev. Phytopathol.* **21**, 363-84

Lisansky, S.G. and Hall, R.A. (1983) Fungal control of insects. In *The filamentous fungi, vol. 4. Fungal Technology* (J.E. Smith, D.R. Berry and B. Kristiansen, eds), pp. 327-45. Edward Arnold, London

Lorz, H. (1984) Variability in tissue culture derived plants. In *Genetic manipulation: impact on man and society* (W. Arber, K. Illmensee, W.J. Peacock, and P. Starlinger, eds), pp. 103-14. ICSU Press, Cambridge, New York and Melbourne

Mantell, S.H., Matthews, J.A. and McKee, R.A. (1985) *Principles of plant biotechnology. An introduction to genetic engineering in plants.* Blackwell Scientific Publications, Oxford

Matzke, A.J.M. and Chilton, M-D. (1981) Site-specific insertion of genes in T-DNA of the *Agrobacterium* tumor-inducing plasmid: an approach to genetic engineering of higher plant cells. *J. Mol. Appl. Genet.* **1**, 39-49

Miller, J.A. (1983) Microbial antifreeze: gene splicing takes to the field. *Science News* **124**, 132
Miller, L.K., Lingg, A.J. and Bulla, L.A. Jr (1983) Bacterial, viral and fungal insecticides. *Science* **219**, 715-21
Moores, J.C., Magazin, M., Ditta, G.S. and Leong, J. (1984) Cloning of genes involved in the biosynthesis of pseudobactin, a high-affinity iron transport agent of a plant growth-promoting *Pseudomonas* strain. *J. Bacteriol.* **157**, 53-8
Neilands, J.B. (1982) Microbial envelope proteins related to iron. *Ann. Rev. Microbiol.* **36**, 285-309
Odell, J.T., Nagy, F. and Chua, N-H. (1985) Identification of DNA sequences required for activity of the cauliflower mosaic virus 35S promoter. *Nature (London)* **313**, 810-12
O'Gara, F. and Shanmugan, K.T. (1976) Regulation of nitrogen fixation. Export of fixed N_2 as NH_4^+. *Biochim. Biophys. Acta* **437**, 313-21
Orser, C.S., Lotstein, R., Lahue, E., Willis, D.K., Panopoulos, N.J. and Lindow, S.E. (1984) Structural and functional analysis of the *Pseudomonas syringae* pv. *syringae ice* region and construction of *ice*$^-$ deletion mutants. *Phytopathol.* **74**, 798
Orser, C., Staskawicz, B.J., Panopoulos, N.J., Dahlbeck, D. and Lindow, S.E. (1985) Cloning and expression of bacterial ice nucleation genes in *Escherichia coli. J. Bacteriol.* **164**, 359-66
Owens, R.A. and Diener, T.O. (1985) Plant disease detection by nucleic acid hybridization. In *Molecular form and function of the plant genome* (L. van Vloten-Doting, G.S.P. Groot and T.C. Hall, eds), pp. 45-53. Plenum Press, New York
Peacock, W.J., Dennis, E.S., Gerlach, W.L., Llewellyn, D., Lorz, H., Pryor, A.J., Sachs, M.M., Schwartz, D. and Sutton, W.D. (1983) Gene transfer in maize: controlling elements and the alcohol dehydrogenase genes. In *Advances in gene technology: molecular genetics of plants and animals.* Miami Winter Symposia — vol. 20 (K. Downey, R.W. Voellmy, F. Ahmad and J. Schultz, eds), pp. 311-25. Academic Press, New York
Peacock, W.J., Dennis, E.S., Gerlach, W.L., Llewellyn, D., Lorz, H., Pryor, A.J., Sachs, M.M., Schwartz, D. and Sutton, W.D. (1984) Gene transfer in maize: controlling elements and the alcohol dehydrogenase genes. In *Genetic manipulation: impact on man and society* (W. Arber, K. Illmensee, W.J. Peacock and P. Starlinger, eds), pp. 115-26. ICSU Press, Cambridge, New York and Melbourne
Pfeiffer, P. and Hohn, T. (1983) Involvement of reverse transcription in the replication of cauliflower mosaic virus: a detailed model and test of some aspects. *Cell* **33**, 781-9
Plazinski, J. and Rolfe, B.G. (1985) Sym plasmid genes of *Rhizobium trifolii* expressed in *Lignobacter* and *Pseudomonas* strains. *J. Bacteriol.* **162**, 1261-9
Plazinski. J., Innes, R.W. and Rolfe, B.G. (1985) Expression of *Rhizobium trifolii* early nodulation genes on maize and rice plants. *J. Bacteriol.* **163**, 812-15
Pühler, A., Aguilar, M.O., Kapp, D., Müller, P., Mohapatra, S., Priefer, U., Reiländer, H., Simon, R. and Weber, G. (1985) Genetic analysis of symbiotic nitrogen fixation in the *Rhizobium meliloti-Medicago sativa* system. In *Molecular form and function of the plant genome* (L. van Vloten-Doting, G.S.P. Groot and T.C. Hall, eds), pp. 429-42. Plenum Press, New York
Riesner, D. and Gross, H.J. (1985) Viroids. *Ann. Rev. Biochem.* **54**, 531-64
Roberts, G.P. and Brill, W.J. (1981) Genetics and regulation of nitrogen fixation. *Ann. Rev. Microbiol.* **35**, 207-35
Ruvkun, G.B. and Ausubel, F.M. (1981) A general method for site directed mutagenesis in prokaryotes: construction of mutations in symbiotic nitrogen fixation

genes of *Rhizobium meliloti. Nature (London)* **289**, 85-8

Schroth, M.N. and Hancock, J.G. (1981) Selected topics in biological control. *Ann. Rev. Microbiol.* **35**, 453-76

Schroth, M.N. and Hancock, J.G. (1982) Disease-suppressive soil and root-colonizing bacteria. *Science* **216**, 1376-81

Sekar, V. and Carlton, B.C. (1985) Molecular cloning of the delta endotoxin gene of *Bacillus thuringiensis* var. *israelensis. Gene* **33**, 151-8

Sequeira, L. (1984) Cross-protection and induced resistance: their potential for plant disease control. *Trends Biotechnol.* **2**, 25-9

Sheperd, R.J. (1979) DNA plant viruses. *Ann. Rev. Plant Physiol.* **30**, 405-23

Shewmaker, C.K., Caton, J.R., Houck, C.M. and Gardner, R.C. (1985) Transcription of cauliflower mosaic virus integrated into plant genomes. *Virology* **140**, 281-8

Simpson, R.B., O'Hara, P.J., Kwok, W., Montoya, A.L., Lichtenstein, C., Gordon M.P. and Nester, E.W. (1982) DNA from the A6S/2 crown gall tumor contains scrambled Ti-plasmid sequences near its junction with plant DNA. *Cell* **29**, 1005-14

Spanò, L., Cardarelli, M., Mauro, M.L., Pomponi, M. and Costantino, P. (1985) Hairy root: molecular and physiological aspects. In *Molecular form and function of the plant genome* (L. van Vloten-Doting, G.S.P. Groot and T.C. Hall, eds) pp. 637-53. Plenum Press, New York

Stachel, S.E., Messens, E., Van Montagu, M. and Zambryski, P. (1985) Identification of the signal molecules produced by wounded plant cells that activate T-DNA transfer in *Agrobacterium tumefaciens. Nature (London)* **318**, 624-9

Stachel, S.E., Nester, E.W. and Zambryski, P.C. (1986) A plant cell factor induces *Agrobacterium tumefaciens vir* gene expression. *Proc. Natl Acad. Sci. USA* **83**, 379-83

Stanley, J. (1983) Infectivity of the cloned geminivirus genome requires sequences from both DNAs. *Nature (London)* **305**, 643-4

Starlinger, P., Courage-Tebbe, U., Doring, H.P., Frommer, W.B., Theres, K., Tillman, E., Weck, E. and Werr, W. (1984) Isolation of transposable elements in maize. In *Genetic manipulation: impact on man and society* (W. Arber, K. Illmensee, W.J. Peacock and P. Starlinger, eds), pp. 67-74. ICSU Press, Cambridge, New York and Melbourne

Sundaresan, V., Jones, J.D.G., Ow, D.W. and Ausubel, F.M. (1983) *Klebsiella pneumoniae nifA* product activates the *Rhizobium meliloti* nitrogenase promoter. *Nature (London)* **301**, 728-32

Takatsuji, H., Hirochika, H., Fukushi, T. and Ikeda, J-E. (1986) Expression of cauliflower mosaic virus reverse transcriptase in yeast. *Nature (London)* **319**, 240-3

Takiguchi, Y., Mishima, H., Okuda, M. and Terao, M. (1980) Milbemycins, a new family of macrolide antibiotics: fermentation, isolation and physico-chemical properties. *J. Antibiotics* **33**, 1120-7

Van den Broeck, G., Timko, M.P., Kausch, A.P., Cashmore, A.R., Van Montagu, M. and Herrera-Estrella, L. (1985) Targeting of a foreign protein to chloroplasts by fusion to the transit peptide from the small subunit of ribulose 1,5-bisphosphate carboxylase. *Nature* **313**, 358-63

Van Haute, E., Joos, H., Maes, M., Warren, G., Van Montagu, M. and Schell, J. (1983) Intergenic transfer and exchange recombination of restriction fragments cloned in pBR322: a novel strategy for the reversed genetics of the Ti-plasmids of *Agrobacterium tumefaciens. EMBO J.* **2**, 411-17

Van Montagu, M. and Schell, J. (1982) The Ti plasmids of *Agrobacterium. Curr. Top. Microbiol. Immunol.* **96**, 237-54

Virts, E.L. and Gelvin, S.B. (1985) Analysis of transfer of tumor-inducing plasmids from *Agrobacterium tumefaciens* to *Petunia* protoplasts. *J. Bacteriol.* **162**, 1030-8

Ward, E.S., Ellar, D.J. and Todd, J.A. (1984) Cloning and expression in *Escherichia coli* of the insecticidal-endotoxin gene of *Bacillus thuringiensis* var. *israelensis. FEBS Letters* **175**, 377-82

White, F.F., Taylor, B.H., Huffman, G.A., Gordon, M.P. and Nester, E.W. (1985) Molecular and genetic analysis of the transferred DNA regions of the root-inducing plasmid of *Agrobacterium rhizogenes. J. Bacteriol.* **164**, 33-44

Willmitzer, L., De Beuckeleer, M., Lemmers, M., Van Montagu, M. and Schell, J. (1980) DNA from Ti plasmid present in nucleus and absent from plastids of crown gall plant cells. *Nature (London)* **287**, 359-61

Wood, H.A., Hughes, P.R., Johnston, L.B. and Langridge, W.H.R. (1981) Increased virulence of *Autographa californica* nuclear polyhedrosis virus by mutagenesis. *J. Invertebrate Pathol.* **38**, 236-41

Yadav, N.S., Vanderleyden, J., Bennett, D.R., Barnes, W.M. and Chilton, M-D (1982) Short direct repeats flank the T-DNA on a nopaline Ti plasmid. *Proc. Natl Acad. Sci. USA* **79**, 6322-6

Yanchinski, S. (1985) Plant engineered to kill insects. *New Scientist* **105**, 25

Yoshikawa, H., Takiguchi, Y. and Terao, M. (1983) Terminal steps in the biosynthesis of herbicidins, nucleoside antibiotics. *J. Antibiotics* **36**, 30-5

9

Environmental Biotechnology

Microorganisms have an important role in the natural cycling of elements. This is due primarily to the vast array of microbial degradative activities. Such activities can be exploited in the control of environmental pollution, a major example being the treatment of sewage. Sewage effluent treatments typically aim to reduce organic matter in wastewaters and to decrease the number of potential pathogens, thereby rendering the water safe for discharge into the environment (for a review see Wheatley, 1984). Interest in the recycling of resources has led to the development of programmes for the conversion of biomass from wastewaters and from other sources to useful products, such as food, fuels and chemicals.

Various recalcitrant xenobiotics are released into the environment in large quantities. Microorganisms engineered for the detoxification/degradation of such persistent compounds may be used to improve the effectiveness of effluent treatment systems and to provide broader environmental control. Microorganisms are also of use in the retrieval of valuable metals from metal-contaminated wastewaters and in the cleaning up of oil spills.

Test systems for the detection and classification of DNA-damaging agents frequently employ specially constructed bacterial strains as indicator organisms. Such microbial systems permit rapid screening of the environment for genotoxic agents. In addition, **biosensors**[1] that comprise biocatalysts, from microbial sources, coupled to physicochemical devices (that monitor the activity of the catalysts in chemical transformation of test substrates) are being developed for the detection of various substances of environmental interest. Such biological systems, which enable identification of compounds with potentially adverse effects on human health and on the

[1] A biosensor consists of a biologically sensitive material (for example an enzyme, an organelle, an antibody, or a bacterial or other cell) immobilised in intimate contact with an appropriate transducing system, which together act to convert a specific biochemical reaction into a quantifiable electrical signal.

environment, and assess toxic thresholds, are of considerable value in environmental control programmes.

Microorganisms can also be exploited for the recovery of oil and valuable minerals from the environment. The application of microbial processes is particularly attractive when conventional recovery methods reach their economic limit.

This chapter describes examples of environmental biotechnologies and indicates the scope of genetic manipulation for their improvement and development.

9.1 BIOMASS CONVERSIONS

Biological raw material (biomass) can be converted to useful materials such as foods, bioplastics, fuels and other chemicals by microbial fermentation (Figure 9.1). Plant biomass is found mostly as starch in corn, wheat, potatoes, cassava and other crops, as soluble sugars in corn syrup, molasses and sulphite waste liquors and as cellulose/lignocellulose in urban refuse, animal manures and industrial wastes. Effective, low-cost production of useful commodities from biomass fermentations depends upon inexpensive and abundant biomass substrates that are readily bioconvertible and upon high yields of product from such substrates. Biopolymer-degrading organisms are of value in the direct fermentation of biomass to useful products and in the provision of enzymes that solubilise biomass polymers to mono- and disaccharides that can be used as feedstock. Starches, soluble sugars and nonlignified celluloses may be fermented directly by bacteria. However, lignocellulosic material, which forms a major component of most biomass substrates, generally requires extensive treatments to remove lignin prior to fermentation. The main constituents of lignocellulose are lignin, α-cellulose and hemicellulose. Of these lignin is recalcitrant to fermentation by most bacteria and limits the activity of microbial cellulases and hemicellulases, which hydrolyse celluloses and hemicelluloses respectively to easily fermentable substrates.

Biomass fermentations may require the combined and coordinated activities of mixed populations of bacteria. Where organic wastes serve as substrate, the organisms involved in the fermentation must be resistant to the various toxic components of the waste, in addition to possessing a diversity of degradative abilities.

Transformation of lignocellulose, cellulose or starch to fermentable sugars is an initial stage in biomass conversion to industrially important products. Genetics can be used to improve this and the subsequent fermentation process. Manipulation of organisms for enhanced biopolymer degradation is considered below.

Figure 9.1: Biomass conversions

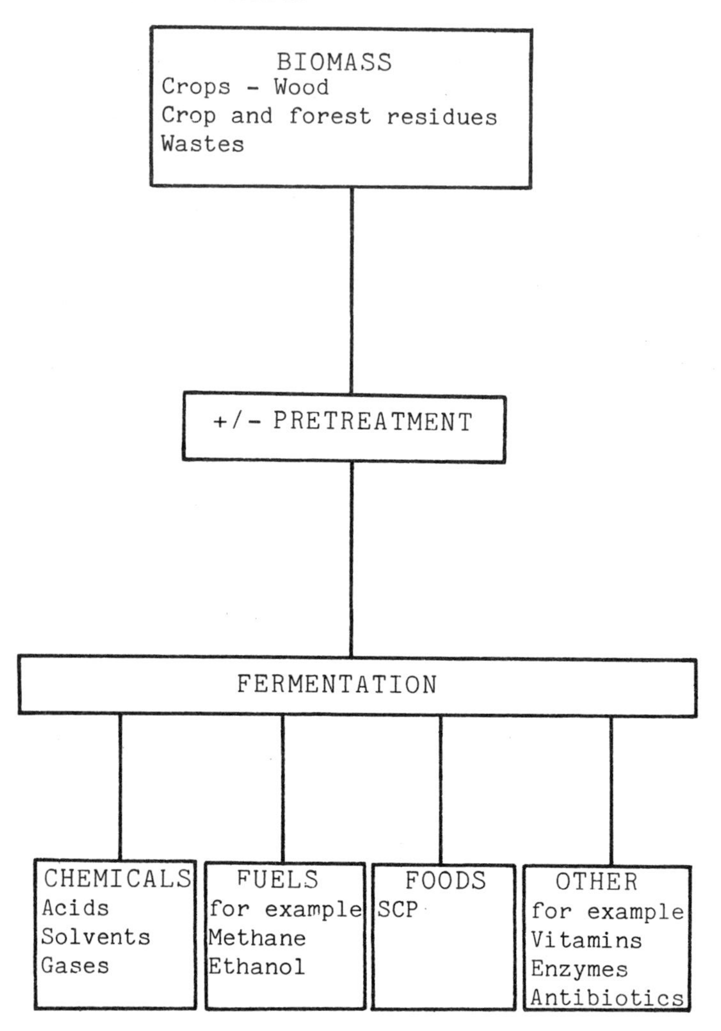

9.1.1 Lignin degradation

The processing of lignocellulosic materials often requires the removal of lignin, which acts as a barrier to the depolymerisation of the cellulose/hemicellulose. Removal of lignin by pulping can be an expensive process and causes pollution. Biodegradation affords an alternative approach to the processing of lignocelluloses. Lignin can be degraded by some filamentous fungi, principally the Basidiomycotina (for example, the white-rot fungi *Phanerochaete chrysosporium* and *Sporotrichum pulverulentum*) and by certain bacteria, such as actinomycetes (for example *Streptomyces viridosporus*) (Paterson *et al.*, 1984). Peroxidase enzymes appear to be involved in the degradative process (Crawford and Crawford, 1984; Schoemaker *et al.*, 1985). Lignin-degrading enzymes are of considerable industrial potential. They may have a role not only in extracting celluloses from plant material, but also in provision of aromatic chemicals from lignin (which is a source of oxygenated aryl rings), in the reclamation of wastewater (since the enzymes can easily oxidise phenols which frequently contaminate water) and in 'cracking' petroleum. Specific lignin-degrading production strains should find utility in the direct bioconversion of lignocellulose to useful products. Ligninolytic strains are thus being manipulated in order to improve ligninolytic activity and to extend the range of lignin-degrading organisms. Intra- and inter-species protoplast fusion techniques and UV-induced mutagenesis have been used to generate improved lignin-degrading streptomycetes (Crawford *et al.*, 1984; Pettey and Crawford, 1984). *Streptomyces viridosporus* T7A degrades lignin to a polyphenolic water-soluble modified polymer, acid precipitable polymeric lignin (APPL), which is useful as an antioxidant and surfactant and possibly as a component in adhesives and resins. The APPL intermediate is probably generated by cleavage of β-ether linkages in lignin. APPL-overproducing strains of *S. viridosporus*, obtained by mutagenesis and protoplast fusion, exhibited increased activity of the enzymes of the β-etherase complex. Further improvements to the fermentation process for production of APPLs may thus be achieved by manipulating the β-etherase complex of these bioconversion strains (Crawford *et al.*, 1984). Gene libraries (section 3.6) of ligninolytic organisms have been constructed (see Paterson *et al.*, 1984) to locate determinants for lignin breakdown. By using specific combinations of appropriately modified genes that specify the enzymes for hydrolysis of lignocellulose (see also sections 9.1.2 and 9.1.3) plant materials should be more efficiently processed and industrial exploitation of bioconversions based on lignocellulosic residues should be made economically more attractive.

9.1.2 Cellulose degradation

Various microorganisms, including *Cellulomonas fimi*, *Clostridium thermocellum*, *Streptomyces* spp., *Thermomonospora* spp. and *Trichoderma reesei*, produce cellulases, which permit the utilisation of cellulose as a carbon source. The cellulase system generally comprises at least three types of enzyme: endo-glucanases, exo-glucanases and cellobiase (β-glucosidase) for the hydrolysis of cellulose to glucose. The proportions of the different enzymes vary for different systems. The enzymes act synergistically in depolymerising cellulose to glucose. Effective use of cellulolytic organisms to convert biomass into substrate suitable for fermentation demands high rates of cellulose degradation. There are several approaches to increasing cellulosis, for example:

(i) Pretreatment of the cellulosic substrate can render it more susceptible to cellulase.
(ii) Mutants with an enhanced ability to degrade cellulose may be used, for example those resistant to catabolite repression or to end-product inhibition[2] (see sections 6.1.1 and 6.2.1.a).
(iii) Appropriate *in vitro* genetic manipulations may be employed to increase production of cellulases and to modify their catalytic properties.

Gene cloning may permit the development of strains possessing either altered proportions of the component enzymes within a cellulase system, or a complement of enzymes originating from different sources. Enzyme production might be altered by manipulating the expression of the genes specifying the cellulase system[3]. This might be achieved by gene amplification, coupling the genes to strong promoters, eliminating promoters that are sensitive to repression, and using efficient translational signals (see Chapter 5). Enzymes might also be restructured for increased stability and enhanced catalytic activity through site-specific mutagenesis (section 4.9). As a preliminary to improving cellulolytic activity, structural genes for the cellulose system, including mesophilic cellulase genes of *C. fimi* (Whittle *et al.*, 1982; Gilkes *et al.*, 1984) and thermophilic cellulase genes of *C. thermocellum* (Cornet *et al.*, 1983) and *Thermomonospora* YX (Collmer and Wilson, 1983), have been cloned and expressed in *E. coli*. The enzymes from thermophiles may be more heat-stable and more resistant to denaturation than those from mesophiles, and may therefore be particularly attractive for industrial applications. By cloning all the determinants of the cellulose system the precise nature of the synergism between the different enzymes should be elucidated.

Highly active cellulolytic strains may be constructed by combining appropriate combinations of cellulase genes. Such strains would serve as good sources of cellulase for efficient conversion of cellulose to soluble sugars for subsequent fermentation. Alternatively specific production strains might be genetically programmed for the direct fermentation of cellulose to useful product. Genes for cellobiohydrolase (exo-glucanase) and endo-glucanase from *Trichoderma reesei* have been introduced into

---

[2] Cellulase activity is subject to end-product inhibition by cellulose and glucose.
[3] Cellulases are inducible enzymes. The inducers include cellulose and cellulose derivatives.

yeast (Shoemaker, 1984) to construct strains for conversion of cellulose to ethanol.

9.1.3 Hemicellulose degradation

Hemicelluloses are polymers of xylose, galactose, mannose, arabinose, other sugars and their uronic acids. The polysaccharide is classified according to the sugar residue present, for example D-xylan, D-galactan, D-mannan and L-arabinan. Xylans constitute a major proportion of hemicelluloses in plants. The xylanase enzyme system (for degradation of xylan) includes endo- and exo-xylanases and β-xylosidases. Xylanases have been found in various bacteria such as *Bacillus* spp., *Bacteroides amylogens, Clostridium* spp. and *Aeromonas* spp., and in fungi including *Myrothecium verrucaria, Aspergillus oryzae* and *Trichoderma reesei*. The hemicellulose hydrolysate contains mainly xylose, which may be used for example as a source of xylitol (sweetener), as an antidiabetic agent and as a substrate for conversion to ethanol (by yeast). Efficient biodegradation of hemicelluloses might be effected by using similar approaches to those described for improving cellulosis (see section 9.1.2). Gene cloning techniques have been employed to manipulate the hemicellulase system. Genes for xylanases of *Bacillus* sp. and *Aeromonas* sp., for example, have been cloned and expressed in *E. coli* (Bernier *et al.*, 1983; Honda *et al.*, 1985; Kudo *et al.*, 1985).

9.1.4 Starch degradation

Utilisation of starch by microorganisms involves its conversion initially to dextrins and oligosaccharides and then to glucose. Microbial hydrolysis of starch to glucose involves several types of enzymes: amylases, glucoamylases and debranching enzymes. Organisms such as *Aspergillus* spp. and *Bacillus* spp. that degrade starch are particularly valuable in the provision of enzymes for starch saccharification. Production of α-amylase from *B. subtilis* has been improved by *in vivo* gene manipulations (see section 6.2.2.c). Furthermore, high yields of a heat-stable α-amylase, for use in high-temperature conversion of starch to glucose, have been obtained by cloning an amylase gene from a thermophile into *B. subtilis* (section 6.3.2). Yeast strains capable of hydrolysing starch completely (by synthesising and secreting α-amylase and glucoamylase with debranching activity) should prove useful in the production of ethanol directly from starch (see, for example, Stewart, 1984; Filho *et al.*, 1986).

9.2 POLLUTION CONTROL

Microbial degradation of hydrocarbons has a part to play in the control of environmental pollution and in hydrocarbon interconversions for industrial purposes. Certain microorganisms, chiefly of the genus *Pseudomonas*, have the capacity to degrade a variety of hydrocarbons and are thus potentially useful in the control of various environmental pollutants. Such organisms often possess **degradative plasmids** (see Table 9.1) that encode all or part of hydrocarbon oxidation pathways that feed into central metabolism (Williams, 1981). Generally these oxidation pathways are inducible and comprise two sets of activities: (i) upper pathway enzymes with specialised activities, for conversion of the initial hydrocarbon to a substrate for (ii) lower pathway enzymes with more generalised activities for conversion of this substrate to intermediates of central metabolism. A well characterised pathway is that for the degradation of toluene and xylenes encoded by the TOL plasmid, pWWO (Figure 9.2) (see Lebens and Williams, 1985; Timmis *et al.*, 1985). In some cases there is overlap between different plasmid-encoded oxidation pathways or between plasmid and chromosome oxidation pathways. Furthermore, there can be different substrate specificities between different enzymes that catalyse the same pathway step. Full expression of particular plasmid-specified catabolic phenotypes requires interaction of chromosomal and plasmid gene products. Chromosomal mutations can effect inhibition of specific plasmid-encoded oxidation steps and/or can block further (intermediary) metabolism of products of initial plasmid-determined reactions. Utilisation of hydrocarbon substrates for growth therefore demands compatibility between host and plasmid-determined metabolic activities. A variety of environmental conditions, including temperature, pH, salinity, oxygen tension and the presence of

Table 9.1: Some degradative plasmids

Plasmid	Substrate	Conjugative ability
TOL	toluene, meta-xylene, para-xylene	+
CAM	camphor	+
OCT	octane, hexane, decane	–
SAL	salicylate	+
NAH	naphthalene	+
NIC	nicotine/nicotinate	+
pJP1	2, 4-dichlorophenoxyacetic acid	+
pAC21	4-chlorobiphenyl	+
pAC25	3-chlorobenzoate	+
pAC27	3- and 4-chlorobenzoate	+

Figure 9.2: The TOL plasmid pWWO and plasmid-specified pathway for the degradation of toluene and xylenes. (a) Map of pWWO showing location of upper and lower (*meta*) pathway genes. Genes for enzymes of the upper pathway are organised in an operon: promoter, *xylC, xylA, xylB*. Genes that have been localised for enzymes of the lower pathway are also arranged in an operon: promoter, *xylD, xylL, xylE, xylG, xylF, xylJ, xylI, xylH*. Tra/Rep, region for conjugal transfer and replication. (b) Pathway for the degradation of toluene and xylenes. XO, xylene oxygenase; BADH, benzylalcohol dehydrogenase; BZDH, benzaldehyde dehydrogenase; TO, toluate/benzoate dioxygenase; DHBDH, dihydrodihydroxybenzoate dehydrogenase; C230, catechol 2,3-dioxygenase; HMSH, hydroxymuconic semialdehyde hydrolase; HMSD, hydroxymuconic semialdehyde dehydrogenase; 4-OT, 4-oxalocrotonate tautomerase; 4-OD, 4-oxalocrotonate decarboxylase; OEH, 2-oxopent-4-enoate hydratase; HOA, 2-oxo-4-hydroxypentonate aldolase. The initial compounds are toluene ($R_1 = R_2 = H$), *m*-xylene ($R_1 = CH_3$; $R_2 = H$) and *p*-xylene ($R_1 = H$; $R_2 = CH_3$). Catabolic genes *xylA* to *xylK* are shown (adapted from Timmis *et al.*, 1985)

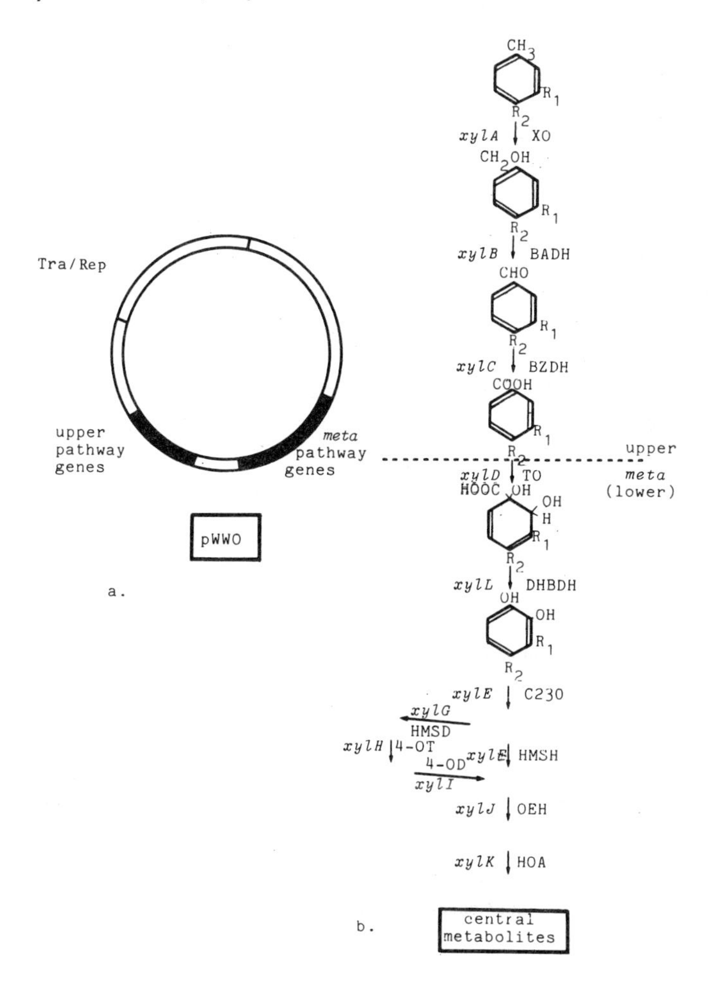

toxicants, also limit the biodegradation of hydrocarbons.

Main approaches to the construction of enhanced hydrocarbon-oxidising strains for use in pollution control involve *in vivo* manipulation of degradative plasmids and/or *in vitro* cloning of genes for specific catabolic enzymes. Substrate range may be extended simply by constructing strains that carry a number of degradative plasmids. Novel hybrid oxidation pathways may be constructed by exploiting the overlap that can occur between various oxidation pathways, encoded by different plasmids or the chromosome. This may be achieved through 'plasmid-assisted molecular breeding' and mutation (Kellogg *et al.*, 1981). A collection of strains, possessing appropriate degradative plasmids, is grown in a chemostat under specific enrichment conditions. The degradative plasmids provide gene pools for the selective evolution of new degradative pathways. This may involve genetic recombination, gene duplications and divergence (see section 9.2.2). By such mechanisms new combinations of upper and lower pathway enzymes can be generated. For example, broad substrate specificity upper pathway enzymes can be recruited to enhance degradative abilities. New combinations of pathway enzymes might also be generated by gene cloning. This methodology permits a more controlled approach to new pathway constructions and enables regulation of expression of specific enzymes. Catabolic genes may be placed under the control of constitutive promoters. In this way an enzyme might be recruited even though the pathway substrate and intermediates do not induce its expression *in vivo*. Furthermore expression of genes for growth-rate-limiting enzymes may be enhanced. However, formation of fully functional novel pathways will depend upon constructing strains with coordinated sets of sequential enzymatic activities. Defective pathways (where, for example, enzymatic activities are mismatched or mutations effect specific pathway blocks) may, however, prove valuable in the development of bioconversion systems for the production of particular metabolic intermediates.

9.2.1 Degradation of petroleum hydrocarbons

Oil pollution can be controlled through the use of specific hydrocarbon-degrading microorganisms. Biological dispersal of oil pollutants may be preferred since many of the detergents that are normally used are not biodegradable. Multiplasmid strains capable of degrading various hydrocarbons in crude oil have been obtained by plasmid transfer. Friello and co-workers (1976) constructed a strain of *Pseudomonas* (termed 'superbug') that harboured a hybrid CAM-OCT, an NAH and a TOL plasmid. This strain was more proficient at degrading crude oil than was a mixture of individual strains each carrying a single plasmid. However, it is questionable whether this multiplasmid strain, which is capable of breaking down camphor, octane, naphthalene, toluene and xylenes, is any more effective in degrading the oil than a mixed microbial population that can break down these and other constituents of the oil. Problems that may be associated with the construction of multiplasmid strains, such as incompatibility, recombination between plasmids and plasmid instability, might be circumvented by cloning genes for a range of degradative functions into a single plasmid. The availability of a broad host range plasmid for the degradation of oil should permit the development of a variety of strains capable of utilising oil wastes from various sources.

9.2.2 Biodegradation of xenobiotics

The release of vast quantities of synthetic chemicals, such as solvents, plasticisers, insecticides, herbicides and fungicides, into the environment, through industrial, agricultural, medical and domestic activities, has created considerable toxicological problems. Some of these chemicals, for example halogenated herbicides and insecticides, persist in nature due to their low rates of biodegradation by the indigenous microbial flora. The development of strains that are capable of enhanced degradation of persistent, toxic chemicals in the environment is thus a desirable objective.

Certain organisms, notably *Pseudomonas* spp., can cometabolise[4] various xenobiotics, including the herbicides 2,4-D (2,4-dichlorophenoxyacetic acid), Dalapon (2,2′-dichloropropionic acid) and 2,4,5-T (2,4,5-trichlorophenoxyacetic acid) in nature. Normally such compounds are oxidised slowly and are not degraded totally.

Genetic determinants for the degradation of various synthetic chemicals, for example simple chlorinated phenoxyacetic or benzoic acids, are encoded on conjugative plasmids (see Table 9.1). By exploiting appropriate combinations of degradative plasmids, strains have been constructed that effect total degradation of specific halogenated aromatics. For example, plasmid pAC25, which permits total degradation of 3-chloro- but not 4-chloro-benzoate, can cooperate with the TOL plasmid (specifying toluene and xylene degradation) for the complete metabolism of both these chlorinated compounds (Reineke and Knackmuss, 1979; Chatterjee *et al.*, 1981). The TOL plasmid encodes a broad substrate-specific benzoate oxygenase, facilitating the conversion of 4-chlorobenzoate to 4-chlorocatechol, which can be degraded by pAC25-encoded enzymes. 4-Chlorobenzoate-degrading *Pseudomonas putida* strains have been isolated by growing cells harbouring pAC25 with TOL-containing cells in a chemostat under

[4] Cometabolism has been used to describe the process in which microorganisms, while growing at the expense of one substrate, transform another compound without deriving any direct benefit from its metabolism (Alexander, 1981).

selective conditions, with 4-chlorobenzoic acid as major carbon and energy source. During enrichment, TOL sequences became transposed to the chromosome, and plasmid pAC27, carrying all the 3-chlorobenzoate degradative genes, was generated from pAC25 by deletion (Chatterjee and Chakrabarty, 1982) (see Figure 9.3). In addition 3,5-dichlorobenzoate-degrading strains could be obtained by introducing the TOL plasmid into cells capable of utilising 4-chlorobenzoate and selecting with 3,5-dichlorobenzoate. These strains harboured both pAC27 and a second plasmid, pAC29, which comprised replication/incompatibility functions of TOL and duplicate copies of a part of pAC27, with modified chlorobenzoate degradative genes specifying utilisation of 3,5-dichlorobenzoate (Figure 9.3).

Development of plasmids specifying novel degradative abilities through the recruitment of genes (that may have undergone duplication and divergence to provide new enzymatic activities) from other plasmids has also formed the basis of a programme for the selective development of a strain capable of degrading the persistent and highly toxic herbicide, 2,4,5-T (a component of Agent Orange) (Chakrabarty *et al.*, 1984). The selective method for breeding such a strain (outlined in Figure 9.4) involved the mixing of strains of *Pseudomonas putida* harbouring a variety of degradative plasmids, such as TOL, SAL and pAC25, with strains from chemical waste-dumping sites, in a chemostat. Selection with 2,4,5-T as sole carbon source led ultimately to the isolation of a single strain of *Pseudomonas cepacia* AC1100 capable of utilising 2,4,5-T and of dechlorinating a number of chlorophenols. Furthermore, this strain could dehalogenate the corresponding fluoro- or bromo-analogues of these compounds. Treatment of contaminated soil with *P. cepacia* AC1100 resulted in a dramatic reduction in the concentration of 2,4,5-T. Moreover, once the 2,4,5-T had been consumed, the strain proved to be competitively inferior to the indigenous microflora (Karns *et al.*, 1984). It should be noted that small amounts of 2,4,5-T (or a breakdown product) remained in the soil despite AC1100 treatments, possibly because some of the herbicide became bound irreversibly to soil particles and was thereby unavailable for microbial degradation. Thus, although microbial systems may be used to clean up contaminated soils, low concentrations of some toxic chemicals may escape effective degradation and remain a hazard.

The extended application of plasmid-assisted breeding under selective conditions in a chemostat should permit the isolation of organisms with an assortment of new degradative abilities for the utilisation of particular toxic compounds. Gene cloning may be necessary for the construction of strains capable of degrading certain recalcitrant molecules that cannot be utilised by strains developed by plasmid-assisted breeding. Furthermore, various pathway constructions generated through plasmid-assisted breeding might be obtained by cloning relevant catabolic genes (see Weightman *et al.*,

Figure 9.3: Development of new degradative functions in pseudomonads. (a) Development of a 4-chlorobenzoate-degrading strain. Plasmid pAC25 specifies degradation of 3-chlorobenzoate, and the TOL plasmid specifies degradation of toluene and xylene. During enrichment with 4-chlorobenzoate, TOL sequences are transposed to the chromosome and pAC27 is formed from pAC25 by deletion. pAC27 contains all the genes for 3-chlorobenzoate degradation. (b) Development of a 3,5-dichlorobenzoate-degrading strain. The TOL plasmid is introduced into a cell harbouring pAC27. Strains selected in the presence of 3,5-dichlorobenzoate contain pAC27 and pAC29. pAC29 contains replication/incompatibility functions of TOL and duplicate copies of portions of pAC27 with mutations in chlorobenzoate genes. ★, Transposable TOL sequences specifying toluate degradation; ●, replication determinants; *x,y,z,* represent chlorobenzoate degradative genes; *x′*, *y′*, *z′*, mutant genes for degradation of 3,5-dichlorobenzoate; 3Cba$^+$. degrades 3-chlorobenzoate; 4Cba$^+$, degrades 4-chlorobenzoate; 3,5-Dcb$^+$, degrades 3,5-dichlorobenzoate (from Chatterjee and Chakrabarty, 1982)

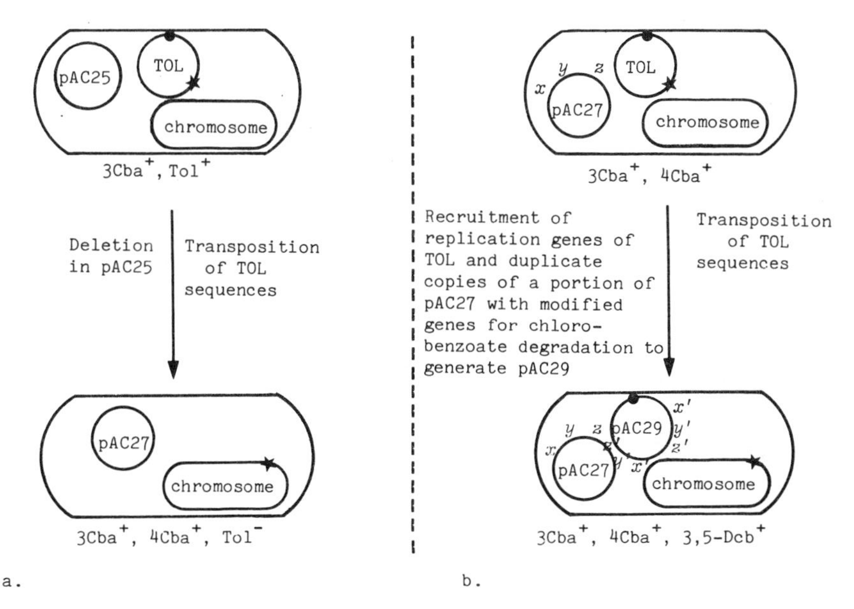

1984). Acquisition of novel degradative functions by a wide range of soil and aquatic microorganisms is likely to enhance the rate of degradation of pollutants and hasten their ultimate removal from the environment. Potential problems associated with this approach include competition by indigenous microflora (for minerals, growth factors and so on) and the presence of innumerable carbon sources in the soil that may be utilised in preference to the toxic chemicals. As an alternative to the use of engineered microorganisms for degradation of specific wastes, suitably manipulated microbial enzymes may be used as sprays or immobilised systems.

Figure 9.4: Selective method for the development of 2,4,5-trichlorophenoxyacetic acid-degrading strain (adapted from Chakrabarty *et al.*, 1984)

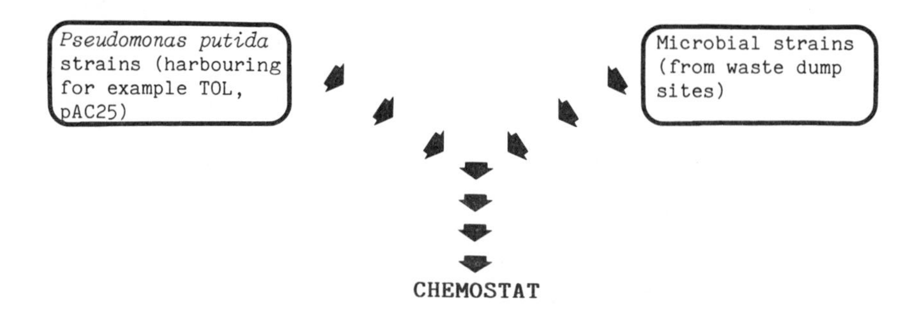

Chemostat operated with substrates such as toluate, chlorobenzoate for 2 weeks. The concentration of these substrates was then gradually reduced, whilst concentration of 2,4,5-T was gradually increased.

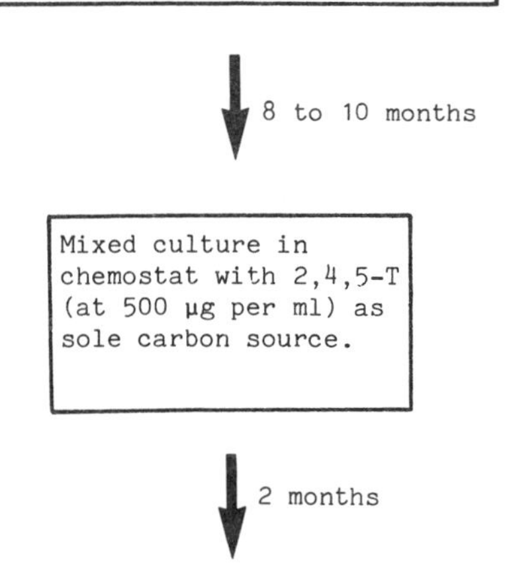

Pure culture of *Pseudomonas cepacia* AC1100 capable of rapidly degrading 2,4,5-T.

9.3 MICROBIAL ENHANCEMENT OF OIL RECOVERY (MEOR)

Only about one-third of the oil in underground reservoirs can be profitably extracted using conventional oil recovery techniques.[5] The development of other, economic methods for recovering extra oil (**enhanced oil recovery, EOR**) is thus desirable. Chemical, physical and biological strategies for EOR are being assessed. Of these the application of microorganisms and/or microbial products to EOR may have certain advantages in terms of cost-effectiveness and improved efficiency (Springham, 1984).

MEOR involves three general approaches:

(1) Stimulation of the indigenous microbial population of a reservoir by injection of requisite nutrients in order to enhance microbial activities related to increased oil production.
(2) Injection of specific microorganisms with beneficial *in situ* activities into a reservoir. Such activities include biopolymer and biosurfactant production, bioleaching of rock matrices and gas production. Gas production effects a partial repressurisation of the reservoir. Carbon dioxide is miscible with the oil and increases oil volume while lowering oil viscosity, thereby making the oil easier to displace. Acid production can effect modification of the rock structure, making the rock more permeable. Inocula of mixed cultures comprising genera such as *Pseudomonas, Escherichia, Arthrobacter, Mycobacterium, Micrococcus, Peptococcus, Bacillus* and *Clostridium* have been used in field trials. Where the appropriate oil type and reservoir conditions prevail, increased oil production has been achieved (see Finnerty and Singer, 1983). Cultures of slime-forming bacteria might be employed for the selective plugging of regions of a reservoir of high permeability.
(3) Use of microbial products, including biopolymers, biosurfactants, fermentation alcohols, acids and ketones, as chemically enhanced oil recovery (CEOR) agents. A major biopolymer used is the extracellular polysaccharide, xanthan gum, which serves to increase the viscosity of injected waters. Surfactants, such as emulsan, reduce the oil-water interfacial tension and enhance oil flow.

Where microorganisms are to be injected into a formation (strategy 2), genetics might be used to combine various desired properties (with respect

---

[5] Conventional methods widely used by oil companies involve primary (water, gas cap or solution gas drives) and secondary (waterflooding and pressure maintenance) processes. Production of oil by these means generally becomes uneconomic when, on average, 30 to 35% of the oil has been recovered (for discussion, see Moses and Springham, 1982).

to oil recovery) into a single strain or a consortium of strains. Such properties include:

(i) small size, to permit efficient penetration of rock strata;
(ii) tolerance to reservoir conditions, such as high temperature, high pressure, high salinity;
(iii) non-fastidious nutritional requirements, including an ability to utilise hydrocarbons at low oxygen tension;
(iv) high yield of products that promote mobilisation of crude oil;
(v) tolerance to substances, for example biocides and corrosion inhibitors, injected with the waterflood used to displace the oil.

The organism(s) of choice for EOR should be the dominant microbial population. Potentially competing organisms could be kept to a minimum by applying biocides to which the chosen organism could be made resistant. In addition, undesirable properties of the bacteria, including plugging of the formation by excessive growth and slime production, promotion of corrosion of pipelines and production of substances having deleterious effects on the oil (such as 'souring'), might be eliminated by appropriate manipulations.

Bioproducts to be used as CEOR agents (strategy 3) may be specifically tailored for increased oil production. Improvements to biopolymers might include; increased shear resistance, higher solution viscosity and resistance to biodegradation in the reservoir. Biosurfactants with an enhanced ability to lower the oil-water interfacial tension would also be desirable.

9.4 MICROBIAL MINING AND METAL RECOVERY

Microorganisms can be used in the extraction of metals from ores and in the recovery of metal ions from solutions (Brierley, 1982, 1984; Curtin, 1983; Monroe, 1985).

9.4.1 Metal extraction from ores

The high operational costs, the limited supply of accessible high-grade metal ores and the environmental pollution associated with conventional mining techniques have stimulated interest in the development of effective, low-cost biological techniques for metal extraction from ores and mine wastes. Metals can be extracted by leaching, which involves cycling an acidic oxidising liquor through the ore to solubilise the metals that can then be recovered from solution. In microbial leaching, bacteria of genera such as *Thiobacillus* and *Sulpholobus* assist directly or indirectly in the solubilisation process. These organisms are able to derive energy from the oxidation of ferrous iron, sulphur and sulphides. Such bacterial activities result in the production of ferric sulphate and sulphuric acid, which constitute the lixiviant that is active in microbial leaching. In addition, various mineral sulphides can be attacked directly by the bacteria, resulting in solubilisation of the metal. Microorganisms may be applied both in the primary recovery of metals from ores and in secondary recovery processes, where the material has already been subjected to conventional processing methods. Microbial leaching processes are generally slower and less efficient than some of the chemical methods of metal extraction, but are less expensive. Increased process efficiency might be attained through appropriate gene manipulations to enhance the leaching capabilities of the bacteria. An increased tolerance to heat, fluctuations in acidity, toxic metals and O_2 deficiency and an improved ability to generate the oxidant, ferric iron, are desired qualities. Implementation of such strain improvements will, however, depend upon the availability of effective gene transfer and cloning systems for the leaching bacteria and on a thorough understanding of the mechanisms involved in the degradation of minerals, which is presently lacking. Although microbial leaching is exploited chiefly for the extraction of copper and uranium, the technique could potentially be applied to the recovery of a variety of metals. Furthermore, since leaching bacteria catalyse the dissolution of inorganic sulphur from coal, pre-combustion desulphurisation of sulphur-bearing coal is possible.

9.4.2 Metal recovery from solution

The ability of microorganisms, such as *Saccharomyces cerevisiae*, *Pseudomonas aeruginosa* and *Rhizopus arrhizus*, to accumulate metal ions, either by adsorption on to the cell surface or by intracellular uptake, can be exploited to recover metals from solution. Such activities serve a dual function of recovering valuable metals (such as gold, silver and platinum) and of removing toxic metals from wastewaters. The accumulation processes involve the electrostatic binding of positively charged ions to the negatively charged cell surface. Such processes are rapid, but relatively nonspecific. Specificity might be increased by introducing the ability to synthesise large quantities of appropriate metal-binding agents (such as metallothioneins) into microorganisms. Metallothioneins, which are found in many eukaryotes and some prokaryotes, have strong affinities for heavy metals such as gold, silver and lead. Genes for metallothioneins have been cloned in bacteria for purposes of recovering heavy metals (see, for example, Curtin, 1983). Ion-exchange resins, which are conventionally used to clean up metal-contaminated industrial wastewater, might thus ultimately be replaced by immobilised recombinant bacteria containing efficient metal-binding proteins or by immobilised metallothioneins themselves.

9.5 MEASUREMENT OF GENOTOXICITY

A number of bacterial tests, including the mutatest (Ames test), inductest (Devoret test) and SOS chromotest may be used for detecting DNA-damaging agents. Such tests can be used in environmental screening programmes for the detection of potentially mutagenic-carcinogenic compounds.

9.5.1 Mutatest (*Salmonella*/mammalian-microsome mutagenicity test)

In the mutatest (Ames test) carcinogens/mutagens may be detected by virtue of their ability to induce reversion of nutritional (His^-) mutants of *Salmonella typhimurium* (Ames *et al.*, 1975). A series of specially constructed *S. typhimurium* strains that are extremely sensitive to the action of mutagens is used in the test. Generally the tester strains are defective in DNA excision repair (*uvrB*$^-$) (see section 4.3.2); are lipopolysaccharide deficient (*rfa*$^-$) (which facilitates entry of chemicals into the cell); possess plasmids such as pKM101 (which exert mutator activity, thereby stimulating conversion of DNA damage into heritable mutations); and are *his*$^-$.[6] These strains contain different types of histidine mutation (such as base substitutions or frameshifts) that permit the detection of different classes of mutagen.

The mutagenic potential of a compound is tested by mixing it with a tester (His^-) strain and determining the numbers of prototrophic (His^+) revertants induced (Figure 9.5). Where the number of induced revertants is significantly greater than the number of spontaneous revertants, the compound is presumed to be acting as a mutagen. (Mutagenic potency of a compound is the histidine-independent revertants per nmol per plate.) Compounds that are potential human carcinogens/mutagens and that require metabolic activation can be detected by incorporating a mammalian microsomal activation preparation (homogenate of rat or human liver) into the test. Various glycosides are potentially mutagenic and carcinogenic. Removal of the sugar groups can render such compounds mutagenic. Incorporation of a cell-free extract of human faeces (fecalase), which contains bacterial glycosidases, into the test system can permit detection of

[6] pKM101 encodes *mucA* and *mucB* functions (analogues of *umuD* and *umuC* functions of *E. coli* (see section 4.3.4)), which enhance the amount of mutagenesis resulting from DNA-damaging treatments.

Figure 9.5: The Ames spot test. The agar plate contains in an overlay of top agar the tester strain, with or without a liver microsomal activation preparation. Test compounds are applied to filter paper discs and placed on the plate. Mutagenicity is assessed from the number of 'mutagen'-induced revertants (A) compared with spontaneous revertants (B)

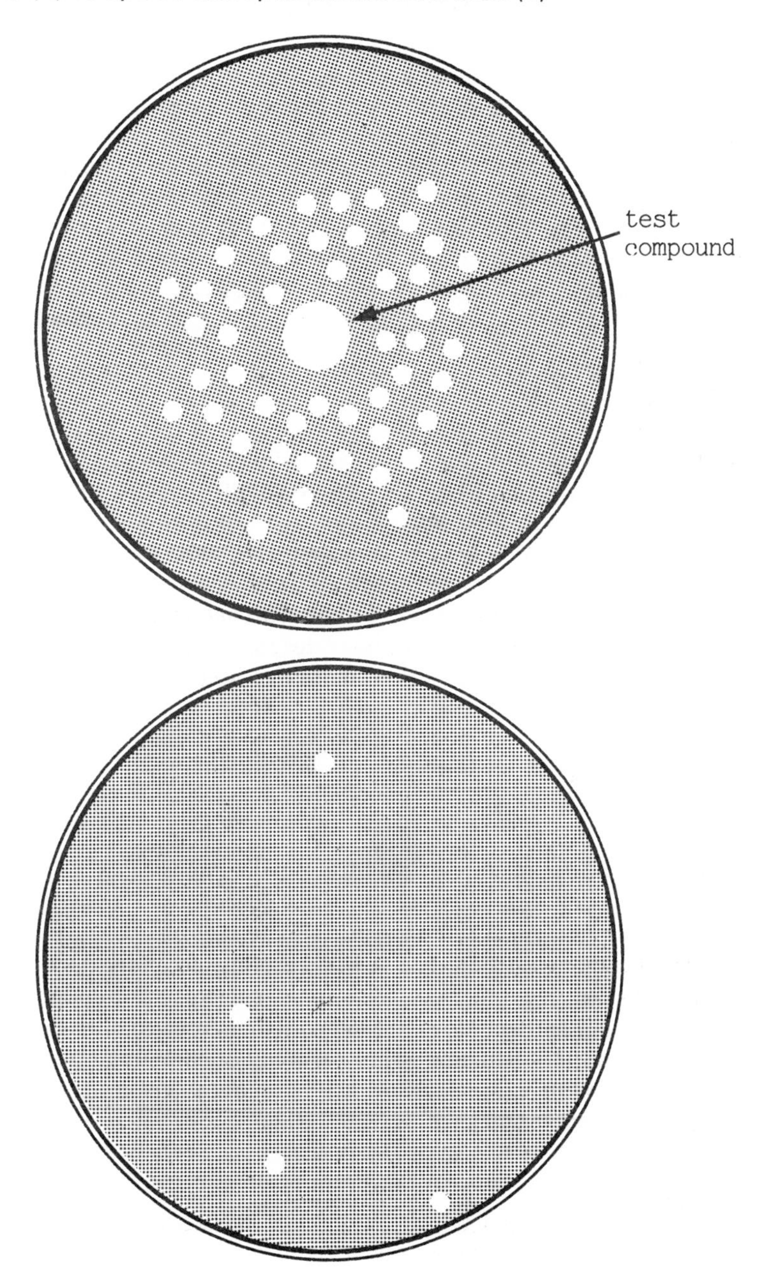

compounds that require deglycosylation for activation. The usefulness of the Ames test as a rapid and economical screen for carcinogens is based on the premise that many carcinogens are also mutagens. The carcinogenicity of compounds that are positive in the Ames test can be determined from their ability to produce tumours in laboratory animals. Good correlation has been found between mutagenicity and carcinogenicity (see Table 9.2; McCann *et al.*, 1975; Quillardet *et al.*, 1982).

9.5.2 Inductest

The inductest (Devoret test) is a phage induction assay utilising λ lysogens of *E. coli.* Detection of carcinogens is based upon the ability of such compounds to induce prophage λ in specific strains of *E. coli* (Moreau *et al.*, 1976). The λ lysogen is *envA*$^-$ (which confers increased permeability to various compounds) and is deficient in excision repair (*uvrB*$^-$) (see section 4.3.2). The test compound can be activated with microsomal enzymes (see section 9.5.1). In the inductest, λ lysogens are plated with indicator bacteria and the test compound is applied. The efficiency of induction is calculated as the ratio of the number of infective centres to the number of original viable cells. (Potency of the compound is taken as the inverse value of amount of the compound in nmol per ml for half maximal induction.) Inclusion of an antibiotic, such as ampicillin (to which the lysogen is sensitive), prevents noninduced bacteria from releasing phages during overnight incubation of the plates (thereby eliminating spontaneous phage production), but does not inhibit the induced lysogens from forming infective centres.

9.5.3 SOS Chromotest

This test relies upon the ability of DNA-damaging agents to induce SOS functions in *E. coli* (see section 4.3.4). One SOS response, which is controlled by the *sulA* (*sfiA*) gene, is inhibition of cell division. A colorimetric assay of induction of this SOS response to DNA damage has been devised by placing the *lacZ* gene for β-galactosidase under the control of *sulA* (Quillardet *et al.*, 1982). Evaluation of the SOS-inducing potency (SOSIP) (induction factor per nmol of compound per assay) of the test compound is based upon its ability to induce *sulA* in *E. coli.* The tester strain carries a *sulA*::*lacZ* operon fusion, a deletion for the normal *lac* region, is deficient in excision repair (*uvrA*$^-$) (see section 4.3.2) and is lipopolysaccharide deficient (*rfa*$^-$) (to facilitate entry of chemicals into the cell). The strain is incubated with the test compound, and β-galactosidase activity (which is strictly dependent upon *sulA* expression) is monitored.

Table 9.2: Responses to bacterial mutagen tests

Compound	Potency*			Carcinogenicity
	SOS chromotest	Inductest	Mutatest	
Fungal toxin: aflatoxin B1	75	0.5	12000	+
Antibiotics: neocarcinostatin	390	0.02	—	Carcinostatic
mitomycin C	70	0.5	—†	Carcinostatic
Esters: ethyl methanesulphonate	0.0004	—	0.03	+
dimethyl sulphate	0.04	—	0.1	+
Nitrosamine: *N*-methyl-*N'*-nitro-*N*-nitrosoguanidine	0.9	0.007	44	+
Benzofuran: 2-nitrobenzofuran	33	0.4	350	
Naphthofuran: 2-nitro-7-methoxynaphtho (2-1-*b*) furan	26000	1200	200000	+
Other: aspirin	—		—	—
caffeine	—		—	?
4-nitroquinoline 1-oxide	71	0.9	2100	+

* Potency for SOS chromotest is induction factor per nmol per assay; for inductest is inverted value of amount of compound in nmol per ml for half-maximal induction; for mutatest is histidine-independent revertants per nmol per plate.

† Mitomycin C is mutagenic in uvr^+ strain.

— Compound had only background activity (data from Quillardet *et al.*, 1982).

The microsomal preparation used to activate carcinogens (see section 9.5.1) may be included in the incubation mixture.

BIBLIOGRAPHY

Alexander, M. (1981) Biodegradation of chemicals of environmental concern. *Science* **211**, 132-8

Ames, B.N., McCann, J. and Yamasaki, E. (1975) Methods for detecting carcinogens and mutagens with the *Salmonella* mammalian microsome mutagenicity test. *Mutat. Res.* **31**, 347-64

Bernier, R., Jr, Driguez, H. and Desrochers, M. (1983) Molecular cloning of a *Bacillus subtilis* xylanese gene in *Escherichia coli. Gene* **26**, 59-65

Brierley, C.L. (1982) Microbiological mining. *Scient. Amer.* **247**, 42-51

Brierley, C.L. (1984) Microbiological mining: technology status and commercial opportunities. In *Biotech 84. The world biotech report 1984, Vol. 1: Europe*, pp. 599-609. Online Publications, Middlesex and New York

Chakrabarty, A.M., Karns, J.S., Kilbane, J.J. and Chatterjee, D.K. (1984) Selective evolution of genes for enhanced degradation of persistent, toxic chemicals. In *Genetic manipulation. Impact on man and society* (W. Arber, K. Illmensee, W.J. Peacock and P. Starlinger, eds), pp. 43-54. ICSU Press, Cambridge, New York and Melbourne

Chatterjee, D.K. and Chakrabarty, A.M. (1982) Genetic rearrangements in plasmids specifying total degradation of chlorinated benzoic acids. *Mol. Gen. Genet.* **188**, 279-85

Chatterjee, D.K., Kellogg, S.T., Watkins, D.R. and Chakrabarty, A.M. (1981) Plasmids in the biodegradation of chlorinated aromatic compounds. In *Molecular biology, pathogenicity and ecology of bacterial plasmids* (S.B. Levy, R.C. Clowes and E.L. Koenig, eds), pp. 519-28. Plenum Press, New York

Collmer, A. and Wilson, D.B. (1983) Cloning and expression of a *Thermomonospora* YX endocellulase gene in *E. coli. Biotechnology* **1**, 594-601

Cornet, P., Millet, J., Beguin, P. and Aubert, J-P. (1983) Characterization of two *cel* (cellulose degradation) genes of *Clostridium thermocellum* coding for endoglucanases. *Biotechnology* **1**, 589-94

Crawford, R.L. and Crawford, D.L. (1984) Recent advances in studies of the mechanisms of microbial degradation of lignins. *Enzyme Microbiol. Technol.* **6**, 434-42

Crawford, D.L., Pettey, T.M., Thede, B.M. and Deobald, L.A. (1984) Genetic manipulation of ligninolytic *Streptomyces* and generation of improved lignin-to-chemical bioconversion strains. In *Proceedings 6th symposium on biotechnology for fuels and chemicals. Biotechnol. Bioeng. symp.* **14**, 241-56

Curtin, M.E. (1983) Microbial mining and metal recovery: corporations take the long and cautious path. *Biotechnology* **1**, 229-35

Filho, S.A., Galembeck, E.V., Faria, J.B. and Frascino, A.C.S. (1986) Stable yeast transformants that secrete functional α-amylase encoded by cloned mouse pancreatic cDNA. *Biotechnology* **4**, 311-15

Finnerty, W.R. and Singer, M.E. (1983) Microbial enhancement of oil recovery. *Biotechnology* **1**, 47-54

Friello, D.A., Mylroie, J.R. and Chakrabarty, A.M. (1976) Use of genetically engineered multi-plasmid microorganisms for rapid degradation of fuel hydrocarbons. In *Proceedings 3rd international biodegradation symposium* (J.M. Sharpley and A. Kaplan, eds), pp. 205-14. Applied Science Publishers, London

Gilkes, N.R., Kilburn, D.G., Langsford, M.L., Miller, R.C. Jr, Wakarchuk, W.W., Warren, R.A.J., Whittle, D.J. and Wong, W.K.R. (1984) Isolation and characterization of *Escherichia coli* clones expressing cellulase genes from *Cellulomonas fimi. J. Gen. Microbiol.* **130**, 1377-84

Honda, H., Kudo, T. and Horikoshi, K. (1985) Molecular cloning and expression of the xylanase gene of alkalophilic *Bacillus* sp. *J. Bacteriol.* **161**, 784-5

Karns, J.S., Kilbane, J.J., Chatterjee, D.K. and Chakrabarty, A.M. (1984) Microbial biodegradation of 2,4,5-trichlorophenoxyacetic acid and chlorophenols. In *Genetic control of environmental pollutants.* Basic Life Sciences, vol. 28 (G.S. Omenn and A. Hollaender, eds), pp. 3-21. Plenum Press, New York

Kellogg, S.T., Chatterjee, D.K. and Chakrabarty, A.M. (1981) Plasmid-assisted molecular breeding — new technique for enhanced biodegradation of persistent toxic chemicals. *Science* **214**, 1133-5

Kudo, T., Ohkoshi, A. and Horikoshi, K. (1985) Molecular cloning and expression of a xylanase gene of alkalophilic *Aeromonas* sp. no. 212 in *Escherichia coli. J. Gen. Microbiol.* **131**, 2825-30

Lebens, M.R. and Williams, P.A. (1985) Complementation of deletion and insertion mutants of TOL plasmid pWWO: regulation implications and location of the *xylC* gene. *J. Gen. Microbiol.* **131**, 3261-9

McCann, J., Choi, E., Yamasaki, E. and Ames, B.N. (1975) Detection of carcinogens as mutagens in the *Salmonella*/microsome test. *Proc. Natl Acad. Sci. USA* **72**, 5135-9

Monroe, D. (1985) Microbial metal mining. *Int. Biotechnol. Lab.* **3**, 19-29

Moreau, P., Bailone, A. and Devoret, R. (1976) Prophage λ induction in *Escherichia coli* K12 *envA uvrB*: a highly sensitive test for potential carcinogens. *Proc. Natl Acad. Sci. USA* **73**, 3700-4

Moses, V. and Springham, D.G. (1982) *Bacteria and the enhancement of oil recovery.* Applied Science Publishers, London

Paterson, A., McCarthy, A.J. and Broda, P. (1984) The application of molecular biology to lignin degradation. In *Microbiological methods for environmental biotechnology* (J.M. Grainger and J.M. Lynch, eds), pp. 33-68. Academic Press, London and New York

Pettey, T.M. and Crawford, D.L. (1984) Enhancement of lignin degradation in *Streptomyces* spp. by protoplast fusion. *Appl. Environ. Microbiol.* **47**, 439-40

Quillardet, P., Huisman, O., D'Ari, R. and Hofnung, M. (1982) SOS chromotest, a direct assay of induction of an SOS function in *Escherichia coli* K-12 to measure genotoxicity. *Proc. Natl Acad. Sci. USA* **79**, 5971-5

Reineke, W. and Knackmuss, H-J. (1979) Construction of haloaromatic utilizing bacteria. *Nature (London)* **277**, 385-6

Schoemaker, H.E., Harvey, P.J., Bowen, R.M. and Palmer, J.M. (1985) On the mechanism of enzymatic lignin breakdown. *FEBS Lett.* **183**, 7-12

Shoemaker, S.P. (1984) The cellulase system of *Trichoderma reesei*: *Trichoderma* strain improvement and expression of *Trichoderma* cellulases in yeast. In *Biotech 84. The world biotech report 1984, vol. 2, USA*, pp. 593-600. Online Publications, Middlesex and New York

Springham, D.G. (1984) Microbiological methods for the enhancement of oil recovery. In *Biotechnology and genetic engineering reviews*, vol. 1 (G.E. Russell, ed.), pp. 187-221. Intercept, Newcastle upon Tyne

Stewart, G.G. (1984) Recent development of genetically manipulated industrial yeast strains. In *Biotech 84. The world biotech report 1984, vol. 2, USA*, pp. 467-88. Online Publications, Middlesex and New York

Timmis, K.N., Lehrbach, P.R, Harayama, S., Don, R.H., Mermod, N., Bas, S., Leppik, R., Weightman, A.J., Reineke, W. and Knackmuss, H.J. (1985) Analysis and manipulation of plasmid-encoded pathways for the catabolism of aromatic compounds by soil bacteria. In *Plasmids in biology.* Basic Life Sciences, vol. 30 (D.R. Helinski, S.N. Cohen, D.B. Clewell, D.A. Jackson and A. Hollaender, eds), pp. 719-39. Plenum Press, New York

Weightman, A.J., Don, R.H., Lehrbach, P.R. and Timmis, K.N. (1984) The identification and cloning of genes encoding haloaromatic catabolic enzymes and the construction of hybrid pathways for substrate mineralization. In *Genetic control of environmental pollutants.* Basic Life Sciences, vol. 28 (G.S. Omenn and A. Hollaender, eds), pp. 47-80. Plenum Press, New York

Wheatley, A.D. (1984) Biotechnology and effluent treatment. In *Biotechnology and genetic engineering reviews,* vol. 1 (G.E. Russell, ed.), pp. 261-309. Intercept, Newcastle upon Tyne

Whittle, D.J., Kilburn, D.G., Warren, R.A.J. and Miller, R.C. Jr (1982) Molecular cloning of a *Cellulomonas fimi* cellulase gene in *Escherichia coli. Gene* **17**, 139-45

Williams, P.A. (1981) Catabolic plasmids. *Trends Biochem. Sci.* **6**, 23-6

10

Conclusion

Most biological processes of use to man are potentially amenable to improvement using genetics. The genes of existing organisms thus represent an important resource that can be exploited in genetic manipulation programmes. The formation of gene libraries for animals or plants that are in danger of becoming extinct may be the only available method for preserving part of that resource. Furthermore, the ability to recover DNA from mummified or fossilised tissue opens up the possibility of acquiring potentially valuable genes from long-extinct species. The isolation of such genes and their combination with other genes would not be possible without modern methods of *in vivo* and *in vitro* genetic manipulation. Nevertheless these newer methods complement but do not necessarily replace conventional genetic techniques.

In vivo and *in vitro* genetic technologies promise numerous benefits in such areas as health care, crop and animal breeding, mineral extraction, pollution control, energy production, chemicals and foods.

The ability to express genes heterologously in microorganisms has made possible the production of many different proteins that are of actual or potential use in human therapy. Such pharmaceutical products include vaccines, interferons, growth hormones, interleukins and neuropeptides, such as endorphins and enkephalins. Many of these peptides would be irksome to purify in quantity by extraction from human or animal cells due to availability of the appropriate tissue. Furthermore, manufacture of biologically active proteins in microorganisms, although not without technical problems, avoids the serious threat of contamination with endogenous viruses present in that tissue. For example, many haemophiliacs treated with the human blood-clotting protein factor VIII extracted from pooled blood products have become contaminated with Human Immune Deficiency Virus (HIV) (previously called Human T-cell Leukemia Virus III (HTLV-III)), the causative agent of acquired immune deficiency syndrome (AIDS).

Molecular genetic techniques, such as those for analysis of RFLPs, have made possible the antenatal diagnosis of an increasing number of genetic defects. Coupled with therapeutic abortions it is likely that some heritable diseases may be greatly reduced in incidence. Furthermore, repair of such defects by gene therapy could be in prospect.

There are many opportunities for protecting crops and enhancing crop yields by *in vitro* gene manipulation. Valuable crop species may be endowed with a variety of desirable properties, including resistance to environmental stresses, that for example, enable growth in entirely new climatic zones. Where microorganisms are used in agricultural programmes, such as those for frost protection and insect, weed and disease control, there is a need to improve the consistency of performance of the organism in relation to different soil and climatic conditions and to tillage operations. It is likely that the application of molecular genetic techniques to crop protection and likewise to health care will effect a shift in emphasis from treatment to prevention.

The creation of novel products may radically alter markets for raw materials. For example, until recently L-phenylalanine was principally used in intravenous feeding and other medically important nutritional products. However, demand for phenylalanine has been dramatically increased by the widespread use of the sweetener aspartame, which is synthesised from this amino acid and aspartic acid. L-phenylalanine can be manufactured by a variety of routes (see Klausner, 1985). The incentive to improve production methods for phenylalanine, to keep pace with demand, is considerable. Genetic methods, ranging from the isolation of bacterial mutants to genetically engineered microorganisms that produce elevated amounts of phenylalanine, are applicable to this problem.

Biosensors (see Chapter 9) have many applications in health care, in agriculture and in monitoring pollution. Implantable biosensors may provide information on specific metabolites in the body, and in conjunction with an electromechanical dispenser effect release of a particular drug. In addition, biosensors may be used to provide rapid and continuous measurements for fermentation control. The applicability of biological sensors depends not only upon the specificity of the biochemical reaction, but also upon the stability and cost of the device employed. All of these properties are potential targets for improvement by genetical means.

Selectivity, reliability and economic viability are also major problems confronting the microbial technologies of ore leaching and of treatment of metal-polluted wastewaters. These and other obstacles, including potential scale-up problems and competition with indigenous organisms, may also be addressed by genetic manipulation.

The potential risks of recombinant DNA technology (most notably the creation of pathogens with novel virulence properties) have received much public attention. In many cases microorganisms can be engineered using strictly *in vivo* (and hence 'natural') methods to possess the same properties as strains manipulated *in vitro*. Moreover, many supposedly unnatural combinations of genes produced by *in vitro* technologies may have been created naturally during the course of evolution and found wanting. It is possible, at some stage, that genetically manipulated organisms grown under contained laboratory or industrial-scale conditions will escape into the environment. Hosts utilised for the manufacture of potentially harmful products may, therefore, be biologically contained by the introduction of mutations that render the organism partially disabled (see section 3.5). The consequences of accidental release of engineered organisms are unknown. It is conceivable that large-scale release could pose a potential biohazard to humans, animals or plants. However, the poor competitive ability of most host microorganisms used for production purposes lessens the possibility of realising any such hazard. Furthermore, studies on laboratory workers engaged for long periods in gene cloning experiments suggest that, thus far, there is no specific risk associated with such activity (Smedley *et al.*, 1982).

The deliberate release into the environment of genetically manipulated microorganisms has raised a number of ethical and safety concerns. For example, fears about the potential consequences of releasing engineered strains of *Pseudomonas syringae* to prevent frost damage in plants (see section 8.2.1) have prompted legal action that delayed field trials in the USA. The effective use of engineered microorganisms in biological control, in improving crop yield and in overcoming the problems of environmental pollution depends upon successful colonisation of appropriate sites. However, the qualities required for survival and propagation (including an ability to tolerate environmental extremes, nutritional flexibility and genetic stability) of such organisms in the wild might include properties that could produce adverse effects on normal ecosystems. There is at present a dearth of information concerning the fate, in the wild, of genetically engineered organisms or the vectors they carry. However, this is perhaps not surprising given that very little is also known about natural gene transfer events in aquatic or terrestrial environments.

BIBLIOGRAPHY

Klausner, A. (1985) Building for success in phenylalanine. *Biotechnology* **3**, 301-7

Smedley, H.M., Sikora, K. and Ciclitira, P.J. (1982) Medical monitoring of genetic engineering research in Cambridge — the first five years. *J. Soc. Occup. Med.* **32**, 167-70

FURTHER READING

In writing this book we have tried, wherever possible, to refer to original references or appropriate reviews. Definitions are generally given in the body of the text. However, for an excellent explanation of a wide variety of terms used in modern genetics the reader is referred to *A dictionary of genetic engineering,* by S.G. Oliver and J.M. Ward (1985), Cambridge University Press, Cambridge and New York. Further information on recombinant DNA technology can be found in *Gene cloning, an introduction,* T.A. Brown (1986), Van Nostrand Reinhold, UK and *Principles of gene manipulation: an introduction to genetic engineering,* 3rd edition, R.W. Old and S.B. Primrose (1985), Blackwell Scientific Publications, Oxford, and on the genetics of *E. coli* in *Gene function: E. coli and its heritable elements,* R.E. Glass (1982), Croom Helm, London/University of California Press, Berkeley and Los Angeles

Index